GOLD

SCIENCE AND APPLICATIONS

GOLD

SCIENCE AND APPLICATIONS

Edited by

Christopher Corti
Richard Holliday

CRC Press
Taylor & Francis Group
Boca Raton London New York

CRC Press is an imprint of the
Taylor & Francis Group, an **informa** business

with the support of
WORLD GOLD COUNCIL

CRC Press
Taylor & Francis Group
6000 Broken Sound Parkway NW, Suite 300
Boca Raton, FL 33487-2742

First issued in paperback 2021

© 2010 by Taylor and Francis Group, LLC
CRC Press is an imprint of Taylor & Francis Group, an Informa business

No claim to original U.S. Government works

ISBN 13: 978-1-03-209945-3 (pbk)
ISBN 13: 978-1-4200-6523-7 (hbk)

Publisher's Note
The publisher has gone to great lengths to ensure the quality of this reprint but points out that some imperfections in the original copies may be apparent.

Library of Congress Cataloging-in-Publication Data

Gold : science and applications / editors, Christopher Corti, Richard Holliday.
    p. cm.
  Includes bibliographical references and index.
  ISBN 978-1-4200-6523-7 (hardcover : alk. paper)
  1. Gold. I. Corti, Christopher. II. Holliday, Richard.

QD181.A9G653 2010
620.1'8922--dc22   2009038080

Visit the Taylor & Francis Web site at
http://www.taylorandfrancis.com

and the CRC Press Web site at
http://www.crcpress.com

In Memoriam

Since writing the preface, we are sad to report that one of the authors, David Thompson, died after a sudden illness. His chapter, co-authored with Sonia Carabineiro, is a testament to his standing as an acknowledged expert in gold catalysis.

Contents

Preface..ix

Contributors xi

Chapter 1 Gold Supply and Demand 1

     Neil Meader

Chapter 2 The Physics and Optical Properties of Gold 13

     Martin G. Blaber, Mike J. Ford, and Michael B. Cortie

Chapter 3 A Primer of Gold Chemistry... 31

     Alan L. Balch and Thelma Y. Garcia

Chapter 4 Surface Electrochemistry of Gold... 51

     L. Declan Burke and Andrew M. O'Connell

Chapter 5 Gold Luminescence... 69

     Vivian Wing-Wah Yam and Chi-Hang Tao

Chapter 6 Gold Catalysis 89

     Sónia Alexandra Correia Carabineiro
     and David Thomas Thompson

Chapter 7 Metallurgy of Gold.. 123

     Jörg Fischer-Bühner

Chapter 8 Gold in Metal Joining... 161

     David M. Jacobson and Giles Humpston

Chapter 9 Jewelry Manufacturing Technology... 191

     Christopher W. Corti

Chapter 10 Biomedical Applications of Gold and Gold Compounds......................... 217

     Elizabeth A. Pacheco, Edward R. T. Tiekink, and Michael W. Whitehouse

Chapter 11 Gold Electroplating 231

     Antonello Vicenzo and Pietro L. Cavallotti

Chapter 12 Gold Thick Film Pastes 279

    Kenichiro Takaoka

Chapter 13 Gold Bonding Wire 287

    Koichiro Mukoyama

Chapter 14 Gold in Dentistry 295

    Paul J. Cascone

Chapter 15 Decorative Gold Materials 317

    Peter T. Bishop and Patsy A. Sutton

Chapter 16 Nanotechnological Applications of Gold 369

    Jonathan A. Edgar and Michael B. Cortie

Chapter 17 Miscellaneous Uses of Gold 399

    Richard Holliday and Christopher Corti

Index 407

Preface

The science and technology of gold is currently going through an exciting period of its history and is leading to new industrial and medical applications of this unique metal. It is unique, not only because of its properties (it is one of only two colored metals, for example, as well as being the most noble and most malleable), but because of its history and its role in society, where it serves as a monetary asset, for adornment, as well as an industrial and decorative metal.

Gold has a unique color and has been prized throughout history for its beauty and value, but it also has an interesting science that we are just beginning to unravel and exploit. Traditional applications have centered mainly on its metallic properties, and its chemistry has been rather overlooked, but that omission is being redressed. The surprising discovery of gold's catalytic properties in recent years and the development of its nanotechnology are leading to some exciting applications with enormous potential in industry and medicine.

There have been many excellent books written on certain aspects of the science and technology of gold, most notably the book of the Hanau gold conference held in 1996 (*Gold: Progress in Chemistry, Biochemistry and Technology*, edited by Hubert Schmidbaur and published in 1999), and there are several new books covering specific sectors of gold science, for example, on catalysis, chemistry, and nanotechnology. However, for those seeking a broader reference source on gold and covering all technical sectors, the most appropriate book on gold, *Gold Usage* by Bill Rapson and Theo Groenewald, was published over 30 years ago in 1978. The world of gold has changed considerably since then, and we believe it is time for an updated book that reflects the more recent developments.

The interest in the science, technology, and applications of gold has grown considerably over recent years, as evidenced by the huge increase in scientific publications in learned journals, the growth in attendance at the series of international gold conferences held since Hanau (it returned to Germany in 2009), and the expansion of readership of the sole journal dedicated to gold, *Gold Bulletin*. It is interesting to note that over \$13 billion worth of gold was used in industrial (and medical) applications in 2007, and consumption in this sector is predicted to grow significantly, much of it in new and green applications. Gold has a significant role to play in our quality of life during the twenty-first century.

With the cooperation of some of the leading experts in their field, we have attempted to produce a book that is a worthy successor to Rapson and to provide an authoritive source of information. We have attempted to cover all the main scientific areas as well as the main areas of application. In some cases this has required some very detailed scientific chapters; other fields require less specific detail and have focused more on the practical application. In addition, to help readers place this science and technology in the context of a much wider gold market, a chapter on the supply, demand and pricing of gold is included. Taken together, our aim for the book is to appeal to those working in both academia and industry. Keen readers might also note that the chapter headings are somewhat different from those of Rapson and reflect the changes that have occurred since it was published. We hope that we have succeeded in producing a reference work that is useful and also enjoyable to read; we have enjoyed editing it, but then we are "aurophiles," so we would, wouldn't we!

We appreciate and acknowledge the first-class contributions of the authors who cooperated in producing this book and the encouragement and support of the team at the publisher, CRC Press. It was a pleasure to work with them. We also acknowledge our employer, World Gold Council, for their support, which has made editing this book so much easier. They, too, have an interest in growing industrial applications for gold, which, in turn, leans on new gold science and technology.

Christopher W. Corti and Richard J. Holliday
International Technology
World Gold Council
London, U.K.

Contributors

Alan L. Balch
Department of Chemistry
University of California, Davis
Davis, California

Peter T. Bishop
Johnson Matthey Technology Centre
Reading, England, United Kingdom

Martin G. Blaber
Institute for Nanoscale Technology
University of Technology Sydney
Sydney, Australia

L. Declan Burke
Department of Chemistry
University College Cork
Cork, Ireland

Sónia Alexandra Correia Carabineiro
Faculty of Engineering
University of Porto
Porto, Portugal

Paul J. Cascone
The Argen Corporation
San Diego, California

Pietro L. Cavallotti
Dipartimento di Chimica Materiali e
    Ingegneria Chimica "Giulio Natta"
Politecnico di Milano
Milano, Italy

Christopher W. Corti
World Gold Council
London, England, United Kingdom

Michael B. Cortie
Institute for Nanoscale Technology
University of Technology Sydney
Sydney, Australia

Jonathan A. Edgar
Institute for Nanoscale Technology
University of Technology Sydney
Sydney, Australia

Jörg Fischer-Bühner
Indutherm Erwärmungsanlagen GmbH
Walzbachtal-Wössingen, Germany
    & Legor Group Srl
Bressanvido (VI), Italy

Mike J. Ford
Institute for Nanoscale Technology
University of Technology Sydney
Sydney, Australia

Thelma Y. Garcia
Department of Chemistry
University of California, Davis
Davis, California

Richard Holliday
World Gold Council
London, England, United Kingdom

Giles Humpston
Tessera Inc.
San Jose, California

David M. Jacobson
Buckinghamshire New University
High Wycombe, England, United Kingdom

Neil Meader
GFMS Limited
London, England, United Kingdom

Koichiro Mukoyama
Tanaka Denshi Kogyo K.K.
Saga, Japan

Andrew M. O'Connell
Department of Chemistry
University College Cork
Cork, Ireland

Elizabeth A. Pacheco
The University of Texas at San Antonio
San Antonio, Texas

Patsy A. Sutton
Johnson Matthey Technology Centre
Reading, England, United Kingdom

Kenichiro Takaoka
Shonan Plant
Tanaka Kikinzoku Kogyo K.K.
Kanagawa, Japan

Chi-Hang Tao
Centre for Carbon-Rich Molecular and
    Nanoscale Metal-Based Materials
    Research, and Department of Chemistry
The University of Hong Kong
Hong Kong, People's Republic of China

David Thomas Thompson
(Deceased)
Newlands, The Village
Reading, England, United Kingdom

Edward R. T. Tiekink
Nanyang Technological University
Singapore

Antonello Vicenzo
Dipartimento di Chimica Materiali e
    Ingegneria Chimica "Giulio Natta"
Politecnico di Milano
Milano, Italy

Michael W. Whitehouse
Griffith University
Brisbane, Australia

Vivian Wing-Wah Yam
Centre for Carbon-Rich Molecular and
    Nanoscale Metal-Based Materials
    Research, and Department of Chemistry
The University of Hong Kong
Hong Kong, People's Republic of China

1 Gold Supply and Demand

Neil Meader

CONTENTS

Introduction 1
Mine Supply 2
Producer Hedging 4
Aboveground Stocks 5
    Central Banks 5
    Scrap 6
Jewelry 7
Other Fabrication 8
Investment 9
Gold Price 11
Disclaimer 12
References 12

INTRODUCTION

To help readers place the information contained in this book in the context of a much wider gold market, a chapter summarizing the supply, demand, and pricing of gold follows. Detailed market terminology has been kept to a minimum, but those readers wishing to further expand their understanding of the market are directed to a separate text [1].

The supply of gold comes essentially from two distinct sources, namely new mine production and previously mined metal now sitting aboveground as stocks in various forms. New mine production provides the greatest amount of gold to the market each year, typically accounting for around two-thirds of total supply. In recent decades, annual production has exploded, up from less than 1,000 tonnes in 1980 to a peak of 2,645 tonnes in 2001, although more recently, output has fallen somewhat.

The bulk of supply from aboveground stocks comes from the recycling of the metal contained in manufactured (or fabricated) products, the vast majority of which comes from jewelry. It is therefore not surprising that the supply from recycling, typically referred to as old scrap, has generally kept to a rising trend, reflecting the addition each year to the pool of available metal from that year's jewelry production.

Other areas of supply from aboveground stocks include sales by central banks, hedging by mining companies (a form of advanced selling), and disinvestment from private holdings (which is reviewed in an upcoming section under demand as investment). Net sales by central banks have been a constant feature of the gold market since 1989, and over the past decade, these provided an average of about 500 tonnes of gold to the market each year. Central banks do purchase gold, but buying globally on a net basis has not been seen since 1988. Hedging by mining companies is, in essence, equivalent to gold producers selling their future production at current prices. Such transactions normally involve gold being borrowed and sold in today's market with producers later delivering their production against these commitments. Hedging was a major source of supply in the late

1980s and the 1990s, but more recently, positive sentiment toward prices has resulted in producers cutting their hedge book commitments. This in essence adds to demand and this position of "de-hedging" on a net basis has been in practice since 2000.

Moving to the demand for gold, the largest portion is accounted for by the fabrication of jewelry. Demand for jewelry can be broadly divided into two categories. The first covers demand in the industrialized world, where adornment is the primary buying motive and purchases are not that responsive to the price of gold. The second concerns demand in the developing world, where quasi-investment motives can dominate, which typically make purchases sensitive to price moves. Fueled by population growth, rising incomes, and favorable gold price conditions, jewelry demand kept to a broadly rising path throughout the 1980s and 1990s. Since then, however, gold jewelry has come under pressure from the diversion of discretionary income to other products (such as electronics), structural changes within the jewelry market, and the long-term rise in gold prices.

The remainder of fabrication demand is mainly accounted for by the uses broadly described in this book, that is, gold's use in electronics (such as gold bonding wires), dentistry, and decorative uses (for example, the plating of costume jewelry). It also includes coins, which are not the subject of this text. Electronics is by some margin the most important area here, having grown comparatively steadily to over 300 tonnes by 2007 or just over 10% of total fabrication demand.

Investment in gold is essentially divided into two main areas, physical and so-called paper products. The former is mainly made up of people buying gold coins or bars while the latter is composed of a myriad of products such as futures, options, shares in exchange-traded funds, or dealings in the "over-the-counter" market. Due to the opaque nature of this area, it is not possible to independently calculate the net impact of investor activity. However, it can be inferred from all the other elements of supply and demand as the balancing item that brings the market into equilibrium. This residual is termed "implied net (dis)investment." Over the years, this has appeared both on the supply side (notably during much of the 1990s) and the demand side (most years so far this decade).

One final aspect of the gold market worth highlighting is the issue of surpluses and deficits. Within most commodities, a deficit, or shortage, of supply translates into higher prices, while surpluses will tend to lead to the opposite. Yet the gold market, and those for the other precious metals, are different in that, since a surplus means an addition to stocks, this reflects investment and this typically accompanies rising prices. Equally at times when investors are selling on a net basis (usually in times of falling prices), this will serve to plug the deficit in the fundamental market.

MINE SUPPLY

There are a number of important points that resources companies need to consider before developing a mine, that will help to determine whether it will be an economically viable venture. At the center of these considerations is the expected long term balance between the price of gold and the cost of production. Once exploration has established the presence of gold minerals at a site, geologists will seek to determine the grade of the ore, that is, how many grams of gold there are per tonne of ore, and the volume of ore in the ground – whether the deposit significant enough in terms of grade and size to be of commercial interest. Positive results in these areas will lead to further technical work, investigating, among other considerations, the rate of recovery (usually expressed as a percentage of the gold within the ore that can be successfully converted to pure gold) and whether a deposit would be more optimally mined as an open pit or underground operation. These investigations generally culminate with a bankable feasibility study that will model projected development and operating costs (and sales revenues) to define whether a deposit can be viably developed as a mine, with a sufficient degree of technical confidence as to be able to raise project finance against its findings.

In conjunction with the price of gold, mining companies will monitor and look to control costs at their operating mines to maintain profitability. These have been steadily increasing over recent years, as a combination of intertwined factors has come into play to increase costs in areas including labor, consumables, and energy. The global resources boom has meant that skilled labor is in high demand, making gold producers face fierce competition in recruiting and retaining staff from both their direct peers and from the wider mining and construction industries. The commodities boom has also intensified competition and inflated prices for key items of plant and consumables. In addition, soaring energy costs have served to ratchet up prices for these items due to the increased cost of their manufacture and delivery, which has subsequently been passed on to the mining companies that consume them. Gold producers have also, of course, faced higher direct energy costs at their operations.

Once economic viability has been established and mining has started, the gold is extracted from the ore. One effective method of extraction is heap leaching. Heap leaching is the practice of obtaining gold by irrigating gold ore with a solvent, most usually sodium cyanide, from which it can be recovered. This method allows gold to be extracted from lower-grade ore at far lower capital and operating costs than is possible with conventional techniques, such as milling. Although cyanidation of gold had been in practice since the nineteenth century, large-scale cyanide heap leaching was first used in the United States, at Carlin, Nevada, in 1970. Thereafter, the technique spread and improved and has been contributing substantially to gold production. According to the industry body, the World Gold Council, its gold mining company members support the International Cyanide Management Code, a voluntary industry program that promotes the responsible management of cyanide used in gold mining, reducing the potential for environmental impacts.

Producers must also be aware of political risk, such as resource nationalism, whereby projects have either been stopped as permits have been rescinded or companies have been forced to surrender mining assets, as has occurred in central Asia. Concerns over security of tenure have also gained prominence in a number of Latin American states in recent years. In addition, mining companies are obliged to fulfill increasingly stringent environmental constraints. Although there is no international environmental standard, most producers come under local legislation covering issues including waste management, the effects on water quality, and mining methods themselves. In many regions, extensive community consultation is now an integral part of projects, and rehabilitation of sites after mining has ceased is standard procedure. The World Gold Council and its members state that they are committed to the principles of the International Council on Mining and Metals (ICMM). These principles seek continuous improvement in sustainable development performance, including environmental performance.

Traditionally, there was a "Big Four" of gold producing countries—South Africa, the United States, Australia, and Canada. For over a century, South Africa was the world's largest gold producer (see Table 1.1). Its peak output reached 1,000 tonnes in 1970, which accounted for about two-thirds of total global production in that year. In 2007, however, it produced 270 tonnes, representing 11% of the world total of 2,476 tonnes. One of the main reasons for the country's declining production is due to the maturity

TABLE 1.1
Geographic Distribution of Mined Gold Supply in 1990 and 2007 (tonnes)

Region	1998	2007
South Africa	496.9	269.9
Australia	310.1	246.3
United States	366.0	239.5
Canada	164.3	101.2
China	165.2	280.5
Peru	92.1	169.6
Russia	127.3	169.2
Indonesia	139.1	146.7
Other	713.00	852.9
Total	2,574	2,475

of many core South African operations; after over a century of intensive mining, shallow deposits have been exploited and mines have become progressively deeper and in many cases seen a fall in ore grades. Deep level mining is labor and technology intensive and, therefore, costly, requiring increasingly rigorous safety measures. Furthermore, the South African industry has experienced a pressing shortage of skilled employees with labor competition from the wider extractives and construction industries.

After reaching peak production in the late 1990s, production from the United States, Australia, and Canada has also been declining. Although output rose during the early part of the decade, low gold prices during this period meant that few new operations were brought into production, as producers were left unable and unwilling to spend capital on exploration, thereby inhibiting output in the latter part of the decade. Many mines were closed as they became uneconomical to maintain. Moreover, many operations were mature by the late 1990s, so the most attractive deposits had already been extracted.

In Latin America, Mexico and Argentina have registered substantial increases in the last decade while Peru is home to one of the two largest gold producing mines in the world, Yanacocha, which started production in 1993. As mining has progressed deeper, however, output has begun to decline there. This is mainly the result of a fall in ore grades and an increase in the stripping ratio—that is, proportion of waste rock that must be mined in the course of extracting ore.

China has now become the world's leading producer, overtaking South Africa in 2007 (Figure 1.1). Market liberalization has had a dramatic impact on China's gold mining sector, and one of the most important changes has been the end to official purchases in 2003, with producers no longer forced to sell all of their production to the People's Bank of China at regulated prices. A further impact of deregulation has been increased participation from foreign investors. Chinese mine production has also shown a remarkable ability to quickly respond to rising gold prices and has risen accordingly with the price. From delivering around 72 tonnes in 1987, it increased production almost fourfold by 2007, when it produced 281 tonnes. Asian production has also been boosted by the contribution from Grasberg in Indonesia, the other of the two largest mines in the world.

PRODUCER HEDGING

A hedge is a transaction that acts to manage the risk of adverse price movements in an asset such as gold. Gold producers typically enter into hedging contracts for one of three reasons. The first is in order to secure a forward premium in price. The second reason is to seek to protect revenues against falling gold prices, while the third is as part of the terms of a financing package. The simplest method of hedging is the forward sale, whereby a producer enters into a contract in which it can receive a fixed payment for a certain amount of gold at a predetermined date in the future. Regardless of any price movement between entering into the contract and its expiration, the producer will realize this guaranteed price on delivery.

There are three main parties involved in a basic gold hedge contract: the producer; the bullion bank, which acts as the producer's dealer; and a central bank. To place a hedge, the producer advises its dealer that it wishes to forward sell, for example, 1 tonne of metal. The dealer then immediately borrows this amount from a central bank (paying a leasing fee), which it immediately sells into the spot market, agreeing to return the 1 tonne by a certain date. The proceeds from this sale are then placed in a high-yielding cash account (subtracting the dealer's fee and the leasing fee on the borrowed gold), usually generating a premium for the producer. The producer is obliged to return 1 tonne of gold to the bullion bank by the preagreed date. At the time of implementation, the action of the bullion bank of borrowing gold and selling it into the spot market increases the supply of gold to the market. However, when the gold is returned, usually by delivering mine production from the producer's account, supply reaching the market is correspondingly constrained. Hedging activity therefore represents an acceleration of supply to the market, but the overall balance of supply and demand is maintained over the lifetime of the transaction.

Hedging and de-hedging activities are thus important in the wider gold market. To give a historical perspective, the 1990s were largely characterized by unrelenting downward pressure on the gold price, which led to steadily increasing levels of hedging undertaken by producers who were seeking revenue protection from further declines. In 1995, hedging had a particularly noticeable impact, reaching 475 tonnes and thus accounting for about 13% of total supply. By September 1999, however, the gold price had made an abrupt change of direction and started to climb. Many hedged producers were caught off-guard and were unable to take advantage of soaring spot prices.

On the back of this crisis, the year 2000 was the first year since the late 1970s when the global producer hedge book did not expand. There has since been a protracted period of de-hedging, in which producers have both delivered into and also prematurely closed out their hedge contracts. This is accomplished either by bullion purchases from the market, delivered to the bullion bank, or through delivering their own production into contracts before they mature. De-hedging posted a record level of 447 tonnes in 2007, representing just over 11% of total gold demand. Expectations for higher gold prices and investors' associated anti-hedging sentiment were the chief reasons for this.

It can be seen that, pre-1999, producer hedging activity was a significant supply component in the gold market, whereas post-1999, hedging activity became a significant demand component of the market. Today, many more exotic and complex financial derivatives and option structures exist, in addition to the humble forward sale, through which producers can hedge. The effect of these more complicated contracts is, however, ultimately the same: hedging activity by gold producers affects the timing of mine supply reaching the market.

ABOVEGROUND STOCKS

In addition to newly mined metal, gold can also be sourced from aboveground stocks, either through the recycling of fabricated products (see section that follows on scrap) or the mobilization of bullion stocks. The latter comprises central bank sales (see following); disinvestment by individuals (covered in the section on investment); and advance sales—the hedging described previously—by mining companies. During the past decade, supply from aboveground stocks has typically accounted for about a third of annual supply.

GFMS estimates put aboveground stocks of gold at the end of 2007 at 161,000 tonnes (the equivalent of over 60 years of current mine production) [1]. It is the sheer scale of these aboveground stocks that sets the yellow metal apart from other commodities. The reason for this buildup is primarily the virtual indestructibility of the metal—almost all of the gold mined throughout history still exists in some form. The metal's durability also allowed gold to become a highly suitable store of value and form of money over the ages, for individuals and state bodies.

The composition of aboveground stocks is also important because it determines the speed and likelihood of the return of the metal to the marketplace. Gold as bullion (usually in bar form) typically comes back the most readily, and stocks in this form stood at about 55,500 tonnes. Just over half of this bullion is held by central banks and other official sector bodies, with the balance being held by private individuals and institutions. More, however, just over half of the total, is held as jewelry items. Lastly, about 19,000 tonnes have been absorbed by other types of fabrication (such as electronics), the least likely area to get recycled back into the market. Just under 4,000 tonnes cannot be accounted for and can be considered irretrievably lost or as true consumption. This would include manufactured items containing gold that have gone to landfill, any metal lost at sea, and so forth.

CENTRAL BANKS

For governments, gold once presided as the dominant asset in the reserves of central banks (commonly referred to as the official sector) and against which all paper currencies were backed. However, with the suspension of the gold standard in 1932 and later dismantling of the Bretton

Woods System in the early 1970s, gold's link to the dollar and the majority of world currencies was severed and the metal's price was allowed to float freely and be determined by the market. The metal still plays valuable roles for central banks, however, such as reserve diversification, instilling public confidence in the central bank, and economic security. It can also generate income through the lending of its bullion.

As of the end of 2007, total official sector gold holdings stood at just under 30,000 tonnes. The largest holdings are in the Western world, with the United States, Germany, France, Italy, and Switzerland making up the top five and mostly for whom gold is far larger in value than other currency reserves. Japan, China, and Taiwan come in respectively at numbers six, eight, and ten, but the value of their gold holdings is tiny compared to their other reserves.

The official sector has been a net supplier of gold to the market for nearly two decades, providing as much as 18% of total annual supply. The bulk of recent sales has come from European banks. These disposals have been conducted under the Central Bank Gold Agreement, which became effective in September 1999 and was renewed in 2004. A key reason behind these five-year agreements was to add an element of certainty to the market. Purchases by central banks have occurred in recent years, such as those by Russia and China but to date their scale remains limited.

Scrap

The recycling of gold takes two forms—so-called process scrap and old scrap. The former concerns gold that never ends up in a finished product and returns unused to a refinery for remelt. This includes areas such as the edgings to a sheet of metal once coins have been stamped out or the filings recovered from the polishing or engraving of jewelry. Because this effectively forms part of a continuous loop, it is of no interest to an economist and is excluded from analysis. Old scrap, in contrast, is the metal recovered from finished items containing gold. The vast bulk comes from jewelry, although small amounts are also received from other areas such as electronic goods or dental alloys.

Old scrap is typically the second largest source of supply after mine production, accounting for approximately a quarter of total supply each year. The amount, however, can prove quite volatile, surging, for example, to a then record of about 1,100 tonnes in 1998 on the back of the East Asian economic crisis before slumping to about 600 tonnes the following year.

Selling by individuals during a time of financial crisis is but one reason for dishoarding. However, perhaps the greatest determinant is changes in the price of gold. One example of this would be the increases in scrap seen emerging from the Middle East at times of high gold prices. Other factors nonetheless mean old scrap does not always track changes in the price. A recession, for example, could trigger an increase in scrap as manufacturers, wholesalers, and retailers remelt unsold jewelry inventory. Changing fashion can even influence scrap, as we are seeing increases in scrap in western Europe as inherited jewelry pieces—considered too dated to be worn by their new owners—are remelted.

With jewelry dominating old scrap, it should come as little surprise that volumes should be greatest in those countries making or buying jewelry. Volumes also tend to be greater in the developing world as pieces there typically have lower markups over the price of the contained gold and as a result of the unsentimental investment motive behind initial purchase. Such pieces are also more likely to be plain, in other words not carry precious or semiprecious stones and be of higher caratage. In industrialized markets, jewelry is more likely to carry stones and be of medium/lower caratage, be branded, and be bought as a gift for adornment—which will lower the propensity for remelt—and the consequent high markups over the contained gold mean a profit can rarely be made. Nonetheless, scrap can still prove significant through the liquidation of poorly selling merchandise or through the previously noted selling of stylistically dated pieces.

Generally, scrap is reused in the markets in which it is generated. However, during times of peak supply (such as after Egypt's currency crisis), amounts surplus to local industry requirements would invariably be exported, often to the major Swiss refineries, where the metal reenters the supply chain.

JEWELRY

The use of gold in jewelry can take many forms but, for the purpose of statistics, it is typically measured only when in "karat" form. This specifically excludes an article of costume jewelry made from, say, brass to which a gold plating has been applied. A karat is a measure of purity of gold, defined as parts per 24. Some jewelry, such as that in China, exists in 24-karat form (essentially pure gold), but the highest caratage typically sold is 22-karat (the norm in India) as other metals such as copper and silver are added to bring sufficient strength to the alloy. Just beneath that sits 21-karat, the norm for many Middle East markets. Purchase of these high-karat grades is often motivated by investment considerations. For this to work, the pieces can carry only a small amount of labor or markup over the value of the contained metal.

The main driver for the purchase of the remaining types of gold jewelry is adornment or fashion. Within this, the next major grade is 18-karat jewelry, the norm in areas such as southern Europe. This is the standard typically used in high-quality pieces, especially those carrying precious stones such as diamonds, because at this concentration of gold, the alloy is sufficiently hard to hold stones and color on an effectively permanent basis. This is also the grade at which "white gold" appears— gold to which a silver-colored metal such as palladium or nickel has been added in sufficient quantity to mask the yellow of the pure gold and render the piece "white" (industry jargon for silver colored). Beneath that is 14-karat gold, the last grade commercially available in which the majority of the alloy is gold; 8, 9, and 10 karat follow. These qualities are the various norms in northern Europe and North America, and in some markets, the United Kingdom, for example, there is legislation to enforce these standards.

In recent years, the lines have blurred between what is considered fine jewelry and costume jewelry. In the past, something like a gold pendant, for example, would be featured only with a gold chain. In today's more informal times, however, the piece could be found on a leather or silk cord. This process has been extended further by recently high prices such that a predominantly steel or brass item might carry a small gold accent, weighing a fraction of a gram.

A common theme to Western markets is the shift from plain gold to pieces with stones and the move to branded or high-fashion designs. All these carry much higher markups over the value of contained gold, and this is important as the trend for sales in terms of total value, number of pieces, or weight of pure gold can go in opposite directions. These trends are also spreading to developing world markets, in particular China where 18-karat gold is gaining market share.

These trends are also important as, in industrialized markets, consumers within the overall jewelry segment devote ever more expenditure proportionately to other materials, such as diamonds, or to the perceived value of design and branding, all of which cuts the amount of money being spent on gold. In addition, all jewelry has been losing market share to other discretionary areas, such as foreign vacations or technology goods (such as cell phones), both of which typically enjoy far greater advertising budgets. This has culminated in the weight of gold sold in jewelry form declining steadily in the industrialized world over the past decade or so.

This trend may not be immediately apparent from the global figures because the bulk of consumption takes place in the developing world, where price is the main determinant of offtake. Global consumption of gold in jewelry in 2007, for instance, came in at just over 2,400 tonnes [2], down by about 800 tonnes since 2000, chiefly as rising gold prices have undermined demand in price-sensitive countries, such as India, Indonesia, or Saudi Arabia. Some countries, however have bucked the trend, most obviously China, whose consumption has been steadily growing since 2002 thanks to strong economic growth. Russian demand has also grown impressively in recent years.

You might be tempted to assume that the trends mentioned previously could mean global consumption might inevitably continue on a downward path. However, a quick review of consumption in per-capita terms implies this may be far from true (as well as highlighting the major cultural differences between countries). Although India is the largest consumer of gold, per-capita consumption is only about 0.5 g and therefore some way below gold-friendly industrialized markets such as Italy and the United States, whose per-capita consumption is about 1 g. China is also comparatively low at a little under 0.3 g and is therefore similar to gold-unfriendly industrialized markets such as Japan and Germany.

As noted previously, the largest jewelry market in the world is India, where consumption has fluctuated roughly between 500 and 600 tonnes a year over the past decade, depending on the price. It is said that Indians hold about 15,000 tonnes of gold, mainly in the form of jewelry. A sizable portion of the demand in India comes from rural parts, where jewelry is foremost a saving instrument.

China has recently moved into second place globally in terms of gold jewelry consumption, after recent growth lifted demand in 2007 to just over 300 tonnes. In contrast, the recent economic and financial crises in the United States, on top of structural change, have meant poor demand for luxury products like jewelry, pushing the United States to third place as demand fell to only about 260 tonnes in 2007. The remaining two in the top five consumers are Turkey and Saudi Arabia. In Turkey, the earthquake of 1999 and the banking crisis of 2001 had adversely impacted domestic demand till 2002. Recovery of the Turkish lira, along with far greater economic and financial stability, has since aided the revival of demand.

If we review global ranking in terms of where the jewelry is produced, the top two stay unchanged, but Turkey moves up to third place while Italy claims fourth. The latter used to be number two, but structural change and market share loss (mostly to Turkey and south-east Asian countries) for its exports in the United States have seriously undermined its output. Fifth place again goes to Saudi Arabia.

Jewelry consumption is critical to the overall gold market as it is the single largest physical user of the yellow metal, typically accounting for about two-thirds of the total demand for gold. Despite this fact, jewelry often plays second fiddle to investment (and on occasion other elements of supply and demand) as the most important determinant of the gold price due to its price sensitivity. Jewelry is best seen as a price taker rather than a price maker, tending to expand or contract depending on the price of gold and therefore acting as a "cushion" for the other supply and demand variables.

OTHER FABRICATION

Gold is used not only in the manufacture of jewelry but also in a wide range of other applications, which form the subject matter for this book. These uses include conductive bonding wires in electronic items, dental alloys, and decorative plating. Along with medals, these other forms of fabrication together account for approximately 15% of total gold demand.

The most important of these nonjewelry applications is electronics, by a significant margin. This industry mainly uses gold in the plating of contacts (see Chapter 11), so that they remain corrosion free, and in bonding wire for the connective wires within electronic items (see Chapter 13). Gold is required for the latter chiefly because of its resistance to corrosion and its ability to be drawn into wires measuring only about 20 μm in diameter. While essentially all electronic goods will contain some gold, those that are more complex, such as cell phones or personal computers, will typically carry far more gold than simple devices such as a toaster.

Gold demand in electronics has been rising steadily in recent years, having recovered from the dot-com bust in 2000, with annual use in 2007 exceeding 300 tonnes. Its ability to grow in the face of much higher gold prices reflects the difficulties in its substitution as well as the underlying growth in sales of electronic goods. By some margin, the largest fabricating country is Japan, followed by the United States and then South Korea (although it should be noted that this ranking refers only to the initial transformation of gold into an intermediate product, such as bonding wire, and not the country of ultimate end use).

The use of gold in dental alloys (see Chapter 14) is one area in which the metal can be readily substituted by other materials (in contrast to say bonding wire). Furthermore, the value of the contained gold makes up a sizable percentage of the total alloy. As a result of these issues, demand in this area can be affected by price. Both factors also mean that despite a rising global population and living standards, gold dental demand now appears to be on a secular downward path (dipping below 60 tonnes in 2007). This is chiefly through substitution by nonprecious metals such as cobalt/chrome, where price is the main concern, or to ceramic fitments, if cosmetic factors are dominant.

One of the main decorative uses of gold is the plating of costume jewelry and the clasps, buckles, and so forth of luxury accessories such as belts or handbags. Another major area is the use of gold in India in jari, a gold thread found in some saris. Together with a host of other uses such as gold pastes for glassware, this decorative and miscellaneous industrial category typically accounts for about 90 tonnes of demand each year.

Lastly, gold is used in coins, whether official or unofficial, and medals. Official coins are best categorized as a form of investment and so have been reviewed in that section. However, unofficial coins and medals are often bought for various reasons such as for gifts or commemorative tokens, although an investment rationale will often still come into play to varying degrees. Official coin fabrication is quite volatile but is invariably the larger of the two: in 2007, for example, this consumed almost 140 tonnes of gold, whereas unofficial coins and medals together stood at only just over 70 tonnes that year (and is normally far lower at about 30 tonnes a year). There is also a marked difference as regards main fabricating countries—the manufacture of unofficial coins and medals is overwhelmingly Indian in origin whereas Turkey and then the United States dominate the production of official coins.

INVESTMENT

Historically, gold was used as the basis for international monetary systems and as a highly effective store of value for investors. In more recent times, the metal has come to meet more modern objectives within the portfolios of both private individuals and institutions. Over the past half-decade in particular, growth in investor interest in gold has arguably made it the single most important influence on the metal's price, certainly as regards day-to-day moves.

Gold is quite commonly purchased as a means of preserving wealth, and one good example of this is as protection against inflation. The 1970s were particularly illustrative of this, as the period was marked by surging U.S. inflation, which resulted in substantial fresh investor inflows into the gold market. Thereafter, as inflation receded as a threat, gold as a form of investment fell out of favor throughout the 1980s and 1990s.

Gold investment can also prove useful as a hedge or form of protection against times of economic instability, which are typically characterized by the poor performance of traditional investments such as equities and bonds. The bursting of the dot-com bubble in 2000 and the recession that followed, for example, all proved to be catalysts and sources of fuel for a turnaround in the gold market. The correlation between the gold price and equities is often reviewed as this typically proves negative, meaning that when returns in one are rising, the other's will generally be moving down.

A third key area is the use of gold as insurance during times of political and military upheaval. Perhaps the most striking example in modern times was the Soviet invasion of Afghanistan in 1979, which contributed significantly to the gold price soaring to a long-standing record (that remained unbroken until 2008). Other more recent events, such as the attacks in the United States on September 11, 2001, or the Iraq War in 2003, also bolstered interest in the gold market.

Applicable to the two points made earlier, geopolitical or financial crises can lead to corporations and governments defaulting on their financial obligations. The fact that gold is no single institution's liability makes it virtually free of the default and credit risk inherent within stocks and bonds and underpins the metal's attributes as a safe haven.

A final major consideration is gold's function as a hedge against a weakening U.S. dollar, again due to the inverse correlation that is normally in place between the two. It was in 2001 that the dollar began to markedly weaken and gold investment was greatly boosted by this. Part of the reason for this relationship is due to gold being priced in U.S. dollars. This means that when the dollar falls (or rises) against other currencies, the metal becomes cheaper (or more expensive) as regards prices in euros, rupees, or whatever. This in turn should boost (or trim) the demand for gold in those countries with currencies not tied to the U.S. dollar, and these changes in demand should then feed through to gold's world price.

Such factors can often overlap, and this tends to make gold investment comparatively volatile. For example, unrest in the Middle East would boost gold's role first as a safe haven during politically unsettled times, second as protection against rising oil prices and general inflation threats, and third as some insurance should such events undermine economic growth.

Given that gold can act as a preserver of wealth, its use as an effective diversifier in investor portfolios has grown dramatically in our more recently troubled times. For instance, while the crisis in the U.S. subprime and global credit markets battered world stock markets and devastated bank balance sheets, the value of gold thrived. Such events were good for sentiment toward the metal since its profile as an effective diversifier was notably enhanced during this period and reaffirmed the assertion that the typical investor portfolio should maintain at least some form of weighting in gold.

Historically, investment was exclusively in a physical form, typically as a bar (sometimes referred to as an ingot) or coins (such as the British sovereign or South Africa's krugerrand). Some might extend the list to include medals, unofficial coins, or even some forms of jewelry; it is quite common in developing world countries, most obviously India, that investment can be the prime motive behind the purchase of gold in jewelry form. Given that motives are blurred, this extended list is typically reviewed as part of fabrication demand, a tradition that this book adheres to.

Recently, the emphasis has swung to a range of products sometimes referred to as "paper gold." These products afford the convenience of exposure to gold prices without the obligation of having to hold and store the physical metal. The decision whether to buy "physical" or "paper" gold (and what sort of paper product) is typically a function of the type of investor involved. The chief factor here is size. An individual (or "retail") buyer would be more focused on physical and those paper products geared to small amounts of capital. In contrast, large financial institutions as well as individuals with high net worth would focus on those paper products appropriate to sizable sums.

One of the most common forms of paper products is gold futures, which trade on exchanges such as New York's Comex. Linked to those are options and other "derivatives." These instruments are usually the mainstay of sophisticated, in their majority institutional, players that often pursue a strategy of short-term gains (as opposed to a longer-term buy-and-hold approach). Such products, however, are also extensively used by commercial participants, namely mining companies and fabricators, to hedge and protect against the underlying gold price risks of their business.

Another key area is the over-the-counter (OTC) markets, which offer a variety of tailor-made products linked to the gold price and have the added advantage of providing greater flexibility and anonymity for participants, as well as lower costs relative to physical holdings and futures. Because of the high entry level of investment required, this arena is largely inaccessible to retail players. A more recent investment product is exchange-traded funds (ETFs), which are similar to equities and trade on stock exchanges. Showing remarkable growth in a short span of time, these vehicles have proved to substantially widen the market's investor base, particularly luring those with long-term investment horizons. Finally, the equities of gold mining firms are another highly popular alternative among all classes of investors and can often yield returns that exceed those of the physical metal, although this lies outside the realm of actual investment in gold.

GOLD PRICE

Gold has been used as a medium of exchange for more than 5,000 years. From 1717 to 1931, it was used, for example, within the British monetary system (and maintained a fixed price against the pound sterling) under what was known as the gold standard, and whereby the sterling price then also acted as the international price of gold. After the World War I, however, the U.S. dollar price was increasingly regarded as the international price. This was confirmed in 1934 when President Roosevelt fixed the price of gold at $35 per ounce, at which level the U.S. Treasury stood ready to buy and sell gold at this fixed ratio. This lasted until 1968, when the gold price was floated freely and determined in the open market. Since that point, the gold price has been quite volatile, rallying to a then all time high of $850 in 1980 before sinking to the mid-$200s by the beginning of this millennium. In recent years, price strength has returned, and in March 2008 the gold price breached the $1,000 mark for the first time, hitting a record $1,011.25. However, this was still well below the 1980 high if adjusted for inflation. Figure 1.1 illustrates the yearly average long-term price of gold.

The most common reference price for gold is the so-called London Fix. This has been set since 1919 by the Gold Fixing members, all of whom are Market Making members of the London Bullion Market Association. The Fixings are a process whereby the members can transact business on the basis of a single quoted price. This price is moved higher or lower until both buyers' and sellers' orders are satisfied. The price is then "fixed," and this is accepted as the average price of the metal. This procedure is carried out twice a day, once at approximately 10:30 a.m. and again at 3:00 p.m.

Even though the international gold price is set in U.S. dollars, the price of gold in other currencies can still have an important impact on the global market, as currency fluctuations can lead to quite divergent price trends within some countries. Recent years, for example, have been characterized by marked dollar weakness, and this means that the price gains in dollar terms have often not been translated into local currency terms. Importantly, from a consumption point of view, the price of gold in Indian rupee terms (basis annual averages) only rose by 5% in 2007 in comparison to the U.S. dollar price jump of 15%. Similarly, but on the producer side, the gold price in Australian dollar terms during the same year rose by only 3%. Had both the rupee and Australian dollar gold prices risen by the same amount as the U.S. dollar price, Indian consumption would have been more depressed and Australian supply more encouraged than was actually the case.

The price of gold and other assets are often closely correlated. The link between gold and the U.S. dollar is perhaps given the most attention, as it is common for gold to strengthen in times of dollar

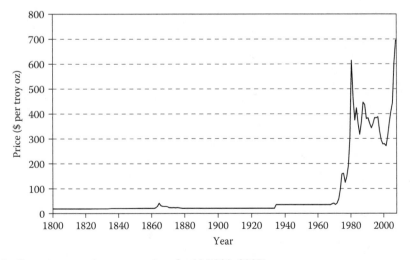

FIGURE 1.1 Long-term yearly average price of gold (1800–2007).

weakness and vice versa. This is largely due to the almost self-fulfilling nature of investment drivers; because speculators tend to believe in a correlation, they act accordingly. Gold prices also generally move in tandem with other precious metals such as silver and platinum, although it is invariably gold that is the leader of the pack. Gold and oil can also enjoy intermittent periods of close correlation, typically positive. This could materialize, for example, at a time of rising Middle Eastern tensions, lifting both the price of oil and the appeal of gold investment as a safe haven. The last correlation of note is with share prices. Here, the link is usually negative, as investors typically shift money from equities into the safe haven of gold during times of crisis and vice versa during times of boom.

DISCLAIMER

While every effort has been made to ensure the accuracy of the information in this document, GFMS Ltd. cannot guarantee such accuracy. Furthermore, the material contained herein has no regard to the specific investment objectives, financial situation, or particular needs of any specific recipient or organization. It is published solely for informational purposes and is not to be construed as a solicitation or an offer to buy or sell any commodities, securities, or related financial instruments. No representation or warranty, either express or implied, is provided in relation to the accuracy, completeness, or reliability of the information contained herein. GFMS Ltd. does not accept responsibility for any losses or damages arising directly, or indirectly, from the use of this document.

REFERENCES

1. *The World of Gold*, Timothy Green, Published by Rosendale Press, London, 1993.
2. *Gold Survey 2008*, Published by GFMS Ltd. London, 2008.

2 The Physics and Optical Properties of Gold

Martin G. Blaber, Mike J. Ford, and Michael B. Cortie

CONTENTS

Introduction...13
Electronic Structure of Gold..13
Crystal Structure and Alloying Behavior 16
Physical Properties..17
Optical Properties..18
    Color of Bulk Gold..18
    Thin Gold Films 21
    Color of Nanoparticles 23
    Gold Sponges and Gold Black...26
Conclusions..26
References...27

INTRODUCTION

Although gold is well down on the periodic table, at position 79, it was the first of the metals to be discovered and exploited by humans. This was almost certainly as a result of it possessing four unique attributes: a bright metallic yellow color, excellent resistance to corrosion, considerable malleability, and a high density (19.32 g/cm³). The high corrosion resistance and density facilitated the concentration of native gold nuggets and powders in the beds of streams, while the yellow color and malleability made it very suitable for the production of jewelry or religious artifacts. A few other metallic elements—such as silver, copper, or platinum—possess color and/or corrosion resistance and/or ductility and/or density, but none to the simultaneous degree exhibited by gold. What are the reasons for this unusual cluster of interesting properties in element 79? The answer, as we will show in this chapter, lies in its special electronic configuration. The electronic configuration controls the optical properties, chemical reactivity, and crystal structure of an element. Of course, the high atomic number of gold does confer a few other attributes that are independent of its electronic configuration. For example, like other relatively heavy elements, it has a large number (48) of isotopes: $_{197}$Au is the common one but $_{198}$Au, which has a half-life of 2.7 days, has been used in radiotherapy [1]. The high atomic number also makes Au relatively opaque to x-rays and ensures that it interacts strongly with electrons in both scanning and transmission electron microscopy. Both of these attributes have generated a few technological applications for gold in their own right [2–5]. Nevertheless, in the present chapter we will focus mainly on the electronic configuration of gold, and on the physical, optical, and crystallographic properties that result from this.

ELECTRONIC STRUCTURE OF GOLD

The high corrosion resistance of gold is a consequence of its first ionization potential being 9.2 eV, which is high compared to those of, for example, silver and copper, at 7.6 and 7.7 eV, respectively [6]. This results in such a large barrier to oxidation that elemental gold is ordinarily free of an oxide

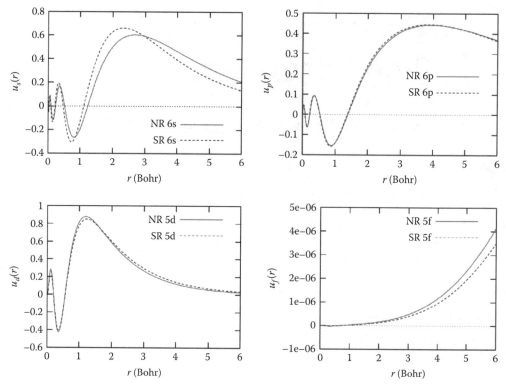

FIGURE 2.1 The radial component of the Au valence wavefunctions u, with no relativistic (NR) corrections and scalar relativistic (SR) corrections. Calculated by M. G. Blaber using all-electron (AE) density functional theory (DFT) within a generalized gradient approximation (GGA) with exchange correlation functional by Perdew, Burke, and Ernzerhof as implemented in the ATOM 3.2.2 pseudopotential generating program distributed in the SIESTA DFT package.

coating. Gold has the electronic configuration $[Xe]4f^{14}5d^{10}6s^1$. In this configuration the 4f electrons underscreen the 5d and the 6s,p electrons from the nuclear charge, resulting in an effect analogous to lanthanide contraction. The lanthanide contraction causes the atomic radius of the lanthanides to decrease across the period as the quality of the shielding per electron decreases; in the 5d metal series this effect gives similar lattice constants to the 4d metals. For some years the stability of the Au_2 dimer and the reduced lattice constant in bulk gold compared to silver were attributed to a similar contraction [7]. However, gold is in a unique spot on the periodic table where the effects of the lanthanide contraction are also superimposed on the onset of relativistic effects. The latter become increasingly important for the heavier elements because as the atomic number increases, the velocity of the 1s electrons approaches the speed of light. This causes these electrons to increase in mass and so their orbital contracts toward the nucleus. To compensate, the higher s and p orbitals, which also have significant electron density in the vicinity of the nucleus, also contract, resulting finally in the outermost 6s and 6p orbitals being smaller than they would otherwise have been without relativistic effects (see Figure 2.1; [10–12]). The contraction of the s and p orbitals causes an increased screening of the core for the 5d and 5f electrons, which do not have much electron density in the vicinity of the nucleus and which are therefore, in the first instance, less susceptible to relativistic effects of their own. Thus, the 5d and 5f orbitals are destabilized and expand outward, with an associated increase in energy [8,9]. The contraction of the 6s orbitals across the lanthanides and 5d metals due to relativistic effects is presented in Figure 2.2. Pyykkö and Desclaux [9] have estimated the average radial velocity, v_r, of the 1s electrons as $v_r = Z$ a.u., where Z is the atomic

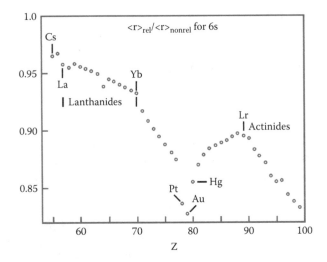

FIGURE 2.2 Fractional contraction of the 6s shell due to relativity. (Redrawn from Relativity and the periodic system of elements, P. Pyykkö and J.-P. Desclaux, *Acc. Chem. Res.*, 1979, **12**, 276. With permission.)

number and a.u. the speed in atomic units. The speed of light in atomic units is 137.036. This means that for gold, $v_r = 79$ a.u., which is about 58% of the speed of light. We can approximate the actual electron mass of a 1s electron as

$$m_e^* = m_e \Big/ \sqrt{1-(v/c)^2} \qquad (2.1)$$

where v and c are the speeds of electron and light, respectively, and m_e is the rest mass of an electron. The mass of a 1s electron in gold then becomes $1.224\ m_e$ from which we can roughly calculate the Bohr radius of the atom using [9]

$$a_0 = 4\pi\varepsilon_0\hbar^2/m_e^* Ze^2 \qquad (2.2)$$

where e is the electron charge, ε_0 is the absolute permittivity, and $\hbar$ is the reduced Planck constant. The result is a radius that is about 20% smaller than would have been the case without the application of relativistic effects.

Both platinum and gold have the electronic configuration $5d^{g-1}6s^1$ where g is the group number, whereas most of the other metals in the 5d period have the configuration $5d^{g-2}6s^2$. The former configuration compounds the 6s orbital contraction and stabilization because the additional 5d electron causes a greater contractive effect on the 6s than would have been the case in the $5d^{g-2}6s^2$ configuration [13]. This effect is calculated to result in the energy of the gold 6s orbital being lowered by approximately 1.63 eV [13].

The hole in the 6s orbital lies at a comparatively low energy as a result of these changes and, when reacted with a very readily ionizable element, gold can actually accept an electron, thus behaving as a pseudohalogen. Not surprisingly, it is the most electronegative metal on the Pauling scale, with a value of 2.44 [14]. The pseudohalogen behavior is manifested in the semiconducting auride [15] compounds RbAu [16] and CsAu [17].

The high first ionization potential of gold and the consequent excellent corrosion resistance of this element are explained by the stabilization of its 6s orbital by the various factors discussed in the preceding paragraphs. The second and third ionization potential, however, follow the opposite trend, giving gold access to higher oxidation states than either silver or copper. The +I , +II, and +III states are readily accessible, and +V states have been reported for fluoride compounds [13,14]. Another consequence of the relativistic effect is that the covalent bond lengths in many gold molecules are shorter than in the equivalent Ag molecules. This is especially marked in the binary molecules Au^-X^+ such as AuCs, AuNa, AuLi, and AuS, whereas the relativistic bond contraction is smaller for the molecules Au^+X^- such as AuF. This is because the destabilization of the 5d orbital increases the radius of the Au^+ ion whereas the contraction of the 6s orbital significantly reduces the size of the Au^- ion [18].

Another unusual attribute of gold is its "aurophilicity" [7], a term used to describe the electron–electron interaction between nonvalence orbitals. The aurophilic effect causes the unusual clustering of gold atoms in a large variety of molecules including dimerization of $ClAuPH_3$ [19] and rearrangement of $C(AuL)_4$ from a tetragonal to a square pyramidal structure so that the Au atoms are closer together [20] (see also reviews by Pyykkö et al. [21,22] and references therein). Formally, the aurophilic effect is quite strong in gold due to the proximity in energy of the 5d electrons to the 6s electron, which makes s-d orbital hybridization easier, resulting in the breaking of the normal $5d^{10}$ configuration which opens the system up to unusual bonding [23].

CRYSTAL STRUCTURE AND ALLOYING BEHAVIOR

In its bulk, metallic form, gold has the face-centered cubic crystal (fcc) structure, an attribute it shares with several other ductile metallic elements including aluminum, silver, nickel, platinum, and copper. All three of the high symmetry planes in gold undergo reconstruction when free of contamination. This is unusual, in most other fcc metals, while the (110) and (100) surfaces may reconstruct the (111) surface does not. The anomaly in gold has also been attributed to relativistic effects [24]. However, the gold (111) can be forced to de-reconstruct quite easily by the adsorption of a self-assembled monolayer or by surface charging caused by the application of a potential [25,26]. The surface tension of liquid Au is comparatively high at about 1.1 to 1.2 J/m^2 [1,27]. It is lowered by alloying with Ag or Zn, but increased in alloys with Cu. Further information on the surface tension, density, and contact angles of liquid Au and its alloys may be found in Ricci and Novakovic [27]. A surface energy of between 1.3 and 1.5 J/m^2 is reported for the solid state [28] but this also depends on which crystal face is being considered (see Table 2.1; [89–91]).

The fcc structure of gold has a multiplicity of close-packed planes on which slip can occur, making it intrinsically more ductile than other crystal structures, such as the body-centered cubic structure of iron and tungsten, or the hexagonal close-packed structure of magnesium or zinc. Interestingly, the fcc structure is not the most stable configuration when gold nanoparticles are smaller than about 10 nm and, instead, icosahedral, decahedral, or defective fcc structures form [29].

TABLE 2.1
Data for the Surfaces of Gold

Surface	Surface Energy meV/$Å^2$	Surface Energy J/m^2	Most Stable Reconstruction	Source
(111)	50	0.80	$\sqrt{3} \times 22 \pm 1$	[89]
(110)	80	1.28	1×2	[90]
(100)	100	1.60	5×20	[91]

Source: Courtesy of M. J. Ford, C. Masens, and M. B. Cortie, The application of gold surfaces and particles in nanotechnology, *Surf. Rev. Lett.*, 2006, **13**, 297.

The physical metallurgical characteristics of gold have been covered in detail elsewhere in Chapter 7. Gold forms a limited range of solid solutions with other elements, a factor that can be also be attributed to significant differences in atomic radius and electronegativity between it and other elements [15]. Although the Au-Cu, Au-Ni, Au-Pd, and Au-Pt systems do form continuous solid solutions at elevated temperatures, the mutual solubility is lost at lower temperatures. In Au-Cu and Au-Pd there are a series of low temperature intermetallic compounds, and in the others a spinodal decomposition occurs. Gold also forms a series of classic Hume-Rothery electron compounds with Zn, Cd, Al, and In. It can be considered to provide one electron per Au atom in the process.

There are a wide variety of other intermetallic compounds containing gold [30] besides the aurides and electron compounds already noted. The compounds $AuAl_2$, $AuGa_2$, and $AuIn_2$ have quite strong colors due to their having a reflectance minimum within the visible range of the spectrum. These minima are caused by strong interband transitions and a well-developed secondary band edge in the visible [31]. However, these substances (like most other intermetallic compounds) are also quite brittle which limits their commercial application in, for example, jewelry. A structural gold nitride has been claimed [32], but this appears to be a metastable material produced by ion implantation. In any case, this "nitride" is of a completely different nature to the explosively unstable gold azide-type compounds. Finally, there are also a few gold-based metallic glass compositions; for example, $Au_{49}Ag_{5.5}Pd_{2.3}Cu_{26.9}Si_{16.3}$ melts at only 371°C and forms a glass-like material if rapidly cooled [33].

PHYSICAL PROPERTIES

The combination of gold's relatively high surface energy (see Table 2.1) and its resistance to oxidation ensures that the element solidifies from the molten state into soft, malleable ingots. Gold has a Vickers hardness of only about 25 HV in the annealed form, but this can be increased to about 60 HV by the application of 60% cold work [14]. Further cold work such as, for example, by application of equal channel angular extrusion (ECAE), will apparently not increase the hardness beyond about 80 HV [34]. The flow stress of pure, annealed gold is similarly very low, in the range of only 25 to 30 MPa [35]. The application of 60 to 70% cold work can raise the flow stress to 220 MPa [14], but it can increased a little further by extreme cold work (as in equal channel angular pressing [ECAP]) or repeated cold rolling to about 245 MPa [35,36] or 280 MPa, respectively [37].

The great malleability of gold has led to its application in the form of decorative gold leaf (Chapter 15). A mere 30 g (roughly a "troy ounce") of gold can be beaten out to a sheet of more than 25 m² in area [1]. With care foils of gold with a thickness of less than 100 nm can be prepared this way [37]. This cannot be achieved by the mechanical working of other metallic elements because the foils of even the most ductile of these fracture when the foil thickness approaches a micron or so. Why is gold so malleable? The reason may be the interplay between its ductile fcc crystal structure and its resistance to oxidation. Nutting and Nuttall proposed [37] that the essential factor is that dislocations in the subgrains of severely deformed gold foil can readily escape to the free surfaces, where they would be annihilated. The presence of an oxide coating on most other metallic elements was said to prevent this from occurring.

A comprehensive tabulated summary of the physical properties of gold was prepared in 1979 by Cohn [14]. Table 2.2 lists a few of the physical parameters that are useful for technological applications, giving updated values where we are aware of them. Note that the melting point of pure, bulk Au was corrected to 1064.18°C in 1990 (from its previously held value of 1064.43°C) and serves as one of the calibration points for the International Temperature Scale ITS-90 [1]. The melting point is lowered in small particles due to the increase in their surface-to-volume ratio. The effect can be expressed as [38]

$$\frac{T_m(r)}{T_m(\infty)} = 1 - \frac{4}{\rho_s L}\left\{\gamma_s - \gamma_l\left(\frac{\rho_s}{\rho_l}\right)^{2/3}\right\}\frac{1}{d} \qquad (2.3)$$

TABLE 2.2
A Compendium of the Values of Some Other Physical Properties of Gold

Property		Notes	Source
Atomic radius	1.44 Å	"Metallic" radius	[14]
Lattice constant	4.07 Å		[14]
Electronegativity	2.44		[14]
Young's modulus	79 GPa	Annealed material at 20°C	[14]
Specific heat	0.1288 J/(g·K)	25°C	[14]
	25.4 J/(mol·K)		[1]
Thermal conductivity	314.4 W/(m·K)	0°C	[14]
Electrical resistivity	2.05×10^{-5} Ω.cm	0°C	[14]

Source: From *CRC Handbook of Chemistry and Physics*, D. R. Lide (ed.) (Taylor and Francis, London, 2008); and Selected properties of gold, J. G. Cohn, *Gold Bull.*, 1979, **12**, 21. With permission.)

where $T_m(r)$ and $T_m(\infty)$ are the melting points of particle of radius r and the bulk material, respectively, ρ_s and ρ_l are the density of solid and liquid, respectively, and γ_s and γ_l are the surface free energy of the solid and liquid. It has been shown that the melting point of gold nanoparticles can be suppressed to 430°C in the naked form [39], and down to 380°C if encased in a silica shell [38]. Surface melting on gold nanoparticles on a substrate has been reported to occur at temperatures of as low as 104°C [40]. However, for very small particles containing only tens of atoms, there is evidence to suggest that melting points can again increase, probably due to the fact that the simple thermodynamic model in Equation 2.3 is no longer valid within the molecular regime [41,42].

The work function of gold, a parameter which is of interest in the context of its use in electronic devices, is usually given in the range 4.8–5.5 eV [14] with 5.3 eV a representative value for use in molecular electronic systems [43]. However, it is worth noting that this figure can be modified by up to 2 eV either way by the attachment of a self-assembled monolayer [43,44].

Gold is an excellent conductor of both heat and electricity, but inferior to both silver and copper. However, the relative immunity of gold toward corrosion has ensured its place in the connectors required for all types of performance-critical electronic [45] devices. Electrical resistance is often expressed as a percentage of the International Annealed Copper Standard (e.g., 100% IACS). On this scale gold comes in at 74% of IACS or less, platinum at 15% or less, and silver at 104% or less [46].

Bulk gold is diamagnetic [47] (i.e., it is weakly repelled by magnetic fields) but extremely small Au nanoparticles (<2 nm in diameter) have recently been reported to have measurable ferromagnetic attributes [47] due to the unpaired electronic configuration of their surfaces.

OPTICAL PROPERTIES

COLOR OF BULK GOLD

The optical properties of all linear optical materials can be described by the complex, frequency-dependent permittivity or dielectric constant:

$$\varepsilon = \varepsilon_1 + i\varepsilon_2 \tag{2.4}$$

which is related to the complex refractive index by the relations

$$m = n + ik = \sqrt{\varepsilon}, \tag{2.5}$$

$$n = \sqrt{\frac{1}{2}\left(\sqrt{\varepsilon_1^2 + \varepsilon_2^2} + \varepsilon_1\right)}, \quad k = \sqrt{\frac{1}{2}\left(\sqrt{\varepsilon_1^2 + \varepsilon_2^2} - \varepsilon_1\right)}, \tag{2.6}$$

$$\varepsilon_1 = n^2 - k^2, \quad \varepsilon_2 = 2nk, \tag{2.7}$$

The permittivity of metals is usually modeled using a Drude-like function of the form:

$$\varepsilon_1(\omega) = \varepsilon_\infty - \frac{\omega_p^2}{\omega^2 + \gamma^2}, \quad \varepsilon_2(\omega) = \frac{\omega_p^2 \gamma}{\omega(\omega^2 + \gamma^2)} \tag{2.8}$$

$$\gamma = \frac{1}{\tau} \tag{2.9}$$

where ε_∞ is the real part of the permittivity at high frequency, ω_p is the bulk plasma frequency which can be approximated by $\omega_p^2 = n.e^2/m_e\varepsilon_0$ where n is the electron density, e the electron charge, m_e the effective electron mass, and ε_0 the permittivity of free space. γ is the damping constant, which is inversely proportional to the relaxation time, τ, which in turn is related to the average time between electron scattering events [48]. The Drude permittivity described in Equation 2.8 is for an idealized metal in the absence of interband transitions (electronic transitions from occupied to unoccupied bands). Interband transitions present themselves in all metals at some point in the UV or visible part of the spectrum. Interband transitions are difficult to accommodate in any simple, analytical form for the dieletric function; the usual method for doing this is to add one or more Lorentz oscillator terms to the Drude form [48].

The real and imaginary components of the dielectric function must preserve causality; this can be expressed in the usual way through the the the Kramers–Kronig relation:

$$\varepsilon_1(\omega) = 1 + \frac{2}{\pi} \int_0^\infty \frac{\Omega \varepsilon_2(\Omega)}{\Omega^2 - \omega^2} \partial\Omega \tag{2.10}$$

where the value of the real part of the permittivity at some frequency ω is simply the integration of the imaginary part of the permittivity at *all* frequencies Ω, weighted by the difference between the two frequencies. The presence of interband transitions will give a relatively large component to the imaginary dielectric function, which through Equation 2.10 will disrupt the Drude-like character of the real part of the permittivity in the vicinity of interband transitions.

The strong yellow color of bulk gold is due to a well-defined band edge at about 2.4 eV [49] (corresponding to roughly 500 nm or green light). The band edge is the energy at which interband transitions from occupied states to unoccupied states become allowed. In gold the band edge is caused by a combination of 5d to 6sp hybrid transitions in the vicinity of the X and L points in the Brillouin zone [50]. The energy of these transitions is very low in comparison to silver (4 eV [51]) due to the previously discussed contractions of the s and p orbitals and the corresponding destabilization of the d and f shells. Calculations of the interband transition spectra of gold, neglecting relativistic effects, put the band edge at approximately 4 eV [52]. Figure 2.3 shows the permittivity for gold, silver, and copper. The curves for the three coinage metals all have a similar shape, and the only real difference is the energy at which the transitions become apparent (the band edge).

The onset of transitions at 500 nm in gold causes the real part of the permittivity to hover around −1 in the region of 2.5–5.0 eV (250 to 500 nm). This results directly in the plasmon resonance "tail" that is visible in gold nanoparticles on the short wavelength side of the resonance, particularly in vacuum where the features are not washed out by solvent or surface effects [53]. This divergence of ε_1 away from a simple Drude-like character causes a sharp rise in the bulk

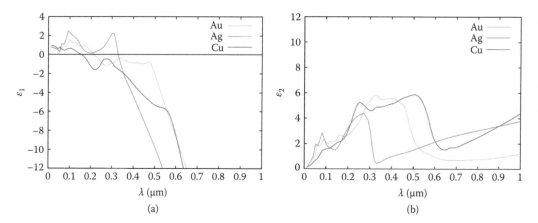

FIGURE 2.3 Real (a) and imaginary (b) components of the complex permittivity of the coinage metals. (Data from Optical properties of selected elements, J. H. Weaver and H. P. R. Frederikse, in *CRC Handbook of Chemistry and Physics,* D. R. Lide (ed.), CRC Press, Boca Raton, 2001.)

reflectivity from around only 35% at 500 nm to 95% at 600 nm. The surface thus reflects red, orange, yellow and some green, but absorbs from the green–blue to the violet parts of the spectrum (see Figure 2.4).

One of the most interesting properties of gold is its reflectivity which increases rapidly in the middle of the visible region of the spectrum and, like most metals, extends into the infrared. This is due to the unique combination of the position of the band edge and the value of the plasma frequency, which defines the properties of the Drude-like tail. In a perfect metal, the value of ε_∞ is 1, and the damping constant γ is negligible. This results in the real part of the permittivity crossing from positive (transmissive) to negative (reflective) at the plasma frequency. Of course, in real metals this is rarely the case, and the plasma frequency calculated from the energy density and reflection edge do not coincide. The interband transitions in gold cause the reflection edge to be much lower in energy than the "plasma frequency" calculated from the electron density. Additionally in gold, the free electron density contributing to the Drude plasma frequency is

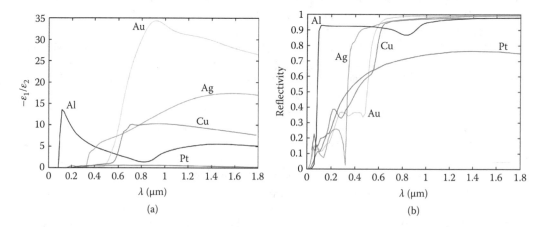

FIGURE 2.4 (a) Ratio of the real part of the permittivity to the imaginary part for selected metals, which is useful for describing the reflectivity spectra and the plasmonic properties of a material. (b) The bulk reflectivity spectra of selected metals. (Data from Optical properties of selected elements, J. H. Weaver and H. P. R. Frederikse, in *CRC Handbook of Chemistry and Physics,* D. R. Lide (ed.), CRC Press, Boca Raton, 2001.)

itself not obvious because of the overlap between 6s and 5d electrons. Values obtained by fitting a Drude model to the dielectric function change depending on what part of the spectrum is fitted. Values can range from $\omega_p = 4.4$ eV [54] to about 9 eV [55–57]. A reasonable value is 9 eV because, if coupled with $\varepsilon_\infty = 7.5$ to account for a large part of the interband transitions, this provides a tolerable Drude approximation to the dielectric function, giving the resonance position for gold nanospheres to within 0.3 eV [55].

THIN GOLD FILMS

Thin films of gold are readily produced on almost any solid substrate by techniques as diverse as electroplating, electroless deposition, and physical vapor deposition. These provide a means by which the desirable optical or surface properties of Au can be exploited without incurring excessive raw material costs. Due to the nobility of gold, such films will generally remain unoxidized, and can be prepared down to a thickness of a few tens of nanometers. Films of less than 80 nm or so in thickness will transmit an appreciable fraction of any blue to green light that falls on them, with red light and the near-infrared being selectively blocked. This is due to the position of the band edge at about 2.4 eV, described previously. This property has stimulated the use of gold films in spectrally selective applications (see Chapter 16). The optical properties of Au thin films can also be significantly varied by control of the deposition conditions. The properties are not only controlled by film thickness, but also by the morphology of the film. In particular, the degree of percolation, or its absence, has a strong effect on the shape of the transmitted and reflected spectra [58,59]. Optimum reflectivity in the infrared requires a continuous coating.

The surface of gold is also able to sustain a surface plasmon polariton (SPP), a nonradiative, propagating surface electromagnetic oscillation that is excited under particular circumstances by incident light. Gold is especially useful for the technological development of this phenomenon because its surfaces can be prepared in an oxide-free condition, and because it can sustain this resonance in the red and near-infrared ranges of the electromagnetic spectrum. This is the basis of an analytical technique known as surface plasmon resonance spectroscopy (SPRS) (see Chapter 16). Control of SPP has also been mooted as a means to achieve subwavelength optoelectronic devices [60].

There are two possible polarizations of a light wave with respect to the surface of a solid. These are the transverse magnetic (TM) mode, in which the electric field is perpendicular to the plane of the surface and in which the magnetic mode is transverse to it, and the transverse electric (TE) mode in which it is the magnetic field oscillation that is perpendicular to the surface and the electric field that is transverse. These are also known as the "p" and "s" modes, respectively [61]. A SPP has the TM characteristic. A further restriction is that it can only propagate if the real parts of the dielectric constants of the materials on the two sides of the interface are of opposite signs. This can easily be satisfied at the interface between metal and insulator (including vacuum), for example, with gold ($\varepsilon_1 < 0$) against air or water ($\varepsilon_1 > 1$).

The behavior of SPPs with a change in frequency, or its "dispersion," is often depicted on a plot of frequency, f, against wave vector, k, where $k = 2\pi/\lambda$, and λ is the wavelength [61,62]. The frequency parameter is more often expressed as an angular frequency, $\omega = 2\pi f$. Light propagating in free space shows no dispersion, that is, all frequencies propagate at the same velocity, the speed of light. Wave vector and frequency are then proportional to each with c being the constant of proportionality. However, generally this is not true when light interacts with matter because the extent to which the wave is retarded in the medium, i.e., its velocity, varies with ω; waves of different frequency propagate at different velocities, their phase velocities. In this case it makes more sense to talk about the group velocity which is the slope of the dispersion curve, i.e., $d\omega/dk$. A wave packet with a relatively narrow spread of frequencies and where the dispersion curve does not vary too rapidly over this frequency range, as a whole, propagates at the group velocity. This is an important concept as all electromagnetic waves will have, in practice, a frequency spread rather than be

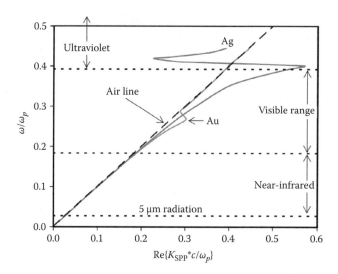

FIGURE 2.5 Dispersion characteristics of gold and silver, plotted using dimensionless parameters (Re means the "real part of").

monochromatic. Obviously k or k_0 can be further normalized if desired by dividing by ω_p, the bulk plasma frequency of the metallic portion, and multiplying by c, where c is the velocity of light in vacuum. As mentioned, a representative value for ω_p of Au is 9 eV, and we will invoke the same value here for Ag, in keeping with the experimental results of Vlasov [63]. These various simplifications permit the construction of a dimensionless dispersion curve for light traveling on an interface (Figure 2.5). The dispersion of light in vacuum, k_0 versus ω, is also shown. It is just a straight line because, as the frequency increases, the wavelength of the light must necessarily decrease in order to satisfy the relationship $\lambda \cdot f = c$. In general light must travel more slowly when it encounters a material of appreciable refractive index, n. This follows because n is defined as c/v.

The dispersion relationship of greatest interest in the present context is that of a TM electromagnetic wave trapped at the gold/insulator interface. It can be shown [61] that this is

$$k_{SPP} = \frac{\omega}{c}\sqrt{\frac{\varepsilon(\omega)\cdot\varepsilon_m}{\varepsilon(\omega)+\varepsilon_m}} \qquad (2.11)$$

where ε_m is the dielectric constant of the insulating medium. In the case of air or vacuum, for which $\varepsilon_m = 1$, this simplifies to

$$k_{SPP} = \frac{\omega}{c}\sqrt{\frac{\varepsilon(\omega)}{\varepsilon(\omega)+1}} \qquad (2.12)$$

This relationship is plotted in Figure 2.5 out to 4 eV (corresponding to UV radiation of 310 nm) using the established values of $\varepsilon(\omega)$ for gold and silver. It can be seen that the velocity of the propagating electromagnetic wave is less than that of c for all but the lowest of ω values, for which it nearly equals c. (Because $k = 2\pi/\lambda$, the value of λ *decreases* from left to right along the horizontal axis of this plot.) Also significant is that the dispersion curve for the SPP lies on or to the right of the light line. Because the momentum of a photon or a SPP is directly proportional to its k value, it can be seen that the SPP of light at a particular frequency carries with it a *greater* momentum than for the corresponding free photon. The increase is due to the manner in which the SPP is bound to the surface [64]. The important consequence of this is that, under normal circumstances, light does not couple to a gold–air surface because it has insufficient momentum. Coupling requires, for example, a Kretschmann prism or some other scheme to bridge the gap [64].

Descriptions of the dispersion curve for SPPs often invoke a Drude model for the dielectric constants of the metal. In these calculations K_{SPP} tends to an asymptote corresponding to the surface plasmon, where the real part of the permittivity $\varepsilon_1 = -1$ and the imaginary part $\varepsilon_2 \approx 0$. When experimentally determined data are used it becomes apparent that the dispersion curve can be quite complex in shape. For example, the dispersion curve for a SPP on silver contains a peculiar nonasymptotic "dogleg" at about 3.6 eV (about 340 nm) [61] due to the onset of strongly absorbtive interband transitions in the metal, with the precise shape and depth of this feature being very sensitive to experimental conditions or values assumed for the dielectric constants [63]. In particular, the value assumed for the plasma frequency ω_p, which is in general not a well-characterized quantity, is quite influential. On the other hand, for gold there is only a blip at 2.4 eV (about 520 nm) due to the much higher value for the imaginary part of the permittivity when the real part is around the value of –1. The momentum-matching requirements also differ between gold and silver, but in a complicated fashion, the significance of which will become apparent when we discuss the propagation lengths possible for SPPs. A SPP on gold has a slightly larger momentum at the slower frequencies of light (i.e., mid to lower portion of the dispersion curves shown); therefore, it is marginally easier to couple light onto a silver surface than a gold one.

Of course, in real metals there is always some attenuation of the electromagnetic energy due to absorptive processes within the metal. Different systems may be compared with reference to their characteristic SPP propagation lengths, L_{SPP}, given by [65]

$$L_{SPP} = [2.\,\mathrm{Im}(K_{SPP})]^{-1} \tag{2.13}$$

However, the following approximation [62] for L is nearly as good

$$L_{SPP} = \lambda_0 \frac{\varepsilon_1^2}{2\pi\varepsilon_2} \tag{2.14}$$

In this case, λ_0 is the wavelength of the light that excited the SPP, and L_{SPP} is the distance over which the power of the SPP falls to $1/e$ of its starting value. Longer L_{SPP}'s are produced by having larger negative values of ε_1 and, to a lesser extent, smaller values of ε_2. In principle SPPs can propagate several hundreds of micrometers along a perfect gold or silver interface with air when excited by near-infrared radiation, whereas propagation of SPPs in metals such as Cu or Al is not as good (Figure 2.6). A platinum interface has virtually no capability to carry a SPP in the near-infrared. The advantages offered by gold become larger at shorter wavelengths, with SPPs in gold potentially able to travel twice as far as those in silver when excited with red light. However, none of the elements examined can sustain an SPP at wavelengths shorter than about 600 nm. It is clear, therefore, that usable SPPs in Au and Ag are restricted to the lower energy portion of the dispersion curve from the origin ($k_{SPP} = 0$, $\omega = 0$) up to the "doglegs" mentioned previously.

In general, exploitation of SPPs depends on their propagating nonradiatively for as long a distance as possible. Any topographical defects on the surface will disrupt and hence attenuate the SPP mode, and in many instances cause it to re-emit or scatter some of its electromagnetic energy as light. Under suitable conditions such re-emitted light, sometimes described as a "plasmon jet," can be imaged in an optical microscope [66].

COLOR OF NANOPARTICLES

It is comparatively well known that nanoparticles of gold have a very different color to bulk gold. This is the localized surface plasmon resonance. Localized surface plasmon resonances in gold

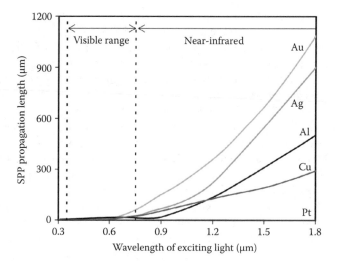

FIGURE 2.6 SPP propagation in selected metal films given by Equation 2.14.

have been the subject of several recent reviews [67–70], and a discussion of their many technological applications is available in Chapter 16. Here we will confine the discussion to a description of the phenomenon itself, and the various factors that influence it.

The localized surface plasmon resonance is caused by a collective oscillation of conduction electrons in a nanoparticle, acting in response to incident light of a suitable frequency. An analytical expression for the simplest resonance in spheres for which the diameter is much smaller than the wavelength of light is available; see, for example, Liz-Marzan [71] or Maier [61]:

$$C_{ext} = 9\frac{\omega}{c}\varepsilon_m^{3/2}V\frac{\varepsilon_2}{(\varepsilon_1 + 2\varepsilon_m)^2 + \varepsilon_2}$$

(2.15)

where $C_{ext} = C_{sca} + C_{abs}$ is the extinction cross-section of the nanoparticle, ω the angular frequency of the light in radians per second, c the velocity of light, ε_m the dielectric constant of the medium surrounding the nanoparticle, V the particle volume, and $\varepsilon = \varepsilon_1 + i\varepsilon_2$ the complex dielectric function of the sphere material (gold in this case). It can be seen that resonance is achieved when the term $(\varepsilon_1 + 2\varepsilon_m) = 0$, i.e., when $\varepsilon_1 = -2\varepsilon_m$. For $\varepsilon_m = 1$, which is the appropriate value for environments of vacuum or air, this means that resonance occurs whenever $\varepsilon_1 = -2$. However, the strength of the resonance can also be seen to depend on the magnitude of ε_2, because the denominator of the equation will be smaller if ε_2 is smaller. This equation describes the dipole resonance for the sphere and the peak position shows no size dependence. For larger particles, or for metals with low plasma frequency, a correction to Equation 2.15 can be introduced [53,72] which accounts for higher order modes and introduces a size dependency. Optical property data for a few well-known metallic elements are shown Figure 2.7 (see also Figure 2.4a). The curves tend from right to left as the wavelength of light is increased, with the complicated behavior at the right-hand side corresponding to the ultraviolet region, while the data pass monotonically off the left-hand side of the graph somewhere in the near-infrared. For a medium with dielectric constant ε_m of 1.3, roughly corresponding to water, the plasmon resonance condition (for a sphere) occurs when $\varepsilon_1 = -2.6$. These points of intersection between dielectric data and resonance conditions are shown on the figure as solid symbols. For gold in water the condition is met for incident light of about 520-nm wavelength, while in silver and copper it occurs at close to 370 nm. The resonance condition for a nanosphere of aluminum or platinum in water is met in the deep ultraviolet at 152 and 105 nm, respectively. Of course, these may or may not be convenient positions from a technological perspective but it is also worth noting the rather different values of ε_2 of each element

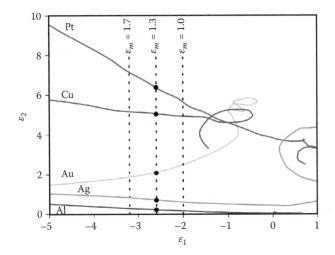

FIGURE 2.7 Complex permittivity data for Al, Ag, Au, Cu, and Pt illustrating also the positions at which a dipole plasmon resonance will be obtained in a nanosphere immersed in a medium of ε_m.

corresponding to these various resonance conditions (the solid symbols). It can be seen that Al and Ag will have stronger plasmon resonances than Au (because at these positions their ε_2 values are smaller), which in turn will be superior in this regard to Pt and Cu. It can also be seen, however, that the situation for gold does improve somewhat if the resonance can be displaced to longer wavelengths (i.e., toward the left-hand side of this figure). This can be done by increasing the dielectric constant of the medium, or by using a less symmetrical nanoparticle with a red-shifted plasmon resonance.

The optical properties of the alkali metals Li, K, or Na are also superior to gold, but of course gold has the great advantage that its nanostructures are resistant to oxidation and are readily fabricated by a wide range of lithographic techniques [73]. This combination of modest strength, mid-visible positioning, and chemical stability is no doubt the reason for the very wide range of technological applications based on the plasmon resonance of gold surfaces or particles (see Chapter 16) rather than, say, on Al or K.

Sols containing gold nanospheres are brownish red when the particles are in the 2- to 6-nm size range, becoming ruby red when the particles are in the 10- to 20-nm size range. As the diameter of the particles increase further the sols become purple then violet in transmitted light. The colors also depend on the shape of the nanospheres and the refractive index of the medium in which they are surrounded [74,75]. The phenomenon has been used to color glass since Roman times [76,77]. These color changes can be fully understood by considering how the plasmon resonance is effected by particle size, shape, and surrounding medium. The red shift in the plasmon resonance of gold nanoparticles is of the order of 15 to 20 nm for a radius increase from 5 to 50 nm [53]). This is not large in itself, but it is important to note that extinction spectra of metallic nanoparticles are due to the sum of two components—absorption and scattering. The ratio of intensity of these two components will shift depending on the size of the nanoparticle. Whereas absorption dominates at low particle volumes (effective radius $\ll \lambda$), scattering dominates at higher volumes (effective radius $\rightarrow \lambda$). Lower limits of this generalization come into play at around 10 nm, below which electron scattering from free surfaces becomes a dominant factor. For particles smaller than 3 nm quantum effects, such as the onset of a band gap, also become important. The net effect of these latter two factors is to attenuate and ultimately wash out the plasmon resonance as the particle size is decreased below 10 nm [78], causing the color of the sol to become brownish rather than red.

On the other hand, as the dimensions of the particle exceed 50 nm or so, optical scattering effects become important and the plasmon resonance is also attenuated. Gold nanospheres of 50-nm diameter or larger scatter a significant proportion of the light falling on them, giving them a violet color when viewed in transmission and a green–brown color in reflection. The light scattered from such large nanoparticles allows them to be located in an optical microscope equipped with "dark-field" optics even though the particles themselves are considerably smaller than the resolving power of the microscope lenses [79]. This phenomenon can be exploited for some types of diagnostic medical testing (see Chapter 16).

Although the optical properties of gold nanospheres have been known and studied for a very long time, it has only been since the 1990s that the possibilities inherent in other nanoparticle shapes became widely appreciated. The simple dipolar resonance of spheres becomes replaced by a variety of other plasmon resonances in these other shapes due to the change in geometric symmetry. In general, these resonances are located in the red end of the visible, and even in the near-infrared. This is of interest in many technological applications, with one reason being that the human body is relatively transparent in the near-infrared region [80]. Although a huge number of interesting shapes such as nanoboxes, nanotriangles, nanocages, nanocubes, nanohexagons, nanocrescents, and nanorings have been produced, most scientific and commercial interest has been focused on only two other shapes: gold nanoshells and gold nanorods. This is because the plasmon resonances of these two shapes are well defined and can be readily tuned into the near-infrared part of the spectrum. The many different shapes of gold nanoparticles and their technological applications are described further in Chapter 16. We also refer the reader elsewhere for head-to-head comparisons of the technological merits of gold nanorods versus gold nanoshells [81,82].

The wavelength of the peak optical extinction in nanorods and nanoshells is controlled by their aspect ratios, although the term has a very different definition for the two shapes. The aspect ratio of a rod is the ratio of its length to its diameter while that of a nanoshell is defined as the ratio of its inner to its outer diameter. Increase of the aspect ratio in either case causes a red shift in the optical extinction. Therefore, control of the position of plasmon resonance can be achieved by simply changing the aspect ratio, if desired. There are also a few other ways in which the resonances can be tuned. For example, the resonances are also red shifted when neighboring nanoparticles are brought into close proximity with one another [83,84]; however, when the particles touch and aggregate on macroscopic scale the resonances collapse and the material becomes black (see gold sponges later). In the case of nanospheres this causes the color of sols to change from red or pink to blue or black. The color change is the basis of diverse analytical or technological applications (see Chapter 16 on nanotechnology). Red shifting can also be achieved by roughening the surface of the nanoparticle [85] or by encasing it in a coating of some dielectric material.

GOLD SPONGES AND GOLD BLACK

The optical properties of gold nanostructures are exceedingly flexible; it is even possible to prepare gold sponges or gold "blacks" which have very high absorptance [86,87]. The essential attribute of such materials is that they possess very well-developed mesoscale porosity and surface roughness so that they trap and absorb incident light. Because the structure is at a size scale that is well below the wavelength of the incident light, it becomes possible to model the optical response using Effective Medium theory. In this scheme the performance of the complex composite structure is modeled as if it were a simple monolithic material, using "effective" values of the refractive indices n and k derived from numerical analysis of experimental data. In a sense the sponge becomes an example of a "metamaterial" with engineered physical properties [88]. Some comments on the applications of such coatings are provided in Chapter 16.

CONCLUSIONS

The unique physical properties of gold may be explained as being the result of its electronic configuration, in particular, the well-developed relativistic contraction that occurs in this element, relative to other elements. The yellow color is the result of the onset of 5d to 6sp hybrid transitions at high symmetry points in the Brillouin zone, resulting in a band edge of approximately 2.4 eV. The ratio of real to imaginary permittivity, which is primarily caused by the position of this band edge, makes for excellent plasmonic characteristics, especially in the upper visible and near-infrared. The electronic configuration has a direct effect on the chemical reactivity of gold, with the relativistically induced strangeness of first and second ionization potentials playing a significant role in the inertness and high-level oxidation states while simultaneously ensuring that this is the most electronegative of the metallic elements. Mechanical, electrical, and other physical properties of gold are primarily influenced by its face centered cubic structure, a factor that provides a plethora of alloying opportunities, high ductility, and relativistically induced surface reconstructions.

REFERENCES

1. *CRC Handbook of Chemistry and Physics*, D. R. Lide (ed.) (Taylor & Francis, London, 2008).
2. The use of gold nanoparticles to enhance radiotherapy in mice, J. F. Hainfeld, D. N. Slatkin, and H. M. Smilowitz, *Phys. Med. Biol.*, 2004, **49**, N309.
3. New frontiers in gold labeling, J. F. Hainfeld and R. D. Powell, *J. Histochem. Cytochem.*, 2000, **48**, 471.
4. *Colloidal Gold: Principles, Methods, and Applications*, M. A. Hayat (ed.) (Academic Press, San Diego, CA, 1989).
5. Impact of gold nanoparticles combined to x-ray irradiation on bacteria, A. Simon-Deckers, E. Brun, B. Gouget, M. Carrière, and C. Sicard-Roselli, *Gold Bull.*, 2008, **41**, 187.
6. Understanding gold chemistry through relativity, H. Schmidbaur, S. Cronje, B. Djordjevic, and O. Schuster, *Chem. Phys.*, 2005, **311**, 151.
7. The fascinating implications of new results in gold chemistry, H. Schmidbaur, *Gold Bull.*, 1990, **23**, 11.
8. Relativistic self-consistent field calculation for mercury, D. F. Mayers, *Proc. R. Soc. London. Ser. A, Math. Phys. Sci. (1934-1990)*, 1957, **241**, 93.
9. Relativity and the periodic system of elements, P. Pyykkö and J.-P. Desclaux, *Acc. Chem. Res.*, 1979, **12**, 276.
10. Generalized gradient approximation made simple, J. P. Perdew, K. Burke, and M. Ernzerhof, *Phys. Rev. Lett.*, 1996, **77**, 3865.
11. A. Garcia, N. Troullier, J. L. Martins, and S. Froyen, ATOM v3.2.2, 2006, http://www.uam.es/siesta/.
12. The SIESTA method for ab initio order-N materials simulation, J. M. Soler, E. Artacho, J. Gale, A. Garcia, J. Junquera, P. Ordejon, and D. Sanchez-Portal, *J. Physics Condensed Matter*, 2002, **14**, 2745.
13. Dependence of relativistic effects on electronic configuration in the neutral atoms of d- and f-block elements, J. Autschbach, S. Siekierski, M. Seth, P. Schwerdtfeger, and W. H. E. Schwarz, *J. Comput. Chem.*, 2002, **23**, 804.
14. Selected properties of gold, J. G. Cohn, *Gold Bull.*, 1979, **12**, 21.
15. The chemistry of gold as an anion, M. Jansen, *Chem. Soc. Rev.*, 2008, **37**, 1826.
16. Electronic properties of alkali-metal--gold compounds, C. Koenig, N. E. Christensen, and J. Kollar, *Phys. Rev. B*, 1984, **29**, 6481.
17. Studies of the semiconducting properties of the compound CsAu, W. E. Spicer, A. H. Sommer, and J. G. White, *Phys. Rev.*, 1959, **115**, 57.
18. Relativistic effects in gold chemistry. I. Diatomic gold compounds, P. Schwerdtfeger, M. Dolg, W. H. E. Schwarz, G. A. Bowmaker, and P. D. W. Boyd, *J. Chem. Phys.*, 1989, **91**, 1762.
19. Ab initio calculations on the (ClAuPh$_3$)$_2$ dimer with relatavistic pseudopotential - Is the autophilic attraction a correlation effect?, P. Pyykkö and Y. F. Zhao, *Angew. Chem. Int. Ed.*, 1991, **30**, 604.
20. The aurophilicity phenomenon: a decade of experimental findings, theoretical concepts and emerging applications, H. Schmidbaur, *Gold Bull.*, 2000, **33**, 3.
21. Strong closed-shell interactions in inorganic chemistry, P. Pyykkö, *Chem. Rev.*, 1997, **97**, 597.
22. Theory of the d(10)-d(10) closed-shell attraction .1. Dimers near equilibrium, P. Pyykkö, N. Runeberg, and F. Mendizabal, *Chem. Eur. J.*, 1997, **3**, 1451.
23. Relativistic effects in inorganic and organometallic chemistry, N. Kaltsoyannis, *J. Chem. Soc. Dalton Trans.*, 1997, **1**, 1.
24. The application of gold surfaces and particles in nanotechnology, M. J. Ford, C. Masens, and M. B. Cortie, *Surf. Rev. Lett.*, 2006, **13**, 297.
25. Reconstruction phenomena at metal-electrolyte interfaces, D. M. Kolb, *Prog. Surf. Sci.*, 1996, **51**, 109.

26. Changes in the surface energy during the reconstruction of Au(100) and Au(111) electrodes, E. Santos and W. Schmickler, *Chem. Phys. Let.*, 2004, **400**, 26.
27. Wetting and surface tension measurements on gold alloys, E. Ricci and R. Novakovic, *Gold Bull.*, 2001, **34**, 41.
28. Calculations of the surface stress tensor and surface energy of the (111) surfaces of iridium, platinum and gold, R. J. Needs and M. Mansfield, *J. Phys: Condensed Matter*, 1989, **1**, 7555.
29. Size- and temperature-dependent structural transitions in gold nanoparticles, K. Koga, T. Ikeshoji, and K. Sugawara, *Phys. Rev. Lett.*, 2004, **92**, 115507.
30. A survey of gold intermetallic chemistry, R. Ferro, A. Saccone, D. Macciò, and S. Delfino, *Gold Bull.*, 2003, **36**, 39.
31. Optical properties of the intermetallic compounds $AuAl_2$, $AuGa_2$ and $AuIn_2$, S. S. Vishnubhatla and J. P. Jan, *Philos. Mag.*, 1967, **16**, 45.
32. X-ray induced decomposition of gold nitride, Y. V. Butenko, L. Alves, A. C. Brieva, J. Yang, S. Krishnamurthy, and L. Šiller, *Chem. Phys. Let.*, 2006, **430**, 89.
33. Gold based bulk metallic glass, J. Schroers, B. Lohwongwatana, W. L. Johnson, and A. Peker, *Appl. Phys. Lett.*, 2005, **87**, 061912.
34. Equal channel angular extrusion of high purity gold solid, W. Y. Yeung, R. Wuhrer, M. Cortie, and M. Ferry, *Mater. Forum*, 2007, **31**, 31.
35. Characteristics of face-centered cubic metals processed by equal-channel angular pressing, N. Q. Chinh, J. Gubicza, and T. G. Langdon, *J. Mater. Sci.*, 2007, **42**, 1594.
36. Microstructural characteristics of pure gold processed by equal-channel angular pressing, J. Gubicza, N. Q. Chinh, P. Szommer, A. Vinogradov, and T. G. Langdon, *Scripta Mater.*, 2007, **56**, 947.
37. The malleability of gold. An explanation of its unique mode of deformation, J. Nutting and J. L. Nuttall, *Gold Bull.*, 1977, **10**, 2.
38. Size-dependent melting of silica-encapsulated gold nanoparticles, K. Dick, T. Dhanasekaran, Z. Zhang, and D. Meisel, *J. Am. Chem. Soc.*, 2002, **124**, 2312.
39. Size effect on the melting temperature of gold particles, P. Buffat and J.-P. Borel, *Phys. Rev. A*, 1976, **13**, 2287.
40. A surface phase transition of supported gold nanoparticles, A. Plech, R. Cerna, V. Kotaidis, F. Hudert, A. Bartels, and T. Dekorsy, *Nano Lett.*, 2007, **7**, 1026.
41. Melting in small gold clusters: a Density Functional molecular dynamics study, B. Soulé de Bas, M. J. Ford, and M. B. Cortie, *J. Phys: Condensed Matter*, 2006, **18**, 55.
42. Ab initio molecular dynamical investigation of finite temperature behaviour of the tetrahedral Au_{19} and Au_{20} clusters S. Krishnamurty, G. S. Shafai, D. G. Kanhere, B. Soulé de Bas, and M. J. Ford, *J. Phys. Chem. A*, 2007, **111**, 10769.
43. Asymmetry in electron transport through molecules studied using ab initio and barrier tunneling models, R. C. Hoft, N. Armstrong, M. J. Ford, and M. B. Cortie, *J. Phys: Condensed Matter*, 2007, **19**, 215206.
44. Toward control of the metal-organic interfacial electronic structure in molecular electronics: a first-principles study on self-assembled monolayers of π-conjugated molecules on noble metal, G. Heimel, L. Romaner, E. Zojer, and J.-L. Brédas, *Nano Lett.*, 2007, **7**, 932.
45. Going for gold, R. Holliday and P. Goodman, *IEE Rev.*, 2002, 15.
46. *Metals Handbook. Desk Edition*, H. E. Boyer and T. L. Gall (ed.) (American Society for Metals, Metals Park, Ohio, 1985).
47. Direct observation of ferromagnetic spin polarization in gold nanoparticles, Y. Yamamoto, T. Miura, M. Suzuki, N. Kawamura, H. Miyagawa, T. Nakamura, K. Kobayashi, T. Teranishi, and H. Hori, *Phys. Rev. Lett.*, 2004, **93**, 116801.
48. *Solid State Physics*, N. W. Ashcroft and N. D. Mermin (ed.) (Harcourt College Publishers, New York, 1976).
49. Spherical perfect lens: Solutions of Maxwell's equations for spherical geometry, S. A. Ramakrishna and J. B. Pendry, *Phys. Rev. B*, 2004, **69**, 241101.
50. The role of relativity in the optical response of gold within the time-dependent current-density-functional theory, P. Romaniello and P. L. de Boeij, *J. Chem. Phys.*, 2005, **122**, 164303.
51. Optical properties of selected elements, J. H. Weaver and H. P. R. Frederikse, in *CRC Handbook of Chemistry and Physics*, D. R. Lide (ed.) (CRC Press, Boca Raton, 2001) 133.
52. Theoretical chemistry of gold, P. Pyykkö, *Angew. Chem. Int. Ed.*, 2004, **43**, 4412.
53. Plasmon absorption in nanospheres: A comparison of sodium, potassium, aluminium, silver and gold, M. G. Blaber, M. D. Arnold, N. Harris, M. J. Ford, and M. B. Cortie, *Phys. B, Condensed Matter*, 2007, **394**, 184.

54. Optical pulse propagation in metal nanoparticle chain waveguides, S. A. Maier, P. G. Kik, and H. A. Atwater, *Phys. Rev. B*, 2003, **67**, 205402.

55. Efficient coupling of high-intensity subpicosecond laser pulses into solids, M. M. Murnane, H. C. Kapteyn, S. P. Gordon, J. Bokor, E. N. Glytsis, and R. W. Falcone, *Appl. Phys. Lett.*, 1993, **62**, 1068.

56. Modelling the optical response of gold nanoparticles, V. Myroshnychenko, J. Rodríguez-Fernández, I. Pastoriza-Santos, A. M. Funston, C. Novo, P. Mulvaney, L. M. Liz-Marzán, and F. J. G. d. Abajo, *Chem. Soc. Rev.*, 2008, **37**, 1792.

57. On geometrical scaling of split-ring and double-bar resonators at optical frequencies, S. Tretyakov, *Metamaterials*, 2007, **1**, 40.

58. Noble-metal-based transparent infrared reflectors: experiments and theoretical analyses for very thin gold films, G. B. Smith, G. A. Niklasson, J. S. E. M. Svensson, and C. G. Granqvist, *J. Appl. Phys.*, 1986, **59**, 571.

59. Infrared-optical properties of vapour-deposited metal films, M. Buskühl and E.-H. Korte, *Anal. Bioanal. Chem.*, 2002, **374**, 672.

60. The New "p-n Junction": Plasmonics enables photonic access to the nanoworld, H. A. Atwater, S. Maier, A. Polman, J. A. Dionne, and L. Sweatlock, *MRS Bull.*, 2005, **30**, 385.

61. *Plasmonics. Fundamentals and Applications*, S. A. Maier (ed.) (Springer, New York, 2007).

62. Surface plasmon–polariton length scales: a route to sub-wavelength optics, W. L. Barnes, *J. Opt. A: Pure Appl. Opt.*, 2006, **8**, S87.

63. Spectroscopy of surface plasmons in metal films with nanostructures, V. Vlasko-Vlasov, A. Rydh, J. Pearson, and U. Welp, *Appl. Phys. Lett.*, 2006, **88**, 173112.

64. Surface plasmon subwavelength optics, W. L. Barnes, A. Dereux, and T. W. Ebbesen, *Nature*, 2003, **424**, 824.

65. Electromagnetic energy transport below the diffraction limit in periodic metal nanostructures S. A. Maier, P. G. Kik, M. L. Brongersma and H. A. Atwater, *Proc. SPIE*, 2001, **4456**, 22.

66. Far-field optical microscopy with a nanometer-scale resolution based on the in-plane image magnification by surface plasmon polaritons, I. I. Smolyaninov, J. Elliott, A. V. Zayats, and C. C. Davis, *Phys. Rev. Lett.*, 2005, **94**, 057401.

67. The optical properties of metal nanoparticles: the influence of size, shape, and dielectric environment, K. L. Kelly, E. Coronado, L. L. Zhao, and G. C. Schatz, *J. Phys. Chem. B*, 2003, **107**, 668.

68. Shape and size dependence of radiative, non-radiative and photothermal properties of gold nanocrystals, S. Link and M. A. El-Sayed, *Inter. Rev. Phys. Chem.*, 2000, **19**, 409.

69. Plasmonics–A route to nanoscale optical devices, S. A. Maier, M. L. Brongersma, P. G. Kik, S. Meltzer, A. A. G. Requicha, and H. A. Atwater, *Adv. Mater.*, 2001, **13**, 1501.

70. Plasmonics: merging photonics and electronics at nanoscale dimensions, E. Ozbay, *Science*, 2006, **311**, 189.

71. Synthesis of nanosized gold-silica core-shell particles, L. M. Liz-Marzán, M. Giersig, and P. Mulvaney, *Langmuir*, 1996, **12**, 4329.

72. *Absorption and Scattering of Light by Small Particles*, C. F. Bohren and D. R. Huffman (ed.) (Wiley, Weinheim, 2004).

73. Preparation of nanoscale gold structures by nanolithography, N. Stokes, A. M. McDonagh, and M. B. Cortie, *Gold Bull.*, 2007, **40**, 310.

74. Surface plasmon spectroscopy of nanosized metal particles, P. Mulvaney, *Langmuir*, 1996, **12**, 788.

75. UV-visible transmission-absorption spectral study of Au nanoparticles on a modified ITO electrode at constant potentials and under potential modulation, A. Toyota, N. Nakashima, and T. Sagara, *J. Electroanalytical Chem.*, 2004, **565**, 335.

76. The Lycurgus Cup – a Roman nanotechnology, I. Freestone, N. Meeks, M. Sax, and C. Higgitt, *Gold Bull.*, 2007, **40**, 270.

77. Gold nanoparticles: assembly, supramolecular chemistry, quantum-size-related properties, and applications toward biology, catalysis, and nanotechnology, M.-C. Daniel and D. Astruc, *Chem. Rev.*, 2004, **104**, 293.

78. Optical absorption spectra of nanocrystal gold molecules, M. M. Alvarez, J. T. Khoury, T. G. Schaaff, M. N. Shafigullin, I. Vezmar, and R. L. Whetten, *J. Phys. Chem. B*, 1997, **101**, 3706.

79. Plasmon resonances in large noble-metal clusters, C. Sonnichsen, T. Franzl, T. Wilk, G. von Plessen, and J. Feldmann, *New J. Phys.*, 2002, **4**, 93.1.

80. Near-infrared optical properties of *ex vivo* human skin and subcutaneous tissues measured using the Monte Carlo inversion technique, C. R. Simpson, M. Kohl, M. Essenpreis, and M. Copey, *Phys. Med. Biol.*, 1998, **43**, 2465.

81. Calculated absorption and scattering properties of gold nanoparticles of different size, shape, and composition: applications in biological imaging and biomedicine, P. K. Jain, K. S. Lee, I. H. El-Sayed, and M. A. El-Sayed, *J. Phys. Chem. B*, 2006, **110**, 7238.

82. Tunable infrared absorption by meta nanoparticles: The case for gold rods and shells, N. Harris, M. J. Ford, P. Mulvaney, and M. B. Cortie, *Gold Bull.*, 2008, **41**, 5.

83. Near infrared optical absorption of gold nanoparticle aggregates, T. J. Norman, J. C. D. Grant, D. Magana, J. Z. Zhang, J. Liu, D. Cao, F. Bridges, and A. V. Buuren, *J. Phys. Chem. B*, 2002, **106**, 7005.

84. Dipole-dipole plasmon interactions in gold-on-polystyrene composites, K. E. Peceros, X. Xu, S. R. Bulcock, and M. B. Cortie, *J. Phys. Chem. B*, 2005, **109**, 21516.

85. Controlled texturing modifies the surface topography and plasmonic properties of Au nanoshells, H. Wang, G. P. Goodrich, F. Tam, C. Oubre, P. Nordlander, and N. J. Halas, *J. Phys. Chem. B*, 2005, **109**, 11083.

86. Optical properties of mesoporous gold films, A. I. Maaroof, M. B. Cortie, and G. B. Smith, *J. Opt. A: Pure Appl. Opt.*, 2005, **7**, 303.

87. Optical and electrical properties of black gold layers in the far infrared, W. Becker, R. Fettig, and W. Ruppel, *Infrared Physics Technol.*, 1999, **40**, 431.

88. Mesoporous gold sponge as a prototype 'meta-material', A. Maaroof, M. B. Cortie, and G. B. Smith, *Phys. B*, 2007, **394**, 167.

89. Relaxation and reconstruction on (111) surfaces of Au, Pt, and Cu, Ž. Crljen, P. Lazić, D. Šokčević, and R. Brako, *Phys. Rev. B*, 2003, **68**, 195411.

90. Reconstruction of charged surfaces: General trends and a case study of Pt(110) and Au(110), A. Y. Lozovoi and A. Alavi, *Phys. Rev. B*, 2003, **68**, 245416.

91. Au(111): A theoretical study of the surface reconstruction and the surface electronic structure, N. Takeuchi, C. T. Chan, and K. M. Ho, *Phys. Rev. B*, 1991, **43**, 13899.

3 A Primer of Gold Chemistry

Alan L. Balch and Thelma Y. Garcia

CONTENTS

Introduction...31
Dissolution of Gold..31
Gold(-I)...33
Gold(0)...33
Gold(I)..35
Gold(II)...39
Gold(III)..40
Gold(V)...42
Clusters with Fractional Oxidation States..43
Utility of Gold Complexes..43
References...46

INTRODUCTION

This chapter is designed to give a short overview of the chemistry of gold with particular, but not exclusive, emphasis on recent developments. There are a number of monographs [1,2], reviews [3,4], and a recent special issue of *Chemical Society Reviews* [5] that give much more comprehensive coverage to this topic. The interested reader will find greater detail in these sources.

This chapter is largely organized around the different oxidation states of gold and focuses on the various structural options that are available in these known oxidation states. Figure 3.1 illustrates the interrelation between oxidation state and coordination geometry.

The drawings in this figure demonstrate how the structure of a gold center coordinated by two trithiacyclononane macrocycles changes as oxidation state varies [6,7]. These ligands present a set of six sulfur atoms as potential donors. In the gold(I) oxidation state the complex has a distorted tetrahedral geometry as seen in part A, and two of the donor sulfur atoms are uncoordinated. In the gold(II) state shown in part B, all of the sulfur atoms are coordinated although the two axial Au-S distances are longer than the other four. Finally, in the gold(III) state again only four sulfur atoms are coordinated but now in a planar arrangement as shown in part C.

Before discussing the available oxidation states and their chemical behavior, we will discuss the methods available to get metallic gold, a notably unreactive metal, into solution.

DISSOLUTION OF GOLD

Many of the chemical reactions of gold compounds are conducted in solution and so it is appropriate to initially consider how soluble gold compounds are obtained from this most attractive but inert metal. Traditionally, the dissolution of gold has involved a strong oxidant and a ligating agent. Thus, gold can be dissolved by treatment with aqua regia (a mixture of nitric acid [the oxidant] and hydrochloric acid [the ligand source], neither of which alone dissolves gold) to form $H[AuCl_4] \cdot n(H_2O)$ or

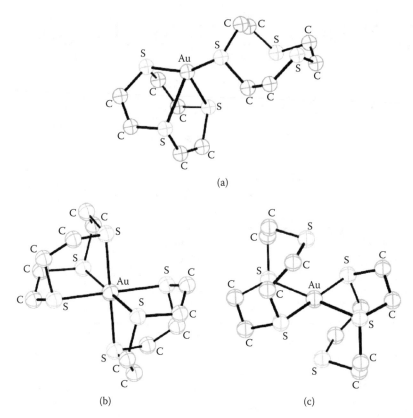

(a)

(b)   (c)

FIGURE 3.1 **(See color insert following page 212.)** The structures of [Au(1,4,7-trithiacyclononane)$_2$]$^{+/2+/3+}$ showing the structural variations in different oxidation states. (a) Monocation Au^I, (b) Dication Au^{II} (c) Trication Au^{III}. For clarity hydrogen atoms are omitted. [Data from Bis(1,4,7-trithiacyclononane)gold dication: a paramagnetic, mononuclear AuII complex, A. J. Blake, J. A. Greig, A. J. Holder, T. I. Hyde, A. Taylor, and M. Schröder, *Angew. Chem. Int. Ed.* 1990, **29**, 197; and Gold thioether chemistry – synthesis, structure, and redox interconversion of [Au(1,4,7-trithiacyclononane)$_2$]$^{+/2+/3+}$, A. J. Blake, R. O. Gould, J. A. Greig, A. J. Holder, T. I. Hyde, and M. Schröder, *J. Chem. Soc., Chem. Commun.*, 1989, 876.]

by oxidation by O$_2$ in the presence of cyanide ion. However, there are reports of metallic gold dissolving under less harsh conditions. For example, thin gold films are attacked by an ethyl alcohol solution of 4-pyridinethiol in air as can be seen in Figure 3.2 [8].

O$_2$ appears to be the oxidant in this dissolution process, which would appear to have significant implications in the handling of gold nanoclusters that frequently are stabilized by coatings of thiolate ligands [9]. In related work, dithoxamide/dihalogen compounds have been demonstrated to act as oxidations that are capable of dissolving gold. Exposure of metallic gold to chloroform solutions of cetyltrimethylammonium bromide also results in dissolution with the formation of [Au^{III}Br$_4$]$^-$ [10]. Again O$_2$ from air is the oxidant. Under rather different anhydrous and anaerobic conditions, Me$_3$AsI$_2$ and Me$_3$PI$_2$ in diethyl ether solution will dissolve metallic gold to produce gold(III) complexes as shown in reactions (3.1) and (3.2) [11]. The products have been isolated as crystalline solids.

$$3Me_3AsI_2 + 2Au^0 \rightarrow 2Au^{III}I_3(AsMe_3) + Me_3As \tag{3.1}$$

$$4Me_3PI_2 + 2Au^0 \rightarrow 2Au^{III}I_3(PMe_3)_2 + I_2 \tag{3.2}$$

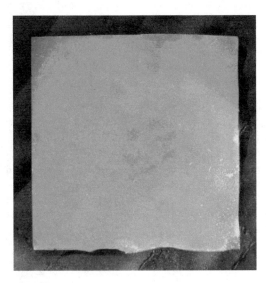

FIGURE 3.2   (See color insert following page 212.) Photograph of a thin gold film that has been briefly immersed into alcohol solution of 4-pyridinethiol. Dissolution of the gold film begins from the corners and edges. (From Oxidation of elemental gold in alcohol solutions, M. T. Risnen, M. Kemell, M. Leskel, and T. Repo, *Inorg. Chem.*, 2007, **46**, 3251. With permission.)

GOLD(–I)

The electron affinity of gold is remarkably high, 222.8 kJ/mol. Of all the elements, only the halogens have higher electron affinities (270–328 kJ/mol). This high electron affinity of gold facilitates the formation of the auride ion, Au(–I) which has the electronic configuration $[Xe]4f^{14}5d^{10}6s^2$ [12,13]. Thus, metallic gold reacts with electropositive metals, cesium and rubidium, to form salt-like compounds such as yellow $CsAu^{-1}$ and $RbAu^{-1}$. These are highly reactive compounds that must be handled under inert atmosphere conditions. The auride ion is isoelectronic with thallium(I), but it is an anion. Consequently, its chemical behavior is far different from that of thallium(I). Indeed, the chemistry of Au(–I) resembles that of the halide ions. Thus $CsAu^{-1}$ can be thought of as $(Cs^+)(Au^-)$.

$CsAu^{-1}$ dissolves in liquid ammonia to give very pale yellow solutions that are highly conductive as expected for an ionic species. Treatment of an ammonia solution of $CsAu^{-1}$ with an ion exchange resin loaded with tetramethylammonium ion or metathesis of $CsAu^{-1}$ with tetramethylammonium chloride in liquid ammonia leads to the formation of the colorless salt $(Me_4N)Au^{-1}$, whose structure is shown in Figure 3.3. $(Me_4N)Au^{-1}$, which consists of isolated ions of the two components, is isostructural with $(Me_4N)Br$.

The salt $[Cs([18]crown-6)(NH_3)_3]Au^{-1}\cdot NH_3$ has been obtained by addition of the crown ether, [18] crown-6, to an ammonia solution of $CsAu^{-1}$ [14]. This salt consists of a cesium cation coordinated by the lone pairs of six-oxygen atoms of the ([18]crown-6) ligand and three ammonia molecules. Figure 3.4 shows a drawing that compares the coordination of Rb^+ (a Lewis acid) and Au^- (a Lewis base). Remarkably, the auride ions are hydrogen bonded to the protons of four ammonia molecules to produce a tetrahedral arrangement of four N-H groups about each auride ion as shown in Figure 3.4. This compound was the first example found in which the auride ion acted as a hydrogen bond acceptor.

GOLD(0)

Complexes containing Au(0), which should be paramagnetic due to their electron configuration ($[Xe]4f^{14}5d^{10}6s^1$), are exceedingly rare. One Au(0) complex that was obtained in solution by electrochemical reduction of the macrocyclic gold(I) complex shown in Figure 3.5 has been detected by

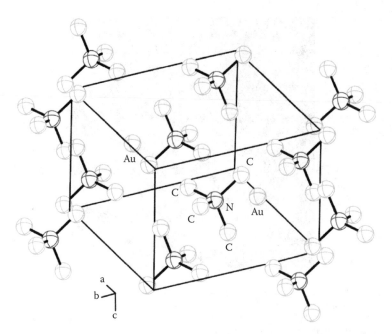

FIGURE 3.3 **(See color insert following page 212.)** The crystal structure of (Me₄N)Au⁻¹. For clarity hydrogen atoms are not shown. (Data from Synthesis and crystal structure determination of tetramethylammonium auride, P. D. C. Dietzel and M. Jansen, *Chem. Commun.*, 2001, 2208.)

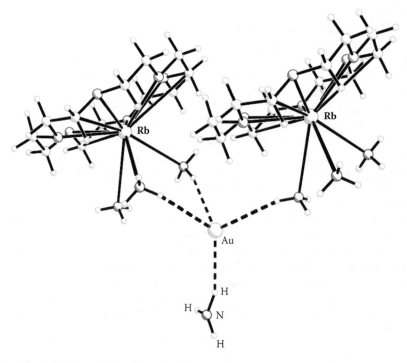

FIGURE 3.4 **(See color insert following page 212.)** A drawing comparing the coordination of Rb⁺ and Au⁻ and emphasizing the hydrogen bonding of ammonia molecules to the auride ion in [Cs([18]crown-6)(NH₃)₃] Au⁻¹·NH₃. (Data from [Rb([18]crown-6)(NH₃)₃]Au·NH₃: gold as acceptor in N-H· · ·Au⁻ hydrogen bonds, H. Nuss and M. Jansen, *Angew. Chem. Int. Ed.*, 2006, **45**, 4369.)

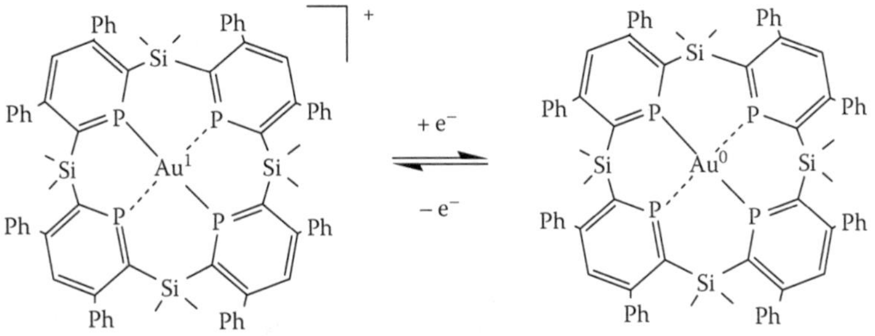

FIGURE 3.5   Formation of a gold(0) complex.

electron paramagnetic resonance (EPR) spectroscopy [15]. Analysis of the EPR spectrum indicates that 24% of the spin resides in a gold s orbital. The EPR spectra of gold(0) atoms deposited in arene and other hydrocarbon matrices at 77 K have also been reported [16].

There is also an old report of a dimeric complex, $Au^0_2(PPh_3)_2$, that was formed by reduction of $ClAu^I(PPh_3)$ [17]. This compound, which has a short Au-Au bond, is an analog of mercury(I) compounds like linear Hg_2Cl_2. Oddly, however, this crystallographic study indicated that this complex had P-Au-Au angles that were strongly bent. Details of this work have not been published, and the compound warrants reinvestigation.

Despite the paucity of well-characterized gold(0) compounds, there are numerous cluster complexes that contain gold in a mixed or fractional oxidation state that lies between Au(0) and Au(I). These clusters are described in a later section on clusters and chains with fractional oxidation states.

GOLD(I)

Gold(I), with an electronic configuration $[Xe]4f^{14}5d^{10}$, forms numerous complexes, many of which are exceedingly stable but undergo ligand exchange reactions easily. These complexes may have coordination numbers ranging from 1 to 4 with coordination number 2 being particularly common.

Examples of one-coordinate complexes are limited by the need to utilize an anion of low nucleophilicity that will not bind to the Au(I). In that regard, $[F_3AsAu^I](SbF_6)$ comes closest to being a one-coordinate complex [18]. Colorless $[F_3AsAu^I](SbF_6)$ is formed by reduction of AuF_3 with AsF_3 in the super acid SbF_5/HF, but it is unstable at room temperature where it decomposes to produce metallic gold. However, $[F_3AsAu^I](SbF_6)$ reacts with xenon in the SbF^5/HF mixture to form room-temperature stable $[F_3AsAu^IXe](Sb_2F_{11})$, which contains a remarkable gold-xenon bond and a linear As-Au-Xe unit [19].

Two-coordinate Au(I) complexes with linear or nearly linear structures are numerous and include anionic ($[Au^I(CN)_2]^-$, $[Au^ICl_2]^-$), neutral $(R_3P)Au^ICl$, $(OC)Au^ICl$, and cationic ($[(R_3P)_2Au^I]^+$, $[(RNC)_2Au^I]^+$) examples.

Attractive interactions between closed shell gold(I) centers are important in determining the solid state structures of many gold(I) complexes and contribute to the properties of such complexes in solution as well [20]. They are particularly important for two-coordinate gold(I) complexes. Thus, in the solid state, two-coordinate Au(I) complexes experience attractive aurophilic interactions that can lead to $Au^I \cdots Au^I$ separations that are shorter than a nominal van der Waal separation of ca 3.6 Å. Theoretical studies have shown that this weakly bonding interaction is the result of correlation effects that are enhanced by relativistic effects [21–25]. Experimental studies of rotational barriers have shown that the strength of the attractive aurophilic interaction is similar to that of hydrogen-bonding: ca. 7–11 kcal/mol [26,27]. Aurophilic interactions have been demonstrated to be sufficiently strong to persist in solution and to play roles in guiding chemical reactions [28].

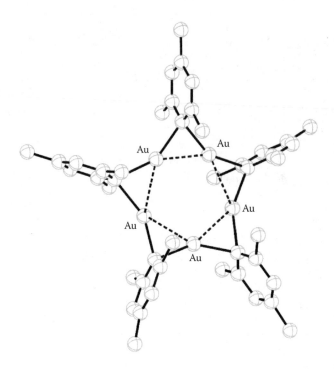

FIGURE 3.6 A picture of {(μ-mesityl)Au^I}$_5$. (Data from A homoleptic arylgold(I) complex: synthesis and structure of pentanuclear mesitylgold(I), S. Gambarotta, C. Floriani, A. Chiesi-Villa, and C. Guastini, *Chem. Commun.*, 1983, 1304.)

Figure 3.6 shows a drawing of {(μ-mesityl)Au^I}$_5$, a cyclic molecule with two-coordinate gold(I) ions that are closely connected due to aurophilic interactions [29]. Notice how the coordination of the gold(I) ions is distorted from linearity.

Aurophilic interactions also play an important role in stabilizing the remarkable hypervalent compounds: [(Ph$_3$PAu)$_6$C]$^{2+}$ with a six-coordinate carbon atom [30], [(Ph$_3$PAu)$_5$C]$^+$ with a five-coordinate carbon atom [31], and [(Ph$_3$PAu)$_5$N]$^{2+}$ with a five-coordinate nitrogen atom [32] as seen in Figure 3.7.

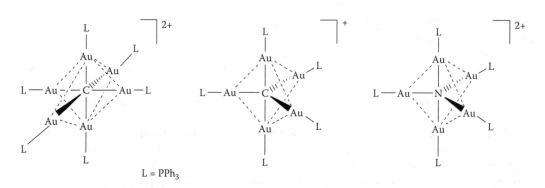

FIGURE 3.7 Gold stabilization of hypervalent carbon and nitrogen.

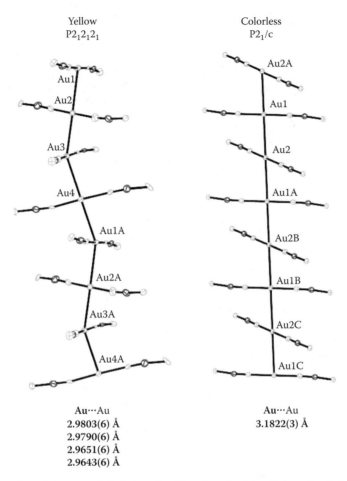

FIGURE 3.8 Drawing of the extended chains of gold cations in the yellow and colorless polymorphs of [(cyclohexyl isocyanide)$_2$Au^I](PF$_6$). Only one carbon atom of each cyclohexyl group is shown. (Data from Aurophilic interactions in cationic gold complexes with two lisocyanide ligands. Polymorphic yellow and colorless forms of [(cyclohexyl isocyanide)$_2$Au^I](PF$_6$) with distinct luminescence, R. L. White-Morris, M. M. Olmstead, and A. L. Balch, *J. Am. Chem. Soc.,* 2003, **125**, 1033.)

Aurophilic interactions are also important in the solid-state structures of many two-coordinate gold(I) complexes. The case of [(cyclohexyl isocyanide)$_2$Au^I](PF$_6$) presents an excellent example [33]. This salt crystallizes in two different fashions. The colorless polymorph contains strictly linear chains of cations with short contacts between the gold atoms as seen in Figure 3.8. In the yellow polymorph, [(cyclohexyl isocyanide)$_2$Au^I](PF$_6$) crystallizes to form bent chains with even shorter separations between the gold ions. The yellow polymorph produces green emission when irradiated with a UV lamp, while under the same conditions the colorless polymorph produces blue emission. The emissive properties of these two salts are a direct consequence of the aurophilic interactions found in the chains of cations [34]. In solution, both salts fully dissociate into monomers that are colorless and nonluminescent.

Aurophilic attractions are also involved in the remarkable behavior of the trinuclear complex, Au$_3$(MeN=COMe)$_3$. The colorless cyclic complex, whose structure is shown in Figure 3.9, crystallizes in three polymorphic forms, one of which contains prismatic stacks of the gold trimers [35]. When irradiated with UV light these crystals produce a long-lived yellow emission. If solvent is dropped on these photoexcited crystals, bright bursts of light are emitted [36,37]. This process

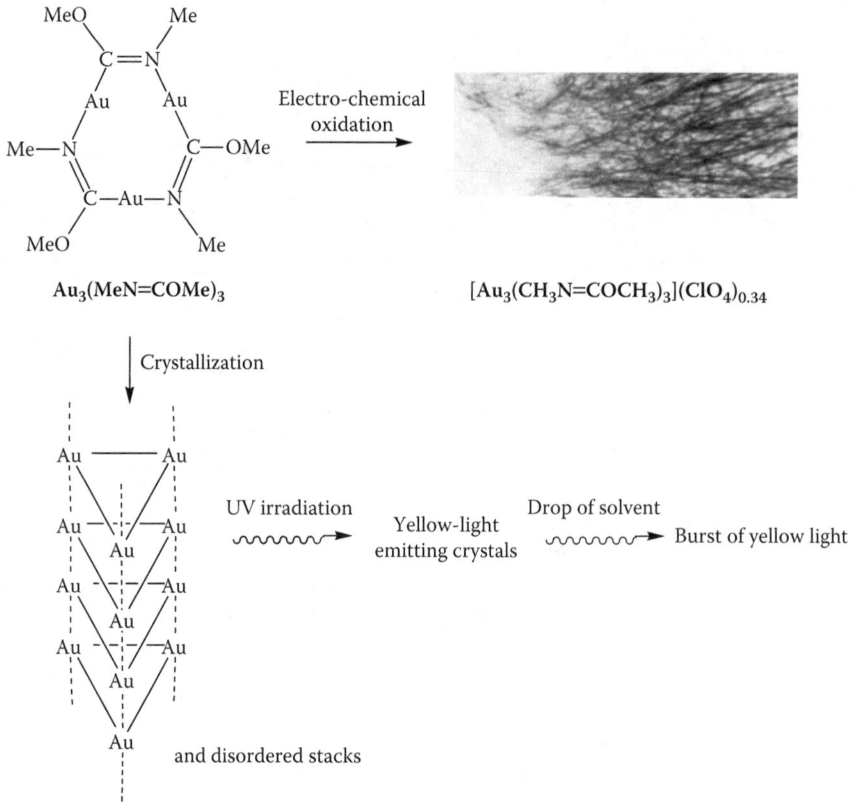

FIGURE 3.9   Solvoluminescence and oxidation of $Au_3(MeN=COMe)_3$.

has been termed solvoluminescence, a property that is unique to this substance. The cyclic trimer, $Au_3(MeN=COMe)_3$, also undergoes electrochemical oxidation to produce fine copper-colored needles of the partially oxidized salt, $[Au_3(CH_3N=COCH_3)_3](ClO_4)_{0.34}$, that are electrically conducting [38]. The abilities of electrons to move along the prismatic stacks of aurophilically connected gold(I) trimers appear to be involved in the solvoluminescence of $Au_3(MeN=COMe)_3$ and the conductivity of $[Au_3(CH_3N=COCH_3)_3](ClO_4)_{0.34}$.

Gold(I) also forms complexes with three- and four-coordinate structures [39]. The three-coordinate complexes may have structures based on a trigonal planar geometry, a planar Y-shaped geometry, or a T-shaped geometry. Four coordinate complexes are generally tetrahedral. An example was shown at the beginning of this chapter in Figure 3.1a.

The relationship between two- and three-coordinate structures for complexes of gold(I) can be very delicate. For example, Figure 3.10 shows drawings of the structures found for three different crystalline forms of $\{(cyclohexyl)_3P\}_2Au^IBr$ [40]. In the α polymorph the complex is present as the linear, two-coordinate salt, $[\{(cyclohexyl)_3P\}_2Au^I]Br$. However, in the γ polymorph the same set of atoms produces a three-coordinate, Y-shaped molecule in which the bromide ion is coordinated to gold and the P-Au-P angle is bent. The β polymorph contains two independent molecules that show changing modes of interaction of the bromide ion with the gold ion and varying P-Au-P angles.

The cation $[(\eta^2\text{-}C_2H_4)_3Au^I]^+$ exhibits a more regular trigonal planar structure as seen in Figure 3.11 [41]. In this complex, the gold(I) ion and the six carbon atoms of the three ethylene ligands lie in a plane. The binding of olefins and acetylenes by gold(I) ions is an important reaction that is involved in many of the processes for which gold(I) complexes serve as catalysts.

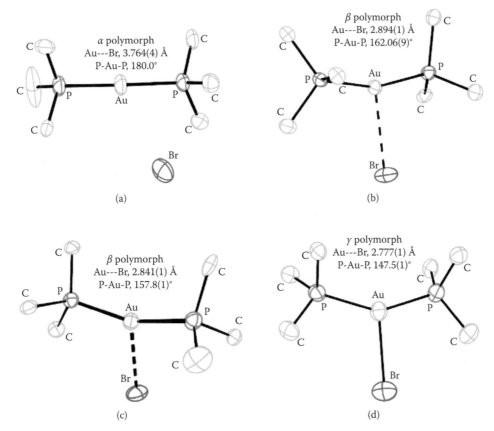

FIGURE 3.10   Drawings of the structures found in the three polymorphs of [(cyclohexyl)$_3$P]$_2$Au^{I}Br. Only one carbon atom of each cyclohexyl group is shown. (Data from Co-ordination and conformational isomerism in bis(tricyclohexyl-phosphine) gold (I) halides, G. A. Bowmaker, C. L. Brown, R. D. Hart, P. C. Healy, C. E. F. Rickard, and A. H. White, *J. Chem. Soc., Dalton Trans.*, 1999, 881.)

GOLD(II)

Compounds of Au(II) have the electronic configuration [Xe]4f^{14}5d^{10} and can exist as paramagnetic monomers or as diamagnetic, dinuclear complexes with Au-Au single bonds [42]. Monomeric complexes are rather rare. Orange Au^{II}(SbF$_6$)$_2$ is formed by dissolution of metallic gold through treatment with F$_2$ in HF/SbF$_5$ [43]. In the crystalline state, the gold atom in Au^{II}(SbF$_6$)$_2$ is situated on a center of symmetry and as an elongated octahedral structure with Au-F bonds to six different anions. The equatorial Au-F bond lengths are 2.09(2) and 2.15(2) Å, while the axial Au-F bond length is 2.64(2) Å. Thus, the odd electron is situated in the d$_{x^2-y^2}$ orbital. The d$_{z^2}$ orbital, which is directed along the axial Au-F bonds, is doubly occupied, which lengthens the axial bonds.

Another monomeric Au(II) compound, [Au^{II}Xe$_4$](Sb$_2$F$_{11}$)$_2$, was produced by treatment of AuF$_3$ with xenon in SbF$_5$/HF solution [44]. The cation, [Au^{II}Xe$_4$]$^{2+}$, has a planar structure with Au-Xe distances ranging from 2.7279(6) to 2.7498(5) Å. The remarkable ability of gold to bind xenon was first discovered with the preparation of [Au^{II}Xe$_4$](Sb$_2$F$_{11}$)$_2$.

With other monomeric gold(II) complexes, there is frequently a question of whether the odd electron resides to a significant extent on the gold atom or whether it is delocalized to such an extent that the compound should be better formulated as a complex of gold(III) with a ligand free radical. These issues are particularly significant when delocalized sulfur donors are present [45]. With thioether ligands as well, gold(II) complexes such as [(1,4,7-trithiacyclononane)$_2$Au^{II}]$^{2+}$, whose

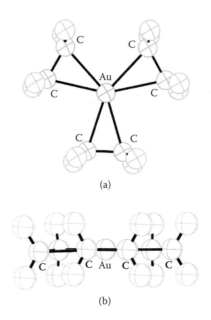

(a)

(b)

FIGURE 3.11 The structure of the cation $[(\eta^2\text{-}C_2H_4)_3Au^I]^+$ in $[(\eta^2\text{-}C_2H_4)_3Au^I][SbF_6]$. A, looking down onto the plane of the gold and carbon atoms; B, looking side-on. (Data from Synthesis and characterization of the gold (I) tris(ethylene) complex $[Au(C_2H_4)_3][SbF_6]$, H. V. R. Dias, M. Fianchini, T. R. Cundari, and C. F. Campana, *Angew. Chem. Int. Ed.*, 2008, **47**, 556.)

structure is shown in Figure 3.1, and [(1-oxa-4,7-dithiacyclononane)$_2$Au^{II}]$^{2+}$ have 27–30% of the spin localized on gold and 62–64% localized on the sulfur donors [46,47].

There are many instances, however, when two gold(II) centers combine to produce binuclear complexes with direct Au^{II}-Au^{II} single bonds. In many cases these bonds can be assembled by an oxidative addition process involving two gold(I) ions that are held in proximity by a suitable bridging ligand. Figure 3.12 shows a classic example, the oxidative addition of I$_2$ to Au$^I_2$[(CH$_2$P(S)Ph$_2$)]$_2$ [48]. Depending on the reaction conditions, this reaction can produce two products, one with a direct Au^{II}-Au^{II} bond (bond length, 2.607(1) Å), the other with a two-coordinate Au^I ion and a separate Au^{III} ion that is bonded to two iodide ligands. Notice that the formation of the Au^{II}-Au^{II} bond results in a pronounced contraction of the separation between the two gold ions.

Amidinate ligands produce complexes with exceedingly short Au^{II}-Au^{II} bonds [49]. An example with an Au^{II}-Au^{II} bond distance of 2.4752(9) Å is shown in Figure 3.13. Currently, the record for the shortest known Au^{II}-Au^{II} bond length is 2.4473(19) Å for a rather complicated compound that involves a pair of amidate ligands bridging the Au^{II}-Au^{II} bond [50].

There are only a few examples of compounds containing Au^{II}-Au^{II} bonds that do not involve bridging ligands [51–53]. Figure 3.14 shows the structure of one of these gold(II) complexes that contains an unsupported Au^{II}-Au^{II} bond.

GOLD(III)

Gold(III), with the electronic configuration [Xe]4f^{14}5d^8, is isoelectronic with platinum(II) and shares many similarities with it. Both form planar, four-coordinate complexes that can be kinetically quite stable. Complexes of gold(III) are numerous and include orange chloroauric acid (H[Au^{III}Cl$_4$]·nH$_2$O), red Au^{III}Cl$_3$ and brown Au^{III}Br$_3$ (which obtain planar four-coordinate geometry by dimerizing through bridging halide ions), and K[Au^{III}(CN)$_4$].

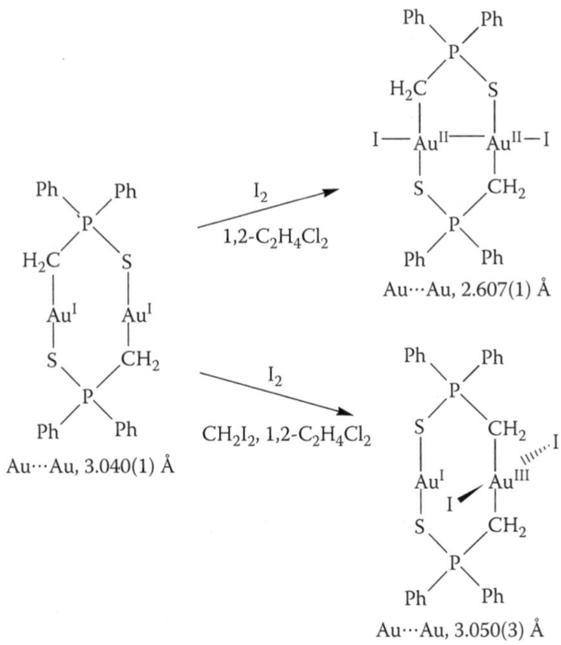

FIGURE 3.12 Binuclear oxidative additions.

Complexes of gold(III) with coordination numbers other than four are rare. Figure 3.15 shows the structure of a well-characterized, five-coordinate gold(III) complex, $(Me_3P)_2Au^{III}I_3$, with a trigonal bipyramidal structure. A report of a six-coordinate gold(III) complex that involves a tetra-aza macrocycle and two axial chloride ligands has appeared [54]. However, the length of the Au-Cl bonds and the fact that the chloride ions dissociate in solution suggests that the true coordination is, as usual, planar, with ion pairing effects bringing the halide ions near the gold(III) center.

Aurophilic interactions between gold(III) ions and gold(I) ions have been observed [55].

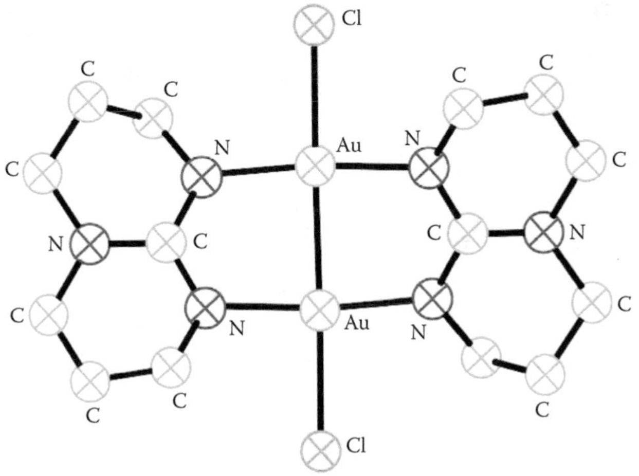

FIGURE 3.13 An amidinate-bridged complex with a short Au^{II}-Au^{II} bond. Hydrogen atoms were omitted. (Data from Synthesis and X-ray structures of silver and gold guanidinate-like complexes. A Au(II) complex with a 2.47 Å Au–Au distance, M. D. Irwin, H. E. Abdou, A. A. Mohamed, and J. P. Fackler, Jr., *Chem. Commun.*, 2003, 2882.)

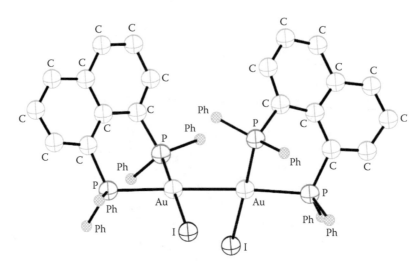

FIGURE 3.14   The structure of a cationic gold(II) complex, [{[Au(dppn)I]}$_2$]$^{2+}$, with an unsupported Au^{II}-Au^{II} bond. Only one carbon atom of each phenyl ring is shown. (Data from synthesis, structural characterization, and photophysics of dinuclear gold(II) complexes [{Au(dppn)Br}$_2$](PF$_6$)$_2$ and [{Au(dppn)I}$_2$](PF$_6$)$_2$ with an unsupported Au^{II}–Au^{II} bond, V. W.-W. Yam, C.-K. Li, C.-L. Chan, and K.-K. Cheung, *Inorg. Chem.*, 2001, **40**, 7054.)

GOLD(V)

Gold(V), with the electronic configuration [Xe]4f^{14}5d^6, is the highest oxidation state of gold for which there is credible evidence. Au^{V}F$_5$, which is a highly reactive oxidant and fluorination agent, has been obtained by a two-stage route [56]. Initially, metallic gold was oxidized by a mixture of fluorine and dioxygen to produce light yellow crystals of (O$_2$)[Au^{V}F$_6$]. Subsequent pyrolysis of (O$_2$)[Au^{V}F$_6$] at 180°C produced red–brown crystals of Au^{V}F$_5$. Au^{V}F$_5$ is dimeric with two fluoride bridges and a nearly octahedral arrangement of six fluoride ions about each gold(V) ion as shown in Figure 3.16. The relatively stable gold(V) anion [Au^{V}F$_6$]$^-$ has a strictly octahedral structure.

Although claims for the formation of a heptafluoride, "AuF$_7$" have been made, recent computational studies indicate that this would not be a stable compound [57,58]. Rather the material formerly believed to be "AuF$_7$" has been reformulated as the adduct Au^{V}F$_5 \cdot$ F$_2$.

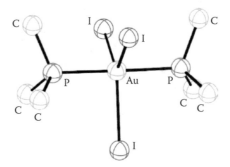

FIGURE 3.15   The structure of a five-coordinate gold(III) complex, (Me$_3$P)$_2$Au^{III}I$_3$ with hydrogen atoms omitted. (Data from The oxidation of gold powder by Me$_3$EI$_2$ (E = P, As) under ambient conditions; structures of [AuI$_3$(PMe$_3$)$_2$], [AuI$_3$(AsMe$_3$)], and [Me$_3$PO)$_2$H][AuI$_2$], S. M. Godfrey, N. Ho, C. A. McAuliffe, and R. G. Pritchard, *Angew. Chem. Int. Ed.*, 1996, **35**, 2344.)

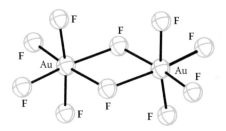

FIGURE 3.16 The structure of a gold(V) complex, Au^VF_5. (Data from Gold pentaflouride: structure and fluoride ion affinity, I-C. Hwang and K. Seppelt, *Angew. Chem. Int. Ed.*, 2001, **40**, 3690.)

CLUSTERS WITH FRACTIONAL OXIDATION STATES

Gold forms an intriguing array of clusters that have oxidations states of gold between 0 and +I. Two major classes can be distinguished: one with phosphine ligands capping the outer surface [59], the other with anionic thiolates on the exterior.

Figure 3.17 shows three cationic examples of clusters with phosphine ligation. The heptanuclear cation $[Au_7(PPh_3)_7]^+$ that consists of a pentagon of gold atoms capped on either side by a $AuPPh_3$ unit is shown in part A [60]. Part B shows a cluster cation $[Au_8(PPh_3)_7]^{2+}$ that involves a central gold atom which is surrounded by seven $AuPPh_3$ units [61]. A cluster, $[Au_{13}(PMe_2Ph)_{10}Cl_2]^{3+}$, that consists of a nearly icosahedral array of 12 gold atoms surrounding a central gold atom is shown in part C [62].

This arrangement of 13 atoms with a single atom surrounded by a nearly icosahedral array of 12 gold atoms is a fundamental building block in gold cluster chemistry. Theoretical and experimental evidence exists for a group of clusters with $W@Au_{12}$ as the prototype. $W@Au_{12}$ consists of a hetero-metal atom (W) surrounded by 12 gold atoms in an icosahedral arrangement [63,64]. These clusters have structures that are reminiscent of endohedral fullerenes.

The core arrangement of the icosahedral Au_{13} unit is also found at the center of the giant $Au_{55}(PPh_3)_{12}Cl_6$ molecule, which has been extensively studied [65].

Gold nanoparticles capped with thiolate groups have received considerable study in recent years. Recently, two such nanoclusters have been isolated in crystalline forms that have proven suitable for single-crystal X-ray diffraction studies. The first of these involved a cluster of 102 gold atoms surrounded by 44 *p*-mercaptobenzoic acid molecules capping the exterior [66].

The second set of reports concerned the structure of the $[Au_{25}(SCHCHPh)_{18}]^+$ [67–69]. The structure of this remarkable cation is shown in Figure 3.18. Part A shows a 13-atom ($Au@Au_{12}$) fragment: the central gold atom and the icosahedral array of 12 gold atoms that surround the central gold atom. The entire cation, without the 18 CH_2CH_2Ph groups, is shown in part B. In this section of the drawing, the 13 core gold atoms are shown in green, and bonds to the central gold atom are omitted for clarity. The cation $[Au_{25}(SCHCHPh)_{18}]^+$ has been reduced to give the neutral cluster, $Au_{25}(SCHCHPh)_{18}$ [70].

In addition to the clusters discussed previously, gold also forms numerous clusters with bridging heteroatoms [71] and heterometallic clusters such as $[Au_{22}\{Fe(CO)_4\}_{12}]^{6-}$, which has a core of gold atoms surrounded by $Fe(CO)_4$ units [72].

UTILITY OF GOLD COMPLEXES

Gold complexes such as those discussed here have found a number of uses. For example, $[AuI(CN)_2]^-$ has been extensively involved in gold mining and gold recovery as well as in gold electroplating. Gold compounds such as those shown in Figure 3.19 have been used in medicine and biology as therapeutic and diagnostic agents [73–75]. Although gold complexes were thought for a long time

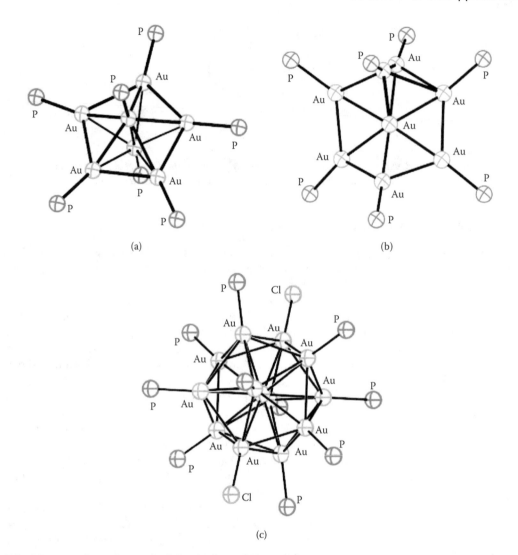

(a)   (b)

(c)

FIGURE 3.17 **(See color insert following page 212.)** The structures of cluster cations. (a) $[Au_7(PPh_3)_7]^+$. (From Intercluster compounds consisting of gold clusters and fullerides: $[Au_7(PPh_3)_7]C_{60} \cdot THF$ and $[Au_8(PPh_3)_8]$ $(C_{60})_2$, M. Schulz-Dobrick and M. Jansen, *Angew. Chem. Int. Ed.*, 2008, **47**, 2256.) (b) $[Au_8(PPh_3)_7]^{2+}$. (From Reactions of cationic gold clusters with Lewis bases. Preparation and X-ray structure investigation of $[Au_8(PPh_3)_7(NO_3)2.2CH_2Cl_2$ and $Au_6(PPh_3)_4[Co(CO)_4]_2$, J. W. A. Van der Velden, J. J. Bour, W. P. Bosman, and J. H. Noordik, *Inorg. Chem.*, 1983, **22**, 1983.) (c) $[Au_{13}(PMe_2Ph)_{10}Cl_2]^{3+}$. (From Synthesis and X-ray structural characterization of the centered icosahedral gold cluster compound $[Au_{13}(PMe_2Ph)_{10}Cl_2](PF_2)_3$; the realization of a theoretical prediction, C. E. Briant, B. R. C. Theobald, J. W. White, L. K. Bell, and D. M. P. Mingos, *Chem. Commun.*, 1981, **5**, 201.) For clarity, phenyl rings were omitted.

to lack the catalytic properties displayed by the platinum metals, there has recently been remarkable growth in the utilization of gold complexes in catalysis [76–82]. In part, this catalytic activity occurs because of the ability of gold to bind olefins and other unsaturated molecules. In that regard $[(\eta^2-C_2H_4)_3Au^I]^+$, whose structure is shown in Figure 3.11, provides a significant model for π-olefinic coordination. Subsequent chapters in this book give further information in the utilization of gold complexes in medicine (Chapter 10) and catalysis (Chapter 6).

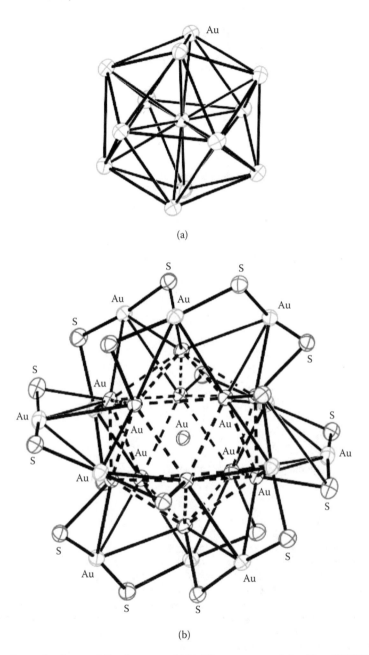

(a)

(b)

FIGURE 3.18   (See color insert following page 212.) The structure of the $[Au_{25}(SCHCHPh)_{18}]^+$. (a) the central gold atom and the icosahedral array of 12 gold atoms at the core; (b) the entire cation without the 18 CH_2CH_2Ph groups. In (b) the 13 core gold atoms are shown in green and bonds to the central gold atom are omitted for clarity. (Data from Crystal structure of the gold nanoparticle $[N(C_8H_{17})_4][Au_{25}(SCH_2CH_2Ph)_{18}]$, M. W. Heaven, A. Dass, P. S. White, K. M. Holt, and R. W. Murray, *J. Am. Chem. Soc.*, 2008, **130**, 3754.)

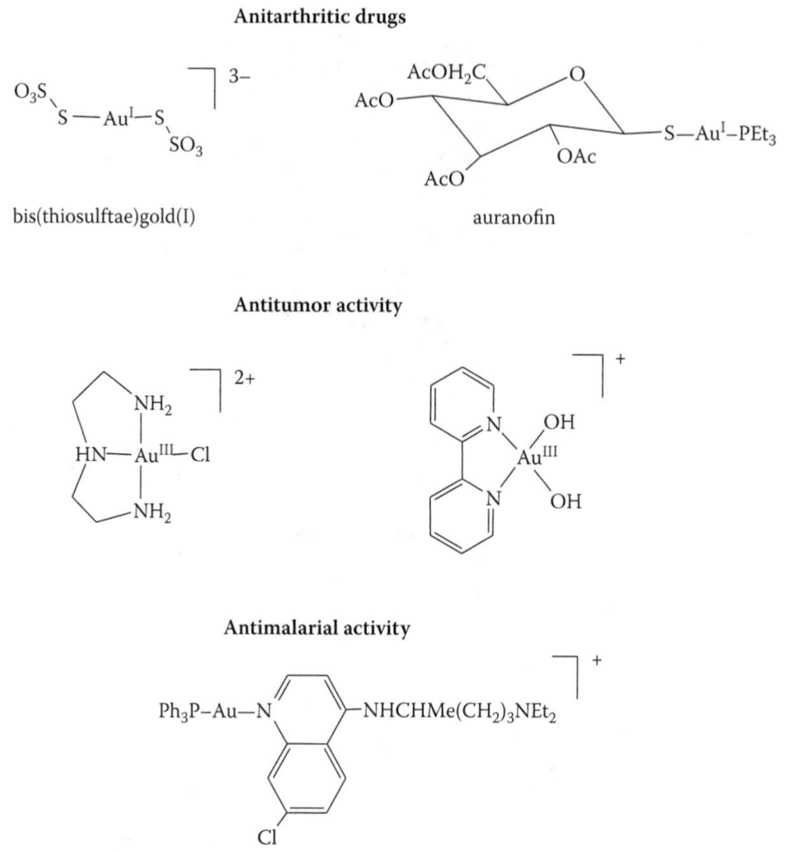

Anitarthritic drugs

bis(thiosulftae)gold(I)   auranofin

Antitumor activity

Antimalarial activity

FIGURE 3.19   Gold compounds with biological activity.

REFERENCES

1. *The Chemistry of Gold*, R. J. Puddephatt, Elsevier, 1978, 1.
2. *Gold Progress in Chemistry, Biochemistry and Technology*, H. Schmidbaur, (ed.), Elsevier, 1999, 1.
3. *Comprehensive Coordination Chemistry II, From Biology to Nanotechnology*, M. C. Gimeno and A. Laguna, Elsevier, 2003, **6**, 911.
4. *Comprehensive Organometallic Chemistry III*, H. Schmidbaur and A. Schier, Elsevier, 2007, **2**, 251.
5. 2008 Gold: chemistry, materials and catalysis issue, *Chem. Soc. Rev.*, 2008, **37**, Special Issue 9.
6. Bis(1,4,7-trithiacyclononane)gold dication: a paramagnetic, mononuclear AuII complex, A. J. Blake, J. A. Greig, A. J. Holder, T. I. Hyde, A. Taylor, and M. Schröder, *Angew. Chem. Int. Ed.*, 1990, **29**, 197.
7. Gold thioether chemistry – synthesis, structure, and redox interconversion of [Au(1,4,7-trithiacyclononane)$_2$]$^{+/2+/3+}$, A. J. Blake, R. O. Gould, J. A. Greig, A. J. Holder, T. I. Hyde, and M. Schröder, *J. Chem. Soc., Chem. Commun.*, 1989, 876.
8. Oxidation of elemental gold in alcohol solutions, M. T. Risnen, M. Kemell, M. Leskel, and T. Repo, *Inorg. Chem.*, 2007, **46**, 3251.
9. Charge transfer complexes of dithioxamides with dihalogens as powerful reagents in the dissolution of noble metals, A. Serpe, F. Artizzu, M. L. Mercuri, L. Pilia, and P. Deplano, *Coord. Chem. Rev.*, 2008, **252**, 1200.
10. Oxidation of solid gold in chloroform solutions of cetyltrimethylammonium bromide, T. Mortier, A. Persoons, and T. Verbiest, *Inorg. Chem. Commun.*, 2005, **8**, 1075.
11. The oxidation of gold powder by Me$_3$EI$_2$ (E = P, As) under ambient conditions; structures of [AuI$_3$(PMe$_3$)$_2$], [AuI$_3$(AsMe$_3$)], and [(Me$_3$PO)$_2$H][AuI$_2$], S. M. Godfrey, N. Ho, C. A. McAuliffe, and R. G. Pritchard, *Angew. Chem. Int. Ed. Engl.*, 1996, **35**, 2344.

12. The chemistry of gold as an anion, M. Jansen, *Chem. Soc. Rev.*, 2008, **37**, 1826.

13. Synthesis and crystal structure determination of tetramethylammonium auride, P. D. C. Dietzel and M. Jansen, *Chem. Commun.*, 2001, 2208.

14. [Rb([18]crown-6)(NH$_3$)$_3$]Au·NH$_3$: Gold as acceptor in N-H· · ·Au⁻ hydrogen bonds, H. Nuss and M. Jansen, *Angew. Chem. Int. Ed.*, 2006, *45*, 4369.

15. Gold(I) and gold(0) complexes of phosphinine-based macrocycles, N. Mézailles, N.Avarvari, N. Maigrot, L. Richard, F. Mathey, P. Le Floch, L. Cataldo, T. Berclaz, and M. Geoffroy, *Angew. Chem. Int. Ed.*, 1999, **38**, 3194.

16. Cryochemical studies. 3. EPR studies of the reaction of group IB metal atoms with arenes at 77 K with a rotating cryostat, A. J. Buck, B. Mile, and J. A. Howard, *J. Am. Chem. Soc.*, 1983, **105**, 3381.

17. Theoretical and structural studies on organometallic cluster molecules, D. M. P. Mingos, *Pure Appl. Chem.* 1980, **52**, 705.

18. Preparation and structure of F$_3$As-Au⁺SbF$_6$⁻, the structures of Au(CO)$_2$⁺ and Au(PF$_3$)$_2$⁺, R. Kuster and K. Seppelt, *Z. Anorg. Allgem. Chem.* 2000, **626**, 236.

19. Gold(I) and mercury (II) xenon complexes, I-C. Hwang, S. Seidel, and K. Seppelt, *Angew. Chem. Int. Ed.*, 2003, **42**, 4392.

20. A briefing on aurophilicity, H. Schmidbaur and A. Schier, *Chem. Soc. Rev.*, 2008, **37**, 1931.

21. Strong closed-shell interactions in inorganic chemistry, P. Pyykkö, *Chem. Rev.* 1997, **97**, 597.

22. Theoretical chemistry of gold, P. Pyykkö, *Angew. Chem. Int. Ed.*, 2004, **43**, 4412.

23. Theoretical chemistry of gold. II, P. Pyykkö, *Inorg. Chim. Acta.*, 2005, **358**, 4113.

24. Theoretical chemistry of gold. III, P. Pyykkö *Chem. Soc. Rev.*, 2008, **37**, 1967.

25. Basis-set limit of the aurophilic attraction using the MP2 method: the examples of [ClAuPH$_3$]$_2$ dimer and [P(AuPH$_3$)$_4$]⁺ ion, P. Pyykkö and P. Zaleski-Ejgierd, *J. Chem. Phys.*, 2008, **128**, 124309-1.

26. Weak intramolecular bonding relationships - the conformation-determining attractive interaction between gold(I) centers, H. Schmidbaur, W. Graf, and G. Müller, *Angew. Chem. Int. Ed.,* 1988, **27**, 417.

27. Auracarboranes with and without Au-Au interactions: An unusually strong aurophilic interaction, D. M. Harwell, M. D. Mortimer, C. B. Knobler, F. A. L. Anet, and M. F. Hawthorne, *J. Am. Chem. Soc.*, 1996, **118**, 2679.

28. Polynuclear (diphenylphosphinomethyl)phenylarsine bridged complexes of gold(I). Bent chains of gold(I) and a role for Au(I)-Au(I) interactions in guiding a reaction, A. L. Balch, E. Y. Fung, and M. M. Olmstead, *J. Am. Chem. Soc.,* 1990, **112**, 5181.

29. A homoleptic arylgold(I) complex: synthesis and structure of pentanuclear mesityl- gold(I), S. Gambarotta, C. Floriani, A. Chiesi-Villa, and C. Guastini, *Chem. Commun.*, 1983, 1304.

30. Aurophilicity as a consequence of relativistic effects - the hexakis(triphenylphosphane-aurio)methane dication [(Ph$_3$PAu)$_6$C]²⁺, F. Scherbaum, A. Grohmann, B. Huber, C. Kruger, and H. Schmidbaur, *Angew. Chem. Int. Ed.*, 1988, **27**, 1544.

31. Synthesis, structure, and bonding of the cation [(C$_6$H$_5$)$_3$PAu)$_5$C]⁺, F. Scherbaum, A. Grohmann, G. Müller, and H. Schmidbaur, *Angew. Chem. Int. Ed.*, 1989, **28**, 463.

32. Electron deficient bonding at pentacoordinate nitrogen, A. Grohmann, J. Riede, and H. Schmidbaur, *Nature*, 1990, **345**, 140.

33. Aurophilic interactions in cationic gold complexes with two lisocyanide ligands. Polymorphic yellow and colorless forms of [(cyclohexyl isocyanide)$_2$Au^I](PF$_6$) with distinct luminescence, R. L. White-Morris, M. M. Olmstead, and A. L. Balch, *J. Am. Chem. Soc.*, 2003, **125**, 1033.

34. Remarkable luminescence behaviors and structural variations of two-coordinate gold(I) complexes, A. L. Balch, *Struct. Bond.*, 2007, **123**, 1.

35. Intermolecular interactions in polymorphs of trinuclear gold(I) complexes: insight into the solvolumi-nescence of Au$^I_3$(MeN=COMe)$_3$, R. L. White-Morris, M. M. Olmstead, S. Attar, and A. L. Balch, *Inorg. Chem.* 2005, **44**, 5021.

36. Solvent-stimulated luminescence from the supramolecular aggregation of a trinuclear gold(I) complex that displays extensive intermolecular Au···Au interactions, J. C. Vickery, M. M. Olmstead, E. Y. Fung, and A. L. Balch*, Angew. Chem. Int. Ed.*, 1997, **36**, 1179.

37. Glowing gold rings: solvoluminescence from planar trigold(I) complexes, E. Y. Fung, M. M. Olmstead, J. C. Vickery, and A. L. Balch, *Coord. Chem. Rev.*, 1998, **171**, 151.

38. Formation of a partially oxidized gold compound by electrolytic oxidation of the solvoluminescent gold(I) trimer, Au$_3$(MeN=COMe)$_3$, K. Winkler, M. Wysocka-Żołopa, K. Rećko, L. Dobrzyński, J. C. Vickery, and A. L. Balch, *Inorg. Chem.* 2009, **48**, 1551.

39. Three- and four coordinate gold (I) complexes, M. Concepcion and A. Laguna, *Chem. Rev.*, 1997, **97**, 511.

40. Co-ordination and conformational isomerism in bis(tricyclohexyl-phosphine) gold(I) halides, G. A. Bowmaker, C. L. Brown, R. D. Hart, P. C. Healy, C. E. F. Rickard, and A. H. White, *J. Chem. Soc., Dalton Trans.*, 1999, 881.

41. Synthesis and characterization of the gold (I) tris(ethylene) complex [Au(C$_2$H$_4$)$_3$][SbF$_6$], H. V. R. Dias, M. Fianchini, T. R. Cundari, and C. F. Campana, *Angew. Chem. Int. Ed.*, 2008, **47**, 556.

42. Coordination chemistry of gold(II) complexes, A. Laguna and M. Laguna, *Coord. Chem. Rev.*, 1999, **193–195**, 837.

43. Synthesis of Au(II) fluoro complexes and their structural and magnetic properties, S. H. Elder, G. M. Lucier, F. J. Hollander, and N. Bartlett, *J. Am. Chem. Soc.*, 1997, **119**, 1020.

44. Xenon as a complex ligand: the tetra xenono gold (II) cation in AuXe$_4^{2+}$ (Sb$_2$F$_{11}^-$)$_2$, S. Seidel and K. Seppelt, *Science*, 2000, **290**, 117.

45. Molecular and electronic structure of square-planar gold complexes containing two 1,2- di(4-*tert*-butylphenyl)ethylene-1,2-dithiolato ligands: [Au(^{2}L)$_2$]$^{1+/0/1-/2-}$. A combined experimental and computational study, S. Kokatam, K. Ray, J. Pap, E. Bill, W. E. Geiger, R. J. LeSuer, P. H. Rieger, T. Weyhermüller, F. Neese, and K. Wieghardt, *Inorg. Chem.*, 2007, **46**, 1100.

46. Redox non-innocence of thioether macrocycles: elucidation of the electronic structures of mononuclear complexes of gold(II) and silver(II), J. L. Shaw, J. Wolowska, D. Collison, J. A. K. Howard, E. J. L. McInnes, J. McMaster, A. J. Blake, C. Wilson, and M. Schröder, *J. Am. Chem. Soc.*, 2006, **128**, 13827.

47. Crystallographic, electrochemical, and electronic structure studies of the mononuclear complexes of Au(I)/(II)/(III) with [9]aneS$_2$O ([9]aneS$_2$O = 1-oxa-4,7-dithiacyclononane), D. Huang, X. Zhang, E. J. L. McInnes, J. McMaster, A. J. Blake, E. S. Davies, J. Wolowska, C. Wilson, and M. Schröder, *Inorg. Chem.*, 2008, **47**, 9919.

48. Isomeric species of [AuCH$_2$P(S)(C$_6$H$_5$)$_2$I]$_2$. Mixed-valent Au(I)/Au(III) and isovalent Au(II)- Au(II) complexes with the same methylenethiophosphinate ligand, A. M. Mazany, and J. P. Fackler, Jr., *J. Am. Chem. Soc.* 1984, **106**, 801.

49. Synthesis and X-ray structures of silver and gold guanidinate-like complexes. A Au(II) complex with a 2.47 Å Au–Au distance, M. D. Irwin, H. E. Abdou, A. A. Mohamed, and J. P. Fackler, Jr., *Chem. Commun.*, 2003, 2882.

50. A silver(I)–gold(II) hexanuclear guanidinate–benzoate cluster with short Au–Au bonds, A. A. Mohamed, H. E. Abdou, A. Mayer, and J. P. Fackler Jr., *J. Clust. Sci.*, 2008, **19**, 551.

51. Synthesis, photophysics and thermal redox reactions of a [{Au(dppn)Cl}$_2$]$^{2+}$ dimer with an unsupported Au-II-Au-II bond, V. W.-W. Yam, S. W.-K. Choi, K.-K. Cheung, *Chem. Commun.*, 1996, 1173.

52. Synthesis, structural characterization, and photophysics of dinuclear gold(II) complexes [{Au(dppn) Br}$_2$](PF$_6$)$_2$ and [{Au(dppn)I$_2$](PF$_6$) $_2$ with an unsupported Au^{II}–Au^{II} bond, V. W.-W. Yam, C.-K. Li, C.-L. Chan, and K.-K. Cheung, *Inorg. Chem.*, 2001, **40**, 7054.

53. Structural studies of gold(I, II, and III) compounds with pentafluorophenyl and tetrahydrothio-phene ligands, J. Coetzee, W. F. Gabrielli, K. Coetzee, O. Schuster, S. D. Nogai, S. Cronje, and H. G. Raubenheimer, *Angew. Chem. Int. Ed.*, 2007, **46**, 2497.

54. Extremely facile template synthesis of gold(III) complexes of a saturated azamacrocycle and crystal structure of a six-coordinate gold(III) complex, M. P. Suh, I. S. Kim, B. Y. Shim, D. Hong, and T.-S. Yoon, *Inorg. Chem.*, 1996, **35**, 3595.

55. Amine-amide equilibrium in gold(III) complexes and a gold(III)-gold(I) aurophilic bond, L. Cao, M. C. Jennings, and R. J. Puddephatt, *Inorg. Chem.*, 2007, **46**, 1361.

56. Gold pentaflouride: structure and fluoride ion affinity, I-C. Hwang and K. Seppelt, *Angew. Chem. Int. Ed.*, 2001, **40**, 3690.

57. Has AuF$_7$ been made?, S. Riedel and M. Kaupp, *Inorg. Chem.*, 2006, **45**, 1228.

58. After 20 years, theoretical evidence that "AuF$_7$" is actually AuF$_5$·F$_2$, D. Himmel and S. Riedel, *Inorg. Chem.*, 2007, **46**, 5338.

59. Gold – a flexible friend in cluster chemistry, D. M. P. Mingos, *J. Chem. Soc., Dalton Trans.*, 1996, 561.

60. Intercluster compounds consisting of gold clusters and fullerides: [Au$_7$(PPh$_3$)$_7$]C$_{60}$•THF and [Au$_8$(PPh$_3$)$_8$] (C$_{60}$)$_2$, M. Schulz-Dobrick and M. Jansen, *Angew. Chem. Int. Ed.*, 2008, **47**, 2256.

61. Reactions of cationic gold clusters with Lewis bases. Preparation and X-ray structure investigation of [Au$_8$(PPh$_3$)$_7$](NO$_3$)2.2CH$_2$Cl$_2$ and Au$_6$(PPh$_3$)$_4$[Co(CO)$_4$]$_2$, J. W. A. Van der Velden, J. J. Bour, W. P. Bosman, and J. H. Noordik, *Inorg. Chem.*, 1983, **22**, 1983.

62. Synthesis and X-ray structural characterization of the centred icosahedral gold cluster compound [Au$_{13}$(PMe$_2$Ph)$_{10}$Cl$_2$](PF$_2$)$_3$; the realization of a theoretical prediction, C. E. Briant, B. R. C. Theobald, J. W. White, L. K. Bell, and D. M. P. Mingos, *Chem. Commun.*, 1981, **5**, 201.

63. Icosahedral WAu_{12}: a predicted closed-shell species, stabilized by aurophilic attraction and relativity and in accord with the 18-electron rule, P. Pyykkö and N. Runeberg, *Angew. Chem. Int. Ed.*, 2002, **41**, 2174.

64. Experimental observation and confirmation of icosahedral $W@Au_{12}$ and $Mo@Au_{12}$ molecules, X. Li, B. Kiran, J. Li, H.-J. Zhai, and L.-S. Wang, *Angew. Chem. Int. Ed.*, 2002, **41**, 4786.

65. The relevance of shape and size of Au_{55} clusters, G. Schmid, *Chem. Soc. Rev.*, 2008, **37**, 1909.

66. Structure of a thiol monolayer-protected old nanoparticle at 1.1 Å resolution, P. D. Jadzinsky, G. Calero, C. A. Ackerson, D. A. Bushnell, and R. D. Kornberg, *Science*, 2007, **318**, 430.

67. Crystal structure of the gold nanoparticle $[N(C_8H_{17})_4][Au_{25}(SCH_2CH_2Ph)_{18}]$, M. W. Heaven, A. Dass, P. S. White, K. M. Holt, and R. W. Murray, *J. Am. Chem. Soc.*, 2008, **130**, 3754.

68. Correlating the crystal structure of a thiol-protected Au cluster and optical properties, M. Zhu, C. M. Aikens, F. J. Hollander, G. C. Schatz, and R. Jin, *J. Am. Chem. Soc.*, 2008, **130**, 5883.

69. On the structure of thiolate-protected Au, J. Akola, M. Walter, R. L. Whetten, H. Häkkinen, and H. Grönbeck, *J. Am. Chem. Soc.*, 2008, **130**, 3756.

70. Conversion of anionic $[Au_{25}(SCH_2CH_2Ph)_{18}]$ cluster to charge neutral cluster via air oxidation, M. Zhu, W. T. Eckenhoff, T. Pintauer, and R. Jin, *J. Phys. Chem. C*, 2008, **112**, 14221.

71. Chalcogenide centred gold complexes, M. C. Gimeno and A. Laguna, *Chem. Soc. Rev.*, 2008, **37**, 1952.

72. An organometallic approach to gold nanoparticles: synthesis and X-ray structure of CO-protected $Au_{21}Fe_{10}$, $Au_{22}Fe_{12}$, $Au_{28}Fe_{14}$ and $Au_{34}Fe_{14}$ clusters, C. Femoni, M. C. Iapalucci, G. Longoni, C. Tiozzo, and S. Zacchini, *Angew. Chem. Int. Ed.*, 2008, **47**, 6666.

73. Gold-based therapeutic agents, C. F. Shaw III, *Chem. Rev.*, 1999, **99**, 2589.

74. Gold(III) compounds as anticancer drugs, C. Gabbiani, A. Casini, and Luigi Messori, *Gold Bulletin*, 2007, **40**, 73.

75. Targeting the mitochondrial cell death pathway with gold compounds, P. J. Barnard and S. J. Berners-Price, *Coord. Chem. Rev.* 2007, **251**, 188.

76. Gold catalysis, A. S. K. Hashmi and G. J. Hutchings, *Angew. Chem. Int. Ed.*, 2006, **45**, 7896.

77. Gold-catalyzed organic transformations, Z. Li, C. Brouwer, and C. He, *Chem. Rev.*, 2008, **108**, 3239.

78. Alternative synthetic methods through new developments in catalysis by gold, A. Arcadi, *Chem. Rev.*, 2008, **108**, 3266

79. Gold-catalyzed cycloisomerizations of enynes: a mechanistic perspective, E. Jiménez-Núñez and A. M. Echavarren, *Chem. Rev.*, 2008, **108**, 3326.

80. Ligand effects in homogeneous Au catalysis, D. J. Gorin, B. D. Sherry, and F. D. Toste, *Chem. Rev.*, 2008, **108**, 3351.

81. Coinage metal catalyzed C–H bond functionalization of hydrocarbons, M. M. Díaz-Requejo and P. J. Pérez, *Chem. Rev.*, 2008, **108**, 3379.

82. Coinage metal-assisted synthesis of heterocycles, N. T. Patil and Y. Yamamoto, *Chem. Rev.*, 2008, **108**, 3395.

4 Surface Electrochemistry of Gold

L. Declan Burke and Andrew M. O'Connell

CONTENTS

Introduction 51
Surface Electrochemistry of Gold 52
    EMS Electrochemistry of Gold 52
    MMS Electrochemistry of Gold 53
    Premonolayer Oxidation of Gold 54
    Monolayer (α) Gold Oxides 54
    Hydrous (β) Oxide Electrochemistry of Gold 55
    FT-AC Voltammetry Investigations 55
Surface Electrocatalysis on Gold 56
Nanoparticle Gold Electrocatalysis 57
    Oxygen Reduction 58
    Carbon Monoxide (CO) Oxidation 59
    Formic Acid and Methanol Oxidation 61
Conclusions 62
References 63

INTRODUCTION

This chapter is concerned mainly with the electrochemistry and, in particular, the electrocatalytic behavior of polycrystalline gold surfaces. Only brief attention will be given to single crystal gold surfaces. Although the study of the latter is quite an active area of research, ordered, single crystal surfaces do not, in theory, contain random defects which are the basis of surface active site [1] and surface catalysis behavior. In reality single crystal metal surfaces generally contain plenty of imperfections [2,3] and it is the low coverage, protruding (or low lattice coordination) surface metal atoms present at the defect sites that dominate the electrocatalytic properties of surfaces. The importance of protruding surface metal atoms is now widely accepted in the surface catalysis area; for example, in a recent account of the mechanism of the interaction of oxygen with gold and its relevance to catalytic oxidation Bond and co-workers [4] stressed the role of such atoms, in very small metal particles, as electron donors that react with oxygen to form ionic species such as Au^+-O_2^- which are the precursors for oxygen dissociation.

In terms of the active site approach to surface catalysis [2,5] only a small fraction—ca. 1.0%—of the surface metal atoms might be involved as reaction centers. Evidently there are two basic types of surface metal atoms: (1) high-coverage, low-energy, well-embedded, equilibrated metal surface (EMS) atoms which often behave as spectator species from a catalytic viewpoint, and (2) low-coverage, high-energy, protruding, metastable metal surface (MMS) atoms which behave as surface active sites. EMS and MMS atoms at the same surface may be chemically identical at the individual level, but because of their different environments at the surface, they differ in chemical, kinetic, and thermodynamic properties [2], i.e., they are effectively different species.

Thus, for gold and related metals there are two types of surface electrochemistries, one relating to the predominant EMS state (this is the conventional electrochemical response of the metal) and the other to the minority MMS state(s). The behavior of the latter is difficult to investigate (at least one author has claimed [6] that catalytic sites remain unidentified) as their coverage and analytical responses are usually quite low, they are intrinsically unstable, and there may well be more than one type of active state or site at a given metal surface.

Electrochemistry has both advantages and disadvantages as a technique for investigating metal surfaces. By varying the electrode potential the electron donor/acceptor ability of the metal surface is altered and reaction, e.g., oxidation of either the metal surface or a dissolved solution species, may be induced. The most common investigation technique used in interfacial electrochemistry is dc cyclic voltammetry. All potential values quoted here refer to the reversible hydrogen electrode (RHE) scale; the pH value in the latter case is arbitary—it is the same as that of the solution surrounding the working electrode. Electrochemical techniques, however, provide no direct information as to the structure or composition of the surface states or species involved and so there is a need for ancillary analytical techniques [7], e.g., surface enhanced Raman spectroscopy (SERS).

The active site (MMS) redox response, which apparently occurs in a fast, quasi-reversible manner, is usually of low magnitude and overlaps with, and is not easily distinguished from, the capacitive double layer response. Furthermore, when an electrocatalytic process takes place, mediated by the active site species, the catalytic process usually dominates the dc response, i.e., the redox behavior of the active site is not at all obvious. In such circumstances advantage may be taken of ac voltammetry techniques to investigate the vital role of the active sites at the interface [8]. Large amplitude Fourier transformed ac (LAFT-ac) voltammetry has been explored, at both a theoretical and experimental level [9–11], by Bond and co-workers, and the application of this technique to the study of gold surfaces in aqueous acid and base was described recently [12].

SURFACE ELECTROCHEMISTRY OF GOLD

EMS Electrochemistry of Gold

The basic (or EMS) electrochemistry of gold in aqueous acid solution is summarized in Figure 4.1. There are two processes involved: (1) double layer charging or discharging, which extends over the entire range of the sweep but is the sole process over the range 0.0 to 1.35 V in the positive sweep, and (2) monolayer (or α) oxide formation and removal. Monolayer oxide formation on gold in acid solution commences in the positive sweep at ca. 1.36 V; however, the process is sluggish, e.g., it does

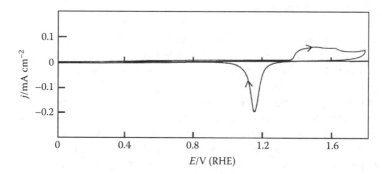

FIGURE 4.1 Cyclic voltammogram (0.0 to 1.80 V, 50 mV s^{-1}) for an unactivated gold wire electrode in 1.0 mol dm^{-3} H$_2$SO$_4$ solution at 25°C. (From Springer Science + Business Media: Generation of active surface states of gold and the role of such states in electrocatalysis, L. D. Burke and A.P. O'Mullane. *J. Solid State Electrochem.*, 2004, 4, 285. With permission.)

not give rise to a sharp voltammetric response. Instead, as illustrated in Figure 4.1, the response extends as a plateau current to at least 1.8 V (oxygen gas evolution commences just above the latter value and occurs quite vigorously above 2.0 V). In the subsequent negative sweep the monolayer oxide reduction response yields a relatively sharp cathodic peak just below 1.2 V and double layer discharging is the sole feature over the range 1.0–0.0 V.

Unlike platinum and the other noble metals [13], gold yields virtually no response for absorbed hydrogen (cathodic evolution of hydrogen gas occurs below 0.0 V and responses due to adsorbed hydrogen are expected just above the latter value). This is in agreement with the assumption that gold surfaces possess very weak chemisorbing and, in terms of the activated chemisorption view of surface catalysis [14], very poor catalytic and electrocatalytic properties. There is an obvious problem here because gold (in oxide-supported nanoparticle form) is an excellent catalyst for certain gas phase reactions [4,15,16] and also the metal, even in massive or extended form, displays quite appreciable surface electrocatalytic activity [17–19] even at potentials within the double layer region where, according to the traditional view summarized in Figure 4.1, the surface gold atoms are inert with respect to both chemisorption and redox behavior.

MMS ELECTROCHEMISTRY OF GOLD

The response shown in Figure 4.1 relates to a low-energy, EMS gold surface. The MMS electrochemistry of gold is concerned with the behavior of unstable, high-energy, protruding surface gold atoms that undergo oxidation to yield a more dispersed (low-density), hydrated, largely amorphous, hydrous (β) gold oxide (a review of metal hydrous oxide electrochemistry was published earlier [20]). Multilayer hydrous oxide deposits may be produced on gold in aqueous media by subjecting the electrode to either severe dc polarization [21–25], e.g., 3 min at 2.2 V, or to repetitive potential cycling using appropriate upper and lower limits [26,27], e.g., 0.90–2.40 V at 50 V s^{-1} for 1200 cycles (the efficiency of the oxide growth reaction is dependent on many factors, including the solution pH). Thick deposits are not produced in a single sweep because the precursor state (MMS gold atoms) exists at the metal surface only at a very low coverage. In the dc procedure oxygen gas evolution occurs quite vigorously initially at an α oxide–coated gold surface, and β oxide formation apparently occurs as a side reaction, i.e., the gas evolution involves repetitive formation and decomposition of an unstable higher oxide (probably AuO_2), and the resulting disturbance of the α oxide film leads to a gradual accumulation of an outer, porous, gold β oxide deposit. In the potential cycling procedure each cycle results in formation and reduction of an α oxide film. Again, there is a side reaction involved, reduction of the place-exchanged α oxide in the negative sweep leads to the formation of some MMS atoms which, in the next positive sweep, are converted to β oxide species. It is important that, at the lower limit used for oxide growth, virtually all of the α, but none of the β, oxide is reduced; in this manner the β oxide deposit can attain multilayer coverage.

Gold hydrous oxide is not particularly stable; according to Pourbaix's thermodynamic data [28] it should undergo reduction to the metal in acid solution at ca. 1.4 V. However, in practice β oxide reduction occurs only under conditions of considerable cathodic overpotential and invariably at a lower potential than the peak for the reduction of the thin inner α oxide film at the same gold surface (after multilayer oxide growth the layer configuration at the interface is usually gold/α oxide/β oxide/aqueous solution; there may be some degree of intermingling of the β oxide and aqueous phase). The β oxide reduction responses in the negative sweeps are often complicated by the presence of several oxide reduction peaks [27,29], plus the fact that (even in terms of the RHE scale) the β oxide reduction peaks tend to occur at a lower potential in base as compared to acid; this effect is highlighted by the observation, Figure 4.2, of a sharp β oxide reduction peak at $E \approx -0.2$ V in the case of gold in base. This type of behavior is not confined to gold. Platinum [30,31] and iridium [32] β oxides also show similar behavior.

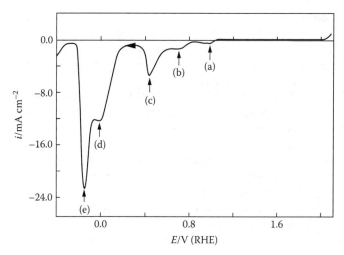

FIGURE 4.2 Reduction sweep (2.1 to –0.4 V, 10 mV s^{-1}) for a multilayer hydrous oxide-coated gold electrode in 1.0 mol dm^{-3} NaOH solution at 25°C; the film was grown in-situ by extended polarization, 90 min. at 2.30 V. (From Springer Science + Business Media: Cyclic voltammetry responses of metastable gold electrodes in aqueous media, L.D. Burke, J.M. Moran, and P.F. Nugent. *J. Solid State Electrochem.*, 2003, **7**, 529. With permission.)

Premonolayer Oxidation of Gold

Premonolayer (or underpotential) oxidation [33] relates to surface active site behavior, i.e., to the presence of low coordination MMS atoms on a predominately high coordination EMS surface. In simple terms the EMS state acts as a support system for the MMS state (or active sites). The MMS state (or states) are attributed to gold atoms present at apical or corner sites at surface protrusions; mobile surface atoms [34] and "magic number" clusters may also be involved. The unusual electronic properties of the latter have been described by Henglein [35]. The coverages of the high energy MMS state are usually quite low and their electrochemical responses (as is evident in Figure 4.1) are often virtually undetectable. There appear to be two options: to increase the sensitivity of the analytical procedure, or to raise the coverage of the MMS state. In practice the latter may be done inadvertently, e.g., when heating a single crystal plane electrode to remove surface defects and contaminants [36]. Rapid thermal quenching is a well-known metallurgical procedure for producing metastable metals in bulk form [37]; the states in question are highly defective and, although thermodynamically unstable, are reasonably persistent (gold electrodes subjected to severe thermal pretreatment [29,38,39] yield very unusual cyclic voltammetry responses).

Monolayer (α) Gold Oxides

The oxide electrochemistry of gold in aqueous media is dominated [28] by auric oxide which can be produced anodically, in thin film form, in either the anhydrous (α) state, Au_2O_3, or the hydrous (β) state, $Au(OH)_3$ or $Au_2O_3 \cdot 3H_2O$. Little attention has been devoted to aurous oxide, Au_2O, or hydroxide, AuOH, as these are not very stable. Gold peroxide, AuO_2, can be produced under highly anodic conditions but is also unstable; it decomposes rapidly with release of oxygen gas and may be an intermediate in the latter reaction. Soluble gold species, e.g., Au^+ and Au^{3+} and auric acid (H_3AuO_3) at low pH and aurate salts in base, may also be involved (usually to a quite limited extent); the presence of complexing species, e.g., CN^- or Cl^- ions, in solution promotes gold dissolution at relatively low potentials.

Generally, nonhydrated oxide films are produced preferentially on gold in acid solution on polarization over the range of ca. 1.36 to 2.0 V. According to Xia and Birss [24] the deposit in question is

AuO at E < 1.5 V, a mixture of AuO and Au_2O_3 at E > 1.5 V, and predominantly Au_2O_3 (up to three monolayers of which may be present) at high potentials, e.g., 2.6 V. Low density, microporous, β oxide deposits, formulated [24] as $Au_2O_3 \cdot nH_2O$ (n = 1 – 10) may be produced (on top of the inner α oxide film) either by extended dc polarization at E > 2.0 V or by extended potential cycling using appropriate upper and lower limits [20,27]. With gold in aqueous media, vigorous oxygen gas evolution is observed at E > 2.0 V.

The mechanism of α oxide film formation on gold in aqueous acid solution had been discussed in detail by Conway [40] and Tian, Pell, and Conway [41]. Monolayer oxide formation on gold in base commences at a lower potential (on the RHE scale) than in acid (ca. 1.20 V in base versus ca. 1.36 V in acid [23,42]). Similar behavior was observed with both Pd [42] and Pt [43], and the effect was attributed to the fact that the initial reaction involves active metal atoms which on oxidation yield significantly hydrated, anionic oxyspecies. Vigorous, steady-state oxygen gas evolution occurs on gold in aqueous media only above ca. 2.0 V [44].

Hydrous (β) Oxide Electrochemistry of Gold

According to Pourbaix [28] the reversible potential for the Au/Au_2O_3 (hydr.) transition at pH = 0 is 1.457 V; with a gold electrode under cyclic voltammetry conditions the oxide formation reaction is likely to occur at a slightly lower potential as surface, rather than bulk, metal atoms are involved. The β oxide formation process tends to be inhibited for two reasons: (1) the α oxide film is formed preferentially and this deposit passivates the surface; (2) to form a low density, porous, β oxide film, water molecules or hydroxide ions must penetrate the outer regions of the compact metal or the α oxide lattice. This apparently is only possible at low coverage defect sites at the gold surface and the resulting response is often minute or negligible and widely ignored. Multilayer β oxide growth may be achieved either by extended dc polarization (at E > 2.0 V) or repetitive potential cycling (the growth mechanisms involved in both cases have been discussed earlier [20]).

In thin film form, gold β oxide deposits undergo reduction in the negative sweep, after α oxide reduction, at ca. 1.0 V in acid and ca. 0.6 V in base [21,23]. This shift in redox potential with solution pH is a common feature of β oxide/oxide and β oxide/metal transitions [20], and is not restricted to gold; it is frequently referred to as a super-Nernstian E/pH shift [17,20,45,46].

A complicating feature of the β oxide electrochemistry of gold is that, for the same thick film, several oxide reduction peaks appear in the negative sweep. The behavior in base is significantly more complex as in this case some of the oxide reduction peaks occur at unusually low potentials—at least one at E < 0.0 V [29]. It was estimated [47] that if the initial product of the reduction of the low density oxide is virtually an isolated gold atom, with zero lattice stabilization energy, the reduction potential should drop to ca. 0.33 V. Furthermore, because the β oxide in alkaline solution is also base stabilized an oxide reduction response at E < 0.0 V for gold in base is not inconceivable, although obviously the peak potential for such a response does not relate to reversible electrode behavior in the thermodynamic sense.

FT-AC Voltammetry Investigations

The use of Fourier transformed alternating current (FT-ac) voltammetry to study premonolayer oxidation and electrocatalysis at gold in aqueous media yields interesting information [12]. The main advantage of this technique is that it achieves almost complete suppression of background capacitive currents, the responses observed in the double layer region, under appropriate conditions, being restricted to faradaic behavior. Some of the more interesting points noted include the following (potential values given in O'Mullane et al. [12] are quoted here in the RHE scale).

With unactivated gold electrodes no dc redox response was evident within the double layer region in either acid or base. However, the more sensitive FT-ac technique showed evidence of premonolayer oxidation, or an active site redox transition, at ca. 0.86 V in acid and ca. 0.80 V in base. After cathodic activation of the gold surface three different types of premonolayer (or active site) redox responses were observed, at ca. 0.48, 0.86, and 1.21 V in acid and ca. 0.10, 0.55, and 0.85 V in base. It is assumed here that the significantly lower transition potentials for gold in base reflect super-Nernstian E/pH behavior [20]).

Substantial FT ac responses were also observed in the initial stages of the monolayer oxide formation region of gold in both acid and base; this may be attributed to the involvement at this stage of the reaction of moderately active surface metal atoms that undergo a quasi-reversible redox, active metal atom/partially hydrated, anionic, oxide transition [42], prior to deposition of the more passivating α oxide deposit at higher levels of surface oxidation.

SURFACE ELECTROCATALYSIS ON GOLD

Because gold displays very weak chemisorbing properties, the activated chemisorption model of electrocatalysis is assumed to be inapplicable in the case of this metal in aqueous media. The alternative, which is well established in the chemically modified electrode [48] and redox sensor [49] area, is the interfacial cyclic redox mediator model which, in the case of gold in aqueous media, is sometimes referred to as the incipient hydrous oxide/adatom mediator (IHOAM) [18,33] model. In the case of the Group 11 metals the mediator systems are unusual in that their redox transitions involve couples with nonequilibrium (or metastable) reduced and oxidized states.

The basis of the IHOAM model of electrocatalysis is summarized in diagrammatic form in Figure 4.3 (a synopsis in chemical terms is given in Scheme 1 in O'Mullane et al. [12]). A low coverage, surface-bonded, interfacial redox mediator system, $M^*/M(OH)_n^{(n-z)-}$, is assumed to undergo a rapid, quasi-reversible redox transition (peaks a and a′ in Figure 4.3) at a potential E_s which is well within the double layer region. In some cases this basic mediator (premonolayer redox) response is not evident (the active state surface coverage being too low) under dc voltammetry conditions. However, several options are available to locate such "hidden" mediator transitions, e.g.,

1. More sensitive analytical techniques, e.g., FT-ac voltammetry [12] or ac impedance [8] may be used.
2. The surface may be deliberately disrupted or superactivated by severe thermal or cathodic pretreatment. Severe thermal activation [38] results in strong promotion of at least one, and probably two, premonolayer oxidation responses in the case of gold in acid at $E < 0.6$ V.
3. The onset of incipient oxidation, curve a in Figure 4.3, provides the interfacial mediator which triggers the oxidation of a dissolved reluctant, $Red_{(aq)}$. Thus, the electrocatalytic oxidation response, curve b in Figure 4.3, has an onset/termination potential coinciding with the interfacial mediator transition potential, and the former may be used to locate

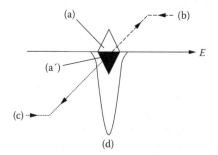

FIGURE 4.3   Schematic representation of the essential features of the IHOAM model of electrocatalysis. (From Springer Science + Business Media: Redox and electrocatalytic activity of copper in base at unusually low, premonolayer potentials, L.D. Burke J.A. Collins, and M.A. Murphy. *J. Solid State Electrochem.*, 1999, **4**, 34. With permission.)

the latter (the same argument applies to an electrocatalytic reduction response, curve c in Figure 4.3).

4. The potential of the maximum of the multilayer hydrous oxide reduction peak, curve d in Figure 4.3, often coincides with (and thus indicates the value of) the interfacial mediator transition potential. Such correlations have been discussed for gold [33], platinum [33], and palladium [50] in acid solution and for copper in base [51].

The reason why a particular electrocatalytic reaction is mediated by a specific interfacial couple is unclear; it may be that the electrocatalytic process induces some surface restructuring [5] so that a particular type of surface active site predominates (as long as the mediated reaction is in progress). Hydrazine oxidation at gold in base has an onset/termination potential [12] at ca. 0.4 V (RHE), i.e., much lower (as expected on the basis of super-Nernstian E/pH behavior [20]) than in acid solution (at ca. 0.8 V).

Oxidation of ethylene glycol at gold in base commences and terminates (see Figure 7(a) in O'Mullane [12]) at ca. 0.9 V (RHE). This is substantially more positive than the onset/termination potential quoted earlier for hydrazine oxidation in base, again suggesting that a different type of interfacial mediator system is involved. Glucose, $CHO(CHOH)_4CH_2OH$, contains a terminal alcohol and a terminal aldehyde group; its oxidation at gold in base under dc conditions [12,52] consists of two distinct responses—one at 0.2 and the other at 0.7 V—indicating that a different mediator system is involved in the oxidation of each group.

In the case of gold, the main mediator system is assumed to involve an Au(0)/Au(III) transition (gold apparently exhibits a number of interfacial couples involving the same oxidation states [19]). Establishing details of the low-coverage mediation species is not easy; the 0 and +3 oxidation states are assumed to be involved as these are by far the most prominent in basic gold electrochemistry [28]. However, it has been suggested [18,19] that a Au(0)/Au(I) mediator system operates catalytically at low potentials, ca. 0.15 V, at gold in base (virtually all the dissolved reductants and oxidants that react in this region exist in, or can readily attain, an anionic state; the anions evidently interact strongly with the cationic form of the mediator). The reduced form of the interfacial couple in question is assumed to involve a gold atom or cluster, Au^*, whereas the oxidized form, AuOH, is considered to be a strong base that exists at the surface mainly in cationic form, i.e., as Au^+_{ads}, the OH^- ion residing in a solvent-separated form at the outer Helmholtz plane.

The oxidation of dimethylamine borane (DMAB), a reducing agent used in electroless plating baths [53], on gold is also assumed to involve the intervention of Au^+_{ads} as a mediator [54]. Similarly, the reduction of anionic oxidants such as the peroxide ion [55], HO_2^-, may involve the intervention of Au^* as a reductant. Other electrocatalytic processes involving anionic species whose responses at gold in base commence and terminate at ca. 0.15 V [19] include polysulfide ion (S_n^{2-}) reduction and formaldehyde (which exists in base in anionic form, $H_2CO^-\cdot OH$) oxidation.

NANOPARTICLE GOLD ELECTROCATALYSIS

Fuel cells continuously convert chemical energy associated with hydrogen or hydrocarbon oxidation to electrical energy and, because they avoid the limitation of the Carnot heat cycle, they have the potential advantages of high energy conversion efficiency plus noise- and pollution-free operation. There are two major areas of potential application of gold in the ambient temperature fuel cell area [56]. First, the bipolar plates in a fuel cell are often made of stainless steel, and to prevent corrosion of the latter and ensure good electrical contact between the plates and the porous electrode material the steel may be plated with a thin layer of corrosion-resistant, highly conducting gold. Despite its obvious inertness from a corrosion viewpoint, gold exhibits surprising electrocatalytic activity and in the second area of application the metal is being investigated as a partial or complete replacement for platinum in the anode or cathode zone in fuel cell stacks. Use was made earlier of gold/platinum cathode electrocatalysts in alkaline H_2/O_2 fuel cells employed in space exploration probes (see p. 350 of Bond et al. [4]). Gold is one of the most active catalysts for oxygen gas reduction in base; however, close to the rest potential, the product with pure gold is predominantly the peroxide ion (HO_2^-) and the function of the platinum in the mixed metal electrocatalyst is to promote decomposition of the peroxide so that oxygen reduction at the mixed metal surface in base occurs via a 4-electron pathway. The following is a brief overview of the use of pure gold and its mixtures with other metals, in extended and nanoparticle form, as electrocatalysts in ambient temperature fuel cells.

OXYGEN REDUCTION

The oxygen reduction reaction (ORR) is the major process at fuel cell cathodes. Typically the cathode electrocatalyst is Pt dispersed on high surface area carbon; however, the performance of Pt is not ideal. The standard potential for the oxygen electrode (H_2O/O_2) in acid solution at 298 K is 1.23 V (RHE). The cathode of the Pt-activated acid fuel cell operating at a moderate current density and an O_2 pressure of 1.0 atm is in the region of 0.8 V—clearly there is considerable scope for improved performance and cost reduction in the fuel cell cathode area. The catalysis of the oxygen reduction reaction, largely at metal surfaces in aqueous media, was surveyed recently by Adzic [57].

Oxygen and hydrogen peroxide undergo reduction at extended gold surfaces in acid solution only below ca. 0.4 V [47]; both the potential range and rate observed are far too low to be of interest in the acid fuel cell cathode area. Similar behavior was observed by Guerin et al. [58] for O_2 reduction at TiO_x-supported gold nanoparticles in acid solution. In the latter work the onset/termination potential for the reaction decreased from ca. 0.55 to 0.3 V as the Au particle size dropped from 4.6 to 1.4 nm. Hayden et al. reported elsewhere [59] that with a similar type of electrocatalyst (TiO_x-supported Au nanoparticles of diameter = 2.8 nm) CO oxidation in acid solution commenced in the positive sweep at ca. 0.2 V (compared with ca. 0.4 V for extended gold). The potential range in question here, 0.2–0.6 V (RHE), for the onset of both electrocatalytic processes coincides with that for the active surface state oxidation responses for both thermally [38] and cathodically [47] pretreated gold in acid solution; again it appears that electrocatalytic behavior even of supported nanoparticles is dominated by active state surface transitions, i.e. the IHOAM model is applicable. The oxygen reduction activity of gold in thin film [60] and supported nanoparticle [61–66] forms, in acid solution, has also been investigated. With regard to the low onset potential for reduction of O_2 there appears to be little advantage in using highly dispersed gold; one apparent exception in this case involves the use of gold nanoparticles embedded in an electroactive polymer [65].

Oxygen reduction at gold in base commences just above 0.9 V [55,57], the initial product being the hydrogen peroxide anion (HO_2^-). There are exceptions to such behavior; for example, if the gold catalyst contains a significant amount of platinum metal the 4-electron pathway will prevail, and this also occurs, over a limited range of potential (ca. 0.9 to 0.7 V), with certain types of single crystal, e.g., Au(100), surfaces [57]. Peroxide ions in oxygen-free base undergo reduction [55] on polycrystalline gold below ca. 0.2 V; evidently the two oxidants, O_2 and HO_2^-, undergo reduction via two different mediating systems at the same interface. Responses for oxygen reduction at supported gold nanoparticles in base have been reported by a number of groups [66–70]. Ring-disk experiments [67,68] indicated that a 2-electron reaction is mainly involved, especially with gold nanorods; with gold nanoparticles more extensive reduction is observed over the range 0.6–0.8 V and below ca. 0.3 V.

Interesting accounts of nanoscale bimetallic electrocatalysts have been published by both Stamenkovic et al. [71] and Hernandez-Fernandez et al. [72]. The latter group prepared samples of Au/Pt bimetallic electrocatalysts on carbon (20 mass % metal; Au/Pt ratio ca. 1/2) by four different techniques, three of which yielded single bimetallic phases. As discussed by Bond [73], the two metals in question display a miscibility gap, the solubilities of Pt in Au (ca.17%) and

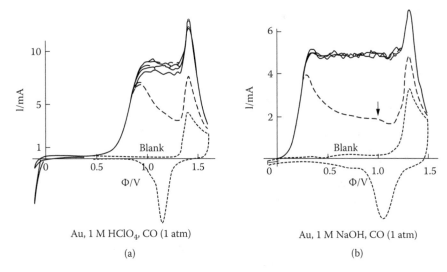

FIGURE 4.4 Voltammograms recorded at 60 mV s^{-1} for gold in (a) acidic and (b) alkaline solution: full lines, CO bubbling; dashed line, CO-saturated, quiescent solution; dotted line, He bubbling. (Reprinted from Electrochemical oxidation of CO on Au in alkaline solution, H. Kita, H. Nakajima, and K. Hayashi, *J. Electroanal. Chem.*, 1985, **190**, 141. With permission.)

Au in Pt (ca. 4%) being relatively low at ambient temperature. However, homogeneous alloys of these two metals can be produced if the particle size is quite small and such alloys often have superior catalytic activity [74]. The preparation and analysis of a range of Au/Pt alloy nanoparticle electrocatalysts and their use for oxygen reduction in both acid and base have also been described [75,76]. In addition, Adzic et al. [77] demonstrated that the stability, with respect to dissolution, of carbon-supported Pt electrocatalysts for oxygen reduction in acid solution was increased by depositing some Au clusters on the Pt nanoparticles; it was suggested that the presence of the Au raises the oxidation potential of the Pt.

CARBON MONOXIDE (CO) OXIDATION

Adsorbed carbon monoxide, CO_{ads}, is a well-known fuel cell anode–deactivating species [78] whose oxidative removal is a topic of major interest. Oxidation responses for dissolved CO at gold in acid and base are shown in Figure 4.4; the reaction in acid commences at ca. 0.5 V, the rate rising to an apparently transport-limited plateau value and then decreasing as the reaction becomes inhibited in the α oxide region above 1.35 V. From a fuel cell anode viewpoint the problem with this CO oxidation response in acid is that the onset potential is rather high; however, Hayden et al. [59,79] observed that with TiO$_x$-supported gold nanoparticles (mean diameter = 2.8 nm) CO oxidation in acid solution commenced at ca. 0.2 V.

The electrocatalytic performance of gold for CO oxidation in base is apparently much better than in acid. According to the data in Figure 4.4b the reaction commences at ca. 0.0 V and rises to a plateau value at ca. 0.3 V, the rate continuing in a transport-limited manner until monolayer oxide inhibition commences at ca. 1.25 V. However, such behavior was observed only with fresh, clean gold, free of CO_{ads}; the latter species accumulates gradually at the gold surface [80] when the electrode is held at low potentials ($E < 0.25$ V), resulting in CO oxidation below 0.5 V being severely impeded. The effect is illustrated in Figure 4.5a. Holding the potential of the gold electrode in CO-stirred base resulted in a decay in oxidation activity below 0.5 V, together with the appearance of a new anodic feature due to oxidation of CO_{ads} at ca. 1.0 V (Figure 4.5b); note also an indication of a quasi-reversible MMS redox response at ca. 0.15 V in the latter diagram. Cathodic pretreatment

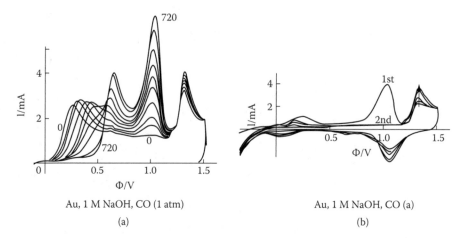

Au, 1 M NaOH, CO (1 atm)   Au, 1 M NaOH, CO (a)

(a)   (b)

FIGURE 4.5   Voltammograms recorded at 60 mV s^{-1} for gold (a) in CO-saturated, quiescent base as a function of electrode holding, or CO adsorption, time (the values for the latter were 0, 10, 20, 40, 60, 80, 120, 180, 360, and 720 s), and (b) in CO-free solution; in this case the surface initially adsorbed CO (for 5 min at –0.25 V) and then He bubbling was carried out for 10 min to remove dissolved CO prior to recording the scans shown here. (Reprinted from Electrochemical oxidation of CO on Au in alkaline solution, H. Kita, H. Nakajima, and K. Hayashi, *J. Electroanal. Chem.*, 1985, **190**, 141. With permission.)

is known to activate gold surfaces [47]; it seems that such activation leads to the gradual loss of the mediator system promoting CO oxidation at $E < 0.5$ V and to the formation of substantial quantities of some type of Au*-CO$_{ads}$ species which undergoes oxidation at ca. 1.0 V.

Studies of CO oxidation on extended gold in aqueous media are too numerous to discuss individually here. Among the more interesting is Weaver's [81,82] surface enhanced Raman spectroscopy (SERS) work which noted the appearance of a response for "terminally bound CO" in CO-saturated acid solution at $E = 0.25$ V. This response disappeared at $E > 0.85$ V; evidently most of the CO$_{ads}$ species were removed electro-oxidatively well within the double layer region.

Spectroscopic investigation of the adsorption and electro-oxidation of CO on gold in aqueous media has been a topic of widespread interest [83–88]. Beltramo et al. [88] and Shubina et al. [89] have discussed the reaction involved from a theoretical viewpoint and suggested that OH plays a different role in CO electro-oxidation on Au and Pt.

There have also been a number of investigations of CO electro-oxidation at gold nanoparticle surfaces [59,79,90,91]. Kumar and Zou [90] used Au nanoparticles attached to indium/tin oxide (ITO)–coated glass slides (the ITO surface itself was catalytically inert). The onset potential for the CO oxidation in base in the positive sweep seemed largely independent of the gold particle size or coverage (the particles diameter ranged from 3 to 50 nm); however, the rate of oxidation in the transport-limited region increased significantly with increasing gold particle coverage. The same authors reported later [91] on the behavior of Pt-coated Au nanoparticles as catalysts for the electro-oxidation of CO and CH$_3$OH in acid solution. Geng and Lu [92] claimed that with regard to CO oxidation on glassy carbon-supported gold nanoparticles in base the electrocatalytic activity was greater at smaller gold particle sizes.

Combinatorial chemistry techniques were used recently to synthesize and automatically test libraries of supported gold nanoparticles as electrocatalysts for CO oxidation. Hayden et al. [59,79] used physical vapor deposition to prepare a range of gold nanoparticles of mean diameter less than 6.5 nm on titania (TiO$_x$; $x \approx 1.96$) and carbon supports which were subsequently tested with regard to CO oxidation activity in acid solution (other authors [93] synthesized a range of Au nanoparticles on TiO$_x$ using pulsed cathodic electrodeposition). It was observed that, under potential sweep conditions, CO oxidation in the positive sweep, at particles of mean diameter 2.8 nm, commenced at ca. 0.2 V (compared with ca. 0.5 V for bulk gold). The support exerted a surprising effect; gold particles (diameter 2.7 nm) on TiO_x remained active, the rate of CO oxidation increasing significantly as the potential was raised (in 0.1 V steps) from 0.2 to 0.6 V. When the experiment was repeated using similar gold nanoparticles supported on carbon, no CO oxidation response was observed over the same potential range (extended gold surfaces have a significant response for CO oxidation in acid solution only above ca. 0.5 V). The influence of the support on the catalytic (and electrocatalytic) activity of gold nanoparticles is still a matter of debate; the nanoparticles lost their catalytic activity as their diameter dropped below ca. 2.8 nm. The authors attributed the latter behavior to changes in electronic properties associated with quantum size effects [79]; in such circumstances the active sites may become overoxidized.

FORMIC ACID AND METHANOL OXIDATION

The anodic oxidation of small (C_2) oxygenated organic molecules in acid solution is of special interest with regard to the operation of ambient temperature fuel cells. Gold is catalytically inactive with regard to the oxidation of methanol and formaldehyde [88] but, as outlined here later, there is significant interest in this area in the use of Pt/Au electrocatalysts. The oxidation behavior of carbon monoxide and formic acid at gold in acid solution has been compared recently by Beltramo et al. [88]. Based on data obtained using a range of computational, electrochemical, and spectroscopic techniques, the authors concluded that the slow step in formic acid oxidation at low potential, 0.7–1.3 V, involved the reaction of adsorbed formate anions, viz.,

$$HCOO^-_{ads} = CO_2 + H^+ + 2e^- \tag{4.1}$$

A curious feature in this work is that, although the mechanisms of reaction are quite different (CO oxidation involves an oxygen insertion step), carbon monoxide and formic acid oxidation at gold in acid yield quite similar cyclic voltammetry responses (see Figure 3 in Beltramo et al. [88]); in particular, the onset/termination potential at ca. 0.5 V is virtually the same for both compounds. Apparently, irrespective of the mechanism of oxidation, it is the redox transition of the interfacial mediator system (and not the nature of the dissolved reductant) that determines the value of the onset/termination potential.

A number of authors have investigated the use of Pt/Au alloy and Pt-decorated Au (shell-core) nanoparticles as catalysts for formic acid oxidation in acid solution. Park et al. [94] described chemical reduction techniques for preparing Pt/Au alloy, pure Au, and Pt-modified Au nanoparticles on carbon supports. The Pt/Au alloy and Pt-modified Au nanoparticles showed higher activities for formic acid oxidation than pure Pt, especially at low potentials in the region of 0.2 V. These results of Park et al. [94] are supported by the data of Kristian et al. [95]. Although the detailed mechanism of operation of these new electrocatalysts remains to be fully clarified, the combination of improved performance, combined with a substantial reduction in the use of Pt, are attractive features of Pt-decorated Au nanoparticle electrocatalysts from a fuel cell viewpoint.

A recent report [96] on the electrocatalytic activity of Pt nanoparticles generated electrochemically on a gold disk electrode for reactions such as carbon monoxide, methanol, and formic acid oxidation in acid solution, while stressing the importance of the structure and morphology of the Pt, gave no indication of a synergistic interaction between Au and Pt; possibly for such an interaction both metals must be in intimate contact in the nanoparticles state. The direct methanol fuel cell (DMFC) involves methanol oxidation at the anode and oxygen/air reduction at the cathode, the two electrodes being separated by a protonic ion-exchange membrane. The concept of a relatively safe, reactive, directly oxidizable liquid fuel is highly attractive but, unfortunately, both electrodes in the DMFC are severely inhibited. With regard to methanol oxidation, an electrocatalyst is required not merely to increase the rate of reaction but, more importantly, to promote this increase at a very low potential, preferably at $E \leq 0.2$ V. Gold is not a very effective catalyst for methanol oxidation, especially at low potential, in acidic media. Hence, attention has been devoted to the use of core-shell (Au/Pt) electrocatalysts for methanol oxidation in both acid [97,98] and base [99–101]. Use of this type of electrocatalyst enhances the Pt utilization but otherwise not the performance of the catalyst; significant rates of methanol oxidation in acid solution in this case were observed [98] only above 0.2 V (SCE), or ca. 0.45 V (RHE); similar behavior for this reaction was reported by Zhoa and Xu [97]. Miyazaki et al. [102] claimed that the activity of $Pt/MoO_x/C$ electrocatalysts for methanol oxidation in acid solution at 0.40 V was enhanced following chemical vapour deposition (CVD) of gold nanoparticles on the surface of the solid; it was suggested that the gold improved the CO tolerance of the anode electrocatalyst.

CONCLUSIONS

The surface chemistry of gold is currently being reassessed in the light of Haruta's discovery of the remarkable catalytic properties of oxide-supported gold nanoparticles [103]. With regard to electrocatalytic behavior the situation is different as high activity for many electrode reactions is observed with extended gold surfaces in aqueous media, especially in base, and in the absence of an oxide support. It is assumed that the important entities at solid metal surfaces from a catalytic viewpoint are surface active sites in the form of protruding metal atoms or clusters which possess unusual thermodynamic and kinetic properties. Such states are more prone to exist in oxidized forms in aqueous media as the resulting ionic states are stabilized by hydration. While the mechanism of surface active site behavior is still a matter of debate, it may be noted that some authors [104] assume that the active ingredient in supported gold nanoparticle catalysts for gas phase reactions such as CO oxidation is a cationic gold state whose generation is assumed to be favored by the presence of moisture.

The basic electrochemistry of gold in aqueous media is usually described in a simple manner as involving a broad potential range exhibiting only double-layer, faradaic reaction–free, behavior, complicated at positive potentials, $E > 1.0$ V (RHE), by monolayer oxide formation/removal reactions. This approach is considered valid for low-energy (EMS) gold surfaces; however, it ignores high-energy (MMS) defect states, the role of the latter as active sites, and hydrous oxide species. The marked electrocatalytic properties of gold appear to be based on quasi-reversible interfacial redox transitions whose low coverage, ill-defined, components function as mediators in oxidation and reduction processes at the gold surface.

Because of its low rate of dissolution in acid and base, its marked resistance to oxidation, and its weak chemisorption capability, gold is an excellent substrate for fundamental research in interfacial electrochemistry. The use of novel techniques in this area, as illustrated by the recent application of FT-ac voltammetry studies [12], is strongly recommended. Applications are not lacking: use of gold electrodes in pulsed electrochemical detector systems [105], amperometric biosensors [49,106] and self-assembled monolayer systems (based for instance on the gold-thiol route) [49,107] are areas of broad interest not discussed here. The intriguing combination of high corrosion resistance and (for some reactions) high catalytic activity suggest useful electrochemical applications of gold in areas such as energy conversion and environmental protection which are in urgent need of attention.

The potential applications of gold in the fuel cell area were surveyed recently by Cameron et al. [108] who emphasized the relatively low cost and greater availability of gold as compared to platinum. According to this group, the most likely initial application of gold with regard to fuel cells is in the fuel conditioning area, e.g., CO removal from the hydrogen feedstock. Use of supported gold catalysts for hydrogen production via the water-gas shift reaction, plus a variety of applications within the cell, are further topics that need to be vigorously explored (see also Chapter 6, Gold Catalysis).

REFERENCES

1. A theory of the catalytic surface, H. S. Taylor, *Proc. R. Soc. London A*, 1925, **108**, 105.
2. Dynamics of reactions at surfaces: the role of deflects: 'active sites', G. Ertl, *Advances in Catalysis*, ed. B. C. Gates and H. Knözinger, 2000, **45**, 49–53. New York: Academic Press.
3. Scanning tunnelling microscopy for metal deposition studies, D. M. Kolb and M. A. Schneeweiss, *Electrochem. Soc. Interface,* 1999, **8**, 26.
4. Chemisorption of simple molecules on gold, G. C. Bond, C. Louis, and D. T. Thompson, *Catalysis by Gold*, 2006, 126–127. London: Imperial College Press.
5. Active sites in heterogeneous catalysts: development of molecular concepts and future challenges, G. A. Somorjai, K. R. McCrea, and J. Zhu. *Topics Catal.*, 2002, **18**, 157.
6. Catalysis on surfaces, B. C. Gates, *Catalytic Chemistry*, 1992, 352. New York: Wiley.
7. R. J. Gale, *Spectroelectrochemistry, Theory and Practice*, 1988. New York: Plenum Press.
8. AC impedance investigation of copper in acid: 1. Role of the high energy surface state, L. D. Burke and R. Sharna, *J. Electrochem. Soc.*, 2008, **155**, D83.
9. Large amplitude Fourier transformed high harmonic alternating current cyclic voltammetry: kinetic discrimination of interfering faradaic processes at glassy carbon and at boron doped diamond electrodes, J. Zhang, S.-X. Guo, and A. M. Bond, *Anal. Chem.*, 2004, **76**, 3619.
10. Theoretical studies of large amplitude alternating current voltammetry for a reversible surface-confined electron transfer process coupled to a pseudo first-order electrocatalytic process, J. Zhang and A. M. Bond, *J. Electroanal. Chem.*, 2007, **600**, 23.
11. Resistance, capacitance, and electrode kinetic effects in Fourier-transformed large amplitude sinusoidal voltammetry: emergence of powerful and intuitively obvious tools for recognition of patterns of behaviour, A. A. Sher, A. M. Bond, D. J. Gavaghan, et al., A*nal Chem.*, 2004, **76**, 6214.
12. Study of the underlying electrochemistry of polycrystalline gold electrodes in aqueous solution and electrocatalysis by large amplitude Fourier transformed alternating current voltammetry, A. P. O'Mullane, A. M. Bond, L. D. Burke, et al., *Langmuir*, 2008, **24**, 2856.
13. Chemisorption at electrodes: hydrogen and oxygen on noble metals and their alloys, R. Woods, *Electroanalytical Chemistry*, ed. A. J. Bard, 1976, **9**, 1. New York: Marcel Dekker.
14. Catalytic activity at surfaces: adsorption and catalysis, P. W. Atkins, *Physical Chemistry*, 4th ed., 1990, 895. Oxford: Oxford University Press.
15. Gold catalysts prepared by coprecipitation for low-temperature oxidation of hydrogen and of carbon monoxide, M. Haruta, N. Yamada, T. Kobayashi, et al., *J. Catal.,* 1989, **115**, 301.
16. Surface chemistry of catalysis by gold, R. Meyer, C. Lemire, S. K. Shaikhutdinov, et al., *Gold Bull.*, 2004, **37**, 72.
17. The electrochemistry of gold: I The redox behavior of the metal in aqueous media, L. D. Burke and P. F. Nugent, *Gold Bull.*, 1997, **30**, 43.
18. The electrochemistry of gold: II The electrocatalytic behaviour of the metal in aqueous media, L. D. Burke and P. F. Nugent, *Gold Bull.*, 1998, **31**, 39.
19. Scope for new applications for gold arising from the electrocatalytic behaviour of its metastable surface states, L. D. Burke, *Gold Bull.*, 2004, **37**, 125.
20. Electrochemistry of hydrous oxide films, L. D. Burke and M. E. G., Lyons. In *Modern Aspects of Electrochemistry*, ed. R. E. White, J. O'M. Bockris, and B.E. Conway, 1986, **18**, 169. New York: Plenum Press.
21. Thick oxide growth on gold in base, L. D. Burke and M. McRann, *J. Electroanal. Chem.*, 1981, **125**, 387.
22. Influence of pH on the reduction of thick anodic oxide films on gold, L. D. Burke, M. E. Lyons, and D. P. Whelan, *J. Electroanal. Chem.*, 1982, **139**, 131.
23. Limit to extent of formation of the quasi-two-dimensional oxide state on Au electrodes, G. Tremilioso-Filho, L. H. Dall'Antonia, and G. Jerkiewicz, *J. Electroanal. Chem.*, 1997, **422**, 149.
24. A multi-technique study of compact and hydrous Au oxide growth in 0.1 M sulphuric acid solutions, S. J. Xia and V. I. Birss, *J. Electroanal. Chem.*, 2001, **500**, 562.
25. Growth of surface oxides on gold electrodes under well defined potential, time and temperature conditions, G. Tremiliosi-Filho, L. H. Dall'Antonia, and G. Jerkiewicz, *J. Electroanal. Chem.*, 2005, **578**, 1.
26. Hydrous oxide formation on gold in base under potential cycling conditions, L. D. Burke and G. P. Hopkins, *J. Appl. Electrochem.*, 1984, **14**, 679.
27. Multicomponent hydrous oxide films grown on gold in acid solution, L. D. Burke and P. F. Nugent, *J. Electroanal. Chem.*, 1998, **444**, 19.

28. Gold, M. Pourbaix, *Atlas of Electrochemical Equilibria in Aqueous Solutions*. 1966, 399. Oxford: Pergamon.
29. Cyclic voltammetry responses of metastable gold electrodes in aqueous media, L. D. Burke, J. M. Moran, and P. F. Nugent, *J. Solid State Electrochem.*, 2003, **7**, 529.
30. Overlap of the oxide and hydrogen regions of platinum electrodes in aqueous acid solution, L. D. Burke and A. J. Ahern, *J. Solid State Electrochem.*, 2001, **5**, 553.
31. Anomalous stability of acid-grown hydrous platinum oxide films in aqueous media, L. D. Burke and D. T. Buckley, *J. Electroanal. Chem.*, 1994, **366**, 239.
32. Use of iridium oxide films as hydrogen gas evolution cathodes in aqueous media, L. D. Burke, N. S. Naser, and B. M. Ahern, *J. Solid State Electrochem.*, 2007, **11**, 655.
33. Premonolayer oxidation and its role in electrocatalysis, L. D. Burke, *Electrochem. Acta*, 1994, **39**, 1841.
34. Determination of step and kink energies on Au(100) electrode in sulphuric acid solutions by island studies with electrochemical STM, S. Dieluweit and M. Giesen, *J. Electroanal. Chem.*, 2002, **524–525**, 194.
35. Physicochemical properties of small metal particles in solution: "microelectrode" reactions, chemisorption, composite metal particles and atom-to-metal transition, A. Henglein, *J. Phys. Chem.*, 1993, **97**, 5457.
36. Flame-annealing and cleaning technique, J. Clavilier. In *Interfacial Electrochemistry*, ed. A. Wieckowski, 1999, 231. New York: Dekker.
37. Processing of metals and alloys: rapid solidification, C. Suryanarayana. In *Material Science and Technology, a Comprehensive Treatment*, 1991, **15**, 57–110, ed. R. W. Cahn, P. Haasen and E. J. Kramer. Weinheim: VCH.
38. The effect of severe thermal pretreatment on the redox behaviour of gold in aqueous acid solution, L. D. Burke, L. M. Hurley, V. E. Lodge, et al., *J. Solid State Electrochem.*, 2001, **5**, 250.
39. An investigation of the electrochemical responses of superactivated gold electrodes in alkaline solution, L. D. Burke and L. M. Hurley, *J. Solid State Electrochem.*, 2002, **6**, 101.
40. Comparative behaviour in surface oxide formation and reduction at Au, B. E. Conway. In *Progress in Surface Science*, ed. S. G. Davidson, 1995, **49**, 407. New York: Pergamon.
41. Nanogravimetry study of the initial stages of anodic surface oxide film growth at Au in aqueous $HClO_4$ and H_2SO_4 by means of EQCN, M. Tian, W. G. Pell, and B. E. Conway, *Electrochim. Acta*, 2003, **48**, 2675.
42. Influence of solution pH on monolayer and multilayer oxide formation processes on gold and palladium, L. D. Burke, M. M. McCarthy, and M. B. C. Roche, *J. Electroanal. Chem.*, 1984, **167**, 291.
43. The possible importance of hydrolysis effects in the early stages of metal surface electrooxidation reactions – with particular reference to platinum, L. D. Burke and M. B. C. Roche, *J. Electroanal. Chem.*, 1983, **159**, 89.
44. Unusual postmonolayer oxide behaviour of gold electrodes in base, L. D. Burke, V. J. Cunnane, and B. H. Lee, *J. Electrochem. Soc.*, 1992, **139**, 399.
45. The oxygen electrode: Part 7 – influence of some electrical and electrolyte variables on the charge capacity of iridium in the anodic region, D. N. Buckley, L. D. Burke, and J. K. Mulcahy, *J. C. S. Faraday 1*, 1976, **72**, 1896.
46. Mediation of oxidation reactions at noble metals anodes by low levels of in situ generated hydroxyl species, L. D. Burke and K. J. O'Dwyer, *Electrochim. Acta*, 1989, **34**, 1659.
47. Generation of active surface states of gold and the role of such states in electrocatalysis, L. D. Burke and A. P. O'Mullane, *J. Solid State Electrochem.*, 2000, **4**, 285.
48. Chemically modified electrodes, R. W. Murray. In *Electroanalytical Chemistry*, ed. A. J. Bard, 1984, **13**, 191–368. New York: Dekker.
49. Electrochemical fuctionalization of a gold electrode with redox-active self-assembled monolayer for electroanalytical application. S. Behera, S. Sampath, and C. R. Raj, *J. Phys. Chem. C*, 2008, **112**, 3734.
50. Anomalous electrochemical behaviour of palladium in aqueous solution, L.D. Burke and L.C. Nagle, *J. Electroanal. Chem.*, 1999, **461**, 52.
51. Redox and electrocatalytic activity of copper in base at unusually low, premonolayer potentials, L. D. Burke J. A. Collins, and M. A. Murphy, *J. Solid State Electrochem.*, 1999, **4**, 34.
52. The role of incipient hydrous oxides in the oxidation of glucose and some of its derivatives in aqueous media, L. D. Burke and T. G. Ryan, *Electrochim. Acta*, 1992, **37**, 1363.
53. Investigation of DMAB oxidation at a gold microelectrode in base, L. C. Nagle and J. F. Rohan, *Electrochem. Solid-State Lett.*, 2005, **8**, C77.
54. Oxidation of some reducing agents used in electroless plating baths at gold anodes in aqueous media, L. D. Burke and B. H. Lee, *J. Appl. Electrochem.*, 1992, **22**, 48.

55. (100)-type behaviour of polycrystalline gold towards O_2 reduction, C. Paliteiro, *Electrochim. Acta*, 1994, **39**, 1633.
56. Uses of gold in fuel cell systems, http://www.utilisegold.com/uses_applications/fuel_cells/.
57. Recent advances in the kinetics of oxygen reduction, R. Adzic. In *Electrocatalysis*, ed. J. Lipkowski and P. N. Ross, 1998, 197–242. New York: Wiley-VCH.
58. A combinatorial approach to the study of particle size effects on supported electrocatalysts: oxygen reduction on gold, S. Guerin, B. E. Hayden, D. Pletcher, et al., *J. Comb.Chem.*, 2006, **8**, 679.
59. Enhanced activity for electrocatalytic oxidation of carbon monoxide on titania-supported gold nanoparticles, B. E. Hayden, D. Pletcher, and J.-P. Suchsland, *Angew. Chem.*, 2007, **119**, 3600.
60. Electrochemical reduction of oxygen on thin-film Au electrodes in acid solution, A. Sarapuu, K. Tammeveski, T. T. Tenno, et al., *Electrochem. Commun.*, 2001, **3**, 446.
61. An extraordinary electrocatalytic reduction of oxygen on gold nanoparticle- electrodeposited gold electrodes, M. S. El-Deab and T. Ohsaka, *Electrochem. Commun.*, 2002, **4**, 288.
62. Oxygen reduction on Au nanoparticles deposited on boron-doped diamond films, Y. Zhang, S. Asahina, S. Yoshihara, et al., *Electrochim. Acta*, 2003, **48**, 741.
63. Electrocatalytic reduction of oxygen to water at Au nanoclusters vacuum-evaporated on boron-doped diamond in acidic solution, I. Yagi, T. Ishida, and K. Uosaki, *Electrochem. Commun.*, 2004, **6**, 773.
64. Electrochemical preparation of a Au crystal with peculiar morphology and unique growth orientation and its catalysis for oxygen reduction, F. Gao, M. S. El-Deab, T. Okajima, et al., *J.Electrochem. Soc.*, 2005, **152**, A1226.
65. Gold nanoparticles dispersed into poly(aminothiophenol) as a novel electrocatalyst – fabrication of a modified electrode and evaluation of its electrocatalytic activity for dioxygen reduction, A. I. Gopalan, K. -P. Lee, K. M. Manesh, et al., *J. Mol. Catal. A*, 2006, **256**, 335.
66. Electrochemical reduction of oxygen on nanostructured gold electrodes. Sarapuu, M. Nurmik, H. Mändar, et al., *J. Electroanal. Chem.*, 2008, **612**, 78.
67. Gold nanoparticles synthesized in water-in-oil microemulsion: electrochemical characterization and effect of surface structure on the oxygen reduction reaction, J. Hernández, J. Solla-Gullón, and E. Herrero, *J. Electroanal. Chem.*, 2004, **574**, 185.
68. Characterisation of the surface structure of gold nanoparticles and nanorods using structure sensitive reactions, J. Hernández, J. Solla-Gullón, E. Herrero, et al., *J. Phys. Chem. B Lett.*, 2005, **109**, 12651.
69. Morphological selection of gold nanoparticles electrodeposited on various substrates, M. S. El-Deab, T. Sotomura, and T. Ohsaka, *J. Electrochem. Soc.*, 2005, **152**, C730.
70. Oxygen reduction at Au nanoparticles electrodeposited on different carbon substrates, M. S. El-Deab, T. Sotomura, and T. Ohsaka, *Electrochim. Acta*, 2006, **52**, 1792.
71. Trends in electrocatalysis on extended and nanoscale Pt-bimetallic alloy surfaces, V. R. Samenkovic, B. S. Mun, M. Arenz, et al., *Nature Materials*, 2007, **6**, 241.
72. Influence of the preparation route of bimetallic Pt-Au nanoparticles electrocatalysts for the oxygen reduction reaction, P. Hernandez-Fernandez, S. Rojas, P. Ocon, et al., *J. Phys. Chem. C*, 2007, **111**, 2913.
73. The electronic structure of platinum-gold alloy particles, G. C. Bond, *Platinum Metals Rev.*, 2007, **51**, 63.
74. Catalysis by gold/platinum group metals, D. T. Thompson, *Platinum Metals Rev.*, 2004, **48**, 169.
75. Electrocatalytic reduction of oxygen: gold and gold-platinum nanoparticle catalysts prepared by two-phase protocol, M. M. Maye, N. N. Kariuki, and J. Luo, *Gold Bull.*, 2004, **37**, 217.
76. Activity-composition correlation of AuPt alloy nanoparticle catalysts in electrocatalytic reduction of oxygen, J. Luo, P. N. Njoki, Y. Lin, et al., *Electrochem. Commun.*, 2006, **8**, 581.
77. Stabilization of platinum oxygen-reduction electrocatalysts using gold clusters, R. R. Adzic, J. Zhang, K. Sasaki, et al., *Science*, 2007, **315**, 220.
78. The oxidation of small organic molecules: a survey of recent fuel cell related research, R. Parsons and T. VanderNoot, *J. Electroanal. Chem.*, 1988, **257**, 9.
79. CO oxidation on gold in acidic environments: particle size and substrate effects, B. E. Hayden, D. Pletcher, M. E. Rendall, et al., *J. Phys. Chem. C*, 2007, **111**, 17044.
80. Electrochemical oxidation of CO on Au in alkaline solution, H. Kita, H. Nakajima, and K. Hayashi, *J. Electroanal. Chem.*, 1985, **190**, 141.
81. The electro-oxidation of carbon monoxide on noble metal catalysts as revisited by real-time surface-enhanced Raman spectroscopy, Y. Zhang and M. J. Weaver, *J. Electroanal. Chem.*, 1993, **354**, 173.
82. Surface-enhanced Raman scattering at gold electrodes: dependence on electrochemical pretreatment conditions and comparison with silver, P. Gao, D. Gosztola, and M. J. Weaver, *J. Electroanal. Chem.*, 1987, **233**, 211.

83. Infrared spectra of carbon monoxide adsorbed on a smooth gold electrode. Part 11: EMIRS and polariza-tion-modulated IRRAS study of the adsorbed CO layer in acidic and alkaline solutions, K. Kunimatsu, A. Aramata, H. Nakajima, et al., *J. Electroanal. Chem.*, 1986, **207**, 293.

84. Study by UV-visible potential-modulated reflectance spectroscopy and cyclic voltammetry of the electroadsorption and electro-oxidation of CO on Au in alkaline media, J. A. Caram and C. Gutierrez, *J. Electroanal. Chem.*, 1991, **314**, 259.

85. Infrared adsorption enhancement for CO adsorbed on Au films in perchloric acid solutions and effects of surface structure studied by cyclic voltammetry, scanning tunnelling microscopy and surface-enhanced IR spectroscopy, S. -G. Sun, W.-B Cai, L.-J. Wan, et al., *J. Phys. Chem. B*, 1999, **103**, 2460.

86. Anion adsorption, CO oxidation, and oxygen reduction reaction at a Au(100) surface: the pH effect, B. B. Blizanac, C. A. Lucas, M. E. Gallagher, et al., *J. Phys. Chem. B,* 2004, **108**, 625.

87. Surface electrochemistry of CO on reconstructed gold single crystal surfaces studied by infrared reflection absorption spectroscopy and rotating disk electrode, B. B. Blizanac, M. Arenz, P. N. Ross, et al., *J. Am. Chem. Soc.*, 2004, **126**, 10130.

88. Oxidation of formic acid and carbon monoxide on gold electrodes studied by surface-enhanced Raman spectroscopy and DTF, G. L. Beltramo, T. E. Shubina, and M. T. M. Koper, *ChemPhysChem,* 2005, **6**, 2597.

89. Density functional theory study of the oxidation of CO by OH on Au (100) and Pt (111) surfaces, T. E. Shubina, C. Hartnig, and M. T. M. Koper, *Phys. Chem. Chem. Phys.,* 2004, **6**, 4215.

90. Electrooxidation of carbon monoxide on gold nanoparticle ensemble electrodes: effects of particle cover-age, S. Kumar and S. Zou, *J. Phys. Chem. B*, 2005, **109**, 15707.

91. Electrooxidation of carbon monoxide and methanol on platinum-overlayer-coated gold nanoparticles: effects of film thickness, S. Kumar and S. Zou, *Langmuir,* 2007, *23*, 7365.

92. Size effect of gold nanoparticles on the electrocatalytic oxidation of carbon monoxide in alkaline solu-tion, D. Geng and G. Lu, *J. Nanopart. Res.*, 2007, **9**, 1145.

93. Automated electrochemical synthesis and characterization of TiO₂ supported Au nanoparticle elec-trocatalysts, S.-H. Baeck, T. F. Jaramillo, A. Kleinman-Shwarsctein, et al., *Meas. Sci. Technol.*, 2005, **16**, 54.

94. Surface structure of Pt-modified Au nanoparticles and electrocatalytic activity in formic acid electro-oxidation, I.-S. Park, K.-S. Lee,Y.-E. Sung, et al., *J. Phys. Chem. C*, 2007, **111**, 19126.

95. Highly efficient submonolayer Pt-decorated Au nano-catalysts for formic acid oxidation, N. Kristian, Y. Yan, and X. Wang, *Chem. Commun.*, 2008, 353.

96. The electrooxidation of small organic molecules on platinum nanoparticles supported on gold: influence of platinum deposition procedure, F. J. E. Scheijen, G. L. Beltramo, M. T. M. Koper, et al., *J. Solid State Electrochem.*, 2008, **12**, 483.

97. Enhancement of Pt utilization in electrocatalysts by using gold nanoparticles, D. Zhao and B.-Q. Xu, *Angew. Chem. Int. Ed,,* 2006, **45**, 4955.

98. Au @ Pt nanoparticles prepared by one-phase protocol and their electrocatalytic properties for methanol oxidation, L. Yang, J. Chen, X. Zhong, et al., *Colloids and Surfaces A: Physicochem. Eng. Aspects,* 2007, **295**, 21.

99. Gold-platinum alloy nanoparticle assembly as catalyst for methanol electrooxidation, Y. Lou, M. M. Maye, L. Han, et al., *Chem. Commun.*, 2001, 473.

100. Catalytic activation of core-shell assembled gold nanoparticles for electrocatalytic methanol oxidation reaction, J. Luo, M. M. Maye, Y. Lou, et al., *Catal. Today*, 2002, **77**, 127.

101. Characterization of carbon-supported AuPt nanoparticles for electrocatalytic methanol oxidation reac-tion, J. Luo, P. N. Njoki, Y. Lin, et al., *Langmuir*, 2006, **22**, 2892.

102. Electro-oxidation of methanol on gold nanoparticles supported on Pt/MO$_x$/C$_1$, K. Miyazaki, K. Matsuoka, Y. Iriyama, et al., *J. Electrochem. Soc.*, 2005, **152**, A1870.

103. Gold as a novel catalyst in the 21st century: preparation, working mechanism and applications, M. Haruta, *Gold Bull.*, 2004, **37**, 27.

104. Role of cationic gold in supported CO oxidation catalysts, J. C. Fierro-Gonzalez, J. Guzman, and B. C. Gates, *Topics Catal.*, 2007, **44**, 103.

105. Liquid chromatography with pulsed electrochemical detection, D. C. Johnson and W. R. LaCourse, *Anal. Chem.*, 1990, **62**, 589A.

106. Gold electrodes modified with 16H, 18H-dibenzo[c,l]-7,9-dithia-16,18-diazapentacene for electrocatalytic oxidation of NADH, V. Rosca, L. Muresan, I. C. Popescu, et al., *Electrochem. Commun.*, 2001, **3**, 439.

107. Self-assembled monolayers of a hydroquinone-terminated alkanethiol onto a gold surface: interfacial electrochemistry and Michael-addition reaction with glutathione, M. Shamsipur, S. H. Kazemi, A. Alizadeh, et al., *J. Electroanal. Chem.*, 2007, **610**, 218.

108. Gold's future in fuel cell systems, D. Cameron, R. Holliday and D. Thompson, J. Power Sources, 2003, **118**, 298.

5 Gold Luminescence

Vivian Wing-Wah Yam and Chi-Hang Tao

CONTENTS

Introduction..69
Luminescent Gold(I) Complexes..70
    Gold(I) Phosphine Complexes..70
    Gold(I) Chalcogenido and Thiolato Complexes 71
    Gold(I) Alkynyl Complexes..76
    Gold (I) Complexes with Dicyanides..79
    Trinuclear Complexes with Exobidentate N^C/N^N Bridging Ligands.................79
Luminescent Gold(III) Complexes 81
Conclusion and Future Prospects..83
References..83

INTRODUCTION

Gold is one of the most inert metals on earth. Despite the stability of gold under ambient conditions, coordination compounds with oxidation numbers of +1, +2, and +3 with various coordination modes and geometries have been discovered.[1] Gold could be made soluble in aqueous solutions upon oxidation by the presence of both a good ligand and an oxidizing agent. For instance, *aqua regia* is needed to convert gold metal to tetrachloroauric acid which is soluble in water, but neither hydrochloric acid nor nitric acid alone could lead to the dissolution. The orange-colored aqueous solution of the trivalent tetrachloroauric acid represents one of the most commonly employed starting materials in the preparation of luminescent gold complexes with different oxidation states.

The chemistry of luminescent gold complexes has attracted increasing attention since the 1990s, a number of which are associated with the weak intermolecular attractive interactions among gold centers resulting from relativistic and correlation effects.[2,3] As a result of the relativistic effect exhibited by gold,[4–6] the 6s orbital and, to a lesser extent, the 6p orbitals will be stabilized while a radial expansion and energetic destabilization of the 5d orbitals would occur. Thus, the closed-shell $5d^{10}$ gold center in gold(I) compounds would have a strong tendency to interact with other elements or other gold atoms. This inter- or intramolecular sub-bonding interaction is widely referred to as "aurophilicity," which was first introduced by Schmidbaur.[2,3] This gold–gold interaction has a bonding energy that is comparable to that of hydrogen bonding,[2] which provides the extra structural stability for polynuclear gold clusters and architectures. These weak attractive interactions, together with the large spin-orbit coupling imposed by the heavy gold metal centers, have made luminescent gold systems fascinating and unique. This chapter aims to give a brief account on the recent development of luminescent gold complexes with different oxidation states, their intriguing luminescence properties, and their potential applications in various fields.

LUMINESCENT GOLD(I) COMPLEXES

Gold(I) Phosphine Complexes

The first example of luminescent gold complex was reported by Dori and co-workers in 1970, in which the photoluminescence properties of $[Au(PPh_3)Cl]$ were described.[7] The annular complex, $[Au_2(dppm)_2]^{2+}$, which possesses fascinating photophysical properties and excited-state redox behavior, was reported by Che[8] and Fackler[9] independently in 1989. These dinuclear complexes were found to exhibit intense, long-lived yellow phosphorescence. The triplet excited state, $[Au_2(dppm)_2]^{2+*}$, was found to be a strong reducing agent, with the excited state reduction potential, $E°[Au_2(dppm)_2^{3+/2+*}]$, of -1.6 (1) V versus saturated sodium chloride calomel electrode (SSCE) determined via oxidative quenching experiments using a series of pyridinium acceptors.[8–10] The electronic absorption and magnetic circular dichroism (MCD) studies of the related systems with $Me_2P\text{-}(CH_2)_n\text{-}PMe_2$ ($n = 1, 2$) as the bridging ligands were reported by Mason and co-workers.[11] The absorption and emission energies of this class of complexes are found to be strongly associated with the extent of both intra- and intermolecular gold–gold interactions in various media. Since then, the study of luminescent Au(I) complexes with gold–gold interactions and the understanding of the effects of such interactions as well as the rational design of luminescent gold complexes for various applications have attracted increasing attention. The exploration of luminescent gold systems has since aroused immense attention. Yam and co-workers have reported the systematic comparison studies of dinuclear complexes, $[Au_2(dmpm)_2]^{2+}$ and $[Au_2(dmpm)_3]^{2+}$, with their trinuclear analogues, $[Au_3(dmmp)_2]^{3+}$ and $[Au_3(dmmp)_3]^{3+}$, in order to investigate and understand the effect of gold–gold interactions and the coordination number and geometry on the luminescence and photophysical properties of gold(I) phosphine systems.[12,13] A red shift in the absorption maximum is observed on going from the dinuclear complexes to the trinuclear analogues. The two-coordinate complexes were found to exhibit dual luminescence in degassed acetonitrile solutions upon excitation at 300–370 nm with the higher emission band assigned as the intraligand phosphorescence while the lower-energy emission band assigned to be originated from the metal-centered $^3[(d\delta^*)^1(p\sigma)^1]$ phosphorescence. The lower-energy emission band was found to follow the same trend as in the electronic absorption spectra. Such a red shift in absorption and emission energies is in line with the reduced HOMO-LUMO gap upon increasing the number of gold atoms in the system as illustrated in Figure 5.1. The three-coordinate complexes exhibited a single emission band in the orange–red region ($[Au_2(dmpm)_3]^{2+}$: $\lambda_{em} = 588$ nm, $\tau_o = 0.85 \pm 0.10$ μs, and $[Au_3(dmmp)_3]^{3+}$: $\lambda_{em} = 625$ nm, $\tau_o = 2.2 \pm 0.20$ μs) which is in contrast to the dual luminescence observed in their dinuclear analogues. The relatively small red shift of 0.13 eV from the dinuclear to trinuclear complexes together with the lifetimes in the microsecond range is supportive of a $^3[(d\delta^*)^1(p\sigma)^1]$ excited state.

Che and co-workers recently reinvestigated a related dinuclear Au(I) system, $[Au_2(dcpm)_2]^{2+}$ with ClO_4^-, PF_6^-, $CF_3SO_3^-$, $[Au(CN)_2]^-$, Cl^-, and I^- counter ions.[14] The emissive states of this system are suggested to be highly affected by the nature of the counter anions and coordinating solvents. The long-lived high-energy emission observed at 368 nm with a lifetime of 4.88 μs for $[Au_2(dcpm)_2](ClO_4)_2$ in the solid state at room temperature upon photoexcitation was assigned as the "intrinsic" metal-centered $^3[d\sigma^*p\sigma]$ emission while the lower-energy emission at 564 nm was attributed to be derived from the exciplex formed between the excited state and the acetonitrile solvent. These assignments have been supported by *ab initio* calculations on the model complex $[Au_2(H_2PCH_2PH_2)_2]^{2+}$.[15] The aurophilic interactions and structural information of the excited states of $[Au_2(dcpm)_2](ClO_4)_2$ have also been probed by resonance Raman studies, in which the ground and $^1[(d\sigma^*)^1(p\sigma)^1]$ excited state vibrational frequencies of 88 and 175 cm^{-1}, were determined, respectively.[16] Thus, the possibility of the origin of the low-energy emission band of $[Au_3(dmmp)_2]^{3+}$ at 580 nm to be derived from exciplex emission cannot be completely excluded.

Gold(I) complexes with bridging phosphine ligands have been used in the design of thin-film oxygen sensor as reported by Mills and co-workers.[17] Three-coordinate gold(I) complexes with phosphine bridging ligands were incorporated into thin plastic films as luminescence quenching

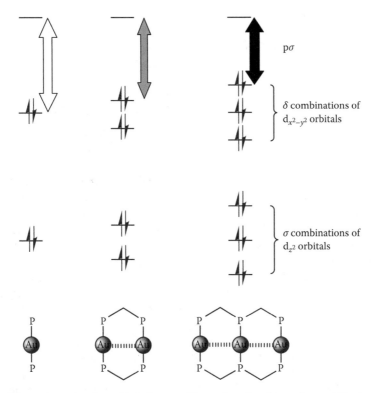

FIGURE 5.1 Schematic molecular orbital diagram illustrating the effect of aurophilic interactions on the absorption and emission energies in Au(I) phosphine systems.

oxygen sensors which are found to be more sensitive than a similar $[Ru(dpp)_3]^{2+}$-based sensor.[18] The luminescence quenching of these sensors by oxygen obeys the Stern-Volmer equation, and Stern-Volmer constants of 5.35×10^{-3} and 0.9×10^{-3} Torr^{-1} are found for the dinuclear and trinuclear complexes, respectively, in a polystyrene polymer matrix.

Other than using bridging phosphines with methylene spacers, bidendate bridging phosphines with aromatic spacers have been employed by various researchers.[19–23] For instance, Yip and Prabhavathy reported the synthesis of a luminescent gold ring complex, $[Au_3(\mu\text{-}Ph_2PAnPPh_2)_3]$ $(ClO_4)_3$, using 9,10-bis(diphenylphosphino)anthracene as the bridging ligand.[22] The trinuclear gold complex was found to have a structure and fluxionality reminiscent of cyclohexane as supported by X-ray crystal analysis and NMR spectroscopic studies. In contrast to the commonly observed vibronically structured emissions from anthracenes, the emission of the complex at 475 nm was structureless and the emission was suggested to be excimeric in nature.

The luminescence studies of four-coordinate tetrahedral Au(I) phosphine complexes using diphosphine with naphthalene spacers were first reported by Yam and co-workers in 2000 (Figure 5.2).[24] The complexes were found to exhibit intense emission in the orange–red region in degassed dichloromethane and in the solid state and the emissions were assigned to be derived from emissive states derived from the $\sigma \rightarrow \pi^*$(naphthalene) transition. Besides exhibiting photoluminescence behavior, the complexes are also found to show electroluminescence (EL) when doped in polymer matrix as the emissive layer in a single-layered EL device.

GOLD(I) CHALCOGENIDO AND THIOLATO COMPLEXES

Apart from gold(I) phosphine systems, gold(I) chalcogenido and thiolato complexes represent another fast growing area of research in luminescent gold complexes. One of the reasons for such

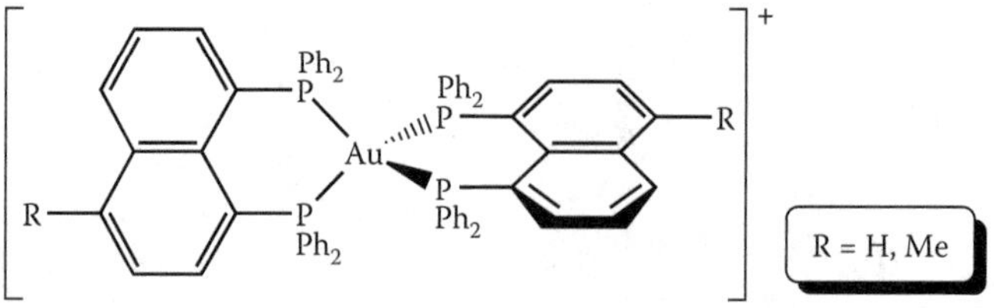

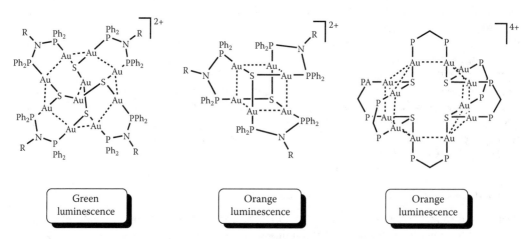

FIGURE 5.2   Structure of four-coordinate tetrahedral luminescent Au(I) complexes.

a growth in research interest in this area is probably due to the immense research interest devoted to the d^{10} metal chalcogenide nanoparticles which possess intriguing intrinsic photoluminescence properties. The metal chalcogenido complexes could serve as model compounds for the understanding of structure and bonding in solid state d^{10} metal chalcogenide nanoparticles. However, the preparation of gold(I) chalcogenido clusters of high nuclearity has been hampered by their tendency to form insoluble binary chalcogenides or polymeric materials.

The employment of phosphine ligands as a "protective coating" and solubilizing groups for the preparation of luminescent hexanuclear,[25] decanuclear,[26] and dodecanuclear[27] gold(I) sulfide complexes has been reported by Yam and co-workers (Figure 5.3). These high-nuclearity luminescent clusters are molecular in nature and their X-ray crystal structures were determined, with short intramolecular aurophilic contacts in the range of 2.939–3.3775 Å. Moreover, these complexes are soluble in common organic solvents and exhibit intense luminescence ranging from green to near IR in degassed solutions and in the solid state upon photoexcitation. For instance, the emission of $[Au_6\{\mu\text{-}Ph_2PN(p\text{-}CH_3C_6H_4)PPh_2\}_3(\mu_3\text{-}S)_2](ClO_4)_2$ recorded in dichloromethane at room temperature was centered at ca. 810 nm while a dual phosphorescence in the green and orange region was observed from a dichloromethane solution of $[Au_{10}(\mu\text{-}PNP)_4(\mu_3\text{-}S)_4](PF_6)_2$ upon photoexcitation. The high-energy emission at ca. 500 nm observed in the dual luminescence of $[Au_{10}(\mu\text{-}PNP)_4(\mu_3\text{-}S)_4]$ $(PF_6)_2$ was attributed to metal-perturbed intraligand phosphorescence. The lower-energy emissions from these clusters with emission maxima ranging from ca. 550 to 800 nm have been tentatively assigned to originate from triplet states of a ligand-to-metal charge transfer character that mixed with metal-centered (ds/dp) states modified by Au···Au interactions (S→Au···Au LMMCT). A series of phosphine-stabilized gold selenide clusters, $[Au_{18}Se_8(dppe)_6]Br_2$ and $[Au_{10}Se_4(dpppe)_4]Br_2$, has been reported by Fenske and co-workers.[28] Short intramolecular Au···Au contacts were found in these clusters and, in particular, a peculiar "multishell" core was noted for $[Au_{18}Se_8(dppe)_6]Br_2$

FIGURE 5.3   Luminescent hexanuclear, decanuclear, and dodecanuclear gold(I) sulfide complexes.

from their X-ray crystal structures. Both the dichloromethane solutions and the solid samples of the $Au_{18}Se_8$ cluster exhibited intense luminescence in the red region on UV-vis excitation at 300–500 nm. The cluster was found to possess high stability in dichloromethane solution and the photo-sensitization of singlet oxygen, $^1O_2$, was observed with a quantum yield of 0.17 ± 0.02 determined in O_2-saturated dichloromethane solutions. The $[Au_{10}Se_4(dpppe)_4]Br_2$ cluster showed a broad near-infrared photoluminescence at 880 nm in dichloromethane which was shifted to ~1020 nm in the solid state at 77 K.

Stable, isolable, water-soluble monolayer-protected Au clusters (MPCs) with thiolate ligands are of great recent research interest.[29–31] Murray and co-workers studied the effect of monolayer and the size of the metal cluster by ligand exchange and galvanic exchange, respectively. These MPCs were found to typically emit in the red to near-infrared (NIR) region with emission maxima of 650–1250 nm on photoexcitation. The authors found that luminescence intensities of the gold MPCs were linearly dependent on the number of gold atoms per cluster and the monolayer, suggesting that the NIR emission is an electronic surface-state phenomenon. An assignment of the low-energy emissions of gold MPC to emissive states associated with a S→Au···Au ligand-to-metal-metal charge transfer (LMMCT) origin was suggested by El-Sayed and Whetten on their studies on the visible to infrared luminescence behavior of a glutathione passivated Au_{28} MPC.[32]

Instead of large clusters with phosphines or thiolate ligands acting as the protective coat, molecu-lar gold(I) complexes of relatively low nuclearity with thiolate ligands have also been reported.[33–39] Fackler and co-workers studied the luminescence properties of a series of monomeric gold(I) com-plexes with phosphine and thiolate as ligands.[33] All the complexes were found to be luminescent at 77 K in the solid state with emission maxima of 413–702 nm and the emissions from these mono-nuclear complexes were found to be affected by both the substituents on the thiolate ligand and the presence of aurophilic interactions between neighboring molecules. In view of the luminescence lifetime in the microsecond range, the emission origins of these complexes have been assigned to be derived from S→Au···Au ligand-to-metal charge transfer (LMCT) triplet excited states. A series of luminescent mononuclear bis(benzenethiolato)gold(I) complexes has been reported by Nakamoto and co-workers.[40] These complexes exhibited emission in the solid state at room temperature with emission maxima of 438–529 nm depending on the substituent on the benzene ring. Unlike most gold(I) thiolate systems, the emission has been suggested to originate from a metal-to-ligand charge transfer (MLCT) or ligand-centered transition.

The emission origin of $[Au(PEt_3)_2]^+[Au(AuPEt_3)_2(i\text{-}mnt)_2]^-$ has been explored and discussed by Zink and co-workers, in which the emission at ca. 510 nm showed weak vibronic structures with average progressional spacings of ~470 cm^{-1} corresponding to the Au–S stretching frequency.[34] Assignment of the emission origin was assisted by the results from the preresonance Raman spec-trum, and the origin was attributed to a [S→Au] LMCT excited state associated with the trinuclear complex anion. Laguna and co-workers reported a series of trinuclear gold(I) dithiocarbamate com-plexes with bridging $Ph_2PCH_2P(Ph)CH_2PPh_2$.[36] The complexes were found to be fluxional in solu-tion, and a red shift in the solid-state emission energy was observed on increasing the number of dithiocarbamate ligands from 1 to 3.

The polymorphic phase change also has substantial influence on the metal···metal interactions and luminescence properties of gold complexes. Balch and co-workers reported a gold cluster complex, $[(\mu_3\text{-}S)\{AuCNC_7H_{13}\}_3](SbF_6)$, with interesting reversible polymorphic phase change on cooling.[41] The cluster in the solid state possesses an arrangement in which the two cations form a pseudo-octahedral array of gold atoms. Significant variations in the aurophilic interactions in the cluster were observed on cooling, which involve a lengthening of the aurophilic contacts in one pair of cations and an induced shortening of such contacts in a second pair of cations (Figure 5.4). On photoexcitation at 355 nm, crystals of $[(\mu_3\text{-}S)\{AuCNC_7H_{13}\}_3](SbF_6)$ were highly emissive at room temperature with an emission maximum at 667 nm. On cooling to 77 K, two emission maxima at 490 and 680 were observed with lifetimes of 10 μs and 25 μs, respectively. The dual luminescence observed at low temperature has been suggested to be associated with the formation of two different

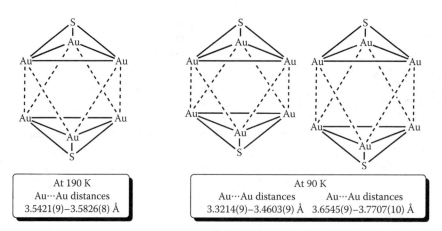

FIGURE 5.4 Lengthening and shortening of the aurophilic contacts in a pair of cations of $[(\mu_3\text{-S})\{AuCNC_7H_{13}\}_3](SbF_6)$ on cooling.

gold clusters, each producing independent emission spectra. It is interesting to note that no changes in the crystal structure and luminescence properties were observed from an analoge of the cluster, $[(\mu_3\text{-S})\{AuCNC_6H_{11}\}_3](PF_6)$, which possesses a similar structure at 190 K.

Laguna and Eisenberg reported a series of $[Au_3(\mu_3\text{-E})Ag(PPh_2py)_3](BF_4)_2$, (E = O, S, Se; PPh_2py = diphenylphosphino-2-pyridine), which was found to possess rich photoluminescence behavior.[42] The structural core of the cluster is a Au_3Ag tetrahedron with an oxygen atom capping the gold atoms in a μ_3-fashion. Intermolecular gold–gold interactions were also observed which link two Au_3Ag tetrahedron together. The emission maxima of the cluster in the solid state are highly dependent on the nature of the μ_3-bridging ligand with values of 466, 554, and 670 nm for E = O, S, and Se, respectively. The room temperature lower-energy emissions at ca. 554 and 670 nm for E = S or Se, respectively, are suggested to be originated from ^{3}LMMCT excited states while an assignment of either a ^{3}LMMCT or metal-centered cluster-based emission is suggested for the high-energy emission at ca. 466 nm.

The idea of ds/dp configurational mixing in the interpretation of the electronic absorption and emission properties of gold(I) complexes which possess weak Au⋯Au interactions was suggested by Vogler and co-workers in the studies of luminescence properties of a tetranuclear gold(I) dithioacetate, $[Au(MeCS_2)]_4$,[43] which was first isolated by Zanazzi and co-workers.[44,45] Other dinuclear gold(I) thiolate complexes bearing phosphine ligands are also known in the literature and found to exhibit interesting luminescence properties. For instance, a series of binuclear gold(I) thiolate complexes with bridging diphosphine ligands was reported by Bruce and Bruce[46] and was shown to display luminescence originated from LMCT excited states. These examples of luminescent gold(I) complexes have demonstrated that the nature of the thiolates and the metal⋯metal separation have significant effects on the luminescence properties of the complexes. Yam and co-workers also reported a series of luminescent dinuclear gold(I) thiolates with aminodiphosphine ligands, $[Au_2(Ph_2PN(R)PPh_2)(SR')_2]$ (R = C_6H_{11}, R' = $C_6H_4F\text{-}p$, $C_6H_4Cl\text{-}p$, $C_6H_4Me\text{-}p$; R = Ph, R' = $C_6H_4Me\text{-}p$; R = ^{n}Pr, R' = $C_6H_4Me\text{-}p$; R = ^{i}Pr, R' = $C_6H_4Me\text{-}p$).[47] Due to the steric requirements of the ligands, very weak to no Au⋯Au interactions were observed in this class of dinuclear complexes, and the solid-state and fluid-solution emission spectra are mainly dominated by a broad intense blue–green emission which was assigned as a metal-perturbed ligand-centered phosphorescence. However, low-energy emission at ca. 600 nm was observed for the solid samples of the complexes and in glass matrices at 77 K which are attributed to a LMCT (S→Au LMCT) origin.

In view of the ability to alter the luminescence properties of Au(I) complexes through a change in the Au⋯Au interactions, the design of luminescence switches or sensors employing aurophilic

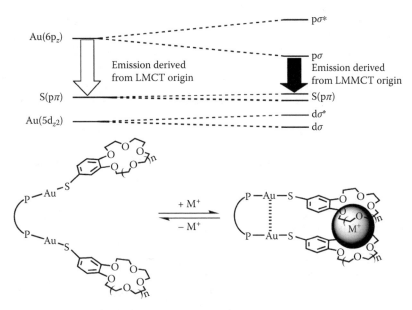

FIGURE 5.5 Schematic diagram showing the switching on of the Au···Au interactions through encapsulation of metal ion by dinuclear gold(I) thiolate complexes.

interactions as the reporter could be envisioned. A series of bis-crown ether–containing dinuclear gold(I) complexes have been shown to exhibit specific binding abilities toward various metal cations according to the ring size of the crown pendants.[48–50] The observation of Au···Au distances of ca. 3.28 (benzo-15-crown-5) and 3.2256(8) Å (benzo-18-crown-6), indicative of fairly weak Au···Au interactions in the solid-state X-ray crystal structures and the green emissions at ca. 492–580 nm observed in dichloromethane solutions at room temperature, are supportive of an emission origin with substantial LMCT(S→Au) character that has no significant Au···Au interactions. On the basis of a judicious choice of an appropriate size match between the crown ether moieties and the metal cations of interest, metal ion-binding in an intramolecular sandwich binding fashion has been demonstrated, which is also supported by electrospray ionization (ESI) mass spectrometry, when the size of the metal cations is larger than that of the crown ether cavities. This intramolecular sandwich binding mode will bring the two Au(I) metal centers closer to each other and subsequently lead to the switching on of the Au···Au interactions that give rise to the LMMCT emission, which is at lower energy than the LMCT emission that is typical of the absence of Au···Au interactions (Figure 5.5).

In addition to reports on the presence of metal ions via Au···Au interactions, there have also been reports on the switching of luminescence properties of gold(I) complexes by mechanical treatment.[51–53] In particular, a series of dinuclear cationic Au(I) complexes bridged by one dppm and one singly deprotonated 2-thiouracil or 6-methyl-2-thiouracil ligand was reported by Eisenberg and co-workers to display luminescence tribochromism.[52,53] In contrast to triboluminescence, in which light is emitted upon grinding or crushing the sample, the term "luminescence tribochromism" represents a sustained change in the emission spectrum upon initial application of pressure. The singly deprotonated complexes are weakly emissive and are suggested to exhibit weak intermolecular Au···Au interactions while the strongly emissive doubly deprotonated forms are shown to exhibit strong intermolecular Au···Au interactions. Through gentle grinding of a solid sample accompanied by liberation of acid vapor or addition of a base, the intermolecular Au···Au interactions will be strengthened which gives rise to the switching on of the blue or cyan photoluminescence. This process has been shown to be reversed through recrystallization of the crushed samples or by addition of an acid.

Considerable attention has been paid to the construction of supramolecular assemblies through noncovalent self-assembly approaches from molecular subunits.[54,55] The Au···Au interactions have bond strengths comparable to those of hydrogen bonding and it appears to be a unique structural motif in the design of novel assemblies.[56,57] Yu, Li and Yam reported a novel chiral gold(I) ring, with a cyclic framework consisting of 16 gold(I) atoms arranged in a closed macrocycle via noncovalent Au(I)···Au(I) interactions, self-assembled from achiral Au_4 units containing piperazine-1,4-dicarbodithiolate and dppm ligands.[58] The Au_{16} ring was found to exist as its tetranuclear monomer in dilute solutions, as revealed by ^{1}H NMR and ESI-MS studies. In addition to the high-energy $^1(d\sigma^* \rightarrow p\sigma)$ transition typical of dinuclear Au(I) diphosphines and the intraligand transitions of the bridging ligands, a low-energy absorption shoulder at ca. 370 nm was observed for the tetranuclear monomer that became more apparent at higher concentrations and did not obey Beer's law. A dimerization process was suggested involving a monomer–dimer equilibrium, with the ε_{abs}(dimer), the equilibrium constant K_{dimer}, and ΔG determined to be 2600 M^{-1} cm^{-1}, 6800 M^{-1}, and 22 kJ mol^{-1}, respectively. It is suggested that the formation of the chiral Au_{16} ring may proceed via the initial dimerization of the tetranuclear monomers through more pronounced Au(I)···Au(I) interactions on increasing concentration, as revealed by the growth of the absorption shoulder at ca. 370 nm in the UV-vis spectra. On photoexcitation, solids of the chiral Au_{16} macrocycle exhibit intense green phosphorescence at 531 nm. A dramatic blue shift in emission energy was observed in the tetranuclear monomer from 605 nm in MeCN at 293 K to 501 nm in ^{n}PrCN glass at 77 K. The emission is assigned to originate from S→Au LMCT excited states, probably modified by the presence of weak Au···Au interactions. The blue shift with decreasing temperature has been attributed to the increased rigidity of the molecular structure at low temperature, which renders a smaller geometrical distortion in the excited states and hence a smaller Stokes shift; this phenomenon has also been reported in the related gold(I) sulfide system.[25] Very recently, a gold macrocyclic ring with 36 gold(I) atoms with dithiocarbamate-type bridging ligands have been studied by Yu, Li and Yam and the macrocyclization was again suggested to be directed by Au···Au interactions.[59] Three bidentate dithiocarbamate units were found to coordinate to six Au(I) centers in a head-to-tail cyclic manner to form a three-bladed propeller-shaped monomer with D_3 symmetry. The two forms of chiral hexanuclear monomers, Δ-Au_6 and Λ-Au_6, were formed in equimolar amounts. Three alternating Δ-Au_6 and Λ-Au_6 units are then linked together by aurophilic interactions to afford a racemic cyclic hexamer of giant Au_{36} ring. The self-assembly process was probed by UV/Vis absorption studies. Dissolution of the Au_{36} compound in dichloromethane under dilute conditions ($\leq 10^{-5}$ M) gave rise to a yellow solution with absorption bands at 280 ($\varepsilon = 86145$), 342 sh (23 270), and 440 nm (10 135 M^{-1} cm^{-1}), which were assigned to absorptions typical of the Au_6 monomer. The yellow solution was found to turn dark red on increasing the concentration, with a low-energy tail growing in at around 500 nm that does not follow Beer's law. This finding was suggestive of the self-assembly of the Au_6 monomer to give higher oligomers or aggregates. This low-energy tail was found to become more obvious and appeared as a shoulder in a less solubilizing solvent (dichloromethane–ethanol mixture) and the low-energy absorptions at longer wavelengths were assigned to LMMCT transitions of the higher oligomers as a result of the more extended Au···Au interactions involving more than one Au_6 unit in these oligomers.

GOLD(I) ALKYNYL COMPLEXES

Rigid wire- or rod-like compounds have attracted recent attention as they possess intriguing electronic properties and well-defined molecular dimensions as well as bear close resemblance to molecular wires. The alkynyl group seems to be a promising candidate for the construction of such entities because of its inherent linearity and rigid structure. In addition, the alkynyl group is also known for its ability to interact with transition metal centers through pπ-dπ overlap which renders its common employment in the preparation of metal-containing carbon-rich materials.[60–62] Linear two-coordination geometries are among the most common coordination modes of Au(I) and are

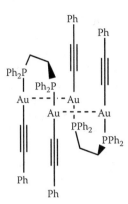

FIGURE 5.6 Structure of [Au$_2$(μ-dppe)(C≡CPh)$_2$].

particularly attractive for the incorporation of alkynyl ligands. This has rendered the gold(I) alkynyls an important class of organometallic gold(I) compounds.

The luminescence properties of Au(I) alkynyl complexes were first reported by Che and co-workers, in which the luminescence behavior of the dinuclear Au(I) alkynyl complex, [Au$_2$(μ-dppe)(C≡CPh)$_2$], was described (Figure 5.6).[63] Short intermolecular Au⋯Au contacts of 3.153(2) Å were observed in the X-ray crystal structure. As discussed in the previous sections, Au⋯Au interactions also play an important role in the photophysical properties of Au(I) complexes. The 550-nm emission of the solid sample of [Au$_2$(μ-dppe)(C≡CPh)$_2$] at room temperature was described as originating from a ^{3}[(dδ*)^1(pσ)^1] triplet excited state arising from the interactions between the two Au atoms.

The photophysical properties of this class of compounds have since been extensively investigated by the groups of Che,[64–66] Yam,[67–69] Mingos,[70] Puddephatt[71–73] and others[74] with emissive states ranging from ^{3}IL to ^{3}MLCT and admixtures of these manifolds. A number of dinuclear Au(I) alkynyl rod-like complexes bridged by an ethynyl or butadiynyl moiety have been synthesized and their photophysical properties reported by Müller et al. and the emission origins have been attributed to excited states derived from either σ(Au−P)→π*(naphthyl) or π→π*(naphthyl) transition. Che and co-workers recently reported the synthesis and photophysical studies of a series of related alkynyl complexes with different alkynyl chain length.[64] The involvement of the π* orbitals of the aryl rings of the phosphine ligands in the excited state was eliminated using an optically transparent tricyclohexylphosphine ligand. In addition, the bulkiness of the phosphine ligand has prevented the formation of Au⋯Au interactions both in solutions and in solid states. Highly vibronic structured emission bands in the near UV-region were observed for the diynyl complex with very sharp single vibronic progression of ~2000 cm^{-1} which is typical of ν(C≡C) modes, and were assigned to an acetylenic ^{3}(π→π*) emission. An extension of the work into the mono- and dinuclear Au(I) complexes with arylalkynyl ligands has also been reported by the same group.[65] The lowest-energy singlet transitions are predominantly intraligand in nature and exhibit both phenyl and acetylenic ^{1}(π→π*) character. Phosphorescence was observed in the emission spectra of these gold(I) arylalkynyl complexes. Delayed fluorescence was observed for complexes bearing long arylalkynyl chains and was suggested to be facilitated through a triplet-triplet annihilation mechanism. The relationship between the phosphorescent emission energy and arylalkynyl chain length was examined by a plot of $\Delta E(S_0–T_1)$ (energy gap between S_0 and T_1) against $1/n$ (n is the number of repeating arylalkynyl units in the mono- and dinuclear complexes). Extrapolation of the lines revealed an estimated $\Delta E(S_0–T_1)$ value for infinite repeating units (i.e., [Cy$_3$PAu]-capped PPE) of ~626 nm (1.98 eV) and ~607 nm (2.04 eV) for the mono and dinuclear species, respectively. In addition, a congenerous mononuclear Au(I) alkynyl complex, [(Cy)$_3$PAu(C≡CC$_6$H$_4$−NO$_2$)], has been reported to exist as two polymorphs with contrasting photophysical properties.[66] At 298 K, the emissive form of the complex is highly phosphorescent with a peak maximum at 504 nm while the other polymorph is nonemissive at 298 K, but emission is detected at 77 K with peak maximum at 486 nm. Crystallographic studies reveal that the major differences between the emissive and nonemissive forms of the complex are the orientations of the molecular dipoles and the dihedral angles between neighboring 4-nitrophenyl moieties and suggest that both the nature of the excited state and the dihedral angle between adjacent [Au(C≡CAr)] moieties determine the phosphorescent properties of these molecular crystals. Besides, the nonlinear optical (NLO) properties of a number of di- and trinuclear Au(I) alkynyl complexes with di- and triethynylbenzene cores have been investigated by Whittall,[75] Powell and Humphrey.[76] These complexes may serve as potential NLO materials with reasonable first- and second-order hyperpolarizabilities. Introduction of a nitro group into these systems was suggested to enhance the optical nonlinearity.

With the recent developments and growing interest in metal-based chemosensors,[77] some works toward the utilization of the luminescent Au(I) system in chemosensing have been described. A series of gold(I) calix[4]crown-5 alkynyl complexes has been reported by Yam et al.[68] These calix[4]crown-5 alkynyl complexes were found to bind K^+ ions preferentially over Na^+ ions with UV–vis absorption spectral changes. It is noteworthy that the crown size of these alkynyl complexes could be fine-tuned by a subtle change in the steric demands on the triarylphosphine ligands, providing a firm basis for the further development of these classes of selective metalloreceptors for various metal ions. These complexes have been found to emit in the yellow region with emission maxima of 578–584 nm in dichloromethane solution at room temperature and was assigned as metal-perturbed intraligand or $\sigma(Au-P) \rightarrow \pi^*$(phosphine) in character.

In contrast to the extensive studies of luminescent gold(I) alkynyl complexes with phosphine ligand, corresponding studies on mononuclear homoleptic dialkynylaurate(I) complexes, [RC≡C–Au–C≡C–R]⁻, are relatively scarce. This class of anionic gold(I) complexes usually shows highly structured absorption and emission bands with vibrational progressional spacings assignable to $\nu(C≡C)$ stretch and the emission origin has been typically assigned as acetylenic $^3IL(\pi \rightarrow \pi^*)$ in nature.[78,79] Mingos et al. reported the synthesis and X-ray structure of a novel catenane of gold alkynyls, where the alkynyl ligands adopt both η^1 and η^2 bonding modes,[80] demonstrating the feasibility of gold(I) alkynyls for the construction of supramolecular assemblies. Yam and co-workers reported a series of tetranuclear gold(I) alkynyl complexes with a novel planar η^2,η^2-coordination mode using diethynylcalix[4]crown-6 in a 1,3-alternate conformation as the building block (Figure 5.7). Short intramolecular Au⋯Au contacts with significant Au⋯Au interactions were observed from the X-ray crystal structure, in which the four gold(I) centers were arranged in a rhomboidal array and capped by the diethynylcalix[4]crown-6 ligands on the two ends. Two Au(I) atoms are σ-bonded to the alkynyl units while the other two Au(I) atoms are each π-coordinated to two alkynyl units in a η^2,η^2-sandwich fashion. On photoexcitation, these tetranuclear complexes are strong emitters in the orange region at room temperature with luminescence quantum yields of up to 0.22. Red shifts

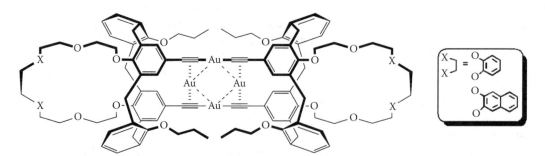

FIGURE 5.7 Luminescent tetranuclear gold(I) alkynyl complexes with a novel planar η^2,η^2-coordination mode.

in emission bands were noted in the solid state (592–611 nm) relative to that recorded in solution (587–588 nm), which may be attributed to the presence of intermolecular Au···Au interactions in the solid state. In view of the large Stokes shifts and microsecond range lifetimes, the emissions in these Au(I) supramolecular assemblies have been tentatively assigned as derived from states of metal-cluster–centered (ds/dp) character that are modified by Au···Au interactions, mixed with metal-perturbed intraligand $\pi \rightarrow \pi^*(C \equiv C)$ states.

GOLD (I) COMPLEXES WITH DICYANIDES

Gold(I) complexes with dicyanides have also been reported to exhibit extensive aurophilic interactions. Patterson and Fackler reported that the solutions of $K[Au(CN)_2]$ in water and methanol exhibit strong photoluminescence at room temperature when the complex concentration is higher than 10^{-2} M.[81] The emission wavelength of the solutions could be tuned ($>15 \times 10^3$ cm^{-1}) in the UV-visible region by systematic variations in concentration, solvent, temperature, and excitation wavelength. Generally, a red shift in emission maximum is observed upon an increase in concentration or a decrease in temperature.[82–84] The emission bands have been assigned to $*[Au(CN)^{2-}]_n$ excimers and exciplexes that differ in "n" and geometry. The energy transfer from these dicyanoaurate(I) moieties, as well as their Ag(I) analogs, to rare earth metal ions has also been reported.[85] The energy transfer efficiency, which is proportional to the spectral overlap between the donor emission and the acceptor absorption, could be tuned by different solvent compositions.

TRINUCLEAR COMPLEXES WITH EXOBIDENTATE N^C/N^N BRIDGING LIGANDS

Homoleptic trinuclear gold(I) complexes bearing exobidentate N^C or N^N ligands represent another important class of luminescent complexes of gold(I). The three gold(I) atoms of this class of complexes are usually arranged in a planar triangular array through intramolecular Au···Au interactions. Pronounced intermolecular Au···Au and $\pi–\pi$ interactions are usually observed due the planar structure, giving rise to intriguing luminescence phenomena.[86] One of the classical examples is the work by Balch and co-workers on their report of solvoluminescence from the crystalline trimer of $[Au_3(MeN=COMe)_3]$ (Figure 5.8).[87] The photoluminescence is triggered by contact of an organic solvent with crystals of the colourless, trimeric compound, $[Au_3(MeN=COMe)_3]$, following photoirradiation. On photoexcitation, solutions of the trinuclear gold complex in chloroform show an emission with maximum at 422 nm, while the polycrystalline solid sample displays dual luminescence with emission maxima at ca. 446 and 552 nm. Interestingly, the lower-energy emission band at ca. 552 nm is greatly enhanced by the addition of solvents, such as chloroform, dichloromethane, toluene, methanol, hexane, and water, in which the complex has greater solubilities.

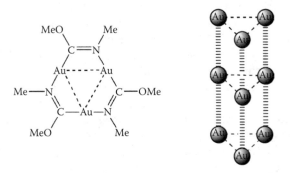

FIGURE 5.8   Structure of $[Au_3(MeN=COMe)_3]$ and a diagram showing the extensive intermolecular aurophilic interactions between the gold atoms in the indefinite columnar stacks.

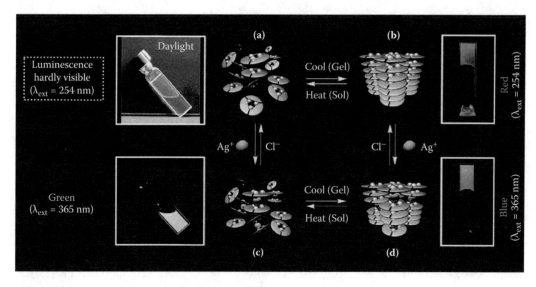

FIGURE 5.9 **(See color insert following page 212.)** Luminescence profiles of the trinuclear gold(I) complex in hexane. Pictures and schematic self-assembling structures of (a) sol, (b) gel, (c) sol containing AgOTf (0.01 equiv.), and (d) gel containing AgOTf (0.01 equiv.). (Reproduced from Phosphorescent organogels via "metallophilic" interactions for reversible RGB-color switching, A. Kishimura, T. Yamashita, T. Aida, *J. Am. Chem. Soc.* 2005, **127**, 179. With permission.)

Crystallographical studies revealed an intramolecular Au···Au separation of 3.308(2) Å and the presence of intramolecular aurophilic interactions. In addition, intermolecular aurophilic interactions with Au···Au distances of 3.346(1) Å are observed in the crystalline sample which facilitated the formation of indefinite columnar stacks of the triangular arrays, both in the eclipsed and staggered conformations in a 2:1 ratio. Extension of this work has led to the identification of three polymorphs with distinct luminescence properties and only the hexagonal polymorph is solvoluminescent that shows a low-energy emission at ~520 nm.[88] Besides, bulkier side groups in the C^N ligand that hamper the formation of columnar stacks of the trinuclear complexes would lead to the absence of solvoluminescence properties.[87]

Aida and co-workers have recently demonstrated the hierarchical self-organization of dendritic macromolecules of trinuclear gold(I) complexes using pyrazole as the bridging ligands.[89] An extension of the work using dendritic pyrazole ligands on trinuclear gold(I) complexes could form phosphorescent organogels via metallophilic interactions for reversible RGB-color switching (Figure 5.9).[90] The complex bearing $C_{18}H_{37}$ chains in hexane self-assembles via a Au···Au metallophilic interaction to form a red-luminescent organogel in hexane with emission maximum of 640 nm. Scanning electron microscopy (SEM) and X-ray diffraction (XRD) analysis of an air-dried metallogel showed the presence of heavily entangled fibers, each consisting of a rectangularly packed columnar assembly. An abrupt change in emission energy to the blue region with an emission maximum of 458 nm was observed on addition of a small amount of Ag^+ ion without disrupting the gel. This process was found to be reversible by the addition of cetyltrimethylammonium chloride, and the original red emission of the gel could be revived. On heating, the red-emitting and blue-emitting gels undergo gel-to-sol transition due to the destabilization of the metallophilic interactions, where the red luminescence of the nondoped mixture becomes hardly visible, while the blue luminescence of the Ag^+-doped mixture turns green with emission maximum of 501 nm. These red, green, and blue long-lived emissions from these materials with lifetimes of 3–6 μs were assigned to electronic transitions from triplet excited states.

LUMINESCENT GOLD(III) COMPLEXES

There has been immense attention to studies of luminescent d^8 metal complexes in recent decades. However, luminescence studies on gold(III) complexes are relatively less explored, in sharp contrast to the isoelectronic platinum(II) systems. This may be attributed to the presence of nonemissive, low-energy d–d ligand field states and the high electrophilicity of the gold(III) center. With the introduction of a strong σ-donating ligand to the gold(III) metal center, the chance of obtaining luminescent gold(III) complexes could be improved through the raising of the energy of the non-emissive ligand field states and the decrease in the electrophilicity of the gold(III) center. This concept has been vividly exemplified and demonstrated by Yam and co-workers in the preparation of a series of luminescent gold(III) diimine complexes, [Au(N^N)R$_2$]ClO$_4$ (N^N = bpy, phen or dpphen; R = mesityl, CH$_2$SiMe$_3$).[91] These complexes emit in the blue–green region in solid state and in fluid solution at room temperature on photoexcitation. Low-temperature solid-state emission spectra of these complexes show well-resolved vibronically structured emission bands with vibrational spacings of ca. $1300-1400 cm^{-1}$, typical of v(C=C) and v(C=N) stretches of the diimine ligands in the ground state. Several possible assignments for the emission have been suggested, including metal-perturbed [$\pi \rightarrow \pi*$] IL, [Au$\rightarrow \pi*$(N^N)] MLCT, [R$\rightarrow \pi*$(N^N)] LLCT, or mixtures of them in origin.

Incorporation of an aryl moiety into the diimine has led to the formation of cyclometalated gold(III) complexes.[92,93] These cyclometalated gold(III) complexes are emissive in fluid solutions and in the solid state at room temperature. In view of the close resemblance of the emission energies to that of the respective ligands, the vibronic-structured emission bands, lifetimes in the microsecond range, and the insensitivity of the emission energies toward different solvents, the emissions are suggested to be of triplet intraligand parentage. The complex [Au(C^N^N-Ph)Cl]$^+$ is a strong photo-oxidant with excited state reduction potential E°*(Au$^{III/II}$) estimated to be 2.2 V (vs. NHE). Recently, a series of cyclometalated gold(III) complexes containing the tridentate C^N^C ligand was reported.[94] The complexes emit only at low temperature from the ^{3}IL excited state. However, red shifts in electronic absorption and emission energies were observed for the dinuclear complexes, which have been attributed to intramolecular π-π stacking of the [Au(C^N^C)] moieties.

As mentioned in the beginning of this section, incorporation of a strong σ-donating ligand to the gold(III) metal center could improve the chance of obtaining luminescent gold(III) complexes through the raising of the energy of the nonemissive ligand field states and the decrease in the electrophilicity of the gold(III) center. A series of cyclometalated gold(III) alkynyl complexes was found to be emissive in various media even at room temperature (Figure 5.10).[95,96] The lowest-energy absorption bands of this class of complexes were assigned as the π–$\pi*$ intraligand (IL) transition of the cyclometalated RC^N(R')^CR ligand with some mixing of a [π(R''C≡C)$\rightarrow \pi*$(RC^N(R')^CR)] ligand-to-ligand charge transfer (LLCT) character. Except for the complexes with an electron-rich amino substituent on the alkynyl ligand, all the complexes showed emission in dichloromethane at room temperature from

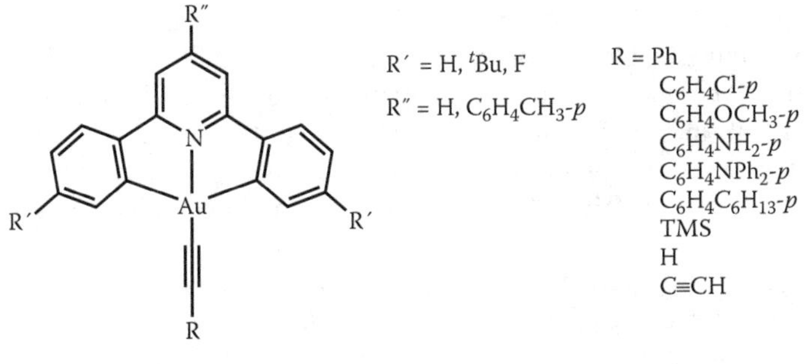

FIGURE 5.10 Structures of a series of cyclometalated gold(III) alkynyl complexes.

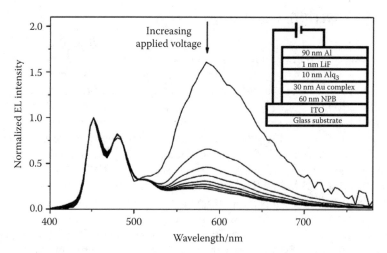

FIGURE 5.11 Tuning of emission color of a device fabricated using [Au(C^N^C)(C≡CC₆H₅)] by different DC voltages applied. (Reproduced from A novel class of phosphorescent gold(III) alkynyl-based organic light-emitting devices with tunable colour, K. M. C. Wong, X. Zhu, L. L. Hung, N. Zhu, V. W. W. Yam, H. S. Kwok, *Chem. Commun.* 2005, 2906. With permission.)

the excited states with predominantly ^{3}IL character of the cyclometalated RC^N(R′)^CR ligand. The amino-substituted complex was found to emit at a much longer wavelength compared to the others, and its emission was tentatively ascribed to a [π(R″C≡C)→π*(RC^N(R′)^CR)] LLCT excited state origin. X-ray crystal structures of these complexes indicated the presence of appreciable π–π stacking interactions in these square planar complexes. The results of the electrochemical and photophysical data have been verified by density functional theory (DFT) and time-dependent density functional theory (TDDFT) computational studies.

There has been recent interest in the exploration of transition metal complexes as electrophosphorescent materials. The luminescent cyclometalated gold(III) alkynyl complexes are neutral compounds that possess the advantage of high thermal stability and are attractive candidates of electrophosphorescent materials. Multilayer organic light emitting diodes (OLEDs) could be fabricated by vacuum deposition employing the gold(III) alkynyl complexes as electrophosphorescent emitters or dopants.[97] They were found to exhibit intense EL derived from the dimeric/oligomeric or excimeric intraligand triplet exciton resulting from the π stacking of the C^N^C ligand on applying a DC voltage. The EL color could be tuned toward the blue with increasing DC voltage. For instance, the EL spectra of a device fabricated using [Au(C^N^C)(C≡CC₆H₅)] (HC^N^CH = 2,6-diphenylpyridine) exhibit a vibronic-structured emission band at 452 nm and a board featureless red–orange band at ca. 585 nm attributed to the electroluminescence of NPB and the Au(III) complex, respectively (Figure 5.11). Upon increasing the DC voltage, the relative EL intensity ratio of the gold complex to the organic compound decreases gradually and the emission color of the device can be tuned from orange to blue, through white. This observation suggested that some holes are blocked by the Au(III) complex leading to light emission from the NPB layer which would become more pronounced with increasing bias. On the other hand, the emission color of the EL devices fabricated from these Au(III) complexes could also be tuned by increasing the concentration of [Au(C^N^C)(C≡CC₆H₄NPh₂)] (HC^N^CH = 2,6-diphenylpyridine) which acts as a dopant in the emitting layer (Figure 5.12). The EL band maxima were found to red shift from 500 to 580 nm on increasing the dopant concentration. It is suggested that a higher dopant concentration leads to a higher order and better packing of the molecules, leading to a stronger π stacking of the C^N^C ligand, and this accounts for the lowering of the energy of the EL band of the device on increasing the dopant concentration.

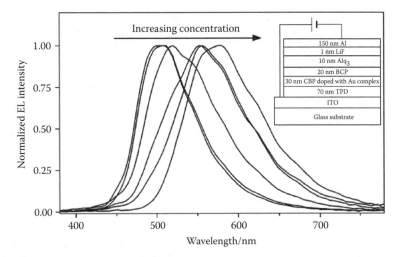

FIGURE 5.12   Tuning of emission colour of a device fabricated using [Au(C^N^C)(C≡CC$_6$H$_4$NPh$_2$)] as a dopant by varing the complex concentration. (Reproduced from A novel class of phosphorescent gold(III) alkynyl-based organic light-emitting devices with tunable colour, K. M. C. Wong, X. Zhu, L. L. Hung, N. Zhu, V. W. W. Yam, H. S. Kwok, *Chem. Commun.* 2005, 2906. With permission.)

CONCLUSION AND FUTURE PROSPECTS

Gold compounds have been widely used in various applications in the field of chemistry, while luminescent gold complexes have attracted immense attention in the last few decades and have emerged as one of the fastest growing sectors in the field of photochemistry and luminescent materials. One of the most probable reasons for such a growth is the interesting and unique phenomenon of forming Au⋯Au interactions which gives rise to luminescent properties and observations that are inaccessible by most other transition metal complexes. This has been exemplified in this chapter by selected works on luminescent gold complexes that highlighted the importance and uniqueness of the presence of Au⋯Au interactions.

The luminescence properties of gold complexes have been shown to be influenced by various factors, such as ligands, coordination modes, microenvironment around the metal centers, temperature, and the extent of metal⋯metal interactions. With the understanding of the underlying principles that govern the diverse luminescence properties of gold complexes, it is envisioned that luminescent gold complexes could find increasing applications in optoelectronics. This unique feature of gold complexes has also opened up novel pathways in sensing applications that are different from the bimolecular quenching mechanism commonly employed in the field of sensing. Moreover, the ability of gold(I) centers to interact with various metal centers has aroused much interest in the employment of gold in the design and synthesis of luminescent mixed-metal complexes and their subsequent applications as sensory materials. In addition to the studies on the catalysis and biomedical applications of gold(III) complexes, recent development of luminescent gold(III) complexes and their applications in OLEDs have represented a new dimension in the research and development of luminescent gold-based molecular materials, which may lead to a promising future in optoelectronic applications.

REFERENCES

1. *Gold: progress in chemistry, biochemistry and technology*, H. Schmidbaur, Wiley, New York, 1999.
2. Ludwig Mond lecture. High-carat gold compounds, H. Schmidbaur, *Chem. Soc. Rev.* 1995, 391.
3. The fascinating implications of new results in gold chemistry, H. Schmidbaur, *Gold Bull.* 1990, **23**, 11.

4. Ab initio calculations on the $(ClAuPH_3)_2$ dimer with relativistic pseudopotential: Is the "aurophilic attraction" a correlation effect?, P. Pyykkö, Y. Zhao, *Angew. Chem., Int. Ed. Engl.* 1991, **30**, 604.

5. Theoretical chemistry of gold, P. Pyykkö, *Angew. Chem., Int. Ed.* 2004, **43**, 4412.

6. The van der Waals forces in the noble metals, J. J. Rehr, E. Zaremba, W. Kohn, *Phys. Rev. B* 1975, **12**, 2062.

7. Photoluminescence of phosphine complexes of d^{10} metals, R. F. Ziolo, S. Lipton, Z. Dori, *J. Chem. Soc., Chem. Commun.* 1970, 1124.

8. Spectroscopic properties and redox chemistry of the phosphorescent excited state of $[Au_2(dppm)_2]^{2+}$ [dppm = bis(diphenylphosphino)methane], C. M. Che, H. L. Kwong, V. W. W. Yam, K. C. Cho, *J. Chem. Soc., Chem. Commun.* 1989, 885.

9. Luminescence and metal-metal interactions in binuclear gold(I) compounds, C. King, J. C. Wang, M. N. I. Khan, J. P. Fackler Jr., *Inorg. Chem.* 1989, **28**, 2145.

10. Spectroscopy and redox properties of the luminescent excited state of $[Au_2(dppm)_2]^{2+}$ (dppm = $Ph_2PCH_2PPh_2$), C. M. Che, H. L. Kwong, C. K. Poon, V. W. W. Yam, *J. Chem. Soc., Dalton Trans.* 1990, 3215.

11. Crystal structures and solution electronic absorption and MCD spectra for perchlorate and halide salts of binuclear gold(I) complexes containing bridging $Me_2PCH_2PMe_2$ (dmpm) or $Me_2PCH_2CH_2PMe_2$ (dmpe) ligands, H. R. C. Jaw, M. M. Savas, R. D. Rogers, W. R. Mason, *Inorg. Chem.* 1989, **28**, 1028.

12. Novel luminescent polynuclear gold(I) phosphine complexes. Synthesis, spectroscopy, and X-ray crystal structure of $[Au_3(dmmp)_2]^{3+}$ [dmmp = bis(dimethylphosphinomethyl)methylphosphine], V. W. W. Yam, T. F. Lai, C. M. Che, *J. Chem. Soc., Dalton Trans.* 1990, 3747.

13. Syntheses, crystal structures and photophysics of organogold(III) diimine complexes, V. W. W. Yam, W. K. Lee, *J. Chem. Soc., Dalton Trans.* 1993, 2097.

14. The intrinsic $^3[d\sigma^*p\sigma]$ emission of binuclear gold(I) complexes with two bridging diphosphane ligands lies in the near UV; emissions in the visible region are due to exciplexes, W. F. Fu, K. C. Chan, V. M. Miskowski, C. M. Che, *Angew. Chem., Int. Ed.* 1999, **38**, 2783.

15. Aurophilic attraction and luminescence of binuclear gold(I) complexes with bridging phosphine ligands: ab initio study, H. X. Zhang, C. M. Che, *Chem. Eur. J.* 2001, **7**, 4887.

16. Resonance Raman investigation of the Au(I)-Au(I) interaction of the $^1[d\sigma^*p\sigma]$ excited state of $Au_2(dcpm)_2(ClO_4)_2$ (dcpm = Bis(dicyclohexylphosphine)methane), K. H. Leung, D. L. Phillips, M. C. Tse, C. M. Che, V. M. Miskowski, *J. Am. Chem. Soc.* 1999, **121**, 4799.

17. Use of luminescent gold compounds in the design of thin-film oxygen sensors, A. Mills, A. Lepre, B. R. C. Theobald, E. Slade, B. A. Murrer, *Anal. Chem.* 1997, **69**, 2842.

18. Oxygen-sensitive luminescent materials based on silicone-soluble ruthenium diimine complexes, I. Klimant, O. S. Wolfbeis, *Anal. Chem.* 1995, **67**, 3160.

19. A coordination polymer of gold(I) with heterotactic architecture and a comparison of the structures of isotactic, syndiotactic, and heterotactic isomers, C. A. Wheaton, R. J. Puddephatt, *Angew. Chem., Int. Ed.* 2007, **46**, 4461.

20. Synthesis and design of novel tetranuclear and dinuclear gold(I) phosphine acetylide complexes. First X-ray crystal structures of a tetranuclear $([Au_4(tppb)(C{\equiv}CPh)_4])$ and a related dinuclear $([Au_2(dppb) (C{\equiv}CPh)_2]$ complex, V. W. W. Yam, S. W. K. Choi, K. K. Cheung, *Organometallics* 1996, **15**, 1734.

21. Synthesis, photophysics and electrochemistry of $[Au_2(dppf)R_2]$ [dppf = $Fe(\eta-C_5H_4PPh_2)_2$; R = alkyl, aryl or alkynyl]. Crystal structure of $[Au_2(dppf)(C_{16}H_9)_2]$ ($C_{16}H_9$ = pyren-1-yl), V. W. W. Yam, S. W. K. Choi, K. K. Cheung, *J. Chem. Soc., Dalton Trans.* 1996, 3411.

22. A luminescent gold ring that flips like cyclohexane, J. H. K. Yip, J. Prabhavathy, *Angew. Chem., Int. Ed.* 2001, **40**, 2159.

23. Self-assembly and molecular recognition of a luminescent gold rectangle, R. Lin, J. H. K. Yip, K. Zhang, L. L. Koh, K. Y. Wong, K. P. Ho, *J. Am. Chem. Soc.* 2004, **126**, 15852.

24. Synthesis, photoluminescent and electroluminescent behaviour of four-coordinate tetrahedral gold(I) complexes. X-Ray crystal structure of $[Au(dppn)_2]Cl$, V. W. W. Yam, C. L. Chan, S. W. K. Choi, K. M. C. Wong, E. C. C. Cheng, S. C. Yu, P. K. Ng, W. K. Chan, K. K. Cheung, *Chem. Commun.* 2000, 53.

25. A novel polynuclear gold-sulfur cube with an unusually large stokes shift, V. W. W. Yam, E. C. C. Cheng, N. Zhu, *Angew. Chem. Int. Ed.* 2001, **40**, 1763.

26. A highly soluble luminescent decanuclear gold(I) complex with a propeller-shaped structure, V. W. W. Yam, E. C. C. Cheng, Z. Y. Zhou, *Angew. Chem. Int. Ed.* 2000, **39**, 1683.

27. A novel high-nuclearity luminescent gold(I)-sulfido complex, V. W. W. Yam, E. C. C. Cheng, K. K. Cheung, *Angew. Chem., Int. Ed.* 1999, **38**, 197.

28. Novel photophysical properties of gold selenide complexes: photogeneration of singlet oxygen by $[Au_{18}Se_8(dppe)_6]Br_2$ and near-infrared photoluminescence of $[Au_{10}Se_4(dpppe)_4]Br_2$, S. Lebedkin, T. Langetepe, P. Sevillano, D. Fenske, M. M. Kappes, *J. Phys. Chem. B* 2002, **106**, 9019.

29. Visible luminescence of water-soluble monolayer-protected gold clusters, T. Huang, R. W. Murray, *J. Phys. Chem. B* 2001, **105**, 12498.

30. Electrochemistry and optical absorbance and luminescence of molecule-like Au_{38} nanoparticles, D. Lee, R. L. Donkers, G. Wang, A. S. Harper, R. W. Murray, *J. Am. Chem. Soc.* 2004, **126**, 6193.

31. Near-IR luminescence of monolayer-protected metal clusters, G. Wang, T. Huang, R. W. Murray, L. Menard, R. G. Nuzzo, *J. Am. Chem. Soc.* 2005, **127**, 812.

32. Visible to infrared luminescence from a 28-atom gold cluster, S. Link, A. Beeby, S. FitzGerald, M. A. El-Sayed, T. G. Schaaff, R. L. Whetten, *J. Phys. Chem. B* 2002, **106**, 3410.

33. Luminescence studies of gold(I) thiolate complexes, J. M. Forward, D. Bohmann, J. P. Fackler Jr., R. J. Staples, *Inorg. Chem.* 1995, **34**, 6330.

34. Synthesis, structure, luminescence, and Raman-determined excited state distortions of a trinuclear gold(I) phosphine thiolate complex, S. D. Hanna, S. I. Khan, J. I. Zink, *Inorg. Chem.* 1996, **35**, 5813.

35. Molecular aggregation of annular dinuclear gold(I) compounds containing bridging diphosphine and dithiolate ligands, S. S. Tang, C. P. Chang, I. J. B. Lin, L. S. Liou, J. C. Wang, *Inorg. Chem.* 1997, **36**, 2294.

36. Synthesis and structural characterization of luminescent trinuclear gold(I) complexes with dithiocarbamates, M. Bardaji, A. Laguna, P. G. Jones, A. K. Fischer, *Inorg. Chem.* 2000, **39**, 3560.

37. Linear chain Au(I) dimer compounds as environmental sensors: a luminescent switch for the detection of volatile organic compounds, M. A. Mansour, W. B. Connick, R. J. Lachicotte, H. J. Gysling, R. Eisenberg, *J. Am. Chem. Soc.* 1998, **120**, 1329.

38. Dinuclear gold(I) dithiophosphonate complexes: synthesis, luminescent properties, and X-ray crystal structures of $[AuS_2PR(OR')]_2$ (R = Ph, R' = C_5H_9; R = 4-C_6H_4OMe, R' = (1S,5R,2S)-(−)-Menthyl; R = Fc, R' = $(CH_2)_2O(CH_2)_2$OMe), W. E. van Zyl, J. M. López-de-Luzuriaga, A. A. Mohamed, R. J. Staples, J. P. Fackler Jr., *Inorg. Chem.* 2002, **41**, 4579.

39. Multiple emissions and brilliant white luminescence from gold(I) O,O'-di(alkyl)dithiophosphate dimers, Y. A. Lee, J. E. McGarrah, R. J. Lachicotte, R. Eisenberg, *J. Am. Chem. Soc.* 2002, **124**, 10662.

40. Solid-state luminescence and crystal structures of novel gold(I) benzenethiolate complexes, S. Watase, M. Makamoto, T. Kitamura, N. Kanehisa, Y. Kai, S. Yanagida, *J. Chem. Soc., Dalton Trans.* 2000, 3585.

41. A reversible polymorphic phase change which affects the luminescence and aurophilic interactions in the gold(I) cluster complex, $[\mu_3$-$S(AuCNC_7H_{13})_3](SbF_6)$, E. M. Gussenhoven, J. C. Fettinger, D. M. Pham, M. M. Malwitz, A. L. Balch, *J. Am. Chem. Soc.* 2005, **127**, 10838.

42. Intensely luminescent gold(I)-silver(I) cluster complexes with tunable structural features, Q. M. Wang, Y. A. Lee, O. Crespo, J. Deaton, C. Tang, H. J. Gysling, M. C. Gimeno, et al., *J. Am. Chem. Soc.* 2004, **126**, 9488.

43. Absorption and emission spectra of tetrameric gold(I) complexes, A. Vogler, H. Kunkely, *Chem. Phys. Lett.* 1988, **150**, 135.

44. Gold(I)-gold(I) interactions. Tetrameric gold(I) dithioacetate, O. Piovesana, P. F. Zanazzi, *Angew. Chem., Int. Ed. Engl.* 1980, **19**, 561.

45. Gold dithiocarboxylates, B. Chiari, O. Piovesana, T. Tarantelli, P. F. Zanazzi, *Inorg. Chem.* 1985, **24**, 366.

46. Solid state EXAFS and luminescence studies of neutral, dinuclear gold(I) complexes. Gold(I)-gold(I) interactions in the solid state, W. B. Jones, J. Yuan, R. Narayanaswamy, M. A. Young, R. C. Elder, A. E. Bruce, M. R. M. Bruce, *Inorg. Chem.* 1995, **34**, 1996.

47. Synthesis and photophysics of dinuclear gold(I) thiolates of bis(diphenylphosphino)-alkyl- and aryl-amines. Crystal structure of $[Au_2\{Ph_2PN(C_6H_{11})PPh_2\}(SC_6H_4F\text{-}p)_2]$, V. W. W. Yam, C. L. Chan, K. K. Cheung, *J. Chem. Soc., Dalton Trans.* 1996, 4019.

48. Proof of potassium ions by luminescence signaling based on weak gold-gold interactions in dinuclear gold(I) complexes, V. W. W., Yam, C. K. Li, C. L. Chan, *Angew. Chem., Int. Ed. Engl.* 1998, **37**, 2857.

49. Molecular design of luminescent dinuclear gold(I) thiolate complexes: from fundamentals to chemosensing, V. W. W. Yam, C. L. Chan, C. K. Li, K. M. C. Wong, *Coord. Chem. Rev.* 2001, **216**, 171.

50. Molecular design of luminescence ion probes for various cations based on weak gold(I)···gold(I) interactions in dinuclear gold(I) complexes, C. K. Li, X. X. Lu, K. M. C. Wong, C. L. Chan, N. Zhu, V. W. W. Yam, *Inorg. Chem.* 2004, **43**, 7421.

51. Syntheses, structure, and photoluminescence properties of the 1-dimensional chain compounds [(TPA)$_2$Au][Au(CN)$_2$] and (TPA)AuCl (TPA = 1,3,5-triaza-7-phosphaadamantane), Z. Assefa, M. A. Omary, B. G. McBurnett, A. A. Mohamed, H. H. Patterson, R. J. Staples, J. P. Fackler Jr., *Inorg. Chem.* 2002, **41**, 6274.

52. Luminescence tribochromism and bright emission in gold(I) thiouracilate complexes, Y. A. Lee, R. Eisenberg, *J. Am. Chem. Soc.* 2003, **125**, 7778.

53. Strong intra- and intermolecular aurophilic interactions in a new series of brilliantly luminescent dinuclear cationic and neutral Au(I) benzimidazolethiolate complexes, J. Schneider, Y. A. Lee, J. Pérez, W. W. Brennessel, C. Flaschenriem, R. Eisenberg, *Inorg. Chem.* 2008, **47**, 957.

54. Chiral spaces: dissymmetric capsules through self-assembly, J. M. Rivera, T. Martin, J. Rebek Jr., *Science* 1998, **279**, 1021.

55. Self-assembly, structure, and spontaneous resolution of a trinuclear triple helix from an oligobipyridine ligand and Ni^{II} ions, R. Kramer, J.-M. Lehn, A. De Cian, J. Fischer, *Angew. Chem., Int. Ed. Engl.* 1993, **32**, 703.

56. Gold(I) macrocycles and topologically chiral [2]catenanes, C. P. McArdle, S. Van, M. C. Jennings, R. J. Puddephatt, *J. Am. Chem. Soc.* 2002, **124**, 3959.

57. Selectivity in the self-assembly of organometallic gold(I) rings and [2]catenanes, N. C. Habermehl, M. C. Jennings, C. P. McArdle, F. Mohr, R. J. Puddephatt, *Organometallics* 2005, **24**, 5004.

58. A chiral luminescent Au$_{16}$ ring self-assembled from achiral components, S. Y. Yu, Z. X. Zhang, E. C. C. Cheng, Y. Z. Li, V. W. W. Yam, H. P. Huang, R. Zhang, *J. Am. Chem. Soc.* 2005, **127**, 17994.

59. Au$_{36}$ crown: a macrocyclization directed by metal-metal bonding interactions, S. Y. Yu, Q. F. Sun, T. K. M. Lee, E. C. C. Cheng, Y. Z. Li, V. W. W. Yam, *Angew. Chem., Int. Ed.* 2008, **47**, 4551.

60. Metal alkynyl σ complexes: synthesis and materials, N. J. Long, C. K. Williams, *Angew. Chem., Int. Ed.* 2003, **42**, 2586.

61. Molecular design of transition metal alkynyl complexes as building blocks for luminescent metal-based materials: structural and photophysical aspects, V. W. W. Yam, *Acc. Chem. Res.* 2002, **35**, 555.

62. Luminescent carbon-rich rhenium(I) complexes, V. W. W. Yam, *Chem. Commun.* 2001, 789.

63. Luminescent gold(I) acetylide complexes. Photophysical and photoredox properties and crystal structure of [{Au(C≡CPh)}$_2$(μ-Ph$_2$PCH$_2$CH$_2$PPh$_2$)], D. Li, X. Hong, C. M. Che, W. C. Lo, S. M. Peng, *J. Chem. Soc., Dalton. Trans.* 1993, 2929.

64. Luminescent μ-ethynediyl and μ-butadiynediyl binuclear gold(I) complexes: observation of $^3(\pi\pi^*)$ emissions from bridging C$_n^{2-}$ units, C. M. Che, H. Y. Chao, V. M. Miskowski, Y. Li, K. K. Cheung, *J. Am. Chem. Soc.* 2001, **123**, 4985.

65. Organic triplet emissions of arylacetylide moieties harnessed through coordination to [Au(PCy$_3$)]$^+$. Effect of molecular structure upon photoluminescent properties, H. Y. Chao, W. Lu, Y. Li, M. C. W. Chan, C. M. Che, K. K. Cheung, N. Zhu, *J. Am. Chem. Soc.* 2002, **124**, 14696.

66. Polymorphic forms of a gold(I) arylacetylide complex with contrasting phosphorescent characteristics, W. Lu, N. Zhu, C. M. Che, *J. Am. Chem. Soc.* 2003, **125**, 16081.

67. Synthesis, structural characterization and photophysical properties of ethyne-gold(I) complexes, T. Müller, S. W. K. Choi, D. M. P. Mingos, D. Murphy, D. J. Williams, V. W. W. Yam, *J. Organomet. Chem.* 1994, **484**, 209.

68. Synthesis, structure, and ion-binding properties of luminescent gold(I) alkynylcalix[4]crown-5 complexes, V. W. W. Yam, S. K. Yip, L. H. Yuan, K. L. Cheung, N. Zhu, K. K. Cheung, *Organometallics* 2003, **22**, 2630.

69. Supramolecular assembly of luminescent gold(I) alkynylcalix[4]crown-6 complexes with planar η^2,η^2-coordinated gold(I) centers, S. K. Yip, E. C. C. Cheng, L. H. Yuan, N. Zhu, V. W. W. Yam, *Angew. Chem. Int. Ed.* 2004, **43**, 4954.

70. Recent studies on alkynyl complexes of the Group 11 and 12 metals, D. M. P. Mingos, R. Vilar, D. Rais, *J. Organomet. Chem.* 2002, **641**, 126.

71. Luminescent gold(I) acetylides: from model compounds to polymers, M. J. Irwin, J. J. Vittal, R. J. Puddephatt, *Organometallics* 1997, **16**, 3541.

72. Angular arenediethynyl complexes of gold(I), M.-A. MacDonald, R. J. Puddephatt, G. P. A. Yap, *Organometallics* 2000, **19**, 2194.

73. Interpenetrating digold(I) diacetylide macrocycles, W. J. Hunks, J. Lapierre, H. A. Jenkins, R. J. Puddephatt, *J. Chem. Soc., Dalton Trans.* 2002, 2885.

74. Some molecular rods: gold(I) complexes of 1,3-diynes. Crystal structures of Au(C≡CC≡CH)(PPh$_3$) and {Cu$_3$(μ-dppm)$_3$}(μ_3-I)(μ_3-C≡CC≡CAuC≡CC≡CH), M. I. Bruce, B. C. Hall, B. W. Skelton, M. E. Smith, A. H. White, *J. Chem. Soc., Dalton Trans.* 2002, 995.

75. Organometallic complexes for nonlinear optics. 14. Syntheses and second-order nonlinear optical properties of ruthenium, nickel and gold σ-acetylides of 1,3,5-triethynylbenzene: X-ray crystal structures of 1-(HC≡C)-3,5-C_6H_3(*trans*-C≡CRuCl(dppm)$_2$)$_2$ and 1,3,5-C_6H_3(C≡CAu(PPh$_3$))$_3$, I. R. Whittall, M. G. Humphrey, S. Houbrechts, J. Maes, A. Persoons, S. Schmid, D. C. R. Hockless, *J. Organomet. Chem.* 1997, **544**, 277.

76. Nonlinear optical properties of transition metal acetylides and their derivatives, C. E. Powell, M. G. Humphrey, *Coord. Chem. Rev.* 2004, **248**, 725.

77. Recent advances in utilization of transition metal complexes and lanthanides as diagnostic tools, V. W. W. Yam, K. K. W. Lo, *Coord. Chem. Rev.* 1998, **184**, 157.

78. Linear ditopic acetylide gold or mercury complexes: synthesis and photophysic studies X-ray crystal structure of PPh$_4$[Au(C≡CC$_5$H$_4$N)$_2$], M. Ferrer, L. Rodríguez, O. Rossell, F. Pina, C. Lima, M. F. Bardia, X. Solans, *J. Organomet. Chem.* 2003, **678**, 82.

79. The first luminescent anionic bis(ethynylphenanthroline)gold(I) complex, Y. Yamamota, M. Shiotsuka, S. Okuno, S. Onaka, *Chem. Lett.* 2004, **33**, 210.

80. A gold(I) [2]catenane, D. M. P. Mingos, J. Yau, S. Menzer, D. L. Williams, *Angew. Chem., Int. Ed. Engl.* 1995, **34**, 1894.

81. Excited-state interactions for [Au(CN)$^{2-}$]$_n$ and [Ag(CN)$^{2-}$]$_n$ oligomers in solution. Formation of luminescent gold-gold bonded excimers and exciplexes, M. A. Rawashdeh-Omary, M. A. Omary, H. H. Patterson, J. P. Fackler Jr., *J. Am. Chem. Soc.* 2001, **123**, 11237.

82. Single-crystal luminescence study of the layered compound potassium dicyanoaurate, N. Nagasundaram, G. Roper, J. Biscoe, J. W. Chai, H. H. Patterson, N. Blom, A. Ludi, *Inorg. Chem.* 1986, **25**, 2947.

83. Tunable energy transfer from dicyanoaurate(I) and dicyanoargentate(I) donor ions to terbium(III) acceptor ions in pure crystals, M. A. Rawashdeh-Omary, C. L. Larochelle, H. H. Patterson, *Inorg. Chem.* 2000, **39**, 4527.

84. Photoluminescence studies of lanthanide ion complexes of gold and silver dicyanides: a new low-dimensional solid state class for nonradiative excited-state energy transfer, Z. Assefa, G. Shankle, H. H. Patterson, R. Reynolds, *Inorg. Chem.* 1994, **33**, 2187.

85. Solvent dependent tunable energy transfer of d^{10} metal dicyanide nanoclusters with Eu^{3+} and Tb^{3+} rare earth ions, Z. Guo, R. L. Yson, H. H. Patterson, *Chem. Phys. Lett.* 2007, **445**, 340.

86. Cyclic trinuclear gold(I) compounds: synthesis, structures and supramolecular acid-base π-stacks, A. Burini, A. A. Mohamed, J. P. Fackler Jr., *Comments Inorg. Chem.* 2003, **24**, 253.

87. Silver-tellurolate polynuclear complexes: from isolated cluster units to extended polymer chains, J. C. Vickery, M. M. Olmstead, E. Y. Fung, A. L. Balch, *Angew. Chem., Int. Ed. Engl.* 1997, **36**, 1179.

88. Intermolecular interactions in polymorphs of trinuclear gold(I) complexes: insight into the solvoluminescence of Au$^I_3$(MeN=COMe)$_3$, R. L. White-Morris, M. M. Olmstead, S. Attar, A. L. Balch, *Inorg. Chem.* 2005, **44**, 5021.

89. Coordination metallacycles of an achiral dendron self-assemble via metal-metal interaction to form luminescent superhelical fibers, M. Enomoto, A. Kishimura, T. Aida, *J. Am. Chem. Soc.* 2001, **123**, 5608.

90. Phosphorescent organogels via "metallophilic" interactions for reversible RGB-color switching, A. Kishimura, T. Yamashita, T. Aida, *J. Am. Chem. Soc.* 2005, **127**, 179.

91. Syntheses, crystal structures and photophysics of organogold(III) diimine complexes, V. W. W. Yam, S. W. K. Choi, T. F. Lai, W. K. Lee, *J. Chem. Soc., Dalton Trans.* 1993, 1001.

92. Gold(III) photooxidants. Photophysical, photochemical properties, and crystal structure of a luminescent cyclometalated gold(III) complex of 2,9-diphenyl-1,10-phenanthroline, C. W. Chan, W. T. Wong, C. M. Che, *Inorg. Chem.* 1994, **33**, 1266.

93. Novel luminescent cyclometalated and terpyridine gold(III) complexes and DNA binding studies, H. Q. Liu, T. C. Cheung, S. M. Peng, C. M. Che, *J. Chem. Soc., Chem. Commun.* 1995, 1787.

94. Application of 2,6-diphenylpyridine as a tridentate [CNC] dianionic ligand in organogold(III) chemistry. Structural and spectroscopic properties of mono- and binuclear transmetalated gold(III) complexes, K. H. Wong, K. K. Cheung, M. C. W. Chan, C. M. Che, *Organometallics* 1998, **17**, 3505.

95. Luminescent gold(III) alkynyl complexes: synthesis, structural characterization, and luminescence properties, V. W. W. Yam, K. M. C. Wong, L. L. Hung, N. Zhu, *Angew. Chem. Int. Ed.* 2005, **44**, 3107.

96. A class of luminescent cyclometalated alkynylgold(III) complexes: synthesis, characterization, and electrochemical, photophysical, and computational studies of [Au(C^N^C)(C≡C–R)] (C^N^C = κ^{3}C,N,C bis-cyclometalated 2,6-diphenylpyridyl), K. M. C. Wong, L. L. Hung, W. H. Lam, N. Zhu, V. W. W. Yam, *J. Am. Chem. Soc.* 2007, **129**, 4350.

97. A novel class of phosphorescent gold(III) alkynyl-based organic light-emitting devices with tunable colour, K. M. C. Wong, X. Zhu, L. L. Hung, N. Zhu, V. W. W. Yam, H. S. Kwok, *Chem. Commun.* 2005, 2906.

6 Gold Catalysis

Sónia Alexandra Correia Carabineiro
and David Thomas Thompson[*]

CONTENTS

History of Development and Unique Aspects of Gold Catalysis..90
Catalyst Preparation Methods 91
   Unsupported Gold 91
   Supported Gold 91
      Deposition Precipitation (DP) 92
      Coprecipitation (CP) 92
      Incipient Wetness (IW)/Impregnation 92
      Use of Colloids...92
      Ion Exchange..92
      Chemical Vapor Deposition (CVD) 93
      Physical Vapor Deposition (PVD)...93
      Liquid-phase Reductive Deposition 93
      Supercritical CO_2 Antisolvent Technique...93
      Chloride-free Preparations 93
   Preparation of Au-PGM "Alloy" Catalysts 94
   Posttreatment...94
Reactions Promoted by Gold Catalysts..95
   Heterogeneous Catalysis 95
      Complete Oxidation 95
      Other Reactions of Environmental Importance..97
      Selective Oxidation 98
      Sugars 98
      Vinyl Chloride..99
      Water–Gas Shift 99
      Selective Hydrogenation 100
   Liquid Phase Homogeneous Catalysis 101
Emerging Commercial Applications..103
   Pollution Control...104
      Air Cleaning 104
      Autocatalysts 104
      Mercury Oxidation 106
      Ethene Oxidation..107
      Volatile Organic Compounds 107
      Water Pollution Control..107
   Chemical Processing 107
      Conversion of Biomass to Platform Chemicals..107
      Vinyl Acetate and Vinyl Chloride Synthesis 108

---

[*] Deceased.

Production of Nylon Precursors 110
Methyl Glycolate 110
Selective Oxidation of Sugars 111
Propene Oxidation 111
Hydrogen Peroxide 113
Petroleum Refining 113
Selective Hydrogenation 113
Fuel Cells and the Hydrogen Economy 113
Sensors 114
Future Prospects 115
Acknowledgments 117
References 117

HISTORY OF DEVELOPMENT AND UNIQUE ASPECTS OF GOLD CATALYSIS

Traditionally gold has been regarded as a poor catalyst compared with other precious metals, but in the 1980s it began to be realized that, when applied to supports as nanoparticles, gold becomes really active [1–3]! Gold particles are readily prepared on the nanoscale and are being used in this form in a wide range of practical applications, including the biomedical monitoring of constituents in body fluids, and in decorative inks, cosmetics, and lubricants (see Chapters 15, 16, 17), and there is a rapidly growing interest in their use as catalysts [4].

In this chapter we focus on an aspect of nanoparticulate gold that is particularly exciting: its use in catalysis. Recent developments in this new technology and its applications are considered. Nanoparticulate gold catalysts are active under mild conditions, even at ambient temperature or below, and this feature makes them quite unique. When nanoparticulate gold, with particles of less than ca. 5 nm, is supported on base metal oxides or activated carbon, very active catalysts are produced. During the 1980s, Haruta and colleagues [5–7] found that gold on oxide can be used to oxidize carbon monoxide at less than 0°C and Hutchings [8,9] showed that gold on carbon is the catalyst of choice for the hydrochlorination of ethyne to give vinyl chloride. These important advances followed the earlier work by Bond et al. [2,10,11], who had made the significant observation that gold on boehmite is selective for the hydrogenation of alkyne in the presence of alkene, and where the unexpected activity of gold on oxide catalysts for unsaturated hydrocarbon hydrogenation under mild conditions was highlighted (although it had been known from as early as 1906 that gold gauze catalyzes the reaction between hydrogen and oxygen between 523 and 673 K [12], and other reactions between hydrogen and unsaturated hydrocarbons had also been studied [13]).

It is particularly noteworthy that nanoparticulate gold or gold alloyed with palladium catalyzes selective oxidation reactions such as alkene epoxidation, selective alcohol oxidation, and direct synthesis of hydrogen peroxide by the hydrogenation of molecular oxygen [14].

These and other examples of the activity of nanoparticulate gold catalysts [1,2,15] imply that suitable investment in tailoring these to commercial requirements will lead to many useful applications, including some with significant commercial value [16,17]. The latter could include catalysts for pollution and emission control and protection from dangerous gases in safety masks, for chemical processing of bulk and speciality chemicals, for clean hydrogen production for the emerging "hydrogen economy" (including fuel cells), and as sensors to detect poisonous or flammable gases, or substances in solution [1,2]. People involved in investigating the development of various applications are making progress toward viable methods of manufacturing significant quantities of catalyst [3,16,17], and devising reliable preparative and storage methods for ensuring suitable durability under required operating conditions.

Until recently gold was also regarded as a poor homogeneous catalyst, with very low turnover numbers (TONs) for the few examples of reactions catalyzed by soluble complexes. Developments in this field have, however, been equally dramatic since the publication of a paper by Teles et al. in

1998 [18], which demonstrated that high turnover numbers and turnover frequencies (TOFs) could be obtained for gold complexes in solution. Since then, it has been found that soluble gold compounds can catalyze some organic reactions that cannot be performed in any other way [13,19]. Gold-catalyzed reactions yielding organic products via homogeneous or heterogeneous catalysis have been recently reviewed by Hashmi [20].

Recently, Angelici published a review describing catalysis on powdered unsupported gold [21]. The author refers, for example, to the reaction between CO, RNH_2, and O_2 to give isocyanates (RNCO) which react with further RNH_2 to give ureas $(RNH)_2CO$. Secondary amines $HN(CH_2R)_2$ can react with oxygen in the presence of gold to undergo dehydrogenation to the imine $RCH=N(CH_2R)$. Attempts to perform similar reactions with copper and silver metal powders were unsuccessful, pointing to the fact that unsupported gold is a catalyst for organic synthesis and this metal is again unique—other metals do not work!

Already, the first practical application for a gold–palladium catalyst within a major industrial process is well established for the manufacture of vinyl acetate monomer (VAM) [22,23], and a pilot plant has been built for the production of methyl glycolate [24].

Investigation of the use of gold catalysts in respirators for carbon monoxide removal and other pollution control applications is well underway [2]. The use of gold catalysts for preferential oxidation (PROX) systems to purify hydrogen by selective oxidation of carbon monoxide has received a lot of attention, and its practical application in supplies of hydrogen for use with PEM fuel cells and chemical processing can confidently be forecast [16,17]. The important role that gold catalysis is likely to play in the transformation of biomass into useful chemicals and fuels is highlighted in a recent article [25], and a very useful broad review of the chemical routes for the transformation of biomass into chemicals has been written by Corma et al. [26]

CATALYST PREPARATION METHODS

In the 20 years since Haruta's initial dramatic breakthrough in observing the low temperature oxidation of CO with oxide-supported nanoparticulate gold [5,6,27,28], there has been much effort in optimizing solution methods for preparing gold catalysts, and alternative CVD and PVD methods [29] have also been devised. The range of methods that can be applied to the preparation of heterogeneous nanogold catalysts is described in our previous review [1] and in Chapter 4 of Bond et al. [2]. The physical properties and characterization methods for small gold particles are discussed in Chapter 3 of that book, and more general aspects of catalyst characterization in review articles [30,31]. Soluble gold complexes for use in homogeneous catalysis, such as methyl(triphenylphosphane)gold are prepared using normal coordination chemistry procedures [18].

Unsupported Gold

Although bulk gold is a comparatively poor catalyst (and the principal reason why the significance of gold catalysis was not generally recognized until some 20 years ago), gold itself in powder form has surprisingly been found to have good catalytic properties for the oxidation of carbon monoxide and glycerol, especially under alkaline conditions [32], and naked gold sols are very active for glucose oxidation in aqueous solution [33]. Nanoporous gold, made by dealloying silver–gold alloys, is a mesoporous metal combining high surface area with high conductivity, which can also be used as an effective catalyst for the reduction of hydrogen peroxide to water. The reaction efficiency is sufficiently high to allow use of the material as a cathode for oxygen reduction in hydrogen PEM fuel cells [34].

Supported Gold

Most of the research carried out on gold catalysis has been done using nanoparticulate gold on oxide or carbon supports. The principal methods for preparing gold catalysts for use in liquid- and gas-phase heterogeneous catalysis are deposition precipitation (DP), coprecipitation (CP), incipient wetness (IW), deposition from colloids, and chemical and physical vapor deposition (CVD and PVD, respectively) [1,2,30]. The most recent detailed review of gold supported on oxide powders is written by Louis [35]. A short summary of all the information available is given in the sections that follow.

Deposition Precipitation (DP)

The gold precursor is precipitated onto a suspension of the preformed support by raising the pH either by the addition of alkali or urea. This method works well with supports having a point of zero charge (PZC) greater than five, e.g., MgO, TiO_2, Al_2O_3, ZrO_2, and CeO_2, but it is not suitable for SiO_2 (PZC ca. 2), SiO_2-Al_2O_3 (PZC ca. 1), tungsta (PZC ca. 1), or activated carbon.

Coprecipitation (CP)

The support and gold precursors are brought out of solution together, most likely as hydroxides, by adding a base such as sodium carbonate to a solution containing relevant gold and base metal precursors; 2–5 wt% Au on oxides such as MgO, TiO_2, Fe_2O_3, Al_2O_3, ZnO, SnO_2, and SiO_2 have been made using this method.

Incipient Wetness (IW)/Impregnation

This is a variation of the traditional impregnation method for preparing precious metals catalysts. In this case the pores of the support are filled with gold solution, usually $HAuCl_4$. Traditional impregnation methods usually lead to the formation of large gold particles (10–35 nm), caused by chloride-promoted agglomeration.

For IW, $HAuCl_4$ is dissolved in a volume of deionized water corresponding to the pore volume of the support, e.g., Al_2O_3. This solution is added to the support with intensive mixing during 15 min. The slightly wet solid is dried immediately for 16 h at 80°C, and then reduced with 5 vol% H_2 in N_2 at 250°C for 2 h. The resulting gold particles were <2 nm and the catalysts were very similar in activity, selectivity, and long-term stability for glucose oxidation as catalysts prepared by DP [36].

Bowker et al. have described an alternative, called the *double impregnation method* (DIM), in which a double impregnation of chloroauric acid and a base (Na_2CO_3) are used to precipitate out gold hydroxide within the pores of the catalyst, followed by washing [37]. The main reason for the enhanced activity of such incipient wetness catalysts is that the double impregnation results in the deposition of Au in the pores of the titania as $Au(OH)_3$, not as gold chloride, which is what usually forms in IW methods. As a result, the Cl is not associated with the Au and is removed from the catalyst by washing, leading to a more active catalyst, with the gold nanoparticles not poisoned or sintered by the presence of chloride (see Chloride-free Preparations later).

Use of Colloids

The fact that gold sols contain small metallic particles was first noted by Faraday in 1857 [38]. Colloids prepared by reduction of chloroauric acid by citric acid and other reducing agents can be used to prepare gold on carbon or oxide supports, by deposition from the colloid, to give good dispersions of gold; 1% Au/C prepared by this method by Michele Rossi's group [39] is now used as a reference catalyst for the scientific community and is distributed by the World Gold Council [40].

Ion Exchange

In this method ions on the surface of the support are replaced by gold ions. Cation exchange can be used for preparing gold on zeolites using $[Au(en)_2]$ (en = 1,2-diaminoethane), and direct anion exchange for preparing gold on alumina, by replacing the hydroxyl groups on the surface of the support by a gold species.

Chemical Vapor Deposition (CVD)

A stream of volatile compound of gold is transported onto a high area support by an inert gas and it reacts chemically with the surface of the support. Dimethylgold acetylacetonate has been used in this way to prepare gold on alumina, titania, silica, MCM-41, and carbon.

Physical Vapor Deposition (PVD)

In this method gold is vaporized from a target under vacuum and deposited on an oxide support or carbon under high vacuum conditions. The resulting catalyst does not need to be washed with water, as do those resulting from solution preparation methods.

The 3M company in Minnesota (USA) has found that very active gold nanocatalysts can be prepared via physical vapor deposition (using equipment originally designed for manufacturing electronic printed circuits), on a wide range of supports, including some that are water soluble, or not suitable for DP because of their unsuitable PZC (e.g., SiO_2). The method provides the advantages of low-cost, superb reproducibility, no need for washing and thermal treatment steps, and has no toxicity hazards. It takes only 2–3 h to make 400 mL of catalyst and there is a minimal loss of gold; 0.7% $Au/TiO_2/C$ catalysts, made this way, are stable for 2 years if kept in a canister for use in respirators. Such a storage life is important for such life-critical applications [29,41,42].

Liquid-phase Reductive Deposition

This new method has recently been reported by Sunagawa et al. [43]. The selective reductive deposition is characteristically performed by the adsorption of metal ion or complexes onto the surfaces where the reduction takes place. Thus, the initial adsorption of metal ions or complexes is the key feature of this technique. Hence, key points of this method are precise control of the metal complex by adjusting solute conditions, such as composition and structure of the metal complex; storing of the suspension until the equilibrium composition is attained, and aging the suspension at a controlled temperature. Using this new method, Au nanoparticles supported on various carriers were successfully obtained.

Supercritical CO$_2$ Antisolvent Technique

Supercritical antisolvent precipitation technique has been used to prepare a new titania catalyst support. The titania precursor was prepared by precipitating $TiO(acac)_2$ from a solution of methanol using supercritical carbon dioxide at 110 bar and 40°C. The titania support was used to prepare a gold catalyst with high activity [44].

Chloride-free Preparations

CVD and PVD can be used to avoid the presence of chloride, which promotes sintering of the gold particles. Preparative methods that avoid the presence of chloride include the use of gold acetate, $Au(OAc)_3$, dimethylgold acetylacetonate, $Me_2Au(acac)$, or gold phosphine complexes such as $[Au^I(PPh_3)]NO_3$ or $[Au^I_3Au^0_6(PPh_3)_8](NO_3)_3]$. These have been used successfully to obtain catalysts with the small gold nanoparticles [1,2]. Unusually, among metal–chloride interactions, the Au-Cl system appears to be covalent [45], and this may be a contributor to the effect of Cl on promotion of sintering gold. Gold's unusually high electronegativity enables it to form a covalent bond with the highly electronegative chlorine. Halide poisoning of catalytic sites is also a cause for deactivation [46].

Additional methods for preparing gold catalysts include deposition of dendrimer-stabilized gold particles, use of a single-step from a gold sol, photochemical deposition, and sonochemical and spray techniques [1,2].

Preparation of Au-PGM "Alloy" Catalysts

Now that gold has been proved to be a very active catalyst on its own, it is being demonstrated that bimetallic Au-PGM and Au-Ag catalysts can be even more active [47,48]. Core-shell Au-Pd catalysts prepared by Edwards et al. [49] for direct synthesis of hydrogen peroxide from hydrogen and oxygen, and the Au-Pd catalyst manufactured for vinyl acetate synthesis from ethene and acetic acid and oxygen [50] are good examples. Improved hydrodechlorination catalysts include metallic Au-Pd nanoparticles made from a mixed metal sol, while those for hydrogen peroxide synthesis were prepared by incipient wetness impregnation, and the vinyl acetate synthesis catalysts included some made by impregnation. Au-Pt/C and SiO_2 catalysts with controlled compositions in the range 10–90% Au with 2–4 nm core sizes and high monodispersity (< 0.5 nm) have been achieved by manipulating the precursor feed ratio [51].

Mixed metal catalysts can also be made using PVD [41,42] and by activation of rapidly quenched metal alloys such as Au_5FeZn_{14} and $Au_5Ag\,Zn_{14}$ under CO oxidation conditions at 553 K [52]. Other methods for preparing mixed metal catalysts include the use of dendrimer-stabilized bimetallic particles, and adsorption of bimetallic molecular clusters [2].

Posttreatment

Having prepared the gold or gold alloy catalyst, a variety of calcination conditions can be used. However, some catalysts are employed effectively without any calcination, as reductive treatments are not always helpful [31]. After drying, the gold is usually in the +3 oxidation state and heating in air or in an inert gas will largely reduce this to Au(0), because Au_2O_3 is unstable under these conditions. Nevertheless, some oxidized gold can remain and this may be an important mechanistic consideration in its use, as suggested in the Bond-Thompson mechanism [31] (Figure 6.1).

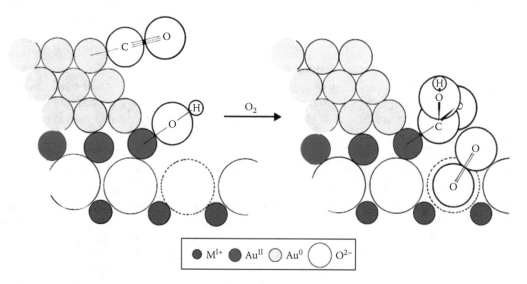

FIGURE 6.1 A representation of the early stages of the oxidation of carbon monoxide at the periphery of an active gold particle. At the left, a carbon monoxide molecule is chemisorbed on a low CN gold atom, and an hydroxyl ion has moved from the support to an Au^{III} ion, creating an anion vacancy. At the right they have reacted to form a carboxylate group, and an oxygen molecule occupies the anion vacancy as O_2^-. This then oxidizes the carboxylate group by extracting an hydrogen atom, forming carbon dioxide, and the resulting hydroperoxide ion HO_2^- then oxidizes a further carboxylate species forming another carbon dioxide and restoring two hydroxide ions to the support surface. This completes the catalytic cycle. No attempt is made to suggest charges carried by the reacting species. (Reprinted from World Gold Council. With permission.)

The extent of the reduction may depend on many parameters, including the nature of the support. The size of the gold particles is also influenced by the thermal treatment, the flow rate and composition of gas through the catalyst (sometimes hydrogen is a component of the gas), and the size of the catalyst sample [1,2]. It is recommended that "as prepared" samples are stored in a refrigerator at <273 K and that calcined catalysts should also be kept cold, and that after drying samples should be kept in a vacuum desiccator in the dark, reduction being performed immediately before use.

REACTIONS PROMOTED BY GOLD CATALYSTS

Bulk gold is chemically inert and is regarded as a poor catalyst, but when gold is in the form of nanoparticles and is deposited on metal oxides or activated carbon it becomes surprisingly active, especially at low temperatures, for such reactions as CO oxidation and propene epoxidation. Oxidation reactions catalyzed by gold have been recently reviewed by Rossi et al. [53]. The catalytic performance of Au is defined by contact structure, support selection, and particle size [54], as well as its oxidation state [31]. Some reactions, such as hydrosilylation, have been reported to be catalyzed by both homogeneous and heterogeneous gold catalysts [55]. Soluble gold species are now established as important catalysts, some of them being uniquely suitable for certain organic syntheses.

The principal reactions catalyzed by gold are indicated in Table 6.1.

HETEROGENEOUS CATALYSIS

Complete Oxidation

Gold catalysts are highly active for the oxidation of carbon monoxide at ambient temperatures and below, as first described by Haruta [5,6] (see Figure 6.2). When gold was coprecipitated with certain

TABLE 6.1
Examples of Reactions Catalyzed by Gold or Gold/PGMs

Reaction	References
Oxidation of CO	[5,6,28,29,31,37,46,52,56–62]
Selective oxidation of CO in hydrogen (PROX)	[63,64]
Complete oxidation of hydrocarbons, including ethene	[58,65]
Cyclohexane to cyclohexanol and cyclohexanone	[66–69]
Selective oxidation of alcohols	[70–72]
Ethene to vinyl acetate	[22,23,50,73]
Ethylene glycol to methyl glycolate	[24,74]
Propene to propene oxide	[75,76]
Sugars and glycerol to organic acids	[33,36,77–85]
Ethyne to vinyl chloride	[8]
Hydrosilylation	[55]
Direct hydrogen peroxide synthesis	[49,86,87]
Water–gas shift	[88–94]
Reaction of hydrogen with oxygen in a fuel cell	[95]
Selective hydrogenation	[10,11,96–101]
Homogeneous catalysis	[13,18–21,102–107]
Autocatalysis	[108–110]
Sensors	[111–115]

Source: Based on An overview of gold-catalysed oxidation processes, D. T. Thompson, *Top. Catal.*, 2006, **38**, 231. (With permission.)

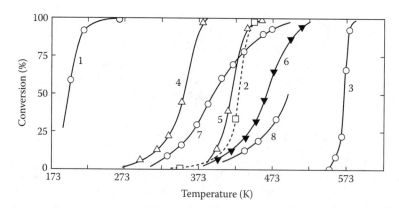

FIGURE 6.2 Oxidation efficiencies for carbon monoxide oxidation for Au/α-Fe$_2$O$_3$ in relation to catalyst temperature in comparison with other catalysts. (1) Au/α-Fe$_2$O$_3$ (Au/Fe =1:19), prepared by coprecipitation; (2) 0.5 wt% Pd/γ-Al$_2$O$_3$, prepared by impregnation; (3) gold fine powder; (4) Co$_3$O$_4$, ex-carbonate; (5) NiO; (6) α-Fe$_2$O$_3$; (7) 5wt% Au/α-Fe$_2$O$_3$, prepared by impregnation; (8) 5 wt% Au/γ-Al$_2$O$_3$, prepared by impregnation. (Reprinted from Catalysis by gold, G. C. Bond and D. T. Thompson, *Cat. Rev. Sci. Eng.*, 1999, **41**, 319. With permission.)

metal oxide supports, active catalysts were produced and the best results were obtained with a 5 wt% Au/α-Fe$_2$O$_3$ catalyst. Even better results in these pioneer experiments were obtained with gold nanoparticles smaller than 5 nm, so that there is sufficient interaction between them and the support. Since then, many other oxide and carbon supports have been shown to be suitable and Au/TiO$_2$ prepared by deposition precipitation is among the most active, particularly when a high pH is used in the catalyst preparation. Examples of their activity are given in Table 6.2.

Differences between the oxidation of carbon monoxide by Au/TiO$_2$ and Pt/SiO$_2$, as a function of the mean diameter of the gold particles, are well illustrated in Figure 6.3. For the gold catalyst the TOF increases sharply as the Au particle size decreases below 4 nm. In contrast, the platinum group metals show a decreasing or steady TOF below this particle size [54]. The much lower temperatures used in the gold case should also be noted.

The gold catalysts tend to lose activity too quickly for some applications but ways of stabilizing them are now being found. For example, addition of iron in the preparation lowers the rate of deactivation when TiO$_2$, SnO$_2$, and CeO$_2$ are used as supports [56]. Improved stability was due to Au particles being in contact with an iron phase such as FeO(OH), but calcination removed this stabilization.

TABLE 6.2
Specific Rates for CO Oxidation for Au/TiO$_2$ Catalysts

Au wt%	Temperature (K)	10^4 mol CO s^{-1}gAu^{-1}	Final pH in Preparation
0.5	300	7.3	8.6–9.0[a]
0.5	300	0.76	7–10
0.06–1.9	300	61	9[b]

[a] Under mass transport control.

[b] Under kinetic control.

Source: Based on Selective oxidation of CO in the presence of H$_2$, H$_2$O and CO$_2$ via gold for use in fuel cells, P. Landon, J. Ferguson, B. E. Solsona, T. Garcia, A. F. Carley, A. A. Herzing, C. J. Kiely, S. E. Golunski, and G. J. Hutchings, *Chem. Commun.*, 2005, 3385; and reprinted from An overview of gold-catalysed oxidation processes, D. T. Thompson, *Top. Catal.*, 2006, **38**, 231. (With kind permission of Springer Science and Business Media.)

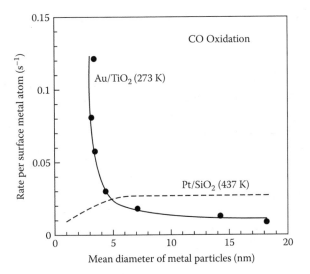

FIGURE 6.3 Relationships between turnover frequency (TOF) for CO oxidation for Au/TiO$_2$ and Pt/SiO$_2$ as a function of the mean diameter of the metal particles at 273 and 437 K, respectively. (Reprinted from When gold is not noble: Catalysis by nanoparticles, M. Haruta, *Chem. Rec.*, 2003, **3**, 75. With permission.)

It is particularly advantageous that gold's catalytic activity is often promoted by moisture [57]; in fact, water can increase the rate by two orders of magnitude. Prototype products that use gold catalysts for low-temperature air quality control are now appearing in the public domain. Among gold catalysts on oxide supports, prepared by CP, DP, and impregnation, Au/CoO$_x$ was the best, maintaining high activity for alkanes, including methane and ethane, at 200°C over 48 h. Interestingly, there was no correlation between CO oxidation and alkane oxidation [58].

Other Reactions of Environmental Importance

The need to preserve the quality of the earth's atmosphere and water is becoming increasingly important [1,2]. Legislation is already in place in many countries throughout the world to reduce pollution levels for gaseous emissions from road vehicles and industrial operations, and also to control impurities in aqueous effluents. This is stimulating efforts to design catalytic processes to meet these requirements, which often involve complete oxidation processes for volatile organic compounds (VOCs), including methane (which is the most difficult) and other hydrocarbons, reduction of nitrogen oxides, ozone decomposition, and removal of halocarbons, sulfur dioxide, dioxins, and VOCs, as well as catalytic wet air oxidation (CWAO) systems for oxidizing organic compounds dissolved in water, and hydrodechlorination of chlorinated organics. Supported gold has been shown to have some merit in most of the oxidation reactions while it also catalyzes reduction of nitrogen oxides with ammonia and ozone decomposition. Au/TiO$_2$ is 5–10 times more active than TiO$_2$ alone for the Claus reaction and for the reduction of SO$_2$ by CO:

$$2\ H_2S + SO_2 \rightarrow 2\ H_2O + 3\ S_{solid} \tag{6.1}$$

$$SO_2\ 2\ CO \rightarrow 2\ CO_2 + S_{solid} \tag{6.2}$$

Sulfur has also been found to be a promoter for the hydrogenation of crotonaldehyde to croton alcohol [96]. When small amounts of thiophene were added, it acted as a promoter, especially when the size of the gold nanoparticles was small. The preconceived notion that sulfur is a poison for gold has therefore been disproved.

Selective Oxidation

The fact that gold is active under very mild conditions means that its potential as a selective oxidation catalyst is high, and many selective oxidation processes are important in the chemical industry [1,2]. Papers have been published on the selective oxidation of propene to propene oxide in the presence of hydrogen, oxidation of sugars and aldehydes to acids, and the oxidation of alcohols and other hydroxyl-compounds. Au-Pd catalysts have been used to oxidize ethene to vinyl acetate in the presence of acetic acid and oxygen and this is a process used by a number of manufacturers worldwide. The selective oxidation of hydrogen to hydrogen peroxide, rather than water, is also catalyzed efficiently by supported Au-Pd catalysts [49].

Gold also has the potential to make commercially important chemical processes more environmentally friendly [25]. This could apply, for example to selective oxidation reactions. Abad et al. has shown that a Au/CeO$_2$ catalyst can be used to selectively oxidize alcohols to aldehydes and ketones under oxygen at atmospheric pressure [70] in the absence of solvent and base; and Enach et al. has shown that supported Au-Pd can increase the rate of alcohol oxidation by a factor of 25, over previously reported values, to give the desired aldehyde in 90% yield [71]. Hutchings and Scurrell have reviewed the design of oxidation catalysts including those containing gold in a series of case studies [116]. Hughes et al. has found that oxygen from the air can be used to make epoxides selectively when Au/C is used as catalyst, using a small quantity of peroxide as initiator, under mild conditions, with no solvent needed [117], whereas hydrogen or another co-reductant was previously found to be necessary to activate oxygen. It is noteworthy that gold particles activate molecular oxygen at such low temperatures and elucidating the mechanism of this reaction will help to progress green sustainable chemistry. In a useful review paper, Fristrup et al. discuss the mechanism of aerobic oxidation of substituted benzyl alcohols and conclude that the rate-determining step involves the generation of a partial positive charge in the benzylic position (i.e., hydride abstraction) [72].

Sugars

The high liquid phase activity for "naked" gold sols compared with similar preparations of other precious metals has been demonstrated by Rossi et al. [39,77] (Figure 6.4). The activity of gold

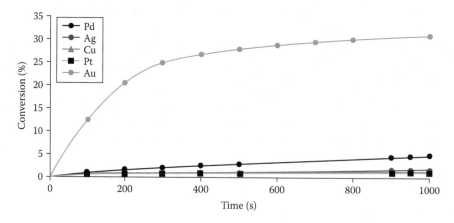

FIGURE 6.4 Comparative activities for "naked" colloidal gold and other metal sols in the oxidation of glucose to gluconic acid. (Adapted from The catalytic activity of "naked" gold particles, M. Comotti, C. Della Pina, R. Matarrese, and M. Rossi, *Angew. Chem. Int. Ed.*, 2004, **43**, 5812; and New perspectives in gold-catalysed oxidation, S. Biella and M. Rossi, Proc. GOLD 2003, Vancouver, Canada, September–October 2003. With permission.)

observed for the first 200 s corresponds to a mean TOF value of 11340 h^{-1} mol gluconic acid per mole Au.

In order to increase the life of the naked colloidal gold, however, it is necessary to support it on carbon. Gold-on-carbon catalysts can be used very effectively to oxidize glucose to gluconic acid [39,77]. The liquid phase oxidation of D-sorbitol has been carried out in water using oxygen as the oxidant and Au/C, Pt/C, and Pd/C as monometallic catalysts. By comparing these with the bimetallic catalysts Au-Pd/C and Au-Pt/C, a strong synergistic effect was observed, producing significant increases in reaction rates. The addition of Au to Pd or Pt also produced systems more resistant to oxygen poisoning [78].

The oxidation of glycerol to glyceric acid, with 100% selectivity, using either 1% Au/charcoal or 1% Au/graphite, has been reported to take place under mild conditions: ca. 55% conversion was obtained in 3 h, at 60°C [79,80]:

$$
\begin{array}{ccc}
\text{glycerol} & \xrightarrow[\substack{1\% \text{ Au/C, } 60°C \\ \text{NaOH}_{aq}, 3 \text{ atm}}]{3 \text{ atm O}_2} & \text{glyceric acid} + \text{etc} \qquad (6.3)
\end{array}
$$

Vinyl Chloride

The reaction between ethyne and hydrogen chloride is efficiently catalyzed by gold to give chloroethene (vinyl chloride). Gold is the most active catalyst for this reaction, as predicted by Hutchings [9]. In fact, $AuCl_3/C$, prepared by incipient wetness impregnation with $HAuCl_4$, is three times more active than the $HgCl_2/C$ used industrially in this reaction catalyzed by oxidized gold [9] (see Figure 6.5).

Water–Gas Shift

The pioneer work of Andreeva et al. with Au/Fe_2O_3 catalysts established that gold is an active catalyst for this reaction [88]. The nature of the support and the preparation technique are clearly of

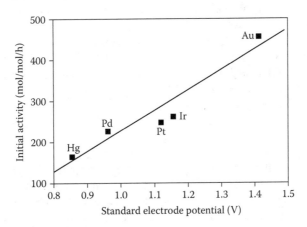

FIGURE 6.5 Correlation of initial hydrochlorination activity (mol HCl mol metal^{-1} h^{-1}) of metal chlorides supported on carbon (453 K, GHSV 1140 h^{-1}) with standard electrode potential (V): catalysts contained 5×10^{-4} mol metal/100 g catalyst. (Adapted from Springer and World Gold Council, Catalytic applications for gold nanotechnology, S. A. C. Carabineiro and D. T. Thompson, in *Nanocatalysis*, ed. U. Heiz and U. Landman, Springer-Verlag, Berlin, 2007.)

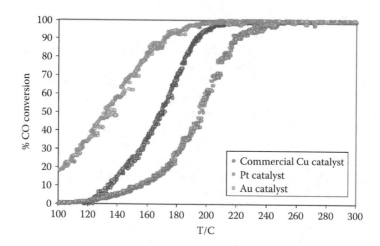

FIGURE 6.6 Comparison between activities of copper, platinum, and gold catalysts in the water–gas shift reaction (MHSV = 40,000 cm^3 g^{-1} h^{-1}). (Reprinted from Springer Science and Business Media, An overview of gold-catalysed oxidation processes, D. T. Thompson, *Top. Catal.,* 2006, **38**, 231. With permission.)

great importance [2] because the mixed oxides Fe$_2$O$_3$-ZnO and Fe$_2$O$_3$-ZrO$_2$ have been reported to be less effective than Fe$_2$O$_3$ itself.

During research aimed at reducing CO levels in hydrogen feeds for fuel cells, to avoid poisoning the platinum anode, a number of Au-based catalysts have been studied for the water–gas shift (WGS). It was recently shown that Au/CeO$_2$-ZrO$_2$ catalysts are highly active for the WGS and produce hydrogen from water at temperatures as low as 100°C [89]. The oxidation state and structure of active gold catalysts have been shown to comprise gold primarily in the zerovalent metallic state, but in intimate contact with the support. This close contact between small metallic gold particles and the support may result in the gold "atoms" having a net cationic charge, but the high activity is associated with metallic gold [90]. Kinetic studies show that the reaction is positive order with respect to CO and H$_2$O, but negative order with respect to CO$_2$ and H$_2$. The reaction mechanism seems significantly different from that for a Pt/CeO$_2$ catalyst [91,92]. In Figure 6.6 the activities for a commercial copper catalyst are compared with that for gold and platinum catalysts, and the advantages for gold at low temperatures can be clearly seen. The fact that the Au/CeO$_2$-ZrO$_2$ catalysts show 20% CO conversion at 100°C is particularly noteworthy.

In a recent Chevron patent [94] it is claimed that a gold catalyst may be used for both high- and low-temperature WGS reactions; 0.1–3.0 wt% Au/sulfated ZrO$_2$ (0.02–2.5 wt% sulfur) were used and 98.2% conversion was obtained at 20,000 GHSV, over 350 h, at 200°C and 30 psig. This patent provides more evidence of advantages for the presence of sulfur in catalysis by gold. This is interesting because another area of doubt in using gold catalysts was thought to be that of potential poisoning by sulfur.

Removal of the carbon monoxide remaining in the hydrogen feeds used for PEM fuel cells can be produced using PROX catalysts. For example, an Au/Fe$_2$O$_3$ coprecipitated catalyst, prepared using a two-stage calcination procedure, achieved target conversion and selectivity for the competitive oxidation of dilute CO, in the presence of moist, excess H$_2$ and CO$_2$, when used at temperatures up to 80°C, and there was no hydrogen conversion to water [63]. A different approach for operating a PROX system, at room temperature, is described in the Applications section.

Selective Hydrogenation

Early investigations in the 1960s and 1970s on hydrogenation using gold and gold alloy catalysts [12,118] led to studies of the potential of supported gold as a very selective hydrogenation catalyst

TABLE 6.3
**Synthesis of Hydrogen Peroxide: Comparison of
Palladium, Gold, and Bimetallic Catalysts in Methanol
under 3.7 MPa Pressure**

Catalyst	Rate of H_2O_2 Formation/mmol $g_{cat}^{-1} h^{-1}$
Au/Al_2O_3	1530
Pd/Al_2O_3	370
$Au-Pd(1:1)Al_2O_3$	4460

Note: At 275 K with mol ratio O_2:H_2 = 1:2, averaged over 30 min.

Source: Adapted from Direct formation of hydrogen peroxide from H_2O_2 using a gold catalysts, P. Landon, P. J. Collier, A. J. Papworth, C. J. Kiely, and G. J. Hutchings, *Chem. Commun.*, 2002, 2058; and An overview of gold-catalysed oxidation processes, D. T. Thompson, *Top. Catal.*, 2006, **38**, 231. (With kind permission of Springer Science and Business Media.)

[10]. In a very significant first paper, published in 1973, after the description of the hydrogenation of buta-1,3-diene and of but-2-yne over 1% Au/boehmite, between 400 and 490 K (H_2 26 kN m^{-2}; hydrocarbon, 13 kN m^{-2}; static system), the products were found to be solely n-butene isomers. The author then writes: "We are at a loss to understand why these catalytic properties of gold have not been reported before, especially since the preparative methods we have used are in no way remarkable. We are sure that the observed activity is due solely to gold, since spectrographic analysis of the $HAuCl_4$ by its supplier failed to detect impurities other than Si (3 ppm) and Fe, Cu and Ag (totalling 1 ppm). We believe that making the gold particles at 380–400 K either introduces a large number of defects which are active sites for hydrogen dissociation, or alternatively that activity resides in some very small particles not sensed by electron microscopy." This paper therefore merits special mention in the recent history of catalysis by gold, although there is literature dating from 1906 on the use of gold as a catalyst [2,30].

More recently, butadiene hydrogenation was found to be almost insensitive to the size of the Au particles and the selection of the support [97]. For crotonaldehyde hydrogenation there was only a slight sensitivity to the selection of the support [97]. The selective removal of alkynes and dienes from the alkene feedstocks used for polymerization has been investigated, so that these impurities do not poison the polymerization catalysts [98].

Theoretical calculations [86] and experimental results have both shown that formation of H_2O_2, from hydrogen and oxygen, is favored over gold surfaces. Au/Al_2O_3 catalysts are particularly effective for this reaction and $Au-Pd/Al_2O_3$ has now been shown to provide a significant improvement over the palladium catalysts used previously [49]. Results are compared in Table 6.3 [87]. These catalysts containing 5 wt% metal were prepared by incipient wetness impregnation of γ-alumina.

It was subsequently shown that calcined $Au-Pd/TiO_2$ catalysts with core-shell Au-Pd particles are very effective with high rates and selectivities of 93% achieved [49], and carbon-supported catalysts are even more active [14]. There is an optimum Au-Pd composition where the rate of hydrogen peroxide production is much higher than for the pure Pd catalyst, and this is gold rich.

LIQUID PHASE HOMOGENEOUS CATALYSIS

Until comparatively recently, i.e., 1998, significant applications-oriented homogeneous catalysis by gold complexes in solution was thought to be unlikely, as the few examples present in the literature had very small turnover numbers (TON = mole product per mole Au). This is in marked contrast to those for the platinum group metals, especially rhodium and palladium, which readily undergo catalytic oxidative-addition/reductive-elimination cycles and can have very high TONs. This has been rationalized by saying that this type of catalysis requires a very delicate balance between the stabilities of the two oxidation states involved, and this has not often been achieved for gold [102]. An additional factor was thought to be the reluctance of gold to form hydride complexes, so that the oxidation of Au^I by dihydrogen or the formation of alkene complexes by β-elimination from Au^{III}-alkyl complexes was virtually unknown. Gold hydrides have, however, recently been postulated as intermediates [103], and eventually a suitable choice of ligand could lead to the isolation of a stable gold hydride.

For members of the platinum group metals, the relative stability of the two critical oxidation states has been successfully adjusted by appropriate choice of ligands, e.g., the inclusion of a good π-bonding ligand, such as carbonyl or a phosphine, increases the stability of the lower oxidation state. With gold, π-bonding has seemed to be of relatively little importance. This kind of thinking was supported experimentally to the extent that the few examples of homogeneous catalysis by gold reported in the literature were associated with very small turnover frequencies (TOF = mole product per mole Au per unit time) and yields of product per mole of catalyst. New thinking is now required to rationalize the dramatically different results initially reported by Teles et al. [18] and now confirmed by a number of other groups of researchers, e.g., methyl(triphenylphosphane)gold(I) catalyzes the addition of methanol to hex-3-yne to give 3,3-dimethoxyhexane:

$$(6.4)$$

Up to 2×10^5 moles of product per mole of catalyst was obtained, with TOFs of up to 5400 h^{-1}. Some of the syntheses are uniquely promoted by gold catalysts, and the reaction mechanisms could also have unique features [13,19,20]. For example, 2,5-disubstituted oxazoles can be synthesized from N-propargylcarboxamides under mild conditions, using 5 mol% AuCl$_3$ in acetonitrile [20,119]. ^{1}H NMR studies were used to detect the intermediate methylene oxazoline, obtained in quantitative yield, and accessible for the first time under the mild conditions of gold catalysis:

quantitative

(TON 1000/TOF 100 h^{-1})

$$(6.5)$$

This is the first direct and catalytic synthesis of such alkylidene oxazolines, and this methodology has been used in synthesis of test substrates in pharmaceutical investigations by Merck [20].

Homogeneous catalysis by gold is not new [13,19], however, and even before 2004 there were about a hundred published papers on this topic, the first being the chlorination of naphthalene to octachloronaphthalene using AuCl or AuCl$_3$, in 1935 [2]. The gold compounds were among the most active catalysts (but it is now known that other Lewis acid catalysts, such as ferric chloride, are superior for this reaction).

Soluble gold compounds are now being used for a wide variety of organic syntheses including the formation of carbon–oxygen, carbon–nitrogen, and carbon–carbon bonds [2,19,20]. One of the milestones in catalytic asymmetric synthesis was the very first example of an asymmetric aldol addition reaction, reported by Ito et al. in 1986 [104]. They used cationic gold(I) complexes of chiral diphosphanes as catalysts and, by so doing, opened up a whole new aspect of catalytic asymmetric synthesis. The results were of fundamental importance and the principle could ultimately be extended to other metals. In these reactions, the chiral information from the ligand is multiplied and induces product chirality.

Toste and co-workers have recently reported that even a chiral counterion with cationic gold–phosphane complexes can be very efficient in inducing chirality [105]. This is a change of paradigm, since traditionally the influence of the ligand, which is close to the substrate, was the dominating one, and little success could be achieved with the more remote counterions. However, Toste's reactions depend on the use of a noncoordinating solvent of low polarity, then the formation of contact ion pairs brings the chiral counterion close to the substrate, because the linear coordination at gold(I) enables it to be even closer than the coordinated ligand. Two remarkable facts are that the influence of the chiral counterion can be more efficient than the influence of a directly coordinated chiral ligand, and that a synergistic influence of the chiral counterion and the chiral ligand is possible. Future investigation will show whether this effect is restricted to gold with its linear coordination, or can be extended to other metals as well (see comments in Hashmi [106]).

Catalytic hydrogenation can be achieved homogeneously with gold complexes. The enantioselective hydrogenation of alkenes and imines has been achieved under mild reaction conditions, using a neutral dimeric Au^I complex {$(AuCl)_2([R,R]$-Me-Duphos)} (($[R,R]$-Me-Duphos) = 1,2-bis[2R,5R]2,5-dimethylpholanebenzene). High TOFs (up to ca. 3900 h^{-1} and high enantiomeric excess (up to 95% with a bulky substrate) were achieved [103]. A gold hydride intermediate was proposed as a possible intermediate in the mechanism.

A significant advance was made in 2001 when the oxidation of sulfides to sulfoxides was achieved using $HAuCl_4/AgNO_3$ in acetonitrile, under 1 atm oxygen or air [107]:

$$R_2S + {}^1/_2 O_2 \rightarrow R_2SO \tag{6.6}$$

This catalyst had activity orders of magnitude higher than the previously used Ru^{II} or Ce^{IV} complexes, and this was a clear indication that sulfur does not poison the catalytic activity of gold. There is in fact clear evidence that sulfur can be a promoter for catalysis by gold [96]. The significant increase of interest in studying homogeneous catalysis by gold has occurred a decade later than the surge of interest in heterogeneous gold catalysis, but again it has its unique features and its versatility is considerable [2,19,20]. Detailed mechanistic understanding awaits further investigation in many cases but it could involve features not experienced when using soluble complexes of the platinum group metals. Increased understanding of the mechanistic aspects of both homogeneous and heterogeneous catalysis by gold could be of mutual benefit to both areas of research.

EMERGING COMMERCIAL APPLICATIONS

Now that nanoparticulate gold catalysts have been found to be active under mild conditions, even at ambient temperature or less, they are seen as having the potential to reduce the running costs of chemical plants and could increase the selectivity of the reactions involved where applicable. There are also indications of ways in which gold can be used to make reactions greener by allowing the absence of solvent. In some cases only gold catalysts have been shown to catalyze certain organic syntheses and they are therefore unique. The use of soluble gold complexes to synthesize pharmaceutical and specialty chemicals is, therefore, very likely. In pollution control applications, such as air cleaning, low light-off autocatalysts and diesel oxidation catalysts, and the purification of hydrogen streams used in fuel cells, heterogeneous gold catalysts have the characteristics to become the catalysts of choice. The reason is that now their durability and resistance to poisons is being shown to be better than anticipated; and as indicated previously, there is clear evidence that sulfur can be a promoter rather than a poison. The Union Chemical Laboratories and Novax Material and

Technology in Taiwan currently market CO-removing masks and escape hoods, designed for fire-fighters, and 3M plans to begin selling a similar product soon [120]. Use of mixed precious metals, especially gold–platinum group metal combinations, can produce even higher activities than the use of gold alone.

POLLUTION CONTROL

Air Cleaning

The use of highly active gold catalysts for protection from carbon monoxide is being developed and appropriate patents are in place [1,2,17]. The major remaining technical hurdle to overcome before widespread application of these technologies is the prevention of deactivation of gold catalysts, caused by the accumulation of contaminants such as carbonates on catalyst surfaces. Gold particle sintering is less important for catalysts operating near ambient temperature. Nonetheless, prototype products such as escape hoods that contain gold catalysts for low temperature air quality control are now being marketed. The demand for this type of product will grow rapidly.

Air-cleaning devices are needed for respiratory protection (gas masks) and for removing carbon monoxide, trace amounts of VOCs and ozone from ambient air indoor office space, and in submarines or space crafts on long missions [1,2]. Gold catalysts are promising materials and their use in respirators for removing carbon monoxide by its conversion to carbon dioxide is being developed. Union Chemical Laboratories (Taiwan) has developed masks for firefighters that last for up to 100 h and operate at room temperature: the gold particle size is 2 nm on oxide supports. Studies on carbon monoxide oxidation over gold catalysts (supported on Fe_2O_3 and TiO_2) in real air confirm that they are useful for the removal of carbon monoxide from both low (10–100 ppm) and high concentrations (10,000 ppm) [59]. Au/Fe_2O_3 is also a catalyst for ozone decomposition and simultaneous elimination of ozone and carbon monoxide at any ratio in the presence of oxygen at ambient temperature [121]. This catalyst is suitable for use in severe conditions such as relatively high ozone concentration and large space velocity. It also shows a high room temperature activity and good resistance to moisture.

Incinerator exhaust gases can contain a huge variety of pollutants, such as dioxins, VOCs, hydrocarbons, nitrogen oxides, carbon oxides, and amine derivatives. Due to the variety, a multicomponent catalyst is required to achieve high catalytic performance for all of the exhaust gas pollutants. However, it has been shown that if several supported single noble metal catalysts are encouraged to work in synergy, the overall catalytic activity can be greatly improved. Multicomponent noble metal catalysts prepared by sequential deposition precipitation for low temperature decomposition of dioxin have been evaluated [122]. Consequently, a ternary component noble metal catalyst (consisting of gold supported on Fe_2O_3-Pt/SnO_2-Ir/La_2O_3) was effective for purifying typical exhaust gases emanating from incinerators even at 423 K [123]. Odor-producing compounds, such as trimethylamine, can be oxidatively decomposed over $Au/NiFe_2O_4$, at temperatures below 373 K [54]. Au/Fe_2O_3, supported on a zeolite wash-coated honeycomb has been used commercially as a deodorizer in Japanese toilets.

Autocatalysts

In the last 60 years the world vehicle fleet has increased from about 40 million to over 700 million, and this figure is projected to increase to 920 million by the year 2010 [124]. Air pollution generated from mobile sources is, consequently, the greatest challenge in the pollution control field. Carbon monoxide, nonmethane hydrocarbons, and nitrogen oxides are the three major pollutants emitted by internal combustion engines [108], and environmental legislation governing the emission of these three types of gas is becoming increasingly stringent. The design of *three-way catalysts* (TWCs) capable of removing these three pollutants simultaneously is continually evolving to meet lower emission requirements. Typically, these catalysts must operate in the presence of 10% water and

10–60 ppm sulfur dioxide at temperatures ranging from 623 to 1273 K and hourly space velocities ranging from 10,000 to over 100,000 h^{-1} for the duration of 100,000 miles of operation [1,2]. Commercial TWCs in use at present are based on platinum, palladium, and rhodium on a support consisting of zirconia-stabilized ceria, zirconia, and α-alumina. Additives include barium oxide and zinc oxide. These PGM-based catalysts perform the task of emission control very well and many aspects of this technology are well established. However, there are areas, such as low light-off and catalysts for diesel exhaust, where gold could play a role in the future. Most of the pollution from automobiles is emitted in the first few minutes after startup, and PGM-based catalysts have significantly higher light-off temperatures than gold. The use of Au-Pt-Pd catalysts in controlling diesel exhaust emissions is currently being developed by Nanostellar Inc., Redwood City, California, USA. A new oxidation catalyst with a flexible Au:Pt:Pd ratio increases hydrocarbon oxidation activity by up to 40%, compared with conventional Pt converters, at equal precious metal cost and also catalyzes the oxidation of carbon monoxide [120,125].

Catalyst efficiency is usually evaluated under simulated driving conditions using the standardized federal test procedure (FTP). A key problem identified in the FTP is the liberation of unburned nonmethane hydrocarbons during the cold start mode of the test when the catalyst monolith is at ambient temperature. As a consequence, the catalyst does not reach the hydrocarbon light-off temperature of about 573 K until approximately 2 min after the start of the test. During this delay, up to 50% of the total unburned hydrocarbons are emitted. Additionally, when the engine operates under prolonged idling conditions, the temperature at the inlet to the converter is typically 553 K, which is an accepted monolith temperature for a catalytic converter mounted approximately 80 cm from the exhaust manifold of a spark ignition engine of 1.8 L displacement [2]. This temperature is again below the light-off for hydrocarbons.

A gold catalyst, with low temperature activity toward carbon monoxide and hydrocarbon oxidation, could be suitable to combat cold start-up emission problems and removal of nitrogen oxides from lean-burn gasoline and diesel engines [109,110]. The justification for developing gold-based technologies is both their promising technical performance and the lower price of gold compared with platinum. There are, however, some technical difficulties to overcome before gold catalysts can be successfully applied in the automotive sphere; these include attaining higher durability and poison resistance, qualities that are currently being evaluated with encouraging results.

A gold-based material has been formulated for use as a TWC in gasoline and diesel applications [108]. This catalyst, developed at Anglo American Research Laboratories, consisted of 1 wt% Au supported on zirconia-stabilized CeO_2, ZrO_2, and TiO_2, and contained 1 wt% CoO_x, 0.1 wt% Rh, 2 wt% ZnO, and 2 wt% BaO as promoters. The catalytically active gold–cobalt oxide clusters were 40–140 nm in size. This catalyst was tested under conditions that simulated the exhaust gases of gasoline and diesel automobiles and survived 773 K for 157 h, with some deactivation.

A significant hurdle for the gold-based TWC is the high operating temperature requirements imposed by gasoline engines. Typically, a catalyst must be able to withstand a temperature of 1373 K for at least 12 h. The gold-based TWC cannot survive under such conditions and it is accepted that gold will not be able to match the high temperature performance of the PGM-based TWCs. However, a relatively simple system in which PGM- and gold-based catalysts operate in parallel or sequentially can be envisaged, where the gold catalyst is in use at low temperatures but is bypassed in favor of the PGM catalyst at higher temperatures, or only sees the gold catalysts when at relatively low temperatures in a second exhaust box. In this way, maximum conversion activity could be maintained both at low temperatures using the gold catalyst and at high temperatures using the PGM catalyst [108].

The formation of ionic gold trapped in an oxide lattice is thought to be responsible for the stability of some Toyota catalysts: there was no reduction in T_{50}% conversion for propene after treatment at 1073 K for 5 h (Table 6.4). A standard Au/Al_2O_3 catalyst under the same conditions suffered significant degradation [126].

TABLE 6.4
Catalyst for Purifying an Exhaust Gas (Toyota)

Catalyst Composition		Temperature at 50% C_3H_6 Conversion (°C)[a]	
Chemical Formula	Au Content (wt%)	Initial	After Durability Test[b]
$Au_2Sr_5O_8$	0.4	345	346
$Au_2Sr_5O_8$	0.4	340	355
$La_2Au_{0.5}Li_{0.5}O_4$	0.2	341	348
$La_2Au_{0.5}Li_{0.5}O_4$	0.2	344	345
Au/Al_2O_3	2	378	433

[a] *Evaluation conditions:* CO 1000 ppm, C_3H_6 670 ppmC, NO 250 ppm, O_2 7.3%, H_2 5%, balance N_2 at 150,000 h^{-1}.
[b] *Durability test conditions:* CO 1000 ppm, C_3H_6 670 ppmC, NO 500 ppm, O_2 6.5%, CO_2 10%, H_2O 10%, balance N_2 at gas temperature of 800°C for 5 h.
Source: Based on Catalyst for purifying an exhaust gas, Y. Miyake and S. Tsuji, Toyota JJK Patent, E. P. 1043059 A1, 2000; and reprinted from An overview of gold-catalysed oxidation processes, D. T. Thompson, *Top. Catal.*, 2006, **38**, 231. (With kind permission of Springer Science and Business Media.)

Under diesel conditions, carbon monoxide and hydrocarbon oxidation is favored. Under the highly oxidizing conditions encountered in the diesel gas stream, reduction of nitric oxide is not expected. A nitric oxide conversion window is observed at temperatures between 493 and 623 K, with a T_{50} value of 523 K. However, large NO absorption bands are observed at temperatures above and below the conversion window [108].

Increased durability is also claimed for gold catalysts prepared by direct anionic exchange on an alumina support [60]. The durability of the catalyst was strongly improved by the complete removal of chloride using an ammonia washing procedure (but this brings with it risks of explosions if fulminating gold is formed from the reaction between soluble gold and ammonia; see Bond et al. [2] (p. 75) and Fisher [127]). This catalyst, tested in various reactions of saturated and unsaturated hydrocarbons from C_1 to C_3 and the oxidation of carbon monoxide, revealed a good activity, which is in an appropriate range of temperature for treatment of automotive exhaust, and longer durability tests may demonstrate further promise. Impregnation has been used for making Au/Al_2O_3, with washing with ammonia to remove chloride [36]; the reactivity for carbon monoxide oxidation at room temperature was comparable with catalysts prepared by deposition-precipitation. This 1 wt% Au/Al_2O_3 catalyst contained 2 nm particles and was stable to hydrothermal sintering, in 10 mol% steam at 873 K for 100 h. This could have important implications for their future use in autocatalyst and other pollution control applications.

Mercury Oxidation

Mercury inhalation has been linked to Alzheimer disease and autism, and limitation to mercury emissions is currently the subject of legislation by the U.S. Environmental Protection Agency (EPA) who will impose limits on mercury emissions from coal-fired boilers in the utilities industry. Mercury control techniques currently used in the industry include the use of flue-gas desulfurization (FGD) units and, as a result of mercury measurements around these units, it is known that oxidized and not elemental mercury is removed by the FGDs. Consequently, one method to increase mercury removal by this type of unit is to introduce a catalyst to promote the oxidation of mercury. Mercury measurement [128,129] led to the discovery that a gold-coated sand sample in a simulated flue-gas environment absorbed elemental mercury until an equilibrium was established and desorption of oxidized mercury began. Individual components of the simulated flue-gas have been evaluated for their effect on the oxidation of mercury, and it was found that nitrogen dioxide and hydrogen chloride were primarily responsible for the mercury oxidation over gold. It is not yet clear whether gold is acting through a truly catalytic mechanism in this instance, but it is the most active of the catalyst materials evaluated to date.

Ethene Oxidation

Gold nanoparticles on a number of supports (alumina, molybdena, and iron, cobalt, and manganese oxides) have been shown to be very active for the complete oxidation of ethene when used on monoliths at atmospheric pressure [65]. Catalysts prepared by deposition precipitation were the most active, followed by coprecipitation and impregnation. The best catalyst of all was Au/Co_3O_4 prepared by DP. The role of the gold nanoparticles was thought to be the promotion of dissociative adsorption of the oxygen which enhances reoxidation of the catalyst. The removal of the ethene produced by fruit on storage is important for controlling the fruit ripening process during both storage and transportation.

Volatile Organic Compounds

Many gaseous compounds are emitted into the atmosphere as a result of industrial and domestic activities. Catalytic combustion is likely to be the best means of destroying these volatile organic compounds (VOCs) [2]. Gold catalysts have the potential of lowering the temperature required for these processes. One of the most promising materials described to date has been Au/mesoporous γ-MnO_2 for removal of toluene, acetaldehyde, and hexane at 25–85°C [130]. A number of other supports have also been investigated, but their use requires higher temperatures [2]. For example, Au/CeO_2 catalyzes the complete conversion of propene at about 200°C [131].

Water Pollution Control

The efficient processing of wastewater is of both industrial and environmental concern. One technique of growing interest is the wet oxidation process, where the oxidation of organic compounds in an aqueous solution or in suspension, by means of oxygen or air, takes place at elevated temperatures (453–588 K) and pressures (2–15 MPa) [2]. The organic material present is first converted into simpler organic compounds, which are then further oxidized to carbon dioxide and water. Catalysts provide the means of using milder conditions, and CWAO processes studied to date using PGM catalysts could benefit from the use of gold or gold/PGM mixed-metal catalysts under milder conditions. Work at the Institut de Recherche sur la Catalyse, Lyon [132], where studies on the CWAO of succinic acid, as a representative organic compound, using Au/TiO_2 at 463 K and 50 bar air pressure, show encouraging performance for gold catalysts.

Trichloroethene (TCE) is one of the most common organic pollutants found in groundwater, deriving from its use as a solvent to degrease metals and electronic parts in the automotive, metals, and electronic industries, but it is a harmful environmental pollutant. Pd/Al_2O_3 catalysts have been used to hydrodechlorinate TCE, but more recent research has demonstrated that Au-Pd nanoparticles are about 70 times more active than Pd/Al_2O_3, on a per-Pd gram basis, for the aqueous phase hydrodechlorination of TCE [48,133]. The gold nanoparticles partially covered with Pd gave the highest activities.

CHEMICAL PROCESSING

Conversion of Biomass to Platform Chemicals

Petroleum-based feedstocks continue to become more expensive and, consequently, chemists are being challenged to devise processes that utilize biomass-derived feedstocks. In one of the latest developments, Claus Christensen and co-workers at the Center for Sustainable & Green Chemistry at the Technical University of Denmark, in Lyngby, have described a gold-catalyzed procedure for selective oxidation of the biomass-derived platform chemicals furfural and hydroxymethylfurfural to form their respective methyl esters [134,135]. Platform molecules are envisaged as key building blocks for the future chemical industry based on renewable chemicals, and methods are being sought to convert biomass into these key chemicals.

Hydroxymethylfurfural (HMF) derived from biomass can be selectively oxidized to furandimethylcarboxylate (FDMC), using Au/TiO_2 catalyst in methanol, in the presence of 8% sodium methoxide, under 4 bar O_2 at 130°C, to give FDMC in 98% yield, in 3 h:

$$(6.7)$$

This efficient conversion is a good example of the potential of gold catalysis to produce useful chemicals from biomass-derived starting materials. Methylfuroate formed from furfural is useful for flavor and fragrance applications and has potential as an industrial solvent. FDMC derived from HMF is a monomer that could replace the terephthalic acid used for making polyester plastics. The course of the reaction is quite different from when a platinum catalyst is used because furan dicarbaldehyde is the product from the Pt-catalyzed oxidation of HMF in water.

According to Christensen there is a great future for producing value-added chemicals from biomass, and this will most likely require intimate integration of biocatalytic and heterogeneous catalytic processes in order to achieve cost-competitive processes that are also environmentally friendly [134]. Just as heterogeneous catalysis can be used to convert petroleum-derived feedstocks into many useful fuels and chemicals, Christensen's group is helping to show that heterogeneous catalysis can be used to convert carbohydrate-based feedstocks into fuels and chemicals, via selective oxidation reactions [135].

The use of Au-Pd catalysts, as recently developed at Cardiff University by Hutchings' group, for hydrogen peroxide synthesis and selective oxidation reactions, is now being used by David Chadwick at Imperial College, London for selective oxidation of sugars and terpenes [136].

Vinyl Acetate and Vinyl Chloride Synthesis

Vinyl acetate monomer (VAM) is an important intermediate used in the production of polyvinyl acetate, polyvinyl butyral, and a variety of other polymers, and there are large-scale uses in emulsion-based paints, wallpaper paste, and wood glue. A gold-palladium catalyst, which includes potassium acetate, is very well established for the production of VAM from ethene, acetic acid, and oxygen in selectivities as high as 96% [1]:

$$(6.8)$$

The presence of Au leads to a significant increase in space–time yield, compared with use of Pd alone (see Table 6.1), and the presence of gold clearly has commercial importance. The bulk alloy is $Pd_{0.8}Au_{0.2}$ but on the surface the composition is richer in gold, $Pd_{0.45}Au_{0.55}$ [22]. Results demonstrating the advantages of using Au-Pd/SiO$_2$ catalysts rather than Pd/SiO$_2$ are given in Table 6.5 [23].

Until the recent emergence of nanoparticulate gold as an important catalyst this was generally regarded as a palladium catalyzed reaction but it is now clear that gold also has a significant role to play. There is likely to be a gas- and liquid-phase component to this process, because water is formed as a by-product.

TABLE 6.5
Yields of Vinyl Acetate Monomer as Reported by Dupont

	VAM Space–Time Yield (gl^{-1}h^{-})*	VAM Selectivity,%*
Pd	124	94.7
Au-Pd	594	91.6
Pd/KOAc	100	95.4
Au-Pd/KOAc	764	93.6

* Fixed bed performance after 40 h on stream: test conditions, 438 K, 115 psig with feed of ethene, acetic acid, oxygen and nitrogen.

Source: Adapted from Discovering the role of Au and KOAc in the catalysis of vinyl acetate synthesis, W. D. Provine, P. L. Mills, and J. J. Lerou, *Stud. Surf. Sci. Catal.*, 1996, **101**, 191; and An overview of gold-catalysed oxidation processes, D.T. Thompson, *Top. Catal.*, 2006, **38**, 231. (With kind permission of Springer Science and Business MediaCatalyst.)

The chemical technologies for these processes were developed in the 1960s and have been operated commercially since the 1970s; today VAM is produced with high selectivity and high space–time yield in many plants around the world. Worldwide VAM capacity is currently about 5 million metric tonnes per year and is expanding, 80% of which is produced via the ethene route. Catalyst consumption for VAM processes worldwide is several hundred tonnes per annum. Potassium acetate is widely used as a co-catalyst and BP has recently introduced a fluidized bed process (as the other processes use fixed beds). The catalyst is durable and typically lasts for 1 to 2 years.

Most of the commercial processes are fixed-bed, but at the end of 2001 BP commissioned a new Leap plant in Hull, UK. This is the world's first fluidized-bed process for VAM, while 80% of today's VAM plants worldwide are more than 20 years old and use a fixed-bed process [2]. BP Chemicals has developed this cost-saving route that allowed process simplification and intensification, requiring only a single reactor compared with the two reactors usually needed in the fixed-bed process. In a fixed-bed reactor, the catalyst which promotes the reaction was in the form of spheres that are packed into tubes. The reaction gases pass through the tubes and around the catalyst particles in the spaces between the spheres, without moving them. In a fluidized-bed reactor, however, the catalyst is in the form of a fine powder, and the gases flow upward through the reactor inducing much better mixing and contact between them and the catalyst, improving heat transfer and allowing the catalyst to be removed and replenished without having to shut down the reactor. In addition, fluidized-bed reactors are cheaper and easier to build (this decision to go to a fluidized-bed process saved 30% in capital costs).

Moving from a fixed- to a fluidized-bed operation also required a new catalyst, and the one selected was a supported gold-palladium system in the form of very fine spheres, prepared in collaboration with Johnson Matthey. Hence, gold-based catalysts are being used for this new fluidized-bed process and are well established in fixed-bed processes for the large-scale manufacture of VAM.

The manufacture of polyvinyl chloride is very important commercially [2,8,9] and the production of the monomer is therefore an important step in this synthesis:

$$HC \equiv CH + HCl \rightarrow 2HC = CHCl \rightarrow PVC \qquad (6.9)$$

The first significant practical demonstration of the commercial relevance of catalysis by gold was carried out by Hutchings, then working in South Africa: gold catalysts supported on activated carbon were found to be about three times more active than commercial mercuric chloride catalysts for vinyl chloride production (see Figure 6.5) and to deactivate much less rapidly than other supported metal catalysts. Deactivation can be minimized if high loadings of gold are used. Also, gold catalysts could be reactivated by treatment off-line with hydrogen chloride or chlorine, and by co-feeding nitric oxide with the reactants from the start of the reaction, deactivation could be virtually eliminated [9]. Gold is thus the catalyst of choice for this reaction, and the economics of its use would be assisted by cost-effectively recycling the gold. This process has potential to be applied in developing countries if market demand for PVC is sustainable.

Production of Nylon Precursors

Gold catalysts can be used in a solventless liquid-phase system to oxidize cyclohexane to cyclohexanol and cyclohexanone using oxygen. Almost all the cyclohexane produced (4.4 million tonnes per annum, and expected to grow at ca. 3%) is converted to cyclohexanol and cyclohexanone, the intermediates in the production of caprolactam and adipic acid, used in the manufacture of nylon-6 and nylon-66 polymers, respectively. The present commercial process for cyclohexane oxidation is carried out at around 423 K and 1–2 MPa over a catalyst such as cobalt naphthenate with approximately 4% conversion and 70–85% selectivity to cyclohexanol and cyclohexanone [66,67,93].

The large demand for these products and the high energy demands for the present process could provide an opening for a more effective catalyst. These results are interesting both for the comparatively high conversion rates of approximately 15%, achieved with high selectivities to cyclohexanol and cyclohexanone, with TONs of up to 3000 h^{-1}, when a zeolite-supported catalyst is used. In addition, the reaction occurs under environmentally benign conditions involving oxygen as the oxidant in a solvent-free system. The catalyst is durable, at least within the limits tried so far. The catalysts used were about 1% Au on ZSM-5. A Solutia Inc. patent describing similar technology has been published [68]. Further investigations of this reaction using organically modified mesoporous silicas as supports have given higher conversions and modified selectivities [69]. Corma Canos et al. have patented a method for preparing cyclohexanone oxime by catalytic hydrogenation of α,β-unsaturated nitro compounds using gold catalysts. [137]. Also Klitgaard et al. showed that bifunctional gold–titania catalysts can be employed to facilitate the oxidation of amines into amides with high selectivity [138]. These results demonstrate that these new methodologies open up new and environmentally benign routes to caprolactam and cyclohexanone oxime, both of which are precursors for nylon-6.

Methyl Glycolate

A liquid-phase air-oxidation process is being developed by Nippon Shokubai in Japan for the one-step production of methyl glycolate from ethylene glycol and methanol using a gold catalyst [2,4,24]:

The catalyst used for this reaction, as described in this paper, was Au-Pb/Al$_2$O$_3$ and the conditions used were 90°C and 0.1–5 MPa [74]. A pilot plant process demonstration, with a capacity of tons per month, has successfully shown that this can be run as a clean and simple continuous process with the product obtained in high purity. Methyl glycolate is used as a solvent for semiconductor manufacturing processes, as a building block for cosmetics, and as a cleaner for boilers and metals. Nippon Shokubai has indicated that the catalyst technology will be used for other syntheses involving one-step esterification of carboxylic acids and lactones. One of the patents claims its use for the synthesis of methyl methacrylate.

Selective Oxidation of Sugars

Colloidal Au and Au/C catalysts can be used to oxidize d-glucose to d-gluconic acid [1,77]. In fact, Au/C catalyst is a valid alternative to most of the investigated multimetallic catalysts based on palladium or platinum. Moreover, gold has the unique property of operating without the external control of pH, thus ensuring total conversion at all pH values and total selectivity to gluconic acid:

$$\text{(6.12)}$$

glucose   gluconic acid

It has been demonstrated that Au/Al_2O_3 catalyst prepared by DP or IW can be an efficient (>99% selectivity) and durable (no loss of activity or selectivity over 110 days) catalyst for this conversion [81] as 3.8 tonnes of gluconic acid can be obtained per gram of gold in 70 days. The starting parameters of 500 mmol L^{-1} glucose feed concentration and 8 h residence time resulted in an average activity of 140 mmol min^{-1} g_{Au}^{-1} and an average conversion of 90%. Gluconic acid is an important food and beverage additive, and is also used as a cleansing agent. It is made on a 60,000 tonnes per annum scale, so there may be further opportunities for gold in the food industry. The catalyst was thus shown to have excellent long-term stability. Kowalczyk et al. [82] have filed a patent covering use of gold catalysts for this conversion.

Oxidation of lactose and maltose with Au/TiO_2 catalysts has been reported to give close to 100% selectivity to lactobionic acid and maltobionic acid, respectively [83], which also have potential uses in the pharmaceutical and detergent industries, as well as in food. The initial activity for maltose was more than twice that for lactose but both reactions gave 100% selectivity to maltobionic acid and lactobionic acid, respectively. Studies of the catalytic conversion of glucose by hydrogenation (with a ruthenium catalyst) and oxidation with a gold-sol/C catalyst to produce sorbitol and gluconic acid, respectively, have been reported by Schimpf et al. [84]. Sorbitol is also manufactured on a 60,000 tonnes per annum scale.

It is pertinent to evaluate whether there could be further opportunities for gold in the food industry: if so, this may have the appeal that any gold residues in the products could be completely harmless since gold is thought to be environmentally benign as it has no known toxic effects.

Propene Oxidation

Propene oxide (methyl oxirane) is used on a large scale for the production of polyurethanes and is usually made using a chlorohydrin process. A greener alternative would be to use the direct gas-phase synthesis of methyloxirane from propene, using molecular oxygen in the presence of hydrogen, thus eliminating chlorine from the production process, as well as reducing water consumption and salt by-products [2,7]:

$$propene + O_2 + H_2 \longrightarrow methyloxirane + H_2O \tag{6.13}$$

Early results gave selectivities above 99%, at low conversions, using a 1 wt% Au/TiO_2 catalyst system at 323 K, when both oxygen and hydrogen are present in the feed gas ($H_2:O_2$: propylene : Ar = 10: 10: 10: 70 vol%) [15]; more recently improved conversions have been obtained while maintaining acceptably high selectivity [139] but improved durability is now required. Patents for direct production of methyloxirane using gold catalysts have been filed by a number of major companies including Bayer, Dow, and Nippon Shokubai [1,2], confirming significant industrial interest in this application, and pilot plants are understood to be operating within the industry. Bayer researchers have claimed an 8% yield of methyl oxirane, with 95% selectivity. Dow reports 92 mol% selectivity at 0.36 mol% propene conversion and a production rate of 8.3 g methyloxirane per kg_{cat} h^{-1}, at 433 K, using a feed stream of 20% propene, 10% hydrogen, 10% oxygen, and the remainder helium over 2 g 0.5 wt% Au on a titania/silica support, which had been calcined at 823 K. The flow rate was 160 cm^3 min^{-1} and olefin GHSV 480 h^{-1} at atmospheric pressure. Methyl oxirane yields of 9% have been obtained when using a silylation treatment and alkaline earth metal salts as promoters. Table 6.6 summarizes some of these results [75,76].

It was concluded that gold particle–support interaction is required together with careful selection of the titania-silica support and control of the gold-particle size. The use of pH 7.0 in the deposition-precipitation method was also recommended, together with calcination at 573 K. Propene oxide has recently been obtained via a two-stage process involving dehydrogenation of propane to propene (selectivity of 57%), followed by a propene to propene oxide oxidation with a selectivity of 8%, and

---

TABLE 6.6
Selectivities and Yields for Propene Epoxidation Using Titanosilicate Supported Catalysts Prepared by a Modified Sol Gel Method

Au %	PP Conv	H_2 Conv	PO Sel	PO Yield
0.42	3.4%	35%	85.4%	4.0%
0.6	4.5%	32.1%	79.6%	5.6%
0.007	1.0%	3.5%	100%	1.0%

TiO-SiO support, space velocity: 4000 h^{-1} cm^3/g_{cat}, feed $Ar/C_3H_6/H_2/O_2$=70/10/10/10, temperature 150°C, Ti/Si = 2/100: prepared by sol-gel method. Catalysts calcined at 300°C.

Source: Based on Propylene epoxidation with O_2 and H_2 over gold nanoparticles supported on mesostructural titanium silicates, M. Haruta, A. K. Sinha, S. Seelan, B. Chowdhury, M. Daté, and S. Tsubota, Proc. 13 International Congress on Catalysis, Paris, July 2004, O1-031; 33; and Catalysis by gold nanoparticles: epoxidation of propene, A. K. Sinha, S. Seelan, S. Tsubota, and M. Haruta, *Top. Catal.*, 2004, **29**, 95; and reprinted from An overview of gold-catalysed oxidation processes, D. T. Thompson, *Top. Catal.*, 2006, **38**, 231. (With kind permission of Springer Science and Business Media.)

a propene oxide space–time yield of 4 g $kg_{cat}^{-1} h^{-1}$, with dual catalyst bed of 1.9% Au/TiO_2 prepared by DP, followed by a Au/TS-1 catalyst. The catalysts showed little deactivation over 12 h at 443 K and 0.1 MPa, in presence of hydrogen and oxygen. Further investigation could lead to an interesting process potential [140].

Hydrogen Peroxide

The market for hydrogen peroxide is very large (ca. 1.9×10^6 tonnes per annum) and is rising by about 10% per annum. This is partly because peroxide is viewed as an environmentally friendly alternative to chlorine as a bacteriocide. Accordingly, there is great incentive to enable hydrogen peroxide to be synthesized where it is to be used, and thus avoid the heavy transport costs for this hazardous material. Also it is only economical to produce it on a large scale, using the sequential hydrogenation and oxidation of alkyl anthraquinone. Often it is required on a much smaller scale which could be manufactured locally with greater efficiency and safety [49,87].

Petroleum Refining

Whereas some of the processes used in the petroleum refining industry may not lend themselves to the use of gold catalysts, there will probably be selected opportunities. There are, for example, continual demands to progressively decrease the levels of sulfur and aromatics in gasoline and diesel distillate fuels. The dual-stage system, which uses a nickel-molybdenum or cobalt-molybdenum catalyst followed by a platinum catalyst, may not exhibit sufficient activity to achieve the final levels of sulfur and aromatics saturation required. The results of experiments using thiophene and dibenzothiophene indicate the possibility of using gold as a component of a hydrodesulfurization and aromatic dehydrogenation catalyst. $Au-Pd/SiO_2$ catalysts are surprisingly more active (by a factor >6) in the hydrodesulfurization of dibenzothiophene than pure palladium catalysts [141]. Such enhanced activity may be explained in terms of the well-known affinity of gold for sulfur, which activates the breakage of the C-S bond without forming stable inactive sulfides.

Selective Hydrogenation

The removal of diene and alkyne impurities from alkene streams via selective hydrogenation is needed to prevent poisoning of the catalysts used for the polymerization of alkenes. Supported gold catalysts offer interesting potential because they can selectively hydrogenate dienes in the presence of monoenes [97,99] and catalyze the conversion of alkynes to alkenes [100,101]. Recent research has shown that hydrogen is dissociatively adsorbed on the gold particles in Au/Al_2O_3 [142]. Gold could also be associated with other metals known to catalyze hydrogenation such as platinum or palladium to modify the chemisorption, activity, or selectivity.

Some time ago the use of Au/SiO_2 or Au/Al_2O_3 for the hydrogenation of canola oil was reported [2 (p. 350),54,143]. The complete reduction of linolenic acid could be achieved at a lower *trans-* isomer content in the products than that obtained using the American Oil Chemists standard nickel catalysts. Nickel catalysts have, of course, been used for over a century for the hardening of natural oils. It may be possible that gold catalysts, using the much more advanced methods of preparation available today, will have a future role in this application. If that is the case, the advantage would be that any gold residues in the products would be completely harmless, as gold is environmentally benign.

FUEL CELLS AND THE HYDROGEN ECONOMY

Nanoparticulate gold on oxide can be used with advantage to catalyze the WGS to produce hydrogen from carbon monoxide and steam [16,90,95]:

$$CO + H_2O \rightarrow H_2 + CO_2 \qquad (6.14)$$

Gold catalyzes this reversible WGS reaction under milder conditions than the commercially used copper–zinc catalysts (see Figure 6.6) [1,2]. For use in a fuel cell, the carbon monoxide still present in the hydrogen, obtained from this or other sources, must be removed to prevent from poisoning the platinum catalyst inside the fuel cell. Gold catalysts have been found to be effective for this PROX reaction at room temperature. Whereas platinum group metals oxidize both the CO and the hydrogen, gold is selective for CO at close to ambient temperatures. For example, this has enabled Project AuTEK in South Africa to develop a new system for hydrogen purification for PEM fuel cells, trade named *AuroPureH$_2$* [4]. This concept is designed to purify on-board cheap hydrogen on vehicles, with the hydrogen feed for the fuel cell being drawn directly from a cylinder. The low (ambient) operating temperature gives high selectivity, and no additional energy is required to heat the reactor. Fuel efficiency is essentially maintained and this simple, low-cost and practical system allows lower Pt loadings for the fuel cell anode, reducing expense while not adding significantly to the weight and volume of the fuel cell system. A 3 wt% Au/TiO$_2$ catalyst removes carbon monoxide to below 1 ppm (from 10 to 2000 ppm CO in H$_2$ with a 1–2 % air bleed) at a space velocity of 850,000 ml.g$_{cat}^{-1}$ h^{-1}, and at a gold cost of less than 1% of the U.S. Department of Energy stipulated target of $45/kW for vehicular fuel cells [64]. Consequently, it outperforms the PtRu and PtMo CO tolerant technologies. An example of the use of this new technology on the performance of a PEM fuel cell is presented in Figure 6.7, for hydrogen feeds containing 1000 ppm carbon monoxide [144]. Table 6.7 and Figure 6.7 show the performance loss for the four technologies investigated.

There is also scope for use of nanoparticulate Au-Pt catalysts in the fuel cells themselves [47]. Work at Brookhaven National Laboratory in New York has shown that platinum oxygen-reduction fuel cell electrocatalysts can be stabilized against dissolution under potential cycling regimes (a continuing problem in vehicle applications) by modifying Pt nanoparticles with gold clusters.

SENSORS

The need for air-quality monitoring demands development of sensors that are selective for detection of individual pollutant gases [1]. Sensors based on gold have therefore been developed for detecting a number of gases, including CO and NOx.

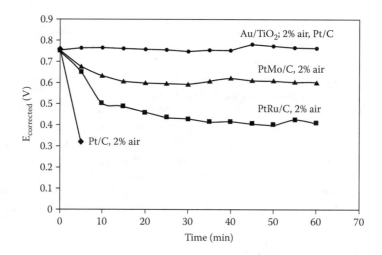

FIGURE 6.7 Plots showing comparison of CO tolerance results (1000 ppm CO, 0.5 A.cm^{-2}, 1.5 times stoichiometric hydrogen flow rate, SV over 3 wt% Au/TiO$_2$ catalyst = 250,000 mol g$_{cat}^{-1}$ h^{-1}, Au-based catalyst chamber at 25°C, fuel cell at 80°C, 30 psi [21]. (Modified from Mintek, Recent developments in the industrial application of gold catalysts, C. W. Corti, R. J. Holliday, D. T. Thompson, and E. van der Lingen, *Science and Technology in Catalysis*, ed. K. Eguchi, M. Machida, and I. Yamanaka, Kodansha, Elsevier, Tokyo, 2007, p. 173. With permission.)

TABLE 6.7
Relative Performance Durability of The Technologies Represented in Figure 6.7

System	Relative Performance Durability
Pt/C	0
PtRu/C	0.55
PtMo/C	0.80
Pt/C + Au/TiO$_2$	1

Source: From Recent developments in the industrial application of gold catalysts, C. W. Corti, R. J. Holliday, D. T. Thompson, and E. van der Lingen, *Science and Technology in Catalysis*, ed. K. Eguchi, M. Machida and I. Yamanaka, Kodansha, Elsevier, Tokyo, 2007, p. 173. (Reproduced from Mintek. With permission.)

For example, *sensors for CO detection* are well established, using Au nanoparticles doped on α-Fe$_2$O$_3$ with particle sizes between 3.2 and 8.8 nm [111]. As expected from previous work on CO oxidation (see section on complex oxidation earlier [57]), water influences the CO sensing characteristics of Au-doped iron oxide sensors: water promotes CO oxidation, probably via a formate intermediate [113]. Also nanogold particle sensors with diameters smaller than 5 nm have been reported on ZnO and on powdered cobalt oxide [1]. Au/CuO/, NiO and Co$_3$O$_4$ composite films (for optical CO sensors) are also reported, as well as a Au-La$_2$O$_3$ loaded SnO$_2$ ceramic [1]. The target of the last example was to develop an alcohol sensitivity–depressed, steadily workable and widely available CO gas sensor (because conventional CO gas sensors had a poor selectivity to ethanol vapor, which coexists very often in kitchens, leading to false alarms, thus ethanol absorbents such as activated carbon had to be used). Results showed that the sensitivity to CO was ten times higher than that to H$_2$, CH$_4$, i-C$_4$H$_{10}$, and C$_2$H$_4$; while the sensitivity to C$_2$H$_5$OH alone remained high [1].

Sensors for NO$_x$ include porous silicon catalyzed by sputtering gold onto the surface, with negligible interference from CO, CH$_4$ or methanol, but humidity appreciably affected the response. WO$_3$ thin films activated by Au layers have been shown to have excellent sensitivity toward NO and NO$_2$ gases. An exhaust gas NO$_x$ sensor that uses a gold–platinum alloy electrode to selectively remove oxygen but not NO has also been reported [1]. A selective NO$_2$ sensor uses a nanogold-sensing electrode based on yttria-stabilized zirconia under dry and wet conditions at 600°C [115].

An amperometric sensor based on gold nanoparticles embedded in a silica-gel matrix electrode coating is available for *detection of hydrogen peroxide* [112], and gold-containing devices have also been studied for detecting dissolved oxygen in water and SO$_2$ in the gas phase [1]. A gold oxide sensor has been used for hydrocarbon monitoring of exhaust gases [1]. Gold nanoparticle embedded in methyltrimethoxysilane sol-gel film-coated electrode can be used as the basis for an amperometric sensor for hydrogen peroxide: it gives good stability and reproducibility [112].

The basis for a very small *H$_2$S sensor* has been described using gold nanoparticles deposited on a silicon wafer. The microelectronic device based on this, where the adsorption of H$_2$S molecules onto the nanoparticles may significantly change the hopping behavior of electrons through the nanoparticles, could be used to develop a novel on-chip H$_2$S sensor [114].

The aforementioned examples are indications of the potential for many useful developments in the use of supported gold nanoparticles for sensing applications [1].

FUTURE PROSPECTS

Now that gold catalysis has become a valuable new part of the field of catalysis as a whole, its potential for applications is beginning to materialize. It has been more than 20 years since the initial breakthroughs of Hutchings and Haruta [7–9,145] showed that gold can be an excellent catalyst, in contrast to the pre-existing perception that it was unlike other precious and transition metals in terms of catalytic activity. The credentials and importance of gold catalysts as part of the green chemistry movement and their use in transformation relevant to the food industry have also been described [146] and have been developed further by others. As a catalyst, gold is unique [19,147] and there are examples of chemical transformations as yet not accomplished by any other metals. Sometimes gold catalysts produce completely different end products from PGM catalysts. The ability of gold to catalyze reactions at ambient or subambient temperatures opens up totally new application opportunities. In fact, this ability to catalyze reactions at low temperatures creates an entirely new field of low temperature catalysis with characteristics not previously encountered [148]. For example, deactivation at room temperature is not due to sintering. When catalysts are operated at higher temperatures, any impurity formation can be burnt off. This does not occur at ambient temperatures but solutions to this problem are available. For example, Au/TiO_2 catalysts for cleaning living atmospheres at ambient temperatures can be regenerated by photoirradiation, and the poisoning of gold catalysts via carbonate deposition caused by interaction of the support with atmospheric carbon dioxide can be avoided by storing the catalyst in a closed canister at room temperature. The catalyst can keep its activity under these conditions for 2 years [61].

The potential of gold catalysts for commercial applications has been reviewed [17,93,95,110,149]. The principle advances include catalysis of a number of reactions of commercial interest and demonstration of stability in the liquid and gas phase for use in the following fields of application:

- Pollution and emission control and safety systems
- Hydrogen economy/fuel cell systems
- Synthesis of bulk and speciality chemical, including platform chemicals from green starting materials
- Use in gas and liquid phase sensors

In addition, gold catalysts are now becoming more commercially available. There have been initiatives, first by the World Gold Council, to have reference catalysts made [40], and then by Project AuTEK to scale up the catalysts preparations to 15 kg and then 65 kg [150]. There are also other industrial companies, such as Johnson Matthey, 3M, Degussa, and Süd Chemie, who have the capability to make gold catalysts on a large scale. The fact that gold is about half the price of platinum, and much more plentiful, is also a significant economic factor, even though the price of both these metals is currently rising.

Gold catalysis is clearly moving into the exploitation phase [17,146]. This is demonstrated by the increasing number of patents being taken by industrial companies, many of which are household names, as well as R&D organizations seeking to commercialize their intellectual property. Patent mapping can be used to give a good indication of forthcoming applications.

The World Gold Council has an active program to promote exploitation of new gold science and technology, and gold catalysis is at the forefront of this activity. We anticipate new industrial applications to be commercialized in the next 2–5 years [17,151]. However, the successful introduction of new commercial applications depends on a number of technical and economic factors. Commercial realization of the promising progress already achieved will occur when the significantly increased levels of investment in appropriate R&D activities are made in ways similar to those which led to successful exploitation of the platinum group metals in autocatalysts and many other commercial applications from the 1970s onward.

Gold catalysis is therefore already established as a new topic in academic science and will make increasing contributions to commercial products and processes in the near future.

ACKNOWLEDGMENTS

The authors are grateful for provision of manuscripts for book chapters on aspects of gold catalysis by Catherine Louis and Michele Rossi et al. We also thank Michele Rossi for a new version of Figure 6.4, and other authors, including Claus Christensen and Robert Angelici, who provided papers in press or newly published.

REFERENCES

1. Catalytic applications for gold nanotechnology, S. A. C. Carabineiro and D. T. Thompson, in *Nanocatalysis*, ed. U. Heiz and U. Landman, Springer-Verlag, Berlin, 2007.
2. *Catalysis by Gold*, G. C. Bond, C. Louis, and D. T. Thompson, Imperial College Press, London, 2006.
3. Status of catalysis by gold following an AURICAT Workshop, G. C. Bond and D. T. Thompson, *Appl. Catal. A*, 2006, **302**, 1.
4. Using gold nanoparticles for catalysis, D. T. Thompson, *Nanotoday*, 2007, **2**, 40.
5. Novel gold catalysts for the oxidation of carbon monoxide at a temperature far below 0°C, M. Haruta, T. Kobayashi, H. Sano, and N. Yamada, *Chem. Lett.*, 1987, **4**, 405.
6. Gold catalysts prepared by coprecipitation for low-temperature oxidation of hydrogen and carbon monoxide, M. Haruta, N. Yamada, T. Kobayashi, and S. Iijima, *J. Catal.*, 1989, **115**, 301.
7. Size- and support dependency in the catalysis of gold, M. Haruta, *Catal. Today*, 1997, **36**, 153.
8. Vapour phase hydrochlorination of acetylene; correlation of catalytic activity of supported metal chloride catalysts, G. J. Hutchings, *J. Catal.*, 1985, **96**, 292.
9. Catalysis: A golden future, G. J. Hutchings, *Gold Bull.*, 1996, **29**,123.
10. Hydrogenation over supported gold catalysts, G. C. Bond, P. A. Sermon, G. Webb, D. A. Buchanan and P. B. Wells, *J. Chem. Soc. Chem. Commun.*, 1973, 444.
11. Gold catalysts for olefin hydrogenation, G. C. Bond and P. A. Sermon, *Gold Bull.*, 1973, **6**, 102.
12. The combination of hydrogen and oxygen in contact with hot surfaces, W.A. Bone and R. V. Wheeler, *Phil. Trans.*, 1906, **206A**, 1.
13. Gold catalysis, A. S. K. Hashmi and G. J. Hutchings, *Angew. Chem. Int. Ed.*, 2006, **45**, 7896.
14. Nanocrystalline gold and gold-palladium alloy catalysts for chemical synthesis, G. J. Hutchings, *Chem. Commun.*, 2008, 1148.
15. New advances in gold catalysis. Parts I and II, D. T. Thompson, *Gold Bull.*, 1998, **31**, 111; and 1999, **32**, 12.
16. Recent developments in the industrial application of gold catalysts, C. Corti, R. Holliday, E. van der Lingen, and D. Thompson, Proc. TOCAT 5, Tokyo, July 2006, Paper IO-A19, p.101.
17. Progress towards the commercial application of gold catalysts, C.W. Corti, R. J. Holliday, and D. T. Thompson, *Top. Catal.*, 2007, **44**, 331.
18. Cationic gold(I) complexes: Highly efficient catalysts for the addition of alcohols to alkynes, J. H. Teles, S. Brode, and M. Chabanas, *Angew. Chem. Int. Ed.*, 1998, **37**, 1415.
19. Homogeneous catalysis by gold, A. S. K. Hashmi, *Gold Bull.*, 2004, **37**, 51.
20. Gold catalysed organic reactions, A. S. K. Hashmi, *Chem. Rev.*, 2007, **107**, 3180.
21. Organometallic chemistry and catalysis on gold metal surfaces, R. J. Angelici, *J. Organomet. Chem.*, 2008, **693**, 847.
22. Pd-Au catalysts for the production of vinyl acetate – review of the patent and scientific literature, H. Lansink-Rotgerink, Proc. GOLD 2006, Limerick, Ireland, Sept. 2006, 34.
23. Discovering the role of Au and KOAc in the catalysis of vinyl acetate synthesis, W. D. Provine, P. L. Mills, and J. J. Lerou, *Stud. Surf. Sci. Catal.*, 1996, **101**, 191.
24. Selective oxidation of alcohol over supported gold catalysts: methyl glycolate formation from ethylene glycol and methanol, T. Hayashi, T. Inagaki, N. Itayama, and H. Baba, *Catal. Today*, 2006, **117**, 210.
25. Gold catalysts: towards sustainable chemistry, T. Ishida and M. Haruta, *Angew. Chem. Int. Ed.*, 2007, **46**, 7154.
26. Chemical routes for the transformation of biomass into chemicals, A. Corma, S. Iborra, and A. Velty, *Chem. Rev.*, 2007, **107**, 2411.
27. Preparation and environmental applications of supported gold catalysts, M. Haruta, *Now and Future*, 1992, **7**, 13.
28. Nanoparticulate gold catalysts for low-temperature CO oxidation, M. Haruta, *J. New Mat. Electrochem. Systems*, 2004, **7**, 163.

29. Highly active CO oxidation catalysts via physical vapour deposition – key differences between solution – derived nanogold catalysts and nanogold by physical vapor methods, T. Wood, C. Chamberlain, A. Siedle, G. Buccellato, D. Fansler, M. Jones, J. Huberty, L. Brey, S.-H. Chou, M. Jain, and B. Veeraraghavan, Proc. GOLD 2006, Limerick, Ireland, Sept. 2006, 156.

30. Catalysis by gold, G. C. Bond and D. T. Thompson, *Cat. Rev. Sci. Eng.*, 1999, **41**, 319.

31. Gold-catalysed oxidation of carbon monoxide, G. C. Bond and D. T. Thompson, *Gold Bull.*, 2000, **33**, 41.

32. Influence of gold particle size on the aqueous-phase oxidation of carbon monoxide and glycerol, W. C. Ketchie, Y.-L. Fang, M. S. Wong, M. Murayama, and R. J. Davis, *J. Catal.*, 2007, **250**, 94.

33. The catalytic activity of "naked" gold particles, M. Comotti, C. Della Pina, R. Matarrese, and M. Rossi, *Angew. Chem. Int. Ed.*, 2004, **43**, 5812.

34. Catalytic reduction of oxygen and hydrogen peroxide by nanoporous gold, R. Zeis, T. Lei, K. Sieradzki, J. Snyder, and J. Erlebacher, *J. Catal.*, 2008, **253**, 132.

35. Gold catalysts, C. Louis, in *Preparation of Heterogeneous Catalysts*, ed. K. P. de Jong, Wiley VCH, Weinheim, Germany, 2008.

36. Preparation of gold catalysts for glucose oxidation by incipient wetness, C. Baatz and U. Prüsse, *J. Catal.*, 2007, **249**, 34.

37. High activity supported gold catalysts by incipient wetness impregnation, M. Bowker, A. Nuhu, and J. Soares, *Catal. Today*, 2007, **122**, 245.

38. Michael Faraday's recognition of ruby gold: the birth of modern nanotechnology, D. T. Thompson, *Gold Bull.*, 2007, **40**, 267.

39. Gold nanoparticles: from preparation to catalytic evaluation, C. Della Pina, E. Falletta, R. Matarrese, and M. Rossi, in *Metal Nanoclusters in Catalysis and Material Science : The Issue of Size-Control - Part II : Methodologies*, ed. B. Corain, N. Toshima, and G. Schmid, Elsevier, New York, 2008.

40. World Gold Council reference catalysts, http://www.utilisegold.com/uses_applications/catalysis/reference_catalysts/ (accessed July 22, 2009).

41. Catalysts, activating agents, support media, and related methodologies useful for making catalyst systems especially when the catalyst is deposited onto the support media using Physical Vapor Deposition, L. Brey, T. E. Wood, G. M. Buccellato, M. E. Jones, C. S. Chamberlain, and A. R. Siedle, WO Patent 2005/030382, 2005, 3M Innovative Properties Co.

42. Heterogeneous, composite carbonaceous catalyst system and methods that use catalytically active gold, J. T. Brady, M. E. Jones, L. A. Brey, G. M. Buccellato, C. S. Chamberlain, J. S. Huberty, A. R. Siedle, and T. E. Wood , WO Patent 2006/074126, 2006, 3M Innovative Properties Co.

43. Liquid-phase reductive deposition as a novel nanoparticle synthesis method and its application to supported noble metal catalyst preparation, Y. Sunagawa, K. Yamamoto, H. Takahashi, and A. Muramatsu, *Catal. Today*, 2008, **132**, 81.

44. Preparation of TiO_2 using supercritical CO_2 antisolvent precipitation (SAS): a support for high activity gold catalysts, Z.-R. Tang, J. K. Bartley, S. H. Taylor, and G. J. Hutchings, *Stud. Surf. Sci. Catal.*, 2006, **162**, 219.

45. Nature of Cl bonding on the Au(111) surface: Evidence of a mainly covalent interaction, T. A. Baker, C. M. Friend, and E. Kaxiras, *J. Am. Chem. Soc.*, 2008, **130**, 3720.

46. Understanding the effect of halide poisoning in CO oxidation over Au/TiO_2, S. M. Oxford, J. D. Heneo, J. H. Yang, M. C. Kung, and H. H. Kung, *Appl. Catal. A*, 2008, **339**, 180.

47. Catalysis by gold-platinum group metals: mixed metal systems displaying increased activity, D. T. Thompson, *Platinum Met. Rev.*, 2004, **48**, 169.

48. Improved Pd-on-Au bimetallic nanoparticle catalysts for aqueous phase trichloroethene hydrodechlorination, M. O. Nutt, K. N. Heck, P. Alvarez, and M. S. Wong, *Appl. Catal. B*, 2006, **69**, 115.

49. Direct synthesis of hydrogen peroxide from H_2 and O_2 using TiO_2-supported Au-Pd catalysts, J. K. Edwards, B. E. Solsona, P. Landon, A. F. Carley, A. Herzing, C. J. Kiely, and G. J. Hutchings, *J. Catal.*, 2005, **236**, 69.

50. Development of high performance catalysts for the production of vinyl acetate monomer, R. Renneke, S.R. McIntosh, V. Arunajatesan, M. Cruz, B.Chen, T. Tacke., H. Lansink-Rotgerink, A. Geisselmann, R. Mayer, R. Hausmann, P. Schinke, U. Rodermerck, and M. Stoyanova, *Top. Catal.*, 2006, **38**, 279.

51. Nanocrystal and surface alloy properties of bimetallic gold-platinum nanoparticles, D. Mott, J. Luo, A. Smith, P. N. Njoki, L. Wang, and C. J. Zhong, *Nanoscale Res. Lett.*, 2007, **2**, 12.

52. Carbon monoxide oxidation over catalysts prepared by in situ activation of amorphous gold-silver-zirconium and gold-iron-zirconium alloys, A. Baiker, M. Maciejewski, S. Tagliaferri, and P. Hug, *J. Catal.*, 1995, **151**, 407.

53. Gold nanoparticles-catalyzed oxidations in organic chemistry, C. Della Pina, E. Falletta, and M. Rossi, in *Nanoparticles and Catalysis*, ed. D. Astruc, Wiley-VCH, Weinheim, Germany, 2008.

54. When gold is not noble: Catalysis by nanoparticles, M. Haruta, *Chem. Rec.*, 2003, **3**, 75.

55. Gold nanoparticles and gold(III) complexes as general and selective hydrosilylation catalysts, A. Corma, C. González-Arellano, M. Iglesias, and F. Sánchez, *Angew. Chem. Int. Ed.*, 2007, **46**, 7820.

56. CO oxidation activity of gold catalysts supported on various oxides and their improvement by inclusion of an iron component, F. Moreau and G. C. Bond, *Catal. Today*, 2006, **114**, 362.

57. Vital role of moisture in the catalytic activity of supported gold nanoparticles, M. Date, M. Okumua, S. Tsubota, and M. Haruta, *Angew. Chem. Int. Ed.*, 2004, **43**, 2129.

58. Supported gold catalysts for the total oxidation of alkanes and carbon monoxide, B. E. Solsona, T. Garcia, C. Jones, S. H. Taylor, A. F. Carley, and G. J. Hutchings, *Appl. Catal. A*, 2006, **312**, 67.

59. Catalytic oxidation of carbon monoxide over gold/iron hydroxide catalyst at ambient conditions, K.-C. Wu, Y.-L. Tung, Y.-L. Chen, and Y. W. Chen, *Appl. Catal. B*, 2004, **53**, 111.

60. Low temperature oxidation of CO and hydrocarbons over supported gold catalysts, R. Cousin, S. Ivanova, F. Ammari, C. Petit, and V. Pitchon, Proc. GOLD 2003, Vancouver, BC, Canada, Sept.–Oct. 2003.

61. Highly active CO oxidation catalysts via physical vapour deposition–key differences between solution-derived nanogold catalysts and nanogold by physical vapor methods, L. Brey and T. Wood, Proc. GOLD 2006, Limerick, September 2006, p. 156.

62. Gold on titania catalyst for the oxidation of carbon monoxide: control of pH during preparation with various gold contents, F. Moreau, G. C. Bond, and A. O. Taylor, *J. Catal.*, 2005, **231**, 105.

63. Selective oxidation of CO in the presence of H_2, H_2O and CO_2 via gold for use in fuel cells, P. Landon, J. Ferguson, B. E. Solsona, T. Garcia, A. F. Carley, A. A. Herzing, C. J. Kiely, S.E. Golunski, and G. J. Hutchings, *Chem. Commun.*, 2005, 3385.

64. Removal of CO from H_2–rich gas streams via selective/preferential oxidation using Au-based catalysts, J. Steyn, G. Pattrick, E. van der Lingen, M. Scurrell, and D. S. Hildebrandt, African Patent Appl. 01120, 2006 + TOCAT 5.

65. Complete oxidation of ethylene over supported gold nanoparticle catalysts, H.-G. Ahn, B.-M. Choi, and D. J. Lee, *J. Nanosci. Nanotechnol.*, 2006, **6**, 3599.

66. A highly efficient oxidation of cyclohexane over Au/ZSM molecular sieve catalyst with oxygen as oxidant, R. Zhao, D. Ji, G. Lu, G. Qian, L. Yan, X. Wang, and J. Suo, *Chem. Commun.*, 2004, **904**, 22.

67. A highly efficient catalyst Au/MCM-41 for selective oxidation of cyclohexane using oxygen, G. Lu, R. Zhao, D. Ji, G. Qian, Y. Qi, X. Wang, and J. Suo, *Catal. Lett.*, 2004, **97**, 115.

68. Gold zeolitic oxidation catalysts for the conversion of cyclohexane into cyclohexanone and cyclohexanol, L. V. Pirutko, A. S. Khatitonov, M. I Khramov, and A. K. Uriate, Solutia, Inc., U.S. Patent 2004158103 A1.

69. Aerobic oxidation of cyclohexane by gold nanoparticles immobilized upon mesoporous silica, K. Zhu, J. Hu, and R. Richards, *Catal. Lett.*, 2005, **100**, 195.

70. A collaborative effect between gold and a support induces the selective oxidation of alcohols, A. Abad, P. Concepcion, A. Corma, and H. Garcia, *Angew. Chem. Int. Ed.*, 2005, **44**, 4066.

71. Solvent-free oxidation of primary alcohols to aldehydes using Au-Pd/ TiO_2 catalysts, D. I. Enache, J. K. Edwards, P. Landon, B. Solsona-Espriu, A. F. Carley, A. A. Herzing, M. Watanabe, C. J. Kiely, D. W. Knight, and G. J. Hutchings, *Science*, 2006, **311**, 362.

72. Mechanistic investigation of the gold-catalysed aerobic oxidation of alcohols, P. Fristrup, L. B. Johansen, and C. H. Christensen, *Catal. Lett.*, 2008, **120**, 184.

73. The nature of the active site for vinyl acetate synthesis over Pd-Au, M. S. Chen, K. Luo, T. Wei, Z. Yan, D. Kumar, C.-W. Yi, and D. W. Goodman, *Catal. Today*, 2006, **117**, 37.

74. Gold catalyses one-step synthesis of methyl glycolate, *Chem. Eng. (New York)*, Sept. 2004, **111**, 20.

75. Propylene epoxidation with O_2 and H_2 over gold nanoparticles supported on mesostructural titanium silicates, M. Haruta, A. K. Sinha, S. Seelan, B. Chowdhury, M. Daté, and S. Tsubota *Proc. 13 Int. Congr. Catalysis*, Paris, July 2004, O1-031; 33.

76. Catalysis by gold nanoparticles: epoxidation of propene, A. K. Sinha, S. Seelan, S. Tsubota, and M. Haruta, *Top. Catal.*, 2004, **29**, 95.

77. New perspectives in gold-catalysed oxidation, S. Biella and M. Rossi, Proc. GOLD 2003, Vancouver, BC, Canada, Sept.–Oct. 2003.

78. Gold-based bimetallic catalysts for liquid phase applications, N. Dimitratos and L. Prati, *Gold Bull.*, 2005, **38**, 73.

79. Selective oxidation of glycerol to glyceric acid using a gold catalyst in aqueous sodium hydroxide, S. Carrettin, P. McMorn, P. Johnston, K. Griffin, and G. J. Hutchings, *Chem. Commun.*, 2002, 696.

80. Oxidation of glycerol using supported gold catalysts, S. Carrettin, P. McMorn, P. Johnston, K. Griffin, C. J. Kiely, G. A. Attard, and G. J. Hutchings, *Top. Catal.*, 2004, **27**, 131.

81. Long-term stability of Au/Al₂O₃ catalyst prepared by incipient wetness in continuous-flow glucose oxidation, N. Thielecke, U. Prüsse, and K.-D. Vorlop, *Catal. Today*, 2007, **122**, 266.

82. Method for selective carbohydrate oxidation using supported gold catalysts, J. Kowalczyk, A. H. Begli, U. Prüsse, H. Bendt, and I. Pitsch, WO Patent 099114 A1 (2004), Südzucker Akt.

83. Selective oxidation of lactose and maltose with gold supported on titania, A. Mirescu, U. Prüsse, and K.-D. Vorlop, Proc. 13 Int. Congr. Catalysis, Paris, July 2004; P5-059.

84. Glucose as renewable feedstock: catalytic conversion of glucose by hydrogenation and oxidation, S. Schimpf, B. Kusserow, Y. Önal, and P. Claus, Proc. 13 International Congress on Catalysis, Paris, July 2004, P5-060.

85. Oxidation of glycerol and propanediols in methanol over heterogeneous gold catalysts, E. Taarning, A. T. Madsen, J. M. Marchetti, K. Egeblad, and C. H. Christensen, *Green Chem.*, 2008, in press.

86. Hydrogen peroxide synthesis over metallic catalysts, P. Paredes Olivera, E. M. Patrito, and H. Sellers, *Surf. Sci.*, 1994, **313**, 25.

87. Direct formation of hydrogen peroxide from H₂O₂ using a gold catalysts, P. Landon, P. J. Collier, A. J. Papworth, C. J. Kiely, and G. J. Hutchings, *Chem. Commun.*, 2002, 2058.

88. Low-temperature water-gas shift reaction over Au/α-Fe₂O₃, D. Andreeva, V. Idakiev, T. Tabakova, and A. Andreev, *J. Catal.*, 1996, **158**, 354.

89. Ceria-zirconia supported Au as highly active low temperature water-gas shift catalysts, A. Amieiro-Fonseca, J. M. Fisher, D.Ozkaya, M. D. Shannon, and D. Thompsett, *Top. Catal.*, 2007, **44**, 223.

90. Gold catalysts for pure hydrogen production in the water-gas shift reaction: activity, structure and reaction mechanism, R. Burch, *Phys. Chem. Chem. Phys.*, 2006, **8**, 5483.

91. Highly active gold-containing water gas shift catalysts for reducing CO levels in hydrogen feeds in fuel cells, J. Breen, R. Burch, J. Gomez-Lopez, A. Amiero, J. Fisher, D. Thompsett, R. Holliday, and D. Thompson, Proc. Fuel Cell Symposium, San Antonio, Texas, November 2004.

92. DFT and in situ EXAFS investigation of gold/ceria-zirconia low-temperature water-gas shift catalysts: identification of the nature of the active form of gold, D. Tibiletti, A. Amiero-Fonseca, R. Burch, Y. Chen, J. M. Fisher, A. Goguet, C. Hardacre, P. Hu, and D. Thompsett, *J. Phys. Chem. B*, 2005, **109**, 22553.

93. An overview of gold-catalysed oxidation processes, D. T. Thompson, *Top. Catal.*, 2006, **38**, 231.

94. Method for making hydrogen using a gold-containing water-gas shift catalyst, A Kuperman and M. E. Moir, Chevron Inc, WO Patent 2005/005032.

95. Gold's future role in fuel cell systems, D. Cameron, R. Holliday, and D. Thompson, *J. Power Sources*, 2003, **118**, 298.

96. Promotion by sulfur of gold catalysts for crotyl alcohol formation from crotonaldehyde hydrogenation, J. E. Bailie and G. J. Hutchings, *Chem. Commun.*, 1999, 2151.

97. The hydrogenation of 1,3-butadiene and crotonaldehyde over highly dispersed Au catalysts, M. Okumura, T. Akita, and M. Haruta, *Catal. Today*, 2002, **74**, 265.

98. Gold based catalysts performances in selective hydrogenation of diene in an excess of alkene, A. Hugon, L. Delannoy, and C. Louis, Proc GOLD 2006, Limerick, Ireland, September 2006, p. 152.

99. Catalysis by group Ib metals .1. Reaction of buta-1,3-diene with hydrogen and with deuterium catalyzed by alumina-supported gold, D. A. Buchanan and G. Webb, *J. Chem. Soc., Faraday Trans.* 1975, **71**,134.

100. Hydrogenation of alkenes over supported gold, P. A. Sermon, G. C. Bond, and P. B. Wells, *J. Chem. Soc., Faraday Trans.*, 1979, **75**, 385.

101. Hydrogenation of acetylene over Au/Al₂O₃ catalyst, J. Jia, K. Haraki, J. N. Kondo, and K. Tamaru, *J. Phys. Chem. B,* 2000, **104**, 11153.

102. Organogold chemistry III: Applications, R. V. Parish, *Gold Bull.*, 1998, **31**, 14.

103. Enantioselective hydrogenation of alkenes and imines by a gold catalyst, C. Gonzalez-Arellano, A. Corma, M. Iglesias, and F. Sanchez, *Chem. Commun.*, 2005, 3451.

104. Catalytic asymmetric aldol reaction: reaction of aldehydes with isocyanoacetate catalyzed by a chiral ferrocenylphosphine-gold(I) complex, Y. Ito, M. Sawamura, and T. Hayashi, *J. Am. Chem. Soc.*, 1986, **108**, 6405.

105. A powerful chiral counterion strategy for asymmetric transition metal catalysis, G. L. Hamilton, E. J. Kang, M. Mba, and F. D. Toste, *Science*, 2007, **317**, 496.

106. Catalysis: Raising the gold standard, A. S. K. Hashmi, *Nature*, 2007, **449**, 292.

107. A homogeneous catalyst for selective O₂ oxidation at ambient temperature. Diversity-based discovery and mechanistic investigation of thioether oxidation by the Au(III)Cl₂NO₃(thioether)/O₂ system, E. Boring, Y. V. Geltii, and C. L. Hill, *J. Am. Chem. Soc.*, 2001, **123**, 1625.

108. The application of supported gold catalysts to automotive pollution abatement, J. R. Mellor, A. N. Palazov, B. S. Grigorova, J. F. Greyling, K. Reddy, M. P. Letsoalo, and J. H. Marsh, *Catal. Today*, 2002, **72**, 145.

109. Recent developments in gold based catalysts systems for automotive applications, G. Pattrick, E. van der Lingen, C. W. Corti, R. J. Holliday, and D. T. Thompson, Preprints CAPoC 6, Brussels, October 2003, O14.

110. The potential for use of gold in automotive pollution control technologies: a short review, G. Pattrick, E. van der Lingen, C. W. Corti, R. J. Holliday, and D. T. Thompson, *Top. Catal.*, 2004, **30/31**, 273.

111. Role of the Au oxidation state in the CO sensing mechanism of Au/iron oxide-based gas sensors, G. Neri, A. Bonavita, C. Milone, and S. Galvagno, *Sens. Actuators B*, 2003, **93**, 402.

112. Gold nanoparticles embedded in silica sol-gel matrix as an amperometric sensor for hydrogen peroxide, G. Maduraiveeran and R. Ramaraj, *J. Electroanal. Chem.*, 2007, **608**, 52.

113. A study of water influence on CO response on gold-doped iron oxide sensors, G. Neri, A. Bonavita, G. Rizzo, S. Galvagno, N. Donato, and L. S. Caputi, *Sens. Actuators B*, 2004, **101**, 90.

114. Suppressed electron hopping in a Au nanoparticle/H_2S system: development towards a H_2S sensor, J. Geng, M. D. R. Thomas, D. S. Shephard, and B. F. G. Johnson, *Chem. Commun.*, 2005, 1895.

115. Selective YSZ-based planar NO_2 sensor using nanogold sensing electrode, V. V. Plashnitsa, T. Ueda, P. Elumalai, and N. Miura, *Chem. Sens.*, 2007, **23**A, 142.

116. Designing oxidation catalysts: are we getting better?, G. J. Hutchings and M. S. Scurrell, *Cattech,* 2003, **7**, 90.

117. Tunable gold catalysts for selective hydrocarbon oxidation under mild conditions, M. D. Hughes, Y.-J. Xu, P. Jenkins, P. McMorn, P. Landon, D. I. Enache, A. F. Carley, G. A. Attard, G. J. Hutchings, F. King, E. H. Stitt, et al., *Nature*, 2005, **437**, 1132.

118. The catalytic properties of gold, G. C. Bond, *Gold Bull.*, 1972, **5**, 11; see also A tribute to Geoffrey Bond on his 80th birthday, D. T. Thompson, *Gold Bull.*, 2007, **40**, 4.

119. Gold 2006 Conference Report, M. Cortie, A. Laguna, and D. T. Thompson, *Gold Bull.*, 2006, **39**, 226.

120. The riches of gold catalysis, R. Burks, *Chem. Eng. News*, September 24, 2007, 87.

121. Supported gold catalysts used for ozone decomposition and simultaneous elimination of ozone and carbon monoxide at ambient temperature, Z. Hao, D. Cheng, Y. Guo, and Y. Liang, *Appl. Catal., B*, 2001, **33**, 217.

122. Multi-component noble metal catalysts prepared by sequential deposition precipitation for low temperature decomposition of dioxin, M. Okumura, T. Akita, M. Haruta, X. Wang, O. Kajikawa, and O. Okada, *Appl. Catal. B*, 2003, **41**, 43.

123. Gold as a novel catalyst in the 21st century: preparation, working mechanism and applications, M. Haruta, Proc. Cat. Gold 2003, Vancouver, BC, Canada, 2003.

124. Review: Automotive catalytic converters: current status and some perspectives, J. Kaspar, P. Fornasiero, and N. Hickey, *Catal. Today*, 2003, **77**, 419.

125. http://www.nanostellar.com/Reports/NS_Gold_Press_Release.doc.

126. Catalyst for purifying an exhaust gas, Y, Miyake and S, Tsuji, Toyota JJK Patent, E.P. 1043059 A1, 2000.

127. Fulminating gold, J. M. Fisher, *Gold Bull.*, 2003, **36**, 155.

128. Application of a gold catalyst to promote mercury control at a coal-fired utility, S. Meischen, Proc. Cat. Gold 2003, Vancouver, BC, Canada, Sept.– Oct. 2003.

129. Pilot testing of mercury oxidation catalysts for upstream of wet FGD systems, G. M. Blythe, Quarterly Technical Progress Report, April–June 2006, URS Corporation, Austin, Texas.

130. Meso structured manganese oxide/gold nanoparticle composites for extensive air purification, A. K. Sinha, K. Suzuki, M. Takahara, H. Azuma, T. Nonaka, and K. Fukumoto, *Angew. Chem. Int. Ed.*, 2007, **46**, 2891.

131. Propene oxidation over gold catalysts: influence of the support, L. F. Liotta, G. Pantaleo, G. Di Carlo, A. M. Venezia, G. Deganello, M. Ousmane, A. Giroir-Fendler, and L. Retailleau, Proc. Catalysis for Society, Krakow, Poland, May 2008.

132. Gold catalysts supported on titanium oxide for catalytic wet air oxidation of succinic acid, M. Besson, A. Kallel, P. Gallezot, R. Zanella, and C. Louis, *Catal. Commun.*, 2003, **4**, 471.

133. Materials synthesis and catalysis of palladium-coated gold nanoparticles, M. S. Wong, M. O, Nutt, K. N. Heck, and P. Alvarez, Proc. GOLD 2006, Limerick, Ireland, Sept. 2006, p.104.

134. Gold catalyst mediates transformation of biomass feedstocks. Aerobic oxidation of furfurals leads to value-added methyl esters, S. K. Ritter, *Chem. Eng. News*, Jan. 3, 2008.

135. Chemicals from renewables: aerobic oxidation of furfural and hydroxymethyl furfural over gold catalysts, E. Taarning, I.S. Nielsen, K. Egeblad, R. Madsen, and C.H. Christensen, *ChemSusChem,* 2008, **1**, 75.

136. Engineering and Physical Sciences Research Council, http://gow.epsrc.ac.uk/ViewGrant.aspx?GrantRef= EP/E009999/1.

137. Method for preparing oximes by catalytic hydrogenation of α,β-unsaturated nitro compounds using gold catalysts, A. Corma Canos and P. Serna-Merino, WO Patent 20077116112, 2007.

138. Oxidations of amines with molecular oxygen using bifunctional gold–titania catalysts, S. K. Klitgaard, K. Egeblad, U. V. Mentzel, A. G. Popov, T. Jensen, E. Taarning, I. S. Nielsen, and C. H. Christensen, *Green Chem.*, 2008, **10**, 419.

139. Gold as a novel catalyst in the 21st century: preparation, working mechanisms and applications, M. Haruta, *Gold Bull.*, 2004, **37**, 27.

140. Oxidation of propane to propylene oxide on gold catalysts, J. J. Bravo-Suarez, K. K. Bando, J. Lu, T. Fujitani, and S. T. Oyama, *J. Catal.*, 2008, **255**, 114.

141. Effect of gold on the HDS activity of supported palladium catalysts, A. M. Venezia, V. La Parola, V. Nicoli, and G. Deganello, *J. Catal.*, 2002, **212**, 56.

142. Hydrogen chemisorption on Al_2O_3-supported gold catalysts, E. Bus, J. T. Miller, and J. A. van Bokhoven, *J. Chem. Phys. B*, 2005, **109**, 14581.

143. Supported gold catalysis in the hydrogenation of canola oil, L. Caceres, L. L. Diosady, W. F. Graydon, and L. J. Rubin, *J. Am. Oil Chem. Soc.*, 1985, **62,** 906.

144. Recent developments in the industrial application of gold catalysts, C. W. Corti, R. J. Holliday, D. T. Thompson, and E. van der Lingen, *Science and Technology in Catalysis*, Ed. K. Eguchi, M. Machida, and I. Yamanaka, Kodansha, Elsevier, Tokyo, 2007, p. 173.

145. M. Haruta and S. Hiroshi, Japanese Patent 60238148 (1985).

146. GOLD 2006 Conference Report, M. Cortie, A. Laguna, and D. T. Thompson, *Gold Bull.*, 2006, **39**, 226.

147. Relations between homogeneous and heterogeneous gold catalysis. Unique catalytic properties of gold for green chemistry, A. Corma, Proc. TOCAT 5, Tokyo, July 2006, Paper O-A3.

148. Deposition of gold nanoparticles on organic polymers, M. Haruta, W. Minagawa, N. Kinoshita, and K. Kuroda, Proc. GOLD 2006, Limerick, September 2006, p. 155

149. Commercial aspects of gold catalysis, C. W. Corti, R. J. Holliday, and D. T. Thompson, *Appl. Catal. A*, 2005, **291**, 253.

150. MINTEK, http://www.mintek.co.za/landing.php?bus_cat=7fc125319d6e1d8b7b496b8953b77777& level=3 (accessed July 22, 2009).

151. World Gold Council, http://www.gold.org (accessed July 22, 2009).

7 Metallurgy of Gold

Jörg Fischer-Bühner

CONTENTS

Introduction..123
Grain Refinement..124
   Solidification..125
   Recrystallization and Grain Growth during Annealing after Cold-Working............126
Strength and Ductility...127
   Mechanisms...127
   Solid Solution Hardening..129
      Conventional Alloying...129
      Microalloying..131
      Disorder–Order Transformation Hardening..131
   Precipitation Hardening...133
      General Aspects...133
      Au-Cu-Ag–based Alloys..136
      Au-Ni- and Au-Pt-based Alloys..139
      Low- and Microalloyed Gold..140
      Multicomponent Alloys...143
   Dispersion Hardening..144
Color Variation...144
   Color Shifts from Yellow to Reddish, Greenish, and Whitish.................................144
   White Gold Alloys...145
   Special Colors of Gold: Black, Blue, and Purple..147
Fluidity, Wetting, and Other Thermal Properties...151
Miscellaneous..152
   Bulk Metallic Glass...152
   Shape Memory Alloys...154
   Metal-Matrix Composites..154
References...155

INTRODUCTION

Gold possesses a unique spectrum of properties. In its pure form, gold has a rich yellow color, is very soft, highly malleable, and ductile at room temperature. It possesses high electrical and thermal conductivities. From the chemical and electrochemical point of view it is the most noble of all metals. It does not corrode, tarnish, or oxidize. It has an unequaled stability toward influences from all atmospheric conditions, be it in a natural, biomedical, technical, or any other environment such as leisure or household. Its stability also persists to higher temperatures, so that it can be molten in air without oxidation despite its high melting temperature of 1064°C. The unique combination of some of these properties is responsible for its use, not only for luxury goods like jewelry, but also in many different technical applications despite its high metal price. In almost all areas of application, however, gold cannot be used in its pure form. It needs to be alloyed with other elements, usually metals, to adjust specific properties. For most applications, pure gold is simply too soft. For several applications, the properties of the pure gold need to be maintained as much as possible, for example, bonding wires in electronics or high karat gold jewelry (karatage is defined as 24kt $\equiv$ 100 w% Au; Xkt = (X*100)/24 wt% Au, for example: 18kt = 75 wt% Au). Here only very small additions (<1000 ppm) of alloying elements are added carefully. This is commonly referred to as *microalloyed* gold. In many other cases, larger modifications of properties by "conventional" alloying with larger additions are necessary. For jewelry applications, alloy compositions are varied over a wide range mainly to achieve a range of colors and mechanical properties, but also to offer products for different price segments (high versus low karat) depending on the gold content (Note: "Karat" is spelled "Carat" in the UK). For both, dental and jewelry applications, investment casting is a widely used manufacturing process, so that alloying additions are used that reduce the melting temperature and increase fluidity of the melt. For gold solders, even lower melting temperatures and especially high wettability are important. For particular solders as well as dental applications, alloying additions are used that lead to easy surface oxidation at elevated temperatures so that a good bond can be obtained with ceramic components.

The basis for any metallurgical approach is provided by phase diagrams. Almost all relevant binary and ternary gold alloy phase diagrams have been reviewed and critically assessed [1–3]. The general alloying behavior of gold, the tendencies to form solid solutions, compounds, and intermetallic phases with different alloying elements, first described in Raub [4], is thoroughly analyzed in Raynor [5], Rapson and Groenewald [6], more recently also in Ferro et al. [7], and is not covered here in such depth.

While peculiarities of different alloys, applications, and respective manufacturing processes are covered in more detail in the corresponding chapters or cited references, the aim of this chapter is to introduce the metallurgical principles of alloying for grain refinement, strengthening, and further property adjustment of gold alloys as indicated above. The corresponding background relevant for standard industrial processing operations like melting, casting, deformation, and heat treatment is indicated wherever appropriate. Furthermore, potentially interesting opportunities using less-common and innovative processing or alloying approaches are briefly covered.

GRAIN REFINEMENT

A fine-grained material provides many advantages over a coarse-grained one. For this reason, grain refinement and grain size control by microalloying additions and proper processing today is applied routinely for gold alloys in dental [8], electronics [9], jewelry [10,11], and other industries. The list of documented benefits related with fine-grained material is long:

- Increase of strength, work-hardening during cold-working, malleability, and ductility
- Improvement of decorative surface characteristics (avoidance of the so-called orange peel effect) and surface-finishing properties
- Increase of chemical homogeneity due to less pronounced segregation of alloying elements during solidification, and related to that:
  - Improvement of corrosion resistance and reduced tendency to stress-corrosion cracking
  - Reduction of susceptibility to embrittlement and hot crack formation in castings

Probably the only disadvantage provided by fine-grained material occurs at high temperatures where a loss of strength and reduction of creep resistance is related with the change of prevailing deformation mechanisms, namely, the onset of diffusion processes along grain boundaries.

During materials processing, two process steps are particularly important for grain size reduction or control, which need to be treated separately as explained in more detail in Ott and Raub [12]: the solidification step after melting and pouring and the annealing step after cold-working.

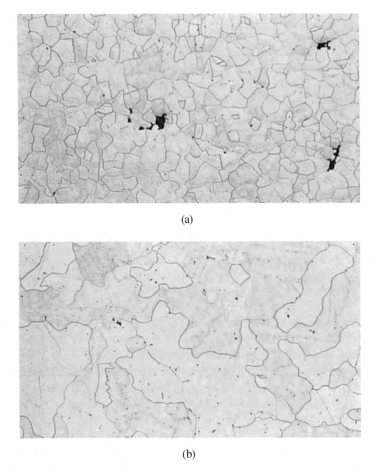

(a)

(b)

FIGURE 7.1   (a) Grain refining effect of 0.0055 wt% iridium in 18 karat yellow gold (50x). (b) 18 kt yellow gold without iridium additions (50x). (From Optimising gold alloys for the manufacturing process, D. Ott, *Gold Technology*, 2002, **34**, 37. With permission.)

SOLIDIFICATION

Grain refinement during solidification mainly is concerned with promotion of the formation of larger numbers of nuclei for crystal growth from the molten state. Based on experience, as well as on thermodynamic considerations [13], this can be achieved by additions of particular elements with comparably high melting point and low solubility in solid gold, such as iridium, ruthenium, rhenium, molybdenum, tungsten, and cobalt. An example of the grain refining effect of Ir in 18kt yellow gold castings is shown in Figure 7.1. The choice of the right addition depends on alloy composition. For instance, it is often reported that the presence of silicon reduces the grain refining effect of iridium in karat gold alloys [10]. The required amount of the additions again depends on alloy composition and can vary between 50 and 1000 ppm. Higher amounts are not meaningful, since no further grain refining effect is usually observed. Furthermore, they can be detrimental, because some additions tend to form clusters or segregate at grain boundaries which reduce the grain refining effect and degrade decorative properties due to formation of hard spots during surface finishing. For this reason, grain refining additions usually are introduced via carefully prepared base metal master alloys, for example, Cu-10% Ir, which need to offer a good liquid and solid solubility for the grain refiner. Therewith a homogeneous distribution in the master alloy is ensured, which provides optimum conditions for the final alloy preparation.

Grain refinement during solidification reportedly can also be obtained by additions of reactive elements like alkaline and rare earth elements, such as calcium, barium, yttrium, or nonmetals like Boron [12]. In these cases, nuclei for crystallization from the melt are provided by formation of high melting compounds (sometimes intermetallic phases) of these reactive additions with other alloying elements or impurities, particularly oxygen.

Apart from alloy additions, grain size during solidification is, to some extent, also influenced by processing [12]. Cold pouring is routinely applied, with low metal and mould temperatures supporting crystal nucleation and retarding crystal growth, leading to more fine-grained castings. Mould agitation or vibration support the homogeneous distribution of nuclei and "mechanically" refine the as-cast structure by breaking up the branches (dendrites) of growing crystals. While its application has been less common in processing of gold castings than for other metals, it is nowadays also successfully applied for jewelry investment casting [14].

RECRYSTALLIZATION AND GRAIN GROWTH DURING ANNEALING AFTER COLD-WORKING

During cold-working by processes such as sheet rolling or wire drawing, the microstructure of a metal becomes heavily distorted, the material is strengthened, and ductility is reduced to very low levels after heavy deformation. Complete or partial soft annealing then is carried out to recover ductility, adjust the final strength level or prepare the material for further deformation processing. For sufficiently high annealing temperature and time, the material recrystallizes, which involves nucleation and growth of new, undeformed grains of regular shape. The control of the annealing parameters, but also of the alloy composition, is essential to ensure a fine-grained recrystallized material. Grain refinement during the annealing stage after deformation is based on two potential mechanisms: increase of the number of recrystallization nuclei which develop in the deformed microstructure and, probably more importantly, the decrease of grain growth velocity.

For gold alloys, some of the additions that have been mentioned above as grain refiners during solidification can also promote the formation of recrystallization nuclei: Particles that are formed by the additions themselves or by reactions with alloying additions or impurities provide obstacles for the deformation process and support the development of deformation inhomogeneities. During annealing, recrystallization nuclei preferentially form and grow in such areas due to energetically favorable conditions. However, if the growth of the nuclei is not inhibited also by the same or other additions, the described mechanism alone does not necessarily lead to a fine-grained recrystallized microstructure. In extreme cases it may even have an adverse effect on grain size, namely, if excessive growth of a few number of preferentially formed nuclei occurs. Furthermore, a fine-grained recrystallized structure energetically is not stable and tends to coarsen during prolonged annealing. In industrial annealing processes, however, the annealing conditions (temperature, time, etc.) often cannot be precisely controlled.

For all these reasons, the decrease of grain growth velocity by specific alloying additions is always of high importance. For gold alloys, it is of even higher importance than for other metals, because usually (and unlike in many other industries) high purity fine gold is used as starting material. Impurities as well as specific alloying additions with low solubility in the metal matrix can reduce the grain boundary mobility drastically [15,16], because they tend to segregate to grain boundaries and therefore need to be able to move together with them during recrystallization and grain growth. This requires time- and temperature-dependent diffusion processes to occur, which slows down recystallization and grain growth kinetics. The more these effects are pronounced, the larger is the mismatch in atomic size between host and foreign atom. Furthermore, grain growth can be inhibited by fine particles that are formed by alloy additions, impurities, or their reaction products like oxides and (intermetallic) compounds and that can effectively pin and stabilize grain boundaries depending on particle size and distribution, a mechanism referred to as Zener pinning.

The influence of impurities on recrystallization kinetics of gold is obvious from the dependence of the recrystallization temperature (defined as the temperature at which the material is 50% recrystallized by 30 min annealing after 90% deformation) on the impurity level; it increases from 112°C for a gold purity of 99.999%, over 160°C for 99.99%, to 200°C for 99.9% [17]. In bonding wires for electronics, grain size control down to grain diameters of ~1 μm is of tremendous importance, because wire diameters nowadays are reduced to below 20 μm. While very high purity fine gold needs to be used to guarantee optimum electrical conductivity, microalloying additions of beryllium and calcium usually together with other rare-earth, alkali or alkaline metals are routinely used in a range of up to only 100 ppm for grain size control and strengthening [9,18]. With regard to larger amounts of alloying additions to gold, a systematic study (with a focus on jewelry alloys) identified cobalt, barium, and zirconium as particularly efficient grain refiners during recrystallization [19]. Today, cobalt additions in a range of 0.1–0.5% probably are most common for grain refining during heat treatment of gold jewelry alloys [10,12]. Lower amounts of Co are required in 14kt compared to 18kt jewelry alloys, due to a reduction of solubility of Co in the base alloy. A synergetic effect of cobalt (as well as zirconium) with boron additions is frequently mentioned.

STRENGTH AND DUCTILITY

Mechanisms

Deformation of any metal takes place by the sliding of crystal planes over each other through the movement of lattice defects, called dislocations. As with many other pure metals, this is particularly easy for pure gold, as characterized by a low hardness of 20–30 HV and a high deformability of over 90% in the as-cast or soft-annealed state. Any distortion of the crystal lattice or any obstacle in the lattice increases the force required to move the dislocations through the lattice. Hence this leads to an increase of hardness or strength but usually also to a decrease of deformability. Alloying additions and impurities in many different ways can lead to such distortions or obstacles to deformation. Figure 7.2 gives an overview of the hardness increase in binary gold alloys depending on the type

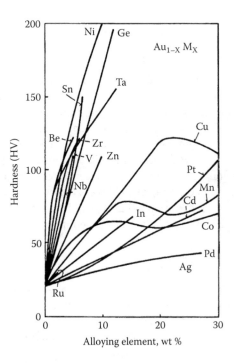

FIGURE 7.2 Hardness trends in binary gold alloys depending on the type and amount of addition in weight % [20]. (From *Edelmetalltaschenbuch* (Precious Metal Pocketbook), eds. Degussa AG Frankfurt, Hüthig-Verlag Heidelberg, Germany, 2nd edition 1995, (in German) . With permission.)

and amount of alloying addition in weight % [20]. The largely different dependencies are striking and related to the different prevailing strengthening mechanisms as well as the particular micro-structural condition the material is in. Among others strengthening is possible by

- **Solid solution hardening:** Alloying elements that dissolve in solid gold can either replace gold atoms in the crystal lattice (substitutional solid solution hardening) or, in case that their atoms are much smaller in size than gold atoms, can fill in the small gaps between gold atoms (interstitial solid solution hardening) in the crystal lattice.
- **Disorder-order transformation hardening:** For particular alloys, usually in the range of stoichiometric alloy compositions, the atoms arrange themselves with lowering of temperature in an ordered solid solution where different atoms occupy strictly defined lattice sites. Strengthening is caused by the related elastic distortions, but also because the movement of dislocations is considerably more difficult because the ordered state needs to be preserved.
- **Precipitation hardening:** Alloying elements for which the solubility in solid gold decreases with temperature can be precipitated by particular heat treatment as finely dispersed particles. This can involve precipitation of ordered, intermetallic compounds with either gold and/or further alloying elements.

Both processes, disorder–order transformation hardening and precipitation hardening, are reversible, that is, the material can be brought back to the nonhardened state by a homogenization heat treatment. With both processes it is mostly referred to as *age-hardening.* As opposed to these,

- **Dispersion hardening** is based on the irreversible formation of finely dispersed particles, mainly oxides (but also carbides, borides), by reaction of alloying elements with oxygen (or carbon, etc.); it often involves special processing routes like internal oxidation or powder metallurgy.

These mechanisms will be discussed in more detail in the following paragraphs. Furthermore strengthening is possible especially by

- **Work hardening:** During plastic deformation the dislocations react not only with alloying elements or particles, but extensively also with each other in many complex ways. While these reactions can transfer the dislocations into immobile obstacles, they also involve a tremendous increase of the number of dislocations, even more mutual interactions and eventually significant strengthening which increases with degree of deformation (until the effect levels off at large deformation levels by a mechanism referred to as dynamic recovery). Work hardening also significantly depends on the type and amount of alloying elements, as shown for some examples in Table 7.1 [20]. This can again be explained by the way these elements are incorporated in the gold matrix or microstructure (see above: solid solution, particles, etc.) and the corresponding interaction mechanisms with the dislocations.
- **Grain size control:** Grain boundaries act as very effective obstacles to dislocation movement. Grain refinement by particular alloying additions and proper processing (discussed previously) therefore contributes to strengthening and work hardening. According to the Hall-Petch relationship, strength is inversely related to grain size squared. It is notable that grain refinement is the only mechanism that at the same time also increases ductility, while all other strengthening mechanisms lead to degradation of ductility.

For a given alloy composition, the final property adjustment by work hardening and grain size control is of particular importance for the production of semifinished material like sheet, rod, tube, and wire. This is achieved either by a final cold-working step, starting from a soft-annealed state, and

TABLE 7.1
Hardness Increase Depending on Amount of Cold-Working and Type and Amount of Alloying Elements

Alloy	Degree of Cold Work (%)		
	0	40	80
Au-X		Hardness HV10	
Au-20Ag	40	95	114
Au-30Ag	42	93	115
Au-25Ag-5Cu	92	160	188
Au-20Ag-10Cu	120	190	240
Au-26Ag-3Ni	83	134	166
Au-25Ag-5Pt	58	106	130
Au-5Co	92	126	154
Au-5Ni	120	162	188
Au-10Pt	78	102	118

(From *Edelmetalltaschenbuch* (Precious Metal Pocketbook), eds. Degussa AG Frankfurt, Hüthig-Verlag Heidelberg, Germany, 2nd edition 1995 (in German). With permission.)

aiming at a desired combination of strength by work-hardening and residual ductility, or by a final annealing treatment, starting from the work-hardened state, during which the deformed microstructure recovers or partially recrystallizes, hence softens to the desired strength level.

SOLID SOLUTION HARDENING

Conventional Alloying

The effect of solid solution hardening increases with the amount of lattice distortion associated with an alloying addition, hence it increases with the difference in atomic size between the host metal and the added element. To some extent this explains the dependence of hardness on the Ag-Cu ratio in the Au-Ag-Cu ternary system, which forms the basis for many common jewelry and dental alloys. Ag atoms are only slightly larger than Au atoms, whereas Cu atoms are ~12% smaller than Au atoms, so that Cu is much more effective in strengthening gold by substititial solid solution (see also Table 7.1). However, while Ag is completely soluble in Au at all temperatures, irrespective of the composition, Cu is soluble in Au only down to a temperature of 410°C (Figure 7.3). Below that temperature a series of intermetallic Au-Cu phases forms, which results in additional hardening effects (see the following sections). Apart from Ag and Cu, Pd is completely soluble in Au and used in large amounts up to ~ 5–20 wt% for jewelry white gold alloys or even ~40 wt% for dental metal-ceramic applications, respectively (see section on white gold alloys later). Also Pt is completely soluble in solid Au at high temperatures and used up to ~25% for dental metal-ceramic applications.

Similar to Ag, the solid solution hardening effect of Pd and Pt in Au is small to moderate, however, so that especially Ag-, Pd-, and Pt-rich gold alloys usually need to be strengthened by further alloying additions or strengthening mechanisms. The effect of small additions up to 0.5 wt% of a large variety of conventional base metals to gold and gold alloys revealed no significant influence on hardness [19], indicating that amounts above ~0.5 wt% are required to recognize substitutional solid solution hardening by those additions. The corresponding influence by additions of 0.5–3 wt% of Ni, Co, Fe, and Sb in 22kt gold jewelry alloys (91.7 wt% Au, Ag:Cu=1:1) is shown in Figure 7.4

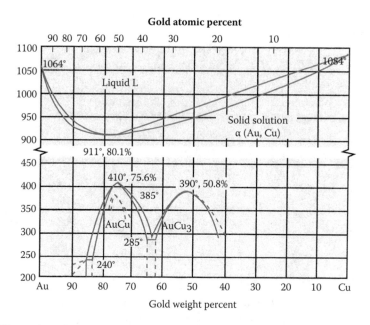

FIGURE 7.3 Binary phase diagram Au-Cu, [1]. (From *Phase Diagrams of Binary Gold Alloys,* eds. H. Okamoto and T. B. Massalski, ASM International, Metals Park, OH, 1987. With permission.)

[21]. With 2 wt% of Co the hardness is raised from ~ 60 HV for a simple ternary 22kt Au-Cu-Ag alloy up to 90–100 HV in the as-cast or soft-annealed state. Large additions of up to 15 wt% Ni are used in 18kt and lower karat white gold jewelry alloys. This leads to more significant strengthening as indicated in Figure 7.1, which is mainly related to precipitation hardening (see section on precipitation hardening later). Base metal additions of Zn, Sn, In, and Ga in the range from 0.1 to 2–3 wt% are frequently used, especially (but not only) in casting alloys and also contribute to strength increase (Figure 7.1) which in part can be attributed to substitutional solid solution hardening. For all these elements, the solid solubility in gold is limited, however, and decreases with temperature, which can result in contributions of precipitation hardening also in an as-cast state.

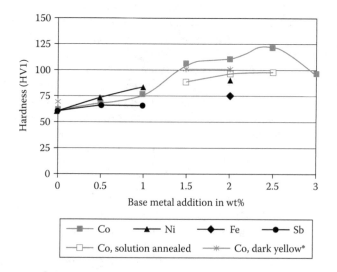

FIGURE 7.4 Hardness trends in as-cast 22kt Au depending on base metal additions [21]. (From Hardening of low-alloyed gold, J. Fischer-Bühner, *Gold Bulletin*, 2005, **38(3)**, 120–131. With permission.)

In order to clearly differentiate between the contributions from solid solution, disorder–order and precipitation strengthening contributions in such alloys, homogenized (i.e., solution annealed) and quickly quenched samples need to be investigated and compared with subsequently aged samples.

Microalloying

As reviewed in Corti [22] and more recently in Corti [27], microalloying additions of rare-earth, light metals, alkali or alkaline metals are used to strengthen 24kt gold jewelry alloys (99.5–99.9 wt% Au) [23– 25], as well as bonding wires in electronics [9,18]. In the as-cast or soft-annealed state, the hardness is raised from 20 to 30 HV for pure gold up to ~60 HV for microalloyed gold. Such additions are also used to strengthen conventional alloys for different applications [18,25]. Rare-earth metals have a significant potential for substitutional solid solution strengthening of gold and gold alloys due to particularly large differences in atomic size to gold [26]. Of the possible rare-earth additions, the light elements like Ce, La, and Nd are the main microalloying metals, often together with some heavy rare-earth metals like Gd. The strengthening potential by microalloying additions of alkali, alkaline or light metals such as Be, Ca, Li, Mg, Sr, Na, or K is reported to be even higher than that of the rare-earth metals [26]. This is related with their particularly small atomic size and weight compared to gold, which in part leads to interstitial rather than substitutional solid solution and a large ratio of dissolved atoms to gold atoms. Comparably little systematic studies are published about the individual effects of the latter additions [28–30], although many of them, especially Ca, are routinely used.

In general, the solid solution strengthening effect of microalloying additions is limited, however, by the usually small and, in part, even neglectible maximum solid solubilities in gold, the formation of eutectics and the related tendencies to segregate during solidification, to accumulate at grain boundaries and to form embrittling phases [26]. Furthermore, many of them show a potential for precipitation hardening if added in amounts exceeding their low temperature solubility in gold. While the addition of 0.5 wt% Gd to gold alloys for potentiometer winding and electric contact applications has been reported, the total concentration of microalloying additions usually is below 100 ppm for bonding wire applications [18], but probably increases up to 0.3 wt% (of mainly Gd and Ca) for jewelry applications [27]. From a practical point of view, the high tendency to oxidation of most of these elements raises the demands for processing in protective gas atmosphere or under vacuum. Especially in the jewelry industry, where scrap is routinely recycled in-house and used again in production, the loss of microalloying additions during processing and the corresponding degradation of properties is a frequent problem and restricts application.

Disorder–Order Transformation Hardening

Strengthening by the disorder-to-order transformation is of particular importance for many Cu-containing gold alloys and is based on the occurrence of the ordered solid solutions (also referred to as intermetallic compounds or intermediate phases) AuCu and $AuCu_3$ in the binary Au-Cu system (see Figure 7.3). The transformations have been extensively researched and reviewed [1–3]. Above 410°C, independent from alloy composition, the solid solution is disordered, hence the atoms are randomly distributed in a face-centered cubic lattice. But below 410°C and 390°C, respectively, the gold and copper atoms occupy specific sites in the crystal lattice. More precisely, in the composition range where the atomic ratio of gold to copper atoms equals 1:1, alternate layers of Au and Cu atoms form the face-centered tetragonal AuCu I phase below 385° (Figure 7.5), with an orthorhombic variation AuCu II being stable between 385°C and 410°C. For atomic ratios approaching 1:3, two versions of $AuCu_3$ I+II of another ordered face-centered phase form ($L1_2$–structure: Cu atoms on lattice faces, Au atoms on lattice edges).

Hardening by the described disorder-to-order transformation in the Au-Cu system is an important feature of many karat gold jewelry alloys [31–34] and dental alloys [35–38]. It is obtained either by slow cooling from elevated temperatures or during aging typically between 150 and 400°C. Different to precipitation hardening (see next section) there is no need for a previous quenching step from elevated temperature for the disorder–order hardening process to occur during low-temperature

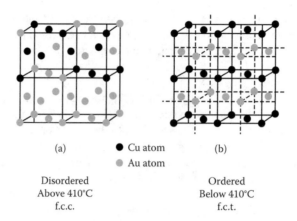

(a)   ● Cu atom   (b)
                 ● Au atom

Disordered   Ordered
Above 410°C   Below 410°C
f.c.c.   f.c.t.

FIGURE 7.5 (a) Crystal structure of a disordered face-centred cubic AuCu – alloy. (b) Crystal structure of the ordered face-centred tetragonal AuCu I – phase. (From 18 Carat yellow gold alloys with increased hardness, R. Süss, E. van der Lingen, L. Gardner, and M. du Toit, *Gold Bulletin*, 2004, **37(3–4)**, 196. With permission.)

aging. An *overaging effect* (softening due to prolonged aging) sometimes is observed and is related to a stronger hardening effect of a partially ordered state if compared to a fully ordered state. In fact, during early stages of the transformation, only nanometer-sized islands of the ordered structure form and grow into the still disordered environment.

A binary alloy with 25 wt% Cu can be hardened to ~350 HV (tensile strength >800 N/mm²) starting from a soft-annealed or as-cast hardness of 165 HV (tensile strength ~500 N/mm²) (Figure 7.6), but ductility then also drops from >40% to <5% tensile elongation. This is why during processing of

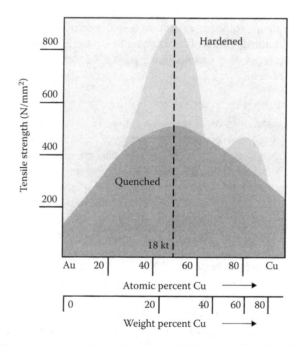

FIGURE 7.6 Effect of heat treatment on the tensile strength of binary Au-Cu alloys [33]. (From A plain man's guide to alloy phase diagrams: their use in jewellery manufacture – Part 2, M. Grimwade, *Gold Technology*, 2000, **30**, 8. With permission.)

such alloys, it is often required to suppress the transformation by quenching from elevated temperatures. However, for the same alloy composition the ductility in the ordered, hardened state can be significantly increased to 20% tensile elongation and even above by appropriate processing routes aiming at microstructural refinement [39].

In ternary and higher-order alloys, the interrelations of alloy composition, processing parameters and effectiveness of the disorder–order hardening mechanism in relation to precipitation hardening are complex, as indicated in Rapson [40]. They are comparably well researched for the ternary Au-Cu-Ag system [32,33], which forms the basis for most of the dental and jewelry alloys. As explained in more detail in the next section, disorder-to-order transformation hardening prevails in Cu-rich 18kt jewelry alloys due to the extension of the AuCu-phases into the ternary system and otherwise overlaps with precipitation hardening. Numerous studies have been carried out with a view to influencing further alloying additions, e.g. Zn, Pd, and Pt which are common alloy components in dental alloys [35–38,41,42]. Accordingly, the disorder–order transformation provides the primary strengthening mechanism in Pd- and Pt-containing Au-Cu-Ag-based dental alloys, with precipitation hardening playing an important role only for alloys with higher Ag and Pt content. Zn additions up to 20 at% equiatomic AuCu-alloys have been systematically investigated and are reported to increase the hardening kinetics and to delay overaging. The acceleration of the AuCu disorder-to-order transformation by Zn additions provides improved hardening properties at relatively low aging temperatures.

Transformations from disordered to ordered solid solutions do also occur in some further binary alloy systems, namely, Au-Cd, Au-Cr, Au-Mn, Au-Nb, Au-Pd, Au-V, and Au-Zn [1–3]. The martensitic transformations associated with the ordering in the Au-Cd and Au-Zn systems are relevant for shape memory applications and are also accompanied with considerable strengthening effects. The transformations in the other alloy systems listed above are, in part, relevant for particular functional applications, but little is known about the impact of the transformations on (mechanical) properties.

PRECIPITATION HARDENING

General Aspects

The prerequisite for precipitation hardening is the reduction of the solid solubility of an alloying element in the host metal lattice. With a view to binary gold alloys, this is the case for alloys of Au with Co, Cr, Fe, Mn, Ni, Pt, Ti, V, and many others [1]. Hence it can occur

- In alloy systems like Au-Ni (Figure 7.7) or Au-Pt with a complete series of solid solutions at high temperature but where a miscibility gap develops with decreasing temperature
- In alloy systems like Au-Co (Figure 7.8) or Au-Cr with a limited mutual solid solubility over the complete temperature range, i.e., where the miscibility gap extends up to the solidus lines
- In alloy systems like Au-Ti (Figure 7.9) or Au-V or many others, with similar features as above, but where one or more intermetallic phases occur.

Obviously, it can also occur in ternary and higher-order systems which can combine some or all of the characteristics indicated above. In all cases, an optimum heat treatment leading to most pronounced age-hardening effects would involve the following:

- Homogenization (or so-called solution annealing) at a temperature high enough to dissolve the alloying element(s) in the gold matrix; for many of the examples previously mentioned typically between 700°C and 900°C for 0.25–2 h depending on the actual alloy composition and dimensions of the sample or product.
- Rapid quenching from the homogenization temperature (usually into water), which suppresses the formation of precipitates. The alloying elements remain dissolved in the crystal lattice and the state is referred to as a supersaturated solid solution.

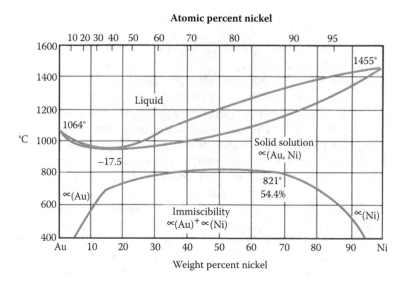

FIGURE 7.7 Binary phase diagram Au-Ni, [1]. (From *Phase Diagrams of Binary Gold Alloys*, eds. H. Okamoto and T. B. Massalski, ASM International, Metals Park, Ohio, 1987. With permission.)

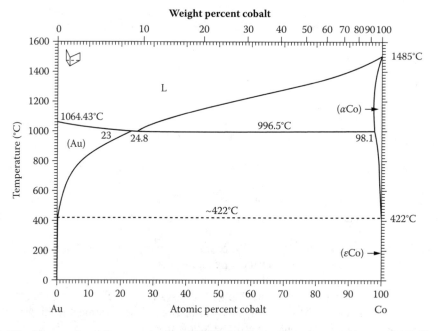

FIGURE 7.8 Binary phase diagram Au-Co, [1]. (From *Phase Diagrams of Binary Gold Alloys*, eds. H. Okamoto and T. B. Massalski, ASM International, Metals Park, Ohio, 1987. With permission.)

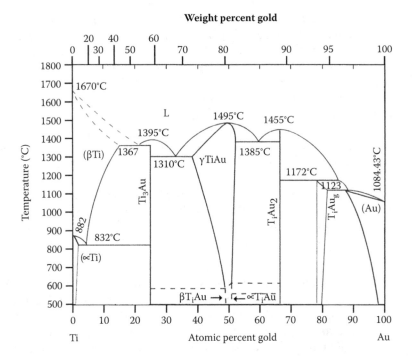

FIGURE 7.9 Binary phase diagram Au-Ti, [1]. (From *Phase Diagrams of Binary Gold Alloys,* eds. H. Okamoto and T. B. Massalski, ASM International, Metals Park, Ohio, 1987. With permission.)

- Aging at low temperatures, typically around 200–400°C, more seldomly up to 600°C, for 0.5 up to several hours, with the annealing time and temperature depending again on alloy composition and many other factors.

During the aging step, very fine precipitates form with dimensions in the submicrometer range, which are very effective in causing hardening. Depending on the alloy system, the precipitates can consist of (following the examples introduced before)

- A second phase highly enriched with the alloying element(s), e.g., a Ni-rich phase with some Au in solid solution (Figure 7.7)
- An almost "pure" metal, e.g., Co due to the extremely low solubility of Au in Co (Figure 7.8), or
- An intermetallic phase, formed either by an compound of Au and the alloying element, e.g., Au_4Ti (Figure 7.9), or formed by compounds of the alloying elements in ternary or higher-order alloy systems.

Thorough analysis of precipitation phenomena by high resolution electron microscopy often reveals that, during aging, not only the volume fraction and size but also the shape, composition, and crystal structure of the precipitates change with aging time, sometimes leading to multistage age-hardening effects. Prolonged aging usually leads to coarsening and coalescence of the precipitates and a related hardness drop, a phenomenon that is referred to as overaging by Ostwald ripening. In an ideal situation, the precipitates are homogeneously distributed in the microstructure. In practice, they do often preferentially form and grow at grain boundaries, leading to coarser precipitates at grain boundaries compared to the grain interiors. Furthermore, the effectiveness of the quenching step is of tremendous importance but often critical to control in industrial processing (which mostly explains significant property discrepancies between research lab results and production trials). Nonoptimum quenching can lead to a gradient in the potential for age-hardening and the related strengthening effect, being highest at the surface and lowest in the center of a sample, semifinished or final product. During (too) slow cooling from the solution annealing temperature, very coarse precipitates, which are comparably ineffective in causing hardening, can already form and therefore reduce the age-hardening potential during the subsequent low-temperature aging step. On the other hand, many industrial processes involve slow to moderately quick cooling steps from high or elevated temperatures, allowing for some age-hardening to occur already during cooling. Although this may not happen in a very defined or controlled way, the properties of a product may well benefit already from the related strength increase. For example, for as-cast products that are manufactured in alloys with a potential for age-hardening, the hardness in the as-cast state can significantly vary depending on product size, process temperatures, and cooling procedures. In severe cases, the formation of precipitates during cooling after casting or after a soft-annealing step can occur to such an extent that ductility is degraded. In such cases, quenching from higher temperature may be inevitable in order to allow for deformation processing of the material.

Au-Cu-Ag–based Alloys

The large miscibility gap in the Ag-Cu system extends widely into the ternary Au-Cu-Ag system (Figure 7.10), leading to age-hardening potential for a wide range of alloys for jewelry and dental applications. The corresponding details have been reviewed [32–34]. Figure 7.11 shows pseudo-binary sections through the ternary phase diagram for constant gold contents of 75, 58.5, and 41.7 wt% (18kt, 14kt, and 10kt alloys, respectively). Figure 7.12 shows the corresponding dependence of hardness on the Ag (and Cu) content for age-hardened, air-cooled and quenched material. Hardening is most pronounced for intermediate Ag contents of roughly 15–30 wt% in 14ct alloys and 20–40 wt%

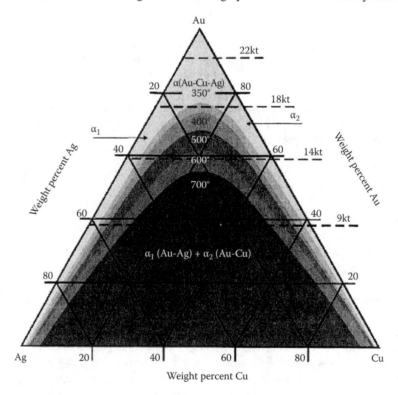

FIGURE 7.10 Isothermal section projections showing extent of miscibility gap with temperature in the Au-Ag-Cu system [33]. (From A plain man's guide to alloy phase diagrams: their use in jewellery manufacture - Part 2, M. Grimwade, *Gold Technology*, 2000, **30**, 8. With permission.)

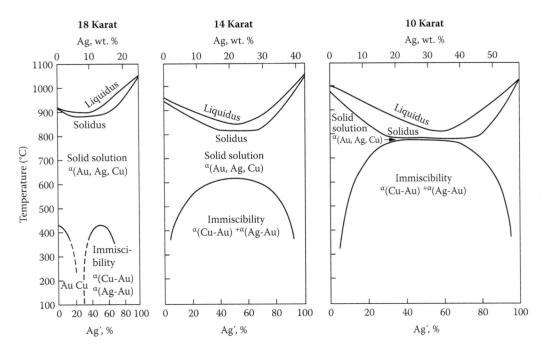

FIGURE 7.11 Pseudo-binary sections through the ternary Au-Cu-Ag phase diagram for constant gold contents of 75, 58.5, and 41.7 wt % (18kt, 14kt, and 10kt alloys, respectively), [20], [32]. (From *Edelmetalltaschenbuch* (Precious Metal Pocketbook), eds. Degussa AG Frankfurt, Hüthig-Verlag Heidelberg, Germany, 2nd edition 1995 (in German). With permission.)

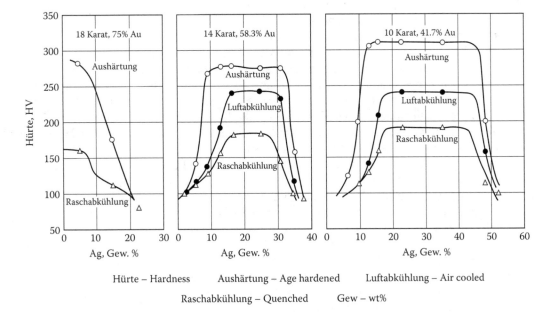

| Hürte – Hardness | Aushärtung – Age hardened | Luftabkühlung – Air cooled |
| Raschabkühlung – Quenched | Gew – wt% | |

FIGURE 7.12 Hardness trends along the pseudo-binary sections shown if Figure 7.11 for age-hardened, air-cooled and quenched material [20]. (From *Edelmetalltaschenbuch* (Precious Metal Pocketbook), ed. Degussa AG Frankfurt, Hüthig-Verlag, Heidelberg, Germany, 2nd edition 1995 (in German). With permission.)

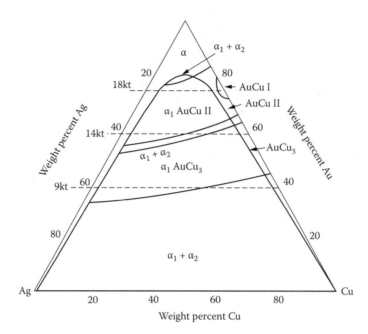

FIGURE 7.13 Isothermal section at 300°C in the Au-Ag-Cu system [2]. (From *Phase Diagrams of Ternary Gold Alloys*, A. Prince, G. V. Raynor, and D. S. Evans, The Institute of Metals, London, UK, 1990. With permission.)

in 10ct alloys. However, during slow cooling and even during quenching from the solid solution at elevated temperatures, they decompose into copper- and silver-rich phases (also referred to as phase separation), leading to already high hardness and difficult workability, for example, of castings or soft-annealed and subsequently air-cooled material. For lower Ag contents down to ~ 5 wt% in 14kt and 10kt alloys as well as higher Ag contents up to roughly 37 wt% in 14kt and 50 wt% in 10kt alloys, single-phased and hence more easy-to-work alloys can be more easily obtained by quenching, with a good potential for subsequent age-hardening. Beyond those boundaries for the Ag content, little to no age-hardening potential is available in the ternary 14kt and 10kt alloys.

The miscibility gap extends much less into the area of 18kt alloys, so that alloys with Ag contents above ~20 wt% are soft and not age hardenable. With lowering of the Ag content the hardness in the aged condition increases steadily, even down to single-phased alloys with low Ag content. This is attributed to the extension of the stability field of the ordered AuCu solid solution into the ternary system and the related hardening effects (see Disorder–Order Transformation Hardening earlier). Hence precipitation hardening and disorder-to-order transformation hardening overlap for a wide range of ternary 18kt Au alloys. In fact, the two mechanisms also overlap for alloys with gold contents down to ~35 wt%, since the Cu-rich precipitates can undergo disorder-to-order transformations during aging or slow cooling as illustrated by the isothermal section shown in Figure 7.13 [33– 36].

From the many further alloying elements that are added to Au-Cu-Ag-based alloys, zinc is of particular importance. It is added in large amounts of 4 wt% and significantly above especially to 14kt and 10kt jewelry alloys for color adjustment, but also because it narrows the volume of the immiscibility gap present in the ternary system. This leads to softer and more workable alloys in both the age-hardened and air-cooled states [32]. The addition of 6 wt% Zn to a 14kt alloy or 9 wt% Zn to a 10kt alloy leads to single-phased alloys that are not age hardenable.

As obvious from Figure 7.12, ternary Au-Ag-Cu–based 18kt jewelry alloys with medium to high Ag content are comparably soft and show little or no age-hardening potential. While the hardness and age-hardening potential of these alloys can be increased by additions of Zn, In, Sn, or Ga for casting applications, there is a demand for alloys with improved age hardenability for applications involving mechanical working, especially in the watch-making industry [43,44]. The influence of

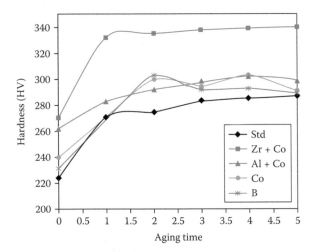

FIGURE 7.14   Age-hardening potential of modified 18kt alloys at 300°C; the time=0h state corresponds to a 50–60% cold-worked state, [43] = Figure 7.12 from [43] 18 Carat yellow gold alloys with increased hardness, R. Süss, E. van der Lingen, L. Gardner, and M. du Toit, *Gold Bulletin*, 2004, **37(3–4)**, 196.

additions of up to 2% of base metals (Co, Ti, V, Cr, Mo, Zr), platinum group metals (PGMs: Pt, Ru, Rh, Ir) as well as Al and B (<0.1%) to an 18ct alloy with equal amounts (in weight percent) of Cu and Ag was investigated in Süss et al. [43]. Additions of the PGM increased the maximum hardness in the aged state to 300–330 HV but the alloys are reported to suffer from PGM inclusions and bad surface finish. More promising is the addition of either 0.08 wt% B, 1 wt% Co, 0.5 wt% Zr + 0.3 wt% Co, or finally 0.8 wt% Zn and Co, each,+ 0.4 wt% Al, which yields increased age-hardening potential if compared to the standard alloy as shown in Figure 7.14.

Au-Ni- and Au-Pt-based Alloys

Although nickel (Ni) is a common alloying element in white gold jewelry alloys, usually together with Cu and Zn (see section on white gold alloys later), comparably little details are published about the age-hardening characteristics of these alloys. The metallurgy of the ternary Au-Ni-Cu system is reviewed in McDonald and Sistare [45] and with a focus to 18kt alloys in Susz and Linker [46]. The miscibility gap present in the binary Au-Ni-system (Figure 7.7) extends largely into the ternary Au-Ni-Cu system as shown in Figure 7.15. For this reason, for many Ni-based white gold alloys, which contain up to ~13–18 wt% Ni for 18kt and 14kt, respectively, phase separation into Au- and Ni-rich second phases and related hardening occurs during slow cooling. This results in comparably high hardness values well above 200 HV, which can be brought near to 300 HV by aging treatment at 350–450°C for 1 h. Several hours of aging at 275–300°C can increase the hardness near to 350 HV for alloys with high Ni content. Even in the solution annealed and quenched condition, alloys are considered to be comparably hard and difficult to work. The addition of Ni (and Zn) to Au-Pd-Ag–based 18kt white gold alloys adds potential for age hardening to these otherwise comparably soft alloys.

In the binary Au-Pt system, an immiscibility gap comparable to the Au-Ni system occurs. Binary alloys containing up to 15 wt% Pt are single phased and comparably soft (<50 HV), but for higher Pt contents the alloys can be age-hardened, for example, at 500°C from a solution-annealed (1000–1150°C) and quenched state: from 120 HV to 300 HV for 30 wt% Pt and from 240 HV to 420 HV for 50 wt% Pt [20]. Accordingly, specific dental Au-based alloys (with ~60–80 wt% Au) in which Pt is used in amounts as high as 15–25 wt% [8], are age hardened by precipitation of Pt-rich particles. In many dental metal-ceramic alloys (with ~70–85 wt% Au) Pt and Pd are added simultaneously, typically 8–10 wt% Pt together with 5–10 wt% Pd. Such alloys are age hardenable, because the

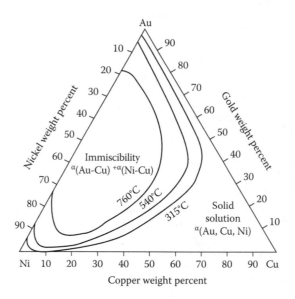

FIGURE 7.15 Isothermal section projections showing the miscibility gap with temperature in the Au-Cu-Ni system [33]. (From A plain man's guide to alloy phase diagrams: their use in jewellery manufacture – Part 2, M. Grimwade, *Gold Technology*, 2000, **30**, 8. With permission.)

miscibility gap in the binary Au-Pt system extends significantly into the ternary Au-Pt-Pd system and broadens toward the Au-Pd system. As a consequence, Au-Pd-Pt–based alloys with as low as 6 wt% Pt are age hardenable [20,47]. Age-hardening mechanisms in dental alloys are continuously researched and discussed controversially [48] (see also Multicomponent Alloys later).

Low- and Microalloyed Gold

The development of strengthened high karat (22kt–24kt) Au alloys with improved hardness and wear resistance and at the same time good suitability for jewelry manufacturing has remained a challenging task to date. It has involved extensive research into precipitation-hardenable alloys.

For 24kt alloys the precipitation hardening potential of a large variety of base metals at alloying levels of max. 1 wt% was reviewed, and titanium was identified as the best candidate in that range [49]. The hardness of a 990 gold alloy containing 1 wt% Ti can be substantially increased from 70 HV in the as-cast state to 170 HV in the age-hardened state (solution annealing for 800°C followed by aging at 500°C for ~1 h) (Figure 7.16). The strengthening effect is based on formation of Au_4Ti-precipitates. Although big efforts were spent on developing and communicating best practices for working with this alloy type [49–51], its usage and spread in jewelry manufacturing has remained very limited up to now due to the high reactivity of titanium and its related tendency to form oxides, nitrides, and carbides. However, binary Au-Ti alloys with ~1.7 wt% Ti (eventually with small Ir-additions for grain refinement) have been introduced with some success in the dental sector as alloys with outstanding biocompatibility [52].

Also considered for wire and foil applications in electronics [53] is 990 AuTi, and it has been discovered that irreversible dispersion hardening by formation of titanium oxides can contribute to the strengthening effects, which is relevant for near-surface regions as well as very thin products (<100 μm).

As an alternative to 990 AuTi in jewelry, a gold alloy with 995 fineness containing 0.2 wt% Co and 0.3 wt% Sb with reportedly good jewelry manufacturing properties was developed [54]. It can be hardened to ~100 HV from the as-cast and homogenized state (~700°C/1 h, water quench) by aging at 300°C for approximately 30 h. The hardening kinetics and effects are significantly accelerated by preceding cold-working (Figure 7.17). According to Fischer-Bühner [21], hardening of 995

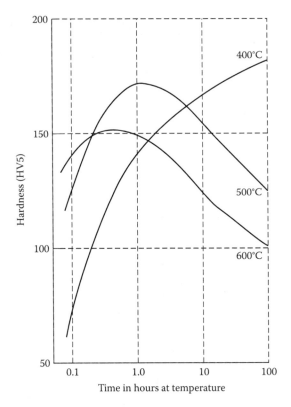

FIGURE 7.16 Age-hardening potential of a 990 Au-Ti alloy [49]. (From The development of 990 Gold-Titanium: Its production, use and properties, G. Gafner, *Gold Bulletin*, 1989, **22**, 112. With permission.)

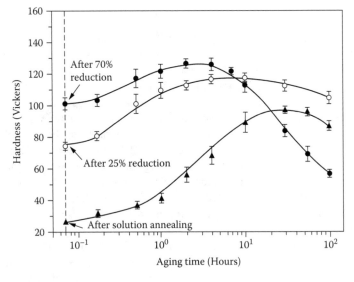

FIGURE 7.17 Age-hardening potential of a 995 Au-Co-Sb alloy at 300°C [54]. (From The development of a novel gold alloy with 995 fineness and increased hardness, M. du Toit, E. van der Lingen, L. Glaner, and R. Süss, *Gold Bulletin*, 2002, **35(2)**, 46. With permission.)

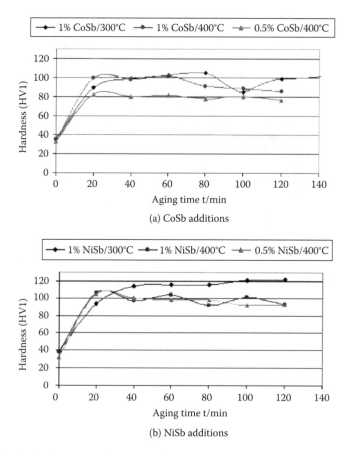

FIGURE 7.18 Age-hardening potential of 990- 995 Au-Co-Sb (a) and Au-Ni-Sb (b) alloys at 300–400°C [21]. (From Hardening of low-alloyed gold, J. Fischer-Bühner, *Gold Bulletin,* 2005, **38(3),** 120–131. With permission.)

and 990 gold alloys to 80–120 HV during aging for 20–60 min at 300–400°C (from the as-cast and homogenized state, 700°C/1 h, water quenching) is achieved by additions of Co+Sb or Ni+Sb in a weight percent ratio of ~1:2. The prevailing hardening mechanism is supposedly based on precipitation of intermetallic CoSb and NiSb compounds (Figure 7.18).

As extensively reviewed [22,27] and described in previous sections, also microalloying additions of rare-earth, light metals, alkali or alkaline metals are used to strengthen 24kt gold jewelry alloys as well as bonding wires in electronics. With a view to the in part very limited solid solubilities of most of these elements in pure gold, and its decrease with temperature, it is commonly assumed that precipitation hardening contributes to the strengthening effects. However, little related in-depth investigations are available. The precipitation hardening potential of different rare-earth elements has been validated [18,26,49] and revealed notable potential for additions of 1 wt% whereas for addition of 0.5 wt% or below age hardening is experimentally confirmed only for Yttrium additions [55]. The expected precipitation phases in the gold-rich Au-RE systems are the gold-rich compounds such as Au_6RE (for La, Ce, Pr, Nd, Pm, Sm, Gd, Tb, Dy, Ho), Au_5RE (for Eu) and Au_4RE (for Er, Tm, Yb, Lu, Sc) . The light REs are expected to have a higher precipitation strengthening potential than the heavy REs, because of their lower solid solubility and because gold-richer precipitate phases exist in Au-light RE systems [18,26].

In 22 karat alloys, significant precipitation hardening is obtained by alloying with conventional base metals like Co and Ni [21,56,57]; 22kt Au alloys with ~1.5–2 wt% Co (rest Ag, Cu) can be

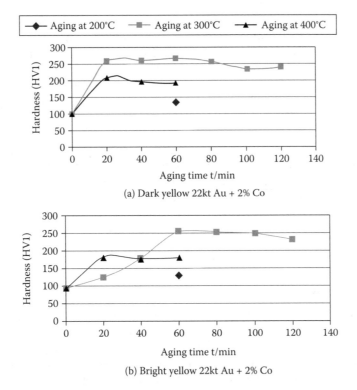

FIGURE 7.19 Age-hardening potential of 22kt Au with 2 wt% Co additions [21], [56]. (From Hardening of low-alloyed gold, J. Fischer-Bühner, *Gold Bulletin*, 2005, **38(3)**, 120–131. With permission.)

hardened by aging for ~1 h at 300–400°C to peak hardness levels between 200 and 260 HV1 (from the as-cast and homogenized state, 850–900°C/1 h, water quenching) (Figure 7.19). Extensive property testing for jewelry manufacturing was carried out with promising results. The required solution temperature, however, is comparably high. Further work showed that the age-hardening behavior of 22kt Au alloyed with Co or Ni can be remarkably influenced by additions of Sb if the weight percent ratio Co:Sb or Ni:Sb is kept ~1:2, similarly to the situation reported for 990–995 Au [21,56]. A total of only 1 wt% of Co+Sb or Ni+Sb additions lead to a peak hardness of ~150 HV1 in 22kt Au by aging 20 min at 400°C after solution annealing at a much lower temperature of only 700°C for 1 h. Precipitation hardening of low-alloyed gold by formation of intermetallic compounds of Ni or Co with base metals like Sb (or Sn, Ga [58]) is an interesting area because comparably low amounts of alloying additions already show an effect and comparably little tendencies to overaging are observed.

Multicomponent Alloys

Most commercial dental and jewelry gold alloys often show a complex combination of five or more alloying elements. Alloying additions of a few percent of Sn, In, Ga, or Zn are frequently added to dental and jewelry casting alloys for reduction of the melting range, improvement of fluidity, or metal-ceramic bonding properties. These elements all have a limited solid solubility in gold and are recognized to improve the strength and sometimes the age-hardening potential of the respective alloys (e.g., Degussa [20]). For example, Sn-additions in Au-Ag-Pd-Pt-Sn dental alloys lead to hardening by precipitation of intermetallic Pd_3Sn, Pt_3Sn, or $(Pd,Pt)_3Sn$ phases [59].

Au-reduced dental alloys with Au contents between 40 wt% and 60 wt% have been developed with the purpose of cost reduction [8,20]. While alloys based on the Au-Ag-Cu system are hardenable by precipitations and disorder-to-order transformation as discussed before, there was a need to develop Cu-free alloys based on the Au-Ag-Pd system, which provides no mutual immiscibility and age-hardening potential. Immiscibility gaps and intermetallic phases occur in the Au-Pd-In ternary and binary systems. Hence, the addition of Indium to Au-Ag-Pd alloys in a certain ratio relative to the Pd-content, for example, 10 wt% Pd and 4 wt% In, allows for age hardening by formation of both Pd- and In-rich precipitates (e.g., from 150 HV in the soft state to 220 HV in the aged state) [20].

The study of age-hardening phenomena in multicomponent alloys is complex and the development of alloys with good age-hardening properties challenging. The use of thermodynamic modeling software allows for calculations on the stability of precipitated phases, in good agreement with experimental results as already demonstrated for multicomponent dental alloys [59]. Hence, the use of such modeling potentially can provide an efficient tool for faster and more direct development of alloys, presupposed that comprehensive thermodynamic property datasets for precious metals, their main alloying elements, and the potential binary, ternary, and higher-order intermetallic phases are available.

DISPERSION HARDENING

Age hardening by disorder-to-order transformation or precipitation hardening is a reversible process, hence the material can be brought back to a soft state by annealing at high temperature followed by quenching. As such, these mechanisms are not efficient in strengthening the material at elevated or high temperature. Contrary to this, dispersion hardening by oxides, carbides or other stable particles is irreversible and can contribute to increased strength at higher temperatures. Gold can be dispersion hardened by amounts as low as 0.42 wt% of TiO_2 particles with a diameter of 0.5 µm, resulting in a ~2.5-fold strength increase at room temperature [55]. Comparable strengthening is also obtained by mixing gold with powders (0.18–0.38 wt%) of cerium-, aluminium-, thorium-, and yttrium-oxide [60]. While powder metallurgical processing is required to obtain these properties, strengthening contributions of dispersion hardening can also be obtained by internal oxidation of gold alloyed with easily oxidizing elements like titanium [53]. Unlike precipitates, oxide particles are stable, i.e., do not dissolve, and show small tendencies to coagulate at elevated temperatures, which explains the irreversibility of the process.

COLOR VARIATION

COLOR SHIFTS FROM YELLOW TO REDDISH, GREENISH, AND WHITISH

A detailed overview on the color properties of gold and gold alloys and an explanation of the mechanisms of color variation are given in Cretu and van der Lingen [61] and Saeger and Rodies [62]. Gold and copper are the only metals that in their pure forms appear yellow and red colored, respectively, whereas all other pure metals appear white or grayish. For pure gold a deep drop of the reflectivity curve occurs within the visible light spectrum, leading to much higher reflectivity for the low-energy end of the visible spectrum than for the rest of the spectrum, hence the yellow color (Figure 7.20). The drop in reflectivity is related to adsorption processes of the incident light energy by the particular electron structure of gold atoms. With copper additions to gold, the drop in the reflectivity curve is shifted further toward the low-energy end of the visible light spectrum, leading to deeper yellow alloys with a pink to reddish hue. With other alloying additions the drop in the reflectivity curve is either shifted toward the high-energy end of the visible light spectrum, as with Ag, as shown also in Figure 7.20, or the shape of the reflectivity curve is flattened as is the case for alloying elements like nickel, palladium or platinum. As a result, gold alloyed with such elements loses its characteristic strong yellow color, and more pale yellow, yellow-greenish, grayish or whitish colored gold alloys result.

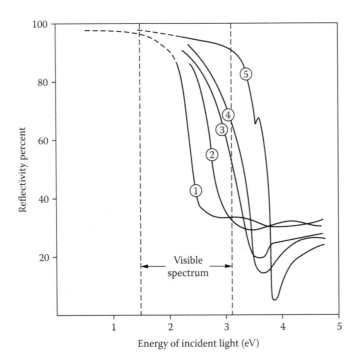

FIGURE 7.20   Reflectivity curves as a function of the energy of incident light for (1) fine gold, (2) Au + 50 at% Ag, (3) Au + 90 at% Ag, (4) Au + 95 at% Ag, (5) fine silver [62]. (The colour of gold and its alloys: The mechanism of variation in optical properties, K. E. Saeger and J. Rodies, *Gold Bulletin*, 1977, **10(1)**, 10. With permission.)

The color spectrum of gold alloys is of special relevance for jewelry applications. For jewelry made in different karatages, the color can be influenced gradually in the range indicated above by variation of the main alloying elements Ag and Cu (Figure 7.21), with zinc as a further significant alloying element, shifting the color of Cu-rich alloys from reddish to dark yellow [32,33]. Standards with defined composition and color have been developed and distributed to allow for qualitative color characterization and comparison, while methods for quantitative color measurement have also been developed which nowadays are routinely applied [61,63,64].

White Gold Alloys

More difficult and sometimes controversially discussed in the literature is the issue of white gold jewelry alloys, which were introduced as lower-cost alternatives to platinum jewelry several decades ago [65]. Gold is almost completely bleached by very high Ag contents, so that low-karat (8/9 kt) white gold alloys usually are based on a high silver content. For white gold jewelry in higher karatages, Ni and Pd still are the most common alloying additions, despite the skin allergy problems related to the usage of Ni on the one hand [66] and the comparably high metal price of Pd on the other hand. This is related to their strong gold bleaching capacity and the comparably good working and casting properties of the corresponding alloys [45,46,65]. These alloys commonly are based on the Au-Ni-Ag-Zn or Au-Pd-Ag-Zn systems, probably with some Cu added to fine-tune workability and strength. The diagram shown in Figure 7.22 illustrates that Pd-based white gold usually is associated with a more grayish-white and warm color, whereas Ni-based white gold is recognized as cold and "steely"-white. Cu additions in both cases degrade the whiteness of the alloys. From the technical point of view, Ni-based white gold is more easy to cast, but alloys with high Ni-content suffer from too high hardness and low workability. Pd-based white gold is comparably soft and easy

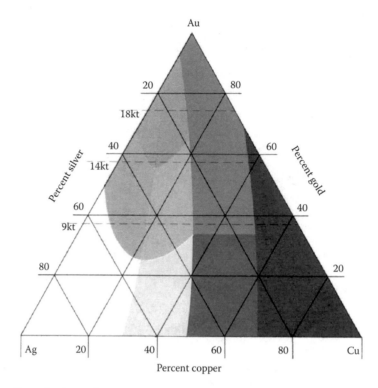

FIGURE 7.21 **(See color insert following page 212.)** Color triangle indicating color variations in the ternary Au-Cu-Ag system [33]. (From A plain man's guide to alloy phase diagrams: their use in jewellery manufacture – Part 2, M. Grimwade, *Gold Technology*, 2000, **30**, 8. With permission.)

to work, but is more difficult to cast due to the higher temperatures involved and higher reactivity with crucibles and mould materials.

Recently the awareness about Ni and its causation of skin allergies have been of concern, and legislation restricting the use of Ni in jewelry alloys has been enacted in Europe. This has focused research on alternatives to Ni and expensive Pd as primary bleachers of gold. A large spectrum of different white gold alloys, in part Ni- and Pd-free, based on additions of, for example, Co, Fe, Mn, Cr, and more recently Ga, has been proposed [65–75]. The actual commercial status of such alloys, however, is that so far none of these alloys has seen big market penetration, probably due to the often comparably poor white color and the degradation of working, casting, tarnishing, and corrosion as well as oxidation properties. As a consequence, the use of conventional alloys but with reduced Pd or Ni content still is most common, whereas only for the high-quality/high-price sector are alloys with a high Pd content in use, sometimes with nickel or platinum addition to even enhance the whiteness [70–73].

The availability of white gold alloys with strongly differing color recently has led to the definition of white gold color, based on the yellowness index [65]. The need for this was enhanced by the fact that most white gold jewelry nowadays is rhodium plated to give a special brilliant white appearance and to effectively hide an often poor, yellowish base material color. Color standards, defined in terms of the yellowness index, representing different white gold categories (premium, standard, off-white, and also nonwhite) and the boundaries between them have been prepared and distributed.

In the dental sector, white colored gold alloys with high Pd or Pt additions are common for metal-ceramic bonding applications. In this case, the resulting strong bleaching is a secondary effect, because Pt and Pd primarily are added to shift the melting temperatures toward higher

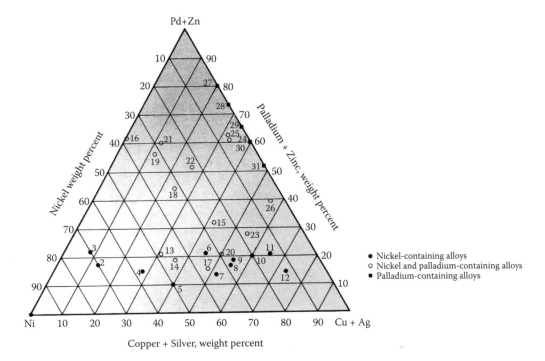

FIGURE 7.22 **(See color insert following page 212.)** Diagram indicating color variations for 18ct white Au depending on Ni, Pd+Zn, and Cu+Ag contents [46]. (From 18ct white gold jewellery alloys: Some aspects of their metallurgy, C. P. Susz and M. Linker, *Gold Bulletin*, 1980, **13(1)**, 15. With permission.)

temperatures, such that the ceramic firing process can be carried out at sufficiently high temperatures without degrading the metal part (see Chapter 14).

The color properties of Au-Pt– and Au-Pt-Pd–based dental white gold alloys have been investigated in detail with a particular focus on the influence of secondary base metal additions [76,77]. Although alloying with both Pt and Pd strongly decolorizes gold, the involvement of a small amount of base metals with a high valence electron concentration (such as In, Sn, Zn) to the parent alloy slightly recovers a gold tinge which, for these applications, is considered useful for the aesthetic appearance. The addition of 2 wt% Sn produces a small amount of a second phase which further increases a gold tinge, while the addition of 4 wt% Sn gives a very light tint of red to the alloy. These findings from the dental sector should also be useful in controlling color of white gold jewelry alloys, because base metal additions of Zn, Sn, In, and Ga in the range from 0.1 to 2–3 wt% are frequently used as secondary additions for improvement of investment casting properties (see Fluidity, Wetting, and Other Thermal Properties).

SPECIAL COLORS OF GOLD: BLACK, BLUE, AND PURPLE

Special colors can be obtained on gold or gold alloys by forming oxides or patinas on the surface by chemical or thermal treatments, or by coating the surface with thin layers of other materials as reviewed in detail [78]. In principle the complete color spectrum can be realized since some of the color effects result from optical interference depending on oxide or coating thickness. While the availability of a larger color spectrum than provided by conventional alloying is of great importance for the jewelry industry, a drawback of most of these coloring approaches is a poor wear resistance which may lead to a loss of the attractive color in short time. Furthermore, coloring by oxide formation requires use of special and difficult-to-process alloys containing comparably large amounts of Co, Mn, Fe, or Cr.

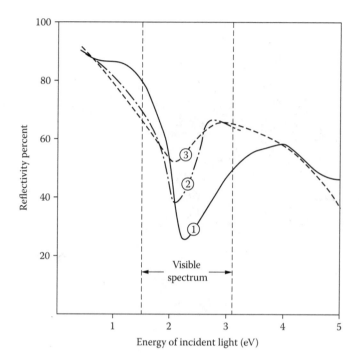

FIGURE 7.23 Reflectivity curves as a function of the energy of incident light for (1) AuAl$_2$, (2) AuIn$_2$, and (3) AuGa$_2$ [62]. (From the colour of gold and its alloys: The mechanism of variation in optical properties, K.E. Saeger and J. Rodies, *Gold Bulletin*, 1977, **10(1)**, 10. With permission.)

As opposed to coloring by surface engineering, some special and intrinsic colors like purple and blue are obtained if gold is alloyed with certain other metals at fixed, stoichiometric compositions. These colors are caused by the formation of intermetallic compounds with particular electron band structures, which make them absorptive for particular parts of the visible light range while staying reflective for the rest of the range, resulting in the special colors.

The three colored gold intermetallic compounds which have attracted the greatest attention are the purple or violet AuAl$_2$ (probably more strictly described as Au$_6$Al$_{11}$), and the blueish AuIn$_2$, and AuGa$_2$ [61,62,78]. The reflectivity curves display characteristic drops in the mid-visible part of the spectrum and strongest reflectivity toward the purple and blue ends of the spectrum (Figure 7.23) [62].

The most pronounced drop in reflectivity and therefore strongest and most attractive color is observed by the purple AuAl$_2$. Unfortunately, like most intermetallic compounds, it suffers from severe intrinsic brittleness which strongly restricts its use. The compound can be formed in the bonds of gold wires and aluminum pads in electronic device manufacture if subjected to temperatures in excess of 250°C, causing joint embrittlement and failure. The purple color of AuAl$_2$ is then usually visible on the fracture surface, so that this phenomenon is known as the "purple plague" in the electronics industry and needs to be avoided by diffusion barrier coatings [79]. While the compound is of some interest for a variety of specialty applications due to some further peculiar properties, as recently reviewed [80], purple gold or so-called amethyst gold has received much attention by the jewelry industry where it can be hallmarked as 18kt due to its gold content of ~78 wt%. In order to overcome the problems associated with the brittleness, a variety of alternative manufacturing routes and alloy additions have been suggested with some success to enable use of the colored compound in jewelry. As reviewed in Corti [78] and Supansomboona et al. [80] this involves application of powder metallurgy (for net-shape manufacturing or simply for grain refinement), composite technology, thermal diffusion treatments, or different surface technologies (thermal spraying, physical vapor deposition). Additions up to ~2–4 wt% of, among others, Pd, Ni, or Cu have been suggested to reduce the brittleness, presumably by formation of a network of ductile second phases in the microstructure. Additions of several weight percent of thorium and tin were claimed by an early patent to eliminate brittleness [81], but it seems that these findings have never been applied or confirmed by later work presumably due to the radioactive properties of thorium. More recently, reduction of brittleness of purple gold reportedly is obtained by a combination of rapid solidification (casting into copper moulds) and alloying with Si+Co (~ 2 wt% total) [82]. The resulting reduction of grain size from usually 50–200 μm down to 2–8 μm leads to ductile fracture behavior as observed on fracture surfaces. A quantification of the obtained ductility level is still lacking, however. For the same composition, slower cooling, as in investment casting, will still lead to coarser-grained microstructure and brittle-like fracture surfaces.

However, the drawback of most alloying additions is that fading of color also occurs by second phase formation [61,83–86]. For binary $AuAl_2$, the color depends strongly on stoichiometry and fades quickly for excess Au or Al additions. This is related to formation of whitish to grayish colored second phases in the microstructure, being another intermetallic AuAl or the aluminum solid solution, respectively. A purple hue is preserved for gold-rich binary Au-Al alloys down to Al contents of 15 wt%.

The quasi-binary section of the ternary Au-Al-Cu phase diagram at 76 wt% Au (Figure 7.24) and the related hardness and color trends were investigated in detail [84,85]. Two ternary intermetallic compounds were identified with approximate stoichiometries of $Au_7Cu_5Al_4$ (around 4 wt% Al, β-phase) and $Au_4(Cu_{0.4}Al_{0.6})_9$ (~7.8 wt% Al, γ'-phase). The corresponding colors are reported as yellow-to-apricot (orange-pink) and gray, respectively. Intermetallic alloys with compositions lying in the β-phase region (5–6 wt% Al) show a colorful glitter on polished surfaces which is the result of a facetting effect caused by quasi-martensitic transformations and related crystal structure

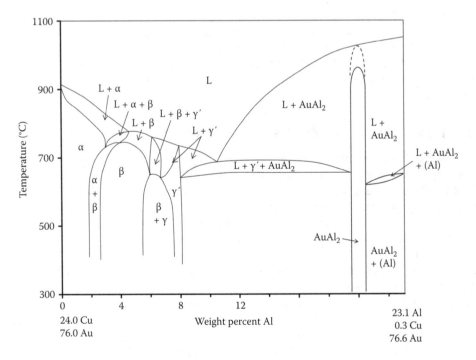

FIGURE 7.24 Pseudo-binary sections through the ternary Au-Cu-Al phase diagram for a constant gold content of 76 wt % (~18kt), [85]. (Modified from Determination of the 76 wt% Au section of the Al-Au-Cu phase diagram, F. C. Levey, M. B. Cortie, and L. A. Cornish, *Journal of Alloys and Compounds*, 2003, **354**, 171.)

changes [61,87,88]; see also Shape Memory Alloys later. A colorful sparkle caused by a comparable mechanism is also observed for binary 23kt AuAl-alloys [89].

For the blueish colored intermetallic compounds $AuIn_2$ and $AuGa_2$, in principle the same restrictions for jewelry applications exist as for $AuAl_2$. Less research into property improvement has been carried out for these compounds due to their much less pronounced color, which together with the intrinsic brittleness and a low hardness make them less attractive. However, electroplating in combination with thermal diffusion treatment as well as investment (bimetal) casting have been successfully applied for manufacturing (multicolored) prototype jewelry [86]. Microalloying additions have been identified that effectively reduce brittleness and increase mechanical shock resistance of $AuGa_2$ and, to somewhat less extent, $AuAl_2$. The particular underlying mechanism has not been studied in detail, but microalloying additions are well known to improve mechanical properties of different intermetallic compounds by a variety of possible mechanisms [86,90].

The three gold intermetallic compounds $AuAl_2$, $AuIn_2$, and $AuGa_2$ all crystallize in the cubic fluorite structure known as CaF_2. The quasi-binary sections $AuAl_2$-$AuIn_2$, $AuAl_2$-$AuGa_2$, and $AuIn_2$-$AuGa_2$ of the corresponding ternary systems were investigated and revealed some limited mutual immiscibilty [91]. However, alloys with compositions lying on the quasi-binary sections display two-phase microstructures composed of purple and bluish phases,. As a result $AuAl_2$ is bleached by Ga and In additions, while Al additions add a purple hue to $AuIn_2$ or $AuGa_2$ [86].

Some further colored intermetallic compounds of gold with alkali metals or light metals have been reported and found to be brittle, difficult to synthesize as single-phased material, and in part unstable in air or in humid atmosphere. Two-colored intermetallic compounds occur in the gold–potassium system, namely, the deep green (olive green) Au_5K and the purple (violet) Au_2K. Au_5Rb crystallizes in the same structure as Au_5K (hexagonal $CaCu_5$-type) and is also deep to olive green colored [92,93]. Single crystals of $Au_7K_4Ge_2$, which crystallizes in a substitution variant of the $MgCu_2$-type, are reported to be intrinsically black and lustrous [94]. Several ternary cubic Zintl phases are described as strikingly colored: Li_2AuSn, orange-reddish; Li_2AuGe, orange; Li_2AuPb, purple; and Li_2AuTl; greenish [95]; Li_2AuIn, greenish; and $Li_2Au_{0.75}In_{1.25}$, purple [96]. On the LiAu-LiIn quasi-binary section, color changes gradually from greenish over yellow, purple to gray-violet as a function of the valence electron concentration of the Zintl phases. The ordered face-centered cubic compound LiAuSb is red-purple; increasing Li gradually up to the formula Li_2AuSb produces gray-black colored compounds with a purple hue [97]. The compound MgAuSn is described as red-purple colored [98].

FLUIDITY, WETTING, AND OTHER THERMAL PROPERTIES

For casting alloys, the improvement of fluidity and wetting properties by appropriate alloying additions is essential in order to support formfilling. For jewelry and dental casting alloys, base metal additions of Zn, Sn, In, and Ga, usually in the range of 0.1 to 2–3 wt%, as well as Si and B, but in a much lower range, are routinely used for these purposes [10]. Notably, the effect of zinc additions was investigated in detail [99]. The surface tension of gold melts is reduced by factors of one-half, one-third, and one sixth by additions of 0.5, 1, and 2 wt% of Zn, respectively, which in turn leads to quantifiable increases in form-filling and reduction of surface roughness of as-cast items (Figure 7.25).

Zn, Si, and B also act as deoxidizers during alloy preparation as well as for obtaining a bright as-cast surface. Due to their affinity for oxygen, part of the added amount of these elements may accumulate in a slag which means a partial loss of these elements in the material, especially during remelting and recasting of scrap. Hence the correct dosage of these additions is complicated. The easy evaporation of zinc, caused by its low vapor pressure, adds to this problem.

All the additions listed above lower the liquidus temperatures and, usually more pronounced, also the solidus temperatures of alloys which effectively means an increase of the width of the melting range. Too much of one or more of the alloying additions therefore increases the risk for formation of hot tears during solidification, due to segregation of alloying elements and development of

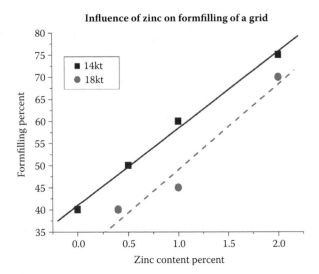

FIGURE 7.25 Relationship between formfilling and zinc content of 14kt and 18kt jewelry casting alloys [99]. (From Gold casting alloys: The effect of Zinc additions on their behaviour, Ch. Raub and D. Ott, *Gold Bulletin*, 1983, **16**, 2. With permission.)

low-melting phases. In severe cases, formation of low-melting compounds can lead to embrittlement during subsequent deformation. The same detrimental effect is also caused by impurities like P, S, Bi, and Pb, which can enter an alloy via the usage of low purity base metals or reusing of poorly cleaned scrap. However, P is also added to alloys deliberately as a deoxidizer (as an alternative to Si or B) during alloy preparation. In summary, the benefits of the alloying additions listed previously have been clearly established so that nowadays they are routinely used in casting alloys. Such alloys need to be particularly well prepared, ensuring a homogeneous distribution of the desired alloying additions in the right dosage adjusted for the particular application and need [10,72,100,101]. For these reasons, it is common practice to use carefully prepared master alloys, which usually also contain balanced amounts of grain refiners (see earlier discussion of grain refinement).

Few reliable and quantified results are available to date on the influence of additions like Zn and Si on further thermal properties important for casting alloys such as the thermal conductivities of molten, partially and completely solidified material, the heat of solidification, and the susceptibility to shrinkage porosity. There is a general lack of such data even for standard alloys without such additions, however, and only recently the determination of those thermal property data has started for the purpose of applying computer simulation to the jewelry investment casting process with the aim of casting defect prevention [102–106]. In the course of this work, it has been revealed that significant differences exist in thermal conductivities for 18ct yellow, red, and white gold alloys near to the respective liquidus temperatures, which together with differences in the heat of solidification and the melting range can explain the characteristic susceptibilities to shrinkage porosity of these standard alloys [105]. While small additions, for example, of Zn and Si supposedly do not significantly alter properties like thermal conductivities and heat of solidification, the marked influence of even small additions on the height and width of the melting range and the solidification morphology should be considered as relevant. Further basic research work is required, however, to improve the understanding of these complex interdependencies for the wide range of dental and jewelry gold casting alloys of different color and caratage.

The availability of high-fluidity Au-based casting alloys is not only of relevance for jewelry and dental applications. The extraordinary flow properties of a Au-based dental casting alloy (with Zn additions, among others) have recently been demonstrated in a study on investment casting of metallic microcomponents for functional applications. Microparts with an aspect ratio of 60 were attained in centrifugal casting even for a relatively low mold temperature of 700°C. Miniature dimensions of wall structures 20 µm wide and 120 µm high were also successfully cast. These results outperformed those of other, non-Au-based casting alloys, and it was concluded that especially Au-based casting alloys offer the best potential for further miniaturization of investment cast components [107].

In dental alloys for metal–ceramic applications, additions of Sn, In, and Zn also serve for formation of very thin surface oxide layers during the firing process, which is a prerequisite for a good bond between the ceramic and the metal surface [8,108]; see also Chapter 14.

Zn, Sn, In, and Ga nowadays are also common constituents of gold soldering and brazing alloys for jewelry and dental applications, while Cd was a main constituent in former times that had to be replaced due to its toxicity. While the characteristics of soldering and brazing alloys are explained in more detail in Chapters 8 and 9, it is worth noting here that, as with casting alloys, the reduction of surface tension and the increase of wettability, together with a lowering of the melting range, are the main reasons for adding these elements to soldering and brazing alloys, although usually in larger amounts. In addition, solder alloys with particularly low melting points have been developed that are not based on these additions but on eutectic compositions in the Au-Si, Au-Ge, and Au-Sn binary and ternary systems. These special solder alloys, which are used for jewelry as well as industrial applications, also form the basis for bulk metallic glass alloys (see the next section).

Thermodynamic modeling of multiple component alloys and their properties reportedly has been applied with success for the development of advanced low-melting point solders and bulk metallic glass alloys [109], as well as for age-hardenable alloys [59]. The possibility to predict important thermal alloy properties like melting ranges, heat of solidification, and prevailing phases in the solid state depending on temperature and compositions should be of high relevance also for the large range of conventional gold alloys, for example the development of improved alloys for casting, soldering or brazing or the optimization of heat treatment procedures for hardening and mechanical working.

MISCELLANEOUS

This final section reviews some special classes of gold alloys with unusual properties that have seen increasing interest in recent research and literature. It is noteworthy that all these speciality alloys have not seen significant application yet which, in part, may be due to a lack of (published) knowledge about further properties relevant for a particular application or the dependence of their outstanding properties on manufacturing conditions, (small) deviations from nominal compositions, and the like.

BULK METALLIC GLASS

The development of metallic materials that do not solidify as usual into a crystalline but an amorphous structure has gained a lot of attention in recent years. The alloy compositions of so-called bulk metallic glasses (BMGs) typically are close to a deep eutectic composition. Freezing-in the atomic arrangement of the molten material, hence avoidance of crystallization during solidification, usually requires fast cooling conditions (Figure 7.26). The glassy state is metastable at room temperature and the material crystallizes if heated above a certain temperature. The large interest in BMGs is based on their unique property spectrum which includes high strength and hardness, large elastic strain limit, low liquidus temperature, large supercooled regions, almost zero shrinkage during solidification, and good processibility for both casting and thermoplastic processing in a manner similar to thermoplastics [110].

It is worth noting here that the first alloy found to exhibit metallic glass formation was the binary Au-Si eutectic composition. It was argued that the abnormal behavior of such alloys is related to

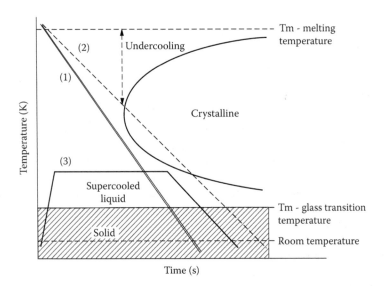

FIGURE 7.26 Schematic time-temperature-transformation diagram for processing of bulk metallic glass alloys; (1) Direct casting, where the liquid is cooled fast enough, in contrast to process (2), to avoid crystallization, (3) Superplastic forming by reheating amorphous "feedstock" material into the supercooled liquid region. [114]. (From Liquidmetal – Hard 18ct and 850 Pt alloys that can be processed like plastics or blown like glass, B. Lohwongwatana, J. Schroers and W. J. Johnson, in *Proceedings of The Sante Fe Symposium on Jewelry Manufacturing Technology 2007*, ed. Eddie Bell, Met-Chem Research, Albuquerque, NM, 2007, 289. With permission.)

their particular atomic structure in the molten state and that the eutectic composition is associated with an ideal ratio of Au and Si atoms stabilizing the liquid structure at the expense of the solid, which accounts for both the extremely low eutectic temperature and the glass-forming ability. The critical cooling rate for glass formation of the binary eutectics Au-Si and Au-Ge and the ternary eutectic Au-Si-Ge is in the order of 10^6 K/s, resulting in a critical casting thickness, d_c, in the range of only 50 μm [111,112]. Great efforts were made to increase the glass-forming ability of these and other candidate alloy systems, which has led to the development of numerous complex Pd-, Pt-, Mg-, Al-, or Zr-based glass-forming alloys [110]. With a view to gold, glass-forming alloys based on Au-Cu-Si were introduced [113]. The maximum casting thickness has been increased from 0.5 mm for the ternary alloy $Au_{55}Cu_{25}Si_{20}$, (all data in atomic percent), over 1 and 2 mm for the quaternary alloys $Au_{46}Ag_5Cu_{29}Si_{20}$ and $Au_{52}Pd_{2.3}Cu_{29.2}Si_{16.5}$, respectively, to 5 mm for the (thus far) best Au-based glassformer, $Au_{49}Ag_{5.5}Pd_{2.3}Cu_{26.9}Si_{16.3}$. In weight percent the Au content of the latter alloy is slightly above 75. Hence, usage for 18kt jewelry was suggested and identified as promising with a view to low casting temperatures, comparably high as-cast hardness (350 HV as opposed to 150–200 HV of most conventional 18kt alloys), improved scratch resistance, high quality surface finish, and brilliant metallic luster [113,114]. For the same alloy, a liquidus temperature of 371°C, a glass transition temperature of 131°C, and a supercooled liquid region of dT = 58 K is reported. The latter is the temperature range above the glass transition temperature, to which the material can be heated after quenching without it crystallizing but in which the amorphous phase first transforms into a highly viscous liquid before eventually crystallizing (Figure 7.26). In this temperature region, BMGs are amenable to superplastic processing using netshape processing methods similar to those employed for thermoplastics, which allows for easy manufacture of highly detailed microcomponents. As a consequence, potential applications apart from jewelry, namely, in the electronic, medical, and other industrial areas have also been suggested.

SHAPE MEMORY ALLOYS

The term "shape memory alloy" is used for a special class of materials that can be plastically severely distorted but will return to their original shape by heating them above a critical temperature. Another peculiarity of the same sort of materials is referred to as *superelasticity* or *pseudoelasticity*. This describes their property to exhibit tremendously large elastic strains if deformed above the critical temperature. Most SMA alloys are intermetallic compounds. Their unusual mechanical properties are caused by reversible, thermoelastic martensitic transformations, which can be activated by temperature modifications as well as stress increase. Once more it was a gold alloy, for which these properties were discovered first, namely, the binary intermetallic alloy AuCd (β-phase) [115]. While the brittleness of AuCd, even in single crystal form, restricted further work and exploitation, the discovery of the ductile intermetallic NiTi and its outstanding shape memory properties initiated further research on gold-based alloys. Consequently, potentially interesting shape memory properties were discovered for ternary Au-Cu-Zn alloys, around $Au_{25}Cu_{30}Zn_{45}$ (in at%) [116,117], and more recently for ternary Au-Ni-Ti alloys, namely, $Au_{13.8}Ni_{36.2}Ti_{50}$ and $Au_{27}Ni_{23}Ti_{50}$ (in at%), where the Ni in the well-researched NiTi-alloy was partially successfully replaced by Au [118]. For AuCuZn-based shape memory alloys, cold workability is still a problem, while hot workability is described as very good. Another problem for AuCuZn-based alloys is the large dependence of properties, especially the critical transformation or processing temperatures, on the zinc content, which is difficult to control during melt-processing of these alloys due to the high zinc content. The processing of the AuNiTi-based alloys is difficult as well, due to high reactivity of Ti with atmosphere and crucible materials. Potential applications that have been discussed for gold-based shape memory alloys range from the dental field over medical instruments or implants to jewelry components with special functional features like gripping of gemstones in mounts.

The martensitic phase transformations of shape-memory alloys are associated with crystallographic shape changes that can be made visible as a surface relief on finished surfaces, leading to a colorful glitter. This has led to the development of the intermetallic alloy family known as Spangold, with compositions around $Au_{43}Cu_{31}Al_{25}$ (in at%) [61,87,88]. The related colorful sparkle is an attractive property for jewelry or decorative applications and is also observed for the intermetallic, high-temperature phase of binary 23kt AuAl-alloys [89]. Little is known about the macroscopic shape memory properties of these alloys as compared to the aforementioned alloy systems [118].

METAL-MATRIX COMPOSITES

Reinforcement of metals with fibers or particles of ceramics, other nonmetallic compounds or other metals, for example, refractory metals that usually are nonsoluble in the host metal, has been a wide research field and has led to many new materials in case of, among others, light metal and Cu- or Fe-based alloys, with a potentially interesting property spectrum for a variety of different applications. With a view to precious metal–based materials, comparably little work related to Au has been carried out (or published) so far, while Ag-matrix [20,119,120] and more recently also Pt-matrix composite materials [121,122] have been developed and are used, in the case of Ag for decades, for functional applications. Ag and Pt are reinforced with oxide particles, which can be realized by internal oxidation or powder metallurgical routes, in order to improve their high-temperature properties, especially the arcing resistance of contact materials or electrode materials, respectively. In contrast, for Au-based contact materials alloying with Ag, Pt, Ni, Cu, Co, or Sn is applied [20,119,120]. Possibilities to incorporate particles into Au coatings by electrochemical means are reviewed in Chapter 11.

In contrast to conventional metal-matrix composites, Au-based metal–metal composites have been developed, in which both the metal and the reinforcing phase are ductile metals that in principle are soluble in each other. They are prepared by powder metallurgy techniques from pure metal powders and subsequently processed by extensive deformation (reductions in area greater than 99.99%), for example, into fine wire. The deformation reshapes the initially equi-axed powder into filaments that are few tenths or hundreds of a nanometer in diameter and millimeters in length. This class of materials is referred to as deformation processed metal-metal composites (DMMCs). Au-based composites with reinforcing phases of 7 vol% Ag, 14 vol% Ag, and 7 vol% Pt with an interesting combination of high strength and high conductivity have been developed and suggested for electronic or electrical applications involving elevated temperature service [123–126]. The strengthening in these composites appears to arise from the minor phase filaments acting as effective barriers to dislocation motion, while the high conductivity can be explained by electrons flowing parallel to the filamentary microstructure aligned with the wire axis. The composites were found to have good thermal stability compared to conventional cold-worked Au interconnection wires. However, these composites will revert to conventional solid solutions if exposed to high temperatures for prolonged times.

REFERENCES

1. *Phase Diagrams of Binary Gold Alloys*, eds. H. Okamoto and T. B. Massalski, ASM International, Metals Park, Ohio, 1987.
2. *Phase Diagrams of Ternary Gold Alloys*, A. Prince, G. V. Raynor, and D. S. Evans, The Institute of Metals, London, 1990.
3. *Ternary Alloy Systems, Subvolume B: Noble Metal Systems*, eds. G. Effenberg and S. Ilyenko, Landolt-Börnstein New Series Group IV Volume 11, Springer-Verlag, Berlin, Germany, 2006.
4. *Die Edelmetalle und ihre Legierungen*, E. Raub, Springer-Verlag, Berlin, 1946.
5. The alloying behaviour of gold, Part I: Solid solutions, G.V. Raynor, *Gold Bulletin*, 1976, **9(1)**, 12; Part II: Compound formation, *Gold Bulletin*, 1976, **9(2)**, 50.
6. *Gold Usage*, W. S. Rapson and T. Groenewald, Academic Press, London, 1978.
7. A survey of gold intermetallic chemistry, R. Ferro, A. Saccone, D. Macciò, and S. Delfino, *Gold Bulletin*, 2003, **36(2)**, 39.
8. Gold in dentistry: alloys, uses and performance, H. Knosp, R. J. Holliday, and C. W. Corti, *Gold Bulletin*, 2003, **36(3)**, 93.
9. Doped and low-alloyed gold bonding wires, Ch. Simons, L. Schräpler, and G. Herklotz, *Gold Bulletin*, 2000, **33(3)**, 89.
10. The effects of small additions and impurities on properties of carat golds, D. Ott, *Gold Technology*, 1997, **22**, 31; Optimising gold alloys for the manufacturing process, D. Ott, *Gold Technology*, 2002, **34**, 37.
11. Basic metallurgy of the precious metals, part II: Development of alloy microstructure through solidification and working, C. W. Corti, in *Proceedings of The Santa Fe Symposium on Jewelry Manufacturing Technology 2008*, ed. E. Bell, Met-Chem Research, Albuquerque, NM, 81.
12. Grain size of gold and gold alloys, D. Ott and Ch. J. Raub, *Gold Bulletin*, 1981, **14(2)**, 69.
13. Grain size in cast gold alloys, J. P. Nielsen and J. J. Tuccillo, *J. Dent. Res.*, 1966, **45**, 964.
14. Vibration technology for the solidification process of investment casting, P. Hofmann, in *Proceedings of the 3rd International Conference on Jewellery Production Technology* (JTF), Legor group Srl, Vicenza, Italy, 2006, 182.
15. A quantitative theory of grain-boundary motion and recrystallization in metals in the presence of impurities, K. Lücke and K. Detert, *Acta Metallurgica*, 1957, **5**, 628; and: Investigation of the dependance of the growth velocity on crystal orientation during the primary crystallisation of aluminium single crystals (in German), G. Masing, K. Lücke, and P. Nolting, *Zeitschrift fur Metallkunde*, 1956, **47**, 64.
16. The mobility and migration of boundaries, in *Recrystallization and Related Annealing Phenomena (Second Edition)*, F. J. Humphreys and M. Hatherly, Elsevier, New York, 2004, 121.
17. Mechanical properties and applications of precious metals at higher temperatures (in German), F. Aldinger and A. Bischoff, in *Festigkeit und Verformung bei hoher Temperatur*, ed. B. Ilschner, DGM Informationsgesellschaft Oberursel, Germany, 1983, 161.
18. Properties and applications of some gold alloys modified by rare earth additions, Y. Ning, *Gold Bulletin*, 2005, **38(1)**, 3.
19. Influence of small additions on the properties of gold and gold alloys (in German), D. Ott and Ch. J. Raub, Part I: *Metall*, 1980, **34**, 629; Part II: *Metall*, 1981, **35**, 543; Part III: *Metall*, 1981, **35**, 1005; Part IV: *Metall*, 1982, **36**, 150.
20. *Edelmetalltaschenbuch* (Precious Metal Pocketbook), ed. Degussa AG Frankfurt, Hüthig-Verlag Heidelberg, Germany, 2nd edition 1995 (in German).

21. Hardening of low-alloyed gold, J. Fischer-Bühner, *Gold Bulletin*, 2005, **38(3)**, 120–131.
22. Metallurgy of microalloyed 24 carat golds, C. W. Corti, *Gold Bulletin*, 1999, **32(2)**, 39 (and references therein); Strong 24 carat golds: the metallurgy of microalloying, C. W. Corti, *Gold Technology*, 2001, **33**, 27; Micro-alloying of 24 ct golds: Update, *Gold Technology*, 2002, **36**, 34.
23. The development of high-strength pure gold, A. Nishio, *Gold Technology*, 1996, **19**, 11.
24. Design opportunities through innovative materials, S. Takahashi, N. Uchiyama, and A. Nishio, *Gold Technology*, 1998, **23**, 12.
25. Jewelry manufacturing with the new high carat golds, J. E. Bernadin, *Gold Technology*, 2000, **30**, 17; Understanding microalloys, J. E. Bernadin, in *Proceedings of The Santa Fe Symposium on Jewelry Manufacturing Technology 2005*, ed. E. Bell, Met-Chem Research, Albuquerque, NM, 53.
26. Alloying and strengthening of gold via rare earth metal additions, Y. Ning, *Gold Bulletin*, 2001, **34(3)**, 77.
27. Microalloying of high carat gold, platinum and silver, C. W. Corti, in *Proceedings of the 2nd International Conference on Jewellery Production Technology* (JTF), Vicenza, Italy, 2005, 141 (and references therein).
28. Gold bonding wire – the development of low loop, long length characteristics, H. Lichtenberger, H. Grohman, G. Lovitz, and M. Zasowski, Proc. IMAPS Conference, San Diego, CA, 1998.
29. The effects of Ca and Pd dopants on gold bonding wire and gold rod, T. S. Sarawati, T. Sritharan, C.I. Pang, Y.H. Chew, C.D. Breach, F. Wulff F., S.G. Mhaisalkar, C.C. Wong, *Thin Solid Films*, 2004, **462–463**, 351, and: Effects of calcium and palladium on mechanical properties and stored energy of hard drawn gold bonding wire, Y. H. Chew, C. C. Wong, C. D. Breach, F. Wulff, S. G. Mhaisalkar, C. I. Pang and T.S. Saraswati, *Thin Solid Films*, 2004, **462–463**, 346.
30. New developments in wire bonding for future packaging, T. Müller, E. Milke, and E. Chung, Proc. IMAPS Conference, San Jose, CA, 2007.
31. Heat treatment of gold alloys: The mechanism of disorder order hardening, C. Chaston, *Gold Bulletin*, 1971, **4**, 70.
32. The metallurgy of some carat gold jewellery alloys Part I: Coloured gold alloys, A. S. McDonald and G. H. Sistare, *Gold Bulletin*, 1978, **11**, 66; Gold-silver-copper alloys, A. S. McDonald and G. H. Sistare, in *Metals Handbook*, American Society for Metals, Materials Park, OH, 1979, **2**, 681.
33. A plain man's guide to alloy phase diagrams: their use in jewellery manufacture – Part 2, M. Grimwade, *Gold Technology*, 2000, **30**, 8.
34. Heat treatment of 14 carat gold jewellery, C. P. Susz, M. Linker, O. Orosz, and D. Sapey, *Aurum*, 1982, **11**, 17.
35. Hardening of gold-based dental casting alloys: Influence of minor additions and thermal ageing, J. -J. Laberge, D. Tréheux, and P. Guiraldeng; *Gold Bulletin*, 1979, **12(2)**, 45.
36. Age-hardening characteristics of a commercial dental gold alloy, K. Yasuda and M. Ohta, *Journal of the Less-Common Metals*, 1980, **70(2)**, 75.
37. Age-hardening and related phase transformations in dental gold alloys, K. Yasuda, *Gold Bulletin*, 1987, **20(4)**, 90.
38. Age-hardening mechanisms in a commercial dental gold alloy containing platinum and palladium, T. Tani, K. Udoh, K. Yasuda, G. van Tendeloo, and J. van Landuyt, *Journal of Dental Research*, 1991, **70(10)**, 1350.
39. Structure and mechanical properties of CuAu and CuAuPd ordered alloys, A. Yu. Volkov, *Gold Bulletin*, 2004, **37(3–4)**, 208.
40. Precipitation hardening and ordering of carat gold jewellery alloys, W. S. Rapson, *Gold Bulletin*, 1978, **11(4)**, 116.
41. Ordering behaviors and age-hardening in experimental AuCu-Zn pseudobinary alloys for dental applications, S. H. Seol, T. Shiraishi, Y. Tanaka, E. Miura, K. Hisatsune, and H. I. Kim, *Biomaterials*, 2002, **23(24)**, 4873–9.
42. High resolution transmission electron microscopy of age-hardenable Au-Cu-Zn alloys for dental applications, S. H. Seol, T. Shiraishi, Y. Tanaka, E. Miura, and K. Hisatsune, *Biomaterials*, 2003, **24(12)**, 2061–6.
43. 18 Carat yellow gold alloys with increased hardness, R. Süss, E. van der Lingen, L. Gardner, and M. du Toit, *Gold Bulletin*, 2004, **37(3–4)**, 196.
44. Gold in Watchmaking, L. F. Trueb, *Gold Bulletin*, 2000, **33(1)**, 11.
45. The metallurgy of some carat gold jewellery alloys Part II: Nickel-containing white gold alloys, A. S. McDonald and G. H. Sistare, *Gold Bulletin*, 1978, **11(4)**, 128.
46. 18ct white gold jewellery alloys: Some aspects of their metallurgy, C. P. Susz and M. Linker, *Gold Bulletin*, 1980, **13(1)**, 15.

47. *Palladium Recovery, Properties and Uses*, E. M. Wise, Academic Press, New York, 1968.
48. Two different types of age-hardening behaviors in commercial dental gold alloys, K. Hisatsune, T. Shiraishi, Y. Takuma, Y. Tanaka, and R. H. Luciano, *Journal of Materials Science: Materials in Medicine*, 2007, **18/4**, 577.
49. The development of 990 gold-titanium: Its production, use and properties, G. Gafner, *Gold Bulletin*, 1989, **22**, 112.
50. Investment casting of gold-titanium alloys, D. Ott, in *Proceedings of The Santa Fe Symposium on Jewelry Manufacturing Technology 1989*, ed. D. Schneller, Met-Chem Research, Albuquerque, NM, 31.
51. *Gold Technology*, 1992, **6**. The complete issue is devoted to 990 Gold.
52. *The Au-Ti – Alloy Esteticor Vision: Materials Basics and Processing in Dental Technology* (in German), J. Fischer, B. Dörfler, and R. Mericske-Stern, Quintessenz-Verlag GmbH, Berlin, 1998.
53. Stable strengthening of 990 gold, D. M. Jacobson, M. R. Harrison, and S. P. S. Sangha, *Gold Bulletin*, 1996, **29(3)**, 95.
54. The development of a novel gold alloy with 995 fineness and increased hardness, M. du Toit, E. van der Lingen, L. Glaner, and R. Süss, *Gold Bulletin*, 2002, **35(2)**, 46.
55. Dispersion hardened gold: A new material of improved strength at high temperatures, M. Poniatowski and M. Clasing, *Gold Bulletin*, 1972, **5**, 34.
56. Hardening of high carat gold alloys, J. Fischer-Bühner, in *Proceedings of The Santa Fe Symposium on Jewelry Manufacturing Technology 2004*, ed. E. Bell, Met-Chem Research, Albuquerque, NM, 2004, 151.
57. Hard gold alloys, R. Süss, E. van der Lingen, M. du Toit, C. Cretu, and L. Glaner, in *Proceedings of Gold 2003: New industrial applications for gold*, Vancouver, Canada, 2003; Hard 22ct gold alloy, C. Cretu, E. van der Lingen, and L. Glaner, *Gold Technology*, 2000, **29**, 25.
58. Leghe d'oro 22 carati induribili; D. Maggian et al., Workshop presentation at Vicenca Jewellery Fair, Summer 2004.
59. Thermodynamic modelling of precious metals alloys, B. Kempf and S. Schmauder, *Gold Bulletin*, 1998, **31(2)**, 51.
60. Dispersion strengthened gold: Improved properties at high temperatures, J. S. Hill, *Gold Bulletin*, 1976, **9**, 76.
61. Coloured gold alloys, C. Cretu and E. van der Lingen, *Gold Bulletin*, 1999, **32(4)**, 115; *Gold Technology*, 2000, **30**, 31.
62. The colour of gold and its alloys: The mechanism of variation in optical properties, K. E. Saeger and J. Rodies, *Gold Bulletin*, 1977, **10(1)**, 10.
63. The colour characteristics of gold alloys, E. F. I. Roberts and K. M. Clarke, *Gold Bulletin*, 1979, **12**, 9.
64. The colour of Au-Ag-Cu alloys: Quantitative mapping of the ternary diagram, R. M. German, M. M. Guzowski, and D. C. Wright, *Gold Bulletin*, 1980, **13**, 113.
65. What is a white gold ? Progress on the issues, C. W. Corti, in *Proceedings of The Santa Fe Symposium on Jewelry Manufacturing Technology 2005*, ed. E. Bell, Met-Chem Research, Albuquerque, NM, 2005, 103.
66. Don't let nickel get under your skin – the European experience!, R. Rushforth, *Gold Technology*, 2000, **28**, 2.
67. Improvement of 18 carat white gold alloys, G. P. O. Connor, *Gold Bulletin*, 1978, **11(2)**, 35.
68. White golds: A review of commercial material characteristics & alloy design alternatives, G. Normandeau, *Gold Bulletin*, 1992, **25(3)**, 94.
69. White golds: A question of compromises, G. Normandeau and R. Roeterink, *Gold Bulletin*, 1994, **27(3)**, 70.
70. Production and characterisation of 18 carat white gold alloys conforming to European Directive 94/27 CE, M. Dabala, M. Magrini, M. Poliero, and R. Galvani, *Gold Technology*, 1999, **25**, 29.
71. White golds – meeting the demands of international legislation, P. Rotherham, *Gold Technology*, 1999, **27**, 34.
72. White gold alloys for investment casting, M. Poliero, *Gold Technology*, 2001, **31**, 10.
73. On Nickel white gold alloys: Problems and possibilities, V. Faccenda, in *Proceedings of The Santa Fe Symposium on Jewelry Manufacturing Technology 2000*, ed. E. Bell, Met-Chem Research, Albuquerque, NM, 2000, 71.
74. Development of new Ni-free, Cr-based white gold alloys – results of a research project, J. Fischer-Buehner and D. Ott, in *Proceedings of The Santa Fe Symposium on Jewelry Manufacturing Technology 2001*, ed. E. Bell, Met-Chem Research, Albuquerque, NM, 2001, 19.

75. Development of 18ct white gold alloys without Ni and Pd, A. Basso, J. Fischer-Buehner, and M. Poliero, in *Proceedings of The Santa Fe Symposium on Jewelry Manufacturing Technology 2008*, ed. E. Bell, Met-Chem Research, Albuquerque, NM, 2008, 31.

76. Optical properties of Au-Pt and Au-Pt-In Alloys, T. Shiraishi, K. Hisatsune, Y. Tanaka, E. Miura, and Y.Takuma, *Gold Bulletin,* 2001, **34(4)**, 129.

77. Optical properties of Au-Pt-Pd-based high noble dental alloys, T. Shiraishi and J. Geis-Gerstorfer, *Gold Bulletin,* 2006, **39(1)**, 9.

78. Blue black and purple! The special colours of gold, C. W. Corti, in *Proceedings of The Santa Fe Symposium on Jewelry Manufacturing Technology 2004*, ed. E. Bell, Met-Chem Research, Albuquerque, NM, 2004, 121; Blue, black, brown and purple: the unusual colours of gold, C. W. Corti, updated version presented at the International Jewellery Symposium, St. Petersburg, Russia, July 3–7, 2006, published in conference proceedings in Russian.

79. Intermetallic formation in gold-aluminum systems, E. Philosky, *Solid State Electronics*, 1970, **13**, 1391.

80. "Purple glory": The optical properties and technology of AuAl2 coatings, S. Supansomboona, A. Maaroof, and M. B. Cortie, *Gold Bulletin*, 2008, **41(4)**, 296.

81. Colored gold-aluminium alloy (in German), L. Weiss and G. Buchenauer, German Patent DRP 710934, 1939.

82. Microstructure of Au-Al systems manipulated by rapid solidification techniques, K. Wongpreedee and P. Ruethaithananon, paper to be presented at *Gold 2009, 5th International Conference on Gold Science, Technology and its Applications*, Heidelberg, 2009.

83. Color technology for jewelry applications, D. P. Agarwal and G. Raykhtsaum, *Proceedings of The Santa Fe Symposium on Jewelry Manufacturing Technology 1988*, ed. D. Schneller, Met-Chem Research, Albuquerque, NM, 1988, 229.

84. Hardness and colour trends along the 76wt% Au (18.2 carat) line of the Au-Cu-Al system, F. C. Levey, M. B. Cortie, and L. A. Cornish, *Scripta Materialia*, 2002, **47**, 95.

85. Determination of the 76 wt% Au section of the Al-Au-Cu phase diagram, F. C. Levey, M. B. Cortie, and L. A. Cornish, *Journal of Alloys and Compounds*, 2003, **354**, 171.

86. J. Fischer-Buehner et al; Ulrich Klotz et al; papers to be published in Proceedings of *The Santa Fe Symposium on Jewelry Manufacturing Technology 2009*, ed. E. Bell, Met-Chem Research, Albuquerque, US, 2009.

87. The development of Spangold, I. M. Wolff and M. B. Cortie, *Gold Bulletin*, 1994, **44(2)**, 141.

88. Spangold - a jewellery alloy with an innovative surface finish, M. Cortie, I. Wolff, F. Levey, S. Taylor, R. Watt, R. Pretorius, T. Biggs, and J. Hurly, *Gold Technology*, 1994, **14(4)**, 30.

89. A 23 Carat Alloy with a Colourful Sparkle, F. C. Levey, M. B. Cortie, and L. A.Cornish, *Gold Bulletin*, 1998, **31(3)**, 75.

90. Intermetallic compounds: Principles and Practice, ed. J.H. Westbrook and R.L. Fleischer, John Wiley and Sons Ltd., Chichester, UK, 1994.

91. The AuAl$_2$-AuGa$_2$-AuIn$_2$ problem: Knight shifts and relaxation times in their pseudobinary alloys, G. C. Carter, I.D. Weisman, L. H. Bennett, and R. E. Watson, *Phys. Rev. B*, 1972, **5(9)**, 3621.

92. On the compounds of Sodium and Potassium with Gold (in German), U. Quadt, F. Weibke and W. Blitz, *Zeitschrift für anorganische und allgemeine Chemie*, 1937, **232**, 298.

93. On the crystal structures of the gold-richest phases in the systems Potassium-Gold and Rubidium-Gold (in German), Ch. J. Raub and V. B. Compton, *Zeitschrift für anorganische und allgemeine Chemie,* 1964, **332**, 5.

94. K$_4$Au$_7$Ge$_2$: a framework structure with Au$_7$-double-tetrahedra and Ge$_2$-dumb-bells (in German), U. Zachwieja, *Zeitschrift für anorganische und allgemeine Chemie*, 1995, **621 (6)**, 975.

95. The crystal structure of the ternary intermetallic phases Li$_2$EX (E=Cu, Ag, Au; X=Al, Ga, In, Tl, Si, Ge, Sn, Pb, Sb, Bi) (in German), H. Pauly, A. Weiss, and H. Witte, *Zeitschrift fur Metallkunde*, 1968, **59(1)**, 47.

96. Phase width and valence electron concentration of the ternary cubic Zintl phases of the NaTl type (in German), H. Pauly, A. Weiss, and H. Witte, *Zeitschrift fur Metallkunde*, 1968, **59(7)**, 554.

97. A new ternary phase in the system Li-Au-Sb (in German), H.U. Schuster and W. Dietsch, *Zeitschrift fur Naturforschung,* 1975, **30b**, 133.

98. Coloured ternary and quaternary Zintl-phases (in German), U. Eberz, W. Seelentag, and H. U. Schuster, *Z. Naturforschung*, 1980, **35b**, 1341.

99. Gold casting alloys: The effect of Zinc additions on their behaviour, Ch. Raub and D. Ott, *Gold Bulletin*, 1983, **16**, 2.

100. Colored carat gold for investment casting, D. Zito, *Gold Technology*, 2001, **31**, 35.

101. The effect of various additions on the performance of an 18ct carat yellow gold casting alloy, G. Normandeau, *Proceedings of The Santa Fe Symposium on Jewelry Manufacturing Technology 1996*, ed. E. Bell, Met-Chem Research, Albuquerque, NM, 1996, 83.

102. Properties of melt and thermal processes during solidification in jewelry casting, D. Ott, in *Proceedings of The Santa Fe Symposium on Jewelry Manufacturing Technology 1999*, ed. E. Bell, Met-Chem Research, Albuquerque, NM, 1999, 487.

103. Computer simulation of jewelry investment casting: What can we expect ?, J. Fischer-Bühner, in *Proceedings of The Santa Fe Symposium on Jewelry Manufacturing Technology 2006*, ed. E. Bell, Met-Chem Research, Albuquerque, NM, 2006, 193.

104. Numerical simulation of the investment casting process: experimental verification, M. A. Grande et al., in *Proceedings of the 2nd International Conference on Jewelry Production Technology*, JTF Vicenca, Italy, 2005, 93.

105. Advances in the prevention of investment casting defects assisted by computer simulation J. Fischer-Bühner, in *Proceedings of The Santa Fe Symposium on Jewelry Manufacturing Technology 2007*, ed. E. Bell, Met-Chem Research, Albuquerque, NM, 2007, 149.

106. Computer simulation of the investment casting process: Widening of the filling step, M. A. Grande, in *Proceedings of The Santa Fe Symposium on Jewelry Manufacturing Technology 2007*, ed. E. Bell, Met-Chem Research, Albuquerque, NM, 2007, 1.

107. Microcasting, G. Baumeister, J. Haußelt, S. Rath, and R. Ruprecht, in *Advanced Micro and Nanosystems Vol. 4. Microengineering of Metals and Ceramics*, ed. H. Baltes, O. Brand, G.K. Fedder, C. Hierold, J. Korvink, and O. Tabata, Wiley-VCH, Weinheim, Germany, 2005, 357.

108. The bonding of gold and gold alloys to non-metallic materials, W. S. Rapson, *Gold Bulletin*, 1979, **12(3)**, 108.

109. Alloys by Design: Knowing the answer before spending the money, B. Lohwongwatana and E. Nisaratanaporn, in *Proceedings of The Sante Fe Symposium on Jewelry Manufacturing Technology 2008*, ed. E. Bell, Met-Chem Research, Albuquerque, NM, 2008, 201.

110. Processing routes, microstructure and mechanical properties of metallic glass and their composites, J. Eckert, J. Das, S. Pauly, and C. Duhamel, *Advanced Engineering Materials*, 2007, **9**, 443.

111. W. Klement, R. H. Willens, and P. Duwez, *Nature*, 1960, **187**, 869.

112. The structure of gold alloys in the liquid state: Novel quench products from some eutectic systems, V. G. Rivlin, R. M. Waghorne, and G.I. Williams, *Gold Bulletin*, 1976, **9(3)**, 84.

113. Gold based metallic glass, J. Schroers, B. Lohwongwatana, W. J. Johnson, and A. Peker, *Applied Physics Letters*, 2005, **87**, 061912.

114. Liquidmetal – Hard 18ct and 850 Pt alloys that can be processed like plastics or blown like glass, B. Lohwongwatana, J. Schroers, and W. J. Johnson, in *Proceedings of The Sante Fe Symposium on Jewelry Manufacturing Technology 2007*, ed. E. Bell, Met-Chem Research, Albuquerque, NM, 2007, 289.

115. Plastic deformation and diffusionless phase changes in metals – The Au-Cd Beta phase, L. C. Chang and T.A. Read, *Transactions AIME, Journal of Metals*, 1951, **189**, 47.

116. Gold alloys with Shape Memory, G. B. Brook, *Gold Bulletin*, 1973, **6(1)**, 8.

117. Gold alloys with Shape Memory, G. B. Brook and R. F. Iles, *Gold Bulletin*, 1975, **8(1)**, 16.

118. Gold with a martensitic transformation: Which opportunities?, S. Besseghini1, F. Passaretti, E. Villa, P. Fabbro, and F. Ricciardi, *Gold Bulletin*, 2007, **40(4)**, 328.

119. *Elektrische Kontakte, Werkstoffe und Anwendungen*, E. Vinaricky, Springer Verlag, Berlin, 2002.

120. *Electrical Contacts – Principles and Applications*, ed. P. G. Slade, M. Dekker, New York, 1999.

121. Platinum ceramic composites as new electrode materials: fabrication, sintering, microstructure and properties, J. Rager, A. Nagel, M. Schwenger, A. Flaig, G. Schneider, and F. Mücklich, *Advanced Engineering Materials*, 2006, **8**, 81.

122. Manufacture of wire from platinum/metal oxide composites, T. Eckhardt, H. Manhardt, and D. F. Lupton, *Materialwissenschaft und Werkstofftechnik*, 2008, **12**, 933.

123. A new method for strengthening gold, A. M. Russell, K. Xu, S. Chumbley, J. Parks, and J. Harringa, *Gold Bulletin*, 1998, **31(3)**, 88.

124. Advances in deformation processed gold composites, V. Gantovnik, A.M. Russell, S. Chumbley, K. Wongpreedee, and D. Field, *Gold Bulletin*, 2000, **33(4)**, 128.

125. Kinetic transformation of nanofilamentary Au metal-metal composites, K. Wongpreedee and A. M. Russell, *Gold Bulletin*, 2004, **37(3–4)**, 174.

126. The stability of Pt nanofilaments in a Au – matrix composite, K. Wongpreedee and A. M. Russell, *Gold Bulletin*, 2007, **40(3)**, 199.

8 Gold in Metal Joining

David M. Jacobson and Giles Humpston

CONTENTS

Introduction.. 161
Solid-State Joining.. 162
Gold in Soldering and Brazing 163
    Gold as a Solderable or Brazeable Metallization.. 163
    Gold-bearing Soldering Alloys 166
        Gold–Tin.. 168
        Gold–Silicon and Gold–Germanium 171
        Gold–Indium 172
        Low Melting Point –High Karat Gold Solders—Options and Limitations......... 173
        22 Karat Gold Solders.. 175
        Diffusion Soldering of Karat Gold... 177
    Gold–based Brazing Alloys... 178
        Karat Gold Brazing Alloys.. 179
        Gold–bearing Brazing Alloys for Industrial Applications................................. 184
Acknowledgments... 188
References.. 188

INTRODUCTION

The three joining processes that will be described here are soldering, brazing, and solid-state diffusion bonding. Soldering and brazing both involve using a filler metal that is heated above its melting point, made to wet the mating surfaces of a joint, with or without the aid of a chemical fluxing agent, leading to the formation of metallurgical bonds between the filler and the respective components. By convention, the joining process is defined as soldering if the filler metal melts below 450°C and as brazing if it melts above this temperature. In both soldering and brazing, it is uncommon for the original surfaces of the components to be eroded by reaction with the filler beyond the microscopic level (<100 μm). Solid-state diffusion bonding involves placing surfaces of two components in contact under a loading that at the least is provided by the weight of the upper component and heating the assembly until the voids at the interface have been removed by diffusion (see next section).

There is another important type of joining process—welding—that involves the fusion of the touching joint surfaces by controlled melting by heat being specifically directed toward the joint. Welding, using a laser or a microplasma torch, is being successfully applied in chain making (see under Karat Gold Brazing Alloys later).

Certain properties of gold are used to advantage in metal joining. In particular,

- Gold does not oxidize when heated in air, neither does it tarnish. This metal possesses excellent corrosion resistance, and in alloys it confers resistance to oxidation.
- Gold forms eutectic alloys with other metals covering a wide range of temperatures, encompassing soldering and brazing temperature regimes.

- Gold is relatively easy to apply as a surface coating by both physical vapor deposition and chemical plating.
- Many gold alloys possess enhanced mechanical properties, especially at elevated temperatures.
- Gold's low elastic modulus and rapid self-diffusion are beneficial for solid-state diffusion bonding.

Gold is unusual in that it is the only element on which both brazes and solders are based. However, because gold is the most expensive constituent of solders and brazes, the use of gold filler alloys tends to be limited to high value applications, such as electronics, photonics, and jewelry manufacture.

SOLID-STATE JOINING

Among metals, gold is one of the most amenable to solid-state diffusion bonding, without requiring a filler metal to melt during the joining operation. Indeed, sound joints can be produced by heating gold joints to temperatures well below the melting point of this metal. Solid-state joining of gold is not new, and examples of gold-based artifacts fabricated using pressure welds have been dated to between 1400 and 1000 BCE [1]. A gold ribbon used as a torch or neck ornament, found in a Celtic grave, was joined by the same method. Other ancient examples of solid-state joining of gold from Egypt and the Black Sea region are cited by Tylecote [2].

Three special characteristics of gold are beneficial for diffusion bonding, namely, the absence of an oxide skin when heated in air which might act as a barrier, the metal's low elastic modulus, and its rapid self-diffusion. As a consequence, diffusion bonding of gold can be achieved at room temperature with plastic deformations of as little as 20%. These favorable properties are widely exploited for gold wire interconnection in microelectronics assembly.

The temperature/pressure curve for a process time of 1 h has been mapped out by Humpston and Baker [3]; see Figure 8.1. Extrapolating this curve indicates that a successful gold–gold bond can be achieved without an applied pressure above about 450°C. This is confirmed by the fact that rods of pure gold placed in contact bond together readily at 500°C.

An extreme form of solid-state bonding under pressure, explosive joining (also referred to as explosive welding), in which a controlled detonation is used to force materials together under a rapid impulse, has been successfully used to bond gold to stainless steel and titanium [4]. However, when the process was applied to bonding gold to aluminum, the formation of the brittle Al_2Au phase

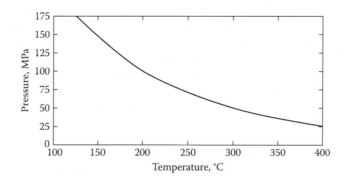

FIGURE 8.1 Temperature/pressure curve for diffusion bonding of gold, for a process time of 1 h. The line on the graph differentiates between joints of acceptable (above) and unsatisfactory (below) tensile joint strength after fabrication.

could not be prevented, which results in a weak interfacial zone. However, a buffer layer of silver between gold and aluminum was found to provide an effective "work around."

Benefits of solid-state joining of gold include:

- The joining operation may be carried out well below the melting point of gold.
- The service temperature of joined assemblies can be higher than the joining temperature without remelting the joints.
- Melting of the components either does not occur or is very slight and highly localized, so that changes to their microstructure are minor.
- The properties of solid-state joints can approach those of the parent materials.
- Because there is very little alloying between components joined in these processes, formation of brittle intermetallic compounds is minimized, if not eliminated, so that a wider range of materials are amenable to joining by nonfusion welding than by fusion welding.
- The joints have no fillets, which can be both an advantage (e.g., cosmetic/aesthetic) and a disadvantage (fillets generally boost the mechanical properties of joints).

On the negative side,

- Solid-state joining is limited in application to specific combinations of materials that provide specific combinations of mechanical or diffusion characteristics.
- Of all the joining methods, solid-state joining is the least tolerant to poor mating of the joint surfaces.
- Joint surfaces need to be scrupulously clean because solid-state joining is a fluxless process.

GOLD IN SOLDERING AND BRAZING

Gold is widely used in soldering and brazing as both a constituent of joining alloys and a tarnish-free finish that reliably maintains its solderability or brazeability.

GOLD AS A SOLDERABLE OR BRAZEABLE METALLIZATION

Gold and some of the platinum group metals do not oxidize when heated in air and can therefore be used as tarnish-free finishes. Although silver will oxidize at air ambient temperature, the oxide dissociates on heating to about 190°C. However, silver is rarely used as a finish on parts to be soldered, especially with ceramics and other insulating materials, because it is prone to electromigration. This is an electrochemical process in which silver is displaced under an applied electric field and in the presence of moisture, which can result in bridging of insulation gaps and lead to a short circuit. Platinum is not often used as a finish for soldering—being twice as expensive as gold—nor is palladium, because it is not as readily wetted as gold by solders, which is outstanding in this respect.

The combination of its chemical inertness, excellent solder wetting properties, and ease of inspection makes gold the natural choice for thin solderable metallizations applied to components for fluxless soldering. It is generally used in combination with an underlayer of another metal to act as a barrier to reaction with the material of the component, which may result in brittle interfaces or dewetting of the solder. Nickel is commonly chosen as the underlayer on metal surfaces, and a reactive metal such as chromium or titanium, applied by sputter deposition, is generally applied to nonmetallic substrates in order to establish good adhesion of the metal. As an example, a sputter-coated film of chromium 0.1 μm thick and overlaid with 0.1 μm of gold has been successfully used for soldering to glass components with smooth surfaces. The chromium represents a metal that is essentially insoluble in most solders while, at the same time, it forms a strong reactive bond to inorganic materials.

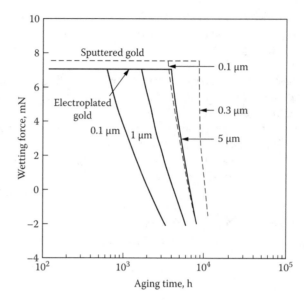

FIGURE 8.2 Solderability shelf-life of gold-coated components. Thicker and denser coatings are more impervious to oxygen and water vapor and therefore confer greater protection to the underlying metal.

The gold layer not only provides a solderable surface but also protects the reactive metal underlayer against oxidation [5]. It needs to be substantially pore free to fully protect the reactive metal underlayer against oxidation over its specified solderability life. Considerably thicker gold layers are needed if they are applied by electroplating rather than by sputtering, especially if the substrate surface is fairly rough. In a situation where a 0.3 μm-thick gold layer deposited by sputtering can be relied on to maintain excellent solderability for a few months, a layer of the same thickness but applied by electroplating may not offer adequate protection to an underlying base metal from atmospheric oxidation for more than a few days, because oxygen will be able to percolate through pores in the gold. This point is illustrated in Figure 8.2, which shows the wettability, in terms of the wetting force measured on a wetting balance, of chromium-metallized coupons coated with different thicknesses of gold by sputtering and electroplating, as a function of the storage time in an open atmosphere.

Gold metallizations are seldom more than a few microns thick, largely on account of the cost of this metal. However, this is not the only reason for limiting the gold layer thickness where tin-based solders, including silver–tin– (copper) and lead–tin are used. The high solubility of tin in gold means that even thick metallizations can completely dissolve in molten solder during the heating cycle. Hard $AuSn_4$ intermetallic phase will form on solidification and, if this phase becomes dominant in the joint, it will be brittle. The embrittlement caused by the $AuSn_4$ intermetallic is a consequence of its intrinsically low fracture toughness and the weak interface between them and the lead phase in the lead–tin eutectic microstructure [6]. To avoid this situation, strict limits are placed on the thickness of the gold layer. For lead–tin solder, $AuSn_4$ becomes the primary phase (the first solid phase to form on solidification and which consequently adopts a massive form) if the concentration of gold exceeds 8% by weight. A maximum gold concentration of 4% is usually taken as a safe working limit in industry. As can be seen from Figure 8.3 [7–9], high lead- and high tin content lead–tin solders can accommodate slightly higher levels of gold before $AuSn_4$ constitutes the primary phase and joint embrittlement occurs. When calculating a safe thickness of gold surface coatings applied to joint surfaces, their surface roughness must be taken into account, because the critical thickness of the gold coating will be reduced by a factor related to the surface roughness.

Low concentrations of the $AuSn_4$ phase are beneficial because they then enhance the mechanical properties of tin-containing solders [10]. At these low levels, the $AuSn_4$ intermetallic phase generated from dissolution of gold by the solders boosts strength hardness, because this phase possesses a

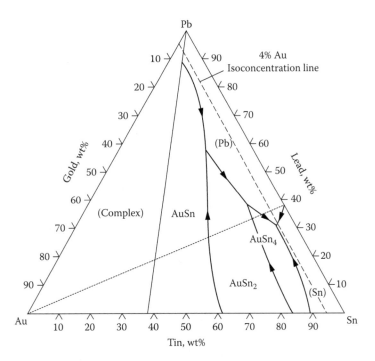

FIGURE 8.3 Liquidus projection of the gold–lead–tin ternary system. The first phase to form on solidification is labeled for each phase field. The 4% gold iso-concentration line is marked on the figure as is the tie line between lead-tin eutectic solder and pure gold. (After Thermal analysis of the AuSn-Pb quasibinary section, G. Humpston and B. L. Davies, *Metal Science*, 1988, **18**, 329–331; and Constitution of the AuSn Pb Sn partial ternary system, G. Humpston and B. L. Davies, *Materials Science and Technology*, 1985, **19**, 433–441; and Constitution of the Au AuSn Pb partial ternary system, G. Humpston and D. S. Evans, *Materials Science and Technology*, 1987, **3**, 621–627. With permission.)

high elastic modulus (i.e., it is much stiffer than the solders). This boost to strength and hardness can be seen, in relation to the Ag-96Sn eutectic solder, in Figure 8.4. From a concentration of about 3.5% to 8% gold, the solder microstructure is characterized by a fine dispersion of the AuSn$_4$ phase, present as a secondary phase. However, when the level of gold in the solder rises toward 10%, there is a sudden change in properties and ingots of the solder become completely unworkable. This change corresponds to the appearance of AuSn$_4$ as the primary phase. It is noteworthy that although the Ag-96Sn alloy contains more tin than lead–tin eutectic solder, it is tolerant to approximately twice the volume fraction of gold before the alloy is embrittled by AuSn$_4$. The critical levels of gold that give rise to AuSn$_4$ as the primary phase are listed in Table 8.1 for a selection of solders.

This restriction on gold coatings does not apply to high-gold and indium-based solders, because they do not react with the gold to form catastrophic embrittling phases. Indium-bearing solders are especially tolerant to gold metallizations because they rapidly form interfacial phases with the gold layer which greatly attenuate further reaction and dissolution. What is more, the interfacial phases that form are comparatively ductile, so that joints are not embrittled by their presence. The low level of gold erosion stems from a combination of the steep slope of the liquidus line on the phase diagram between indium and gold (see Figure 8.5) and the formation of a thin, continuous intermetallic compound AuIn$_2$ between the molten solder and the gold metallization. This layer of compound then acts as an effective barrier against significant further gold dissolution from taking place and results in the profile of the erosion curves shown in Figure 8.6. Thus, indium containing solders can be reliably used in conjunction with very thin gold metallizations as well as thicker ones.

An electroplated layer of gold applied to titanium renders it brazeable with low melting point brazes, without flux. As applied, the gold layer is not adherent and will readily flake off. However, after heat

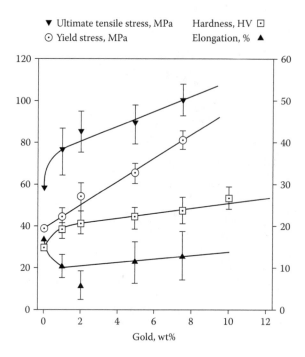

FIGURE 8.4 A selection of mechanical properties of Ag-96Sn solder as a function of gold addition.

treatment at 750°C for 30 min in an inert atmosphere, the gold diffuses into the titanium, giving it a yellowish tinge and a readily brazeable surface. This surface enrichment with gold does not alter the solubility of the parent metal in the filler alloy, as compared with untreated titanium.

Gold-bearing Soldering Alloys

Most gold-bearing solders are gold-rich alloys of eutectic composition and have melting points between 278°C and 363°C [11]. They are based on the gold–tin, gold–antimony, gold–germanium, and gold–silicon binary alloys; see Table 8.2 and Figures 8.7 through 8.10 for the associated phase diagrams. These are the highest melting-point solders that are widely used. Being relatively hard and strong, they produce joints of superior mechanical properties compared with other solders. At first sight, it might be expected that the gold-bearing solders, having a high weight percentage of gold, would have properties not dissimilar to pure gold. However, the large difference in density

TABLE 8.1
Effect of Gold on the Solidus Temperature of Common, Binary Tin-based Eutectic Solders

Solder Composition	wt % of Gold to Form AuSn$_4$ as the Primary Phase	wt % of Gold That Will Dissolve in Solder at 50°C Superheat	Change in Melting Point of Solder by Dissolving 1% Gold, °C
In-48Sn	AuIn$_2$ formed	<1	−1
Bi-43Sn	>1	4	−2
Pb-63Sn	>8	13	−6
Sb-95Sn	>10	30	−15
Ag-96Sn	>11	30	−15

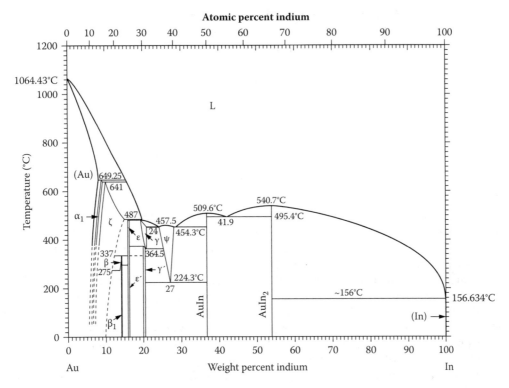

FIGURE 8.5 The gold–indium phase diagram (Courtesy of ASM International).

between gold, on the one hand, and the alloying elements, on the other, means that in terms of volume fraction the gold–silicon eutectic alloy, for example, possesses close to 20% silicon by volume, even though it contains only 3.16 wt% of this element.

Three of these solders (the gold–antimony alloys being the exception) are widely used in the electronics industry as high-temperature solders for attaching semiconductor devices into packages and building hermetic enclosures for sensitive compound semiconductors and optical devices. Gold–antimony alloys are brittle, even when prepared using rapid solidification technology and, as far as the authors are aware, are not much employed in soldering.

The principal gold-based solders are as follows:

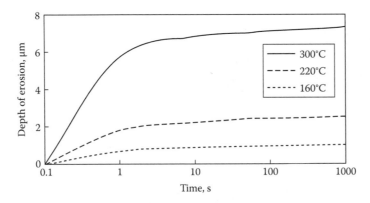

FIGURE 8.6 Erosion of a gold metallization by molten indium as a function of reaction time and temperature. Similar results are obtained for indium-based solders, including gold–indium, silver–indium, indium–lead, and indium–tin.

TABLE 8.2
Principal Gold–bearing Solders

Solder Composition	Eutectic Temperature,°C
Au-3Si	363
Au-12Ge	361
Au-25Sb	356
Au-20Sn	278

Gold–Tin

The Au-20Sn eutectic solder, melting at 280°C, is hard and moderately brittle. These mechanical properties derive from the fact that its constituent phases are two gold tin intermetallic compounds, namely, AuSn (δ) and Au_5Sn (ζ). The first of these phases is very hard, but the ζ phase, being stable over a wide range of composition, has some limited ductility that is imparted to the solder alloy (approximately 2% at room temperature). Although difficult, this solder can be hot rolled to foil and preforms stamped from it. By using rapid solidification casting technology it is possible to produce thin ductile foil, up to about 75 μm thick and having an amorphous microstructure. However, this state is somewhat unstable, and within about 30 min at room temperature the rapidly solidified strip is indistinguishable in its mechanical properties from foil prepared by conventional fabrication methods [12]. Nevertheless, this crystallization can be suppressed for about a year if the quench cooled material is stored under liquid nitrogen (–196°C), and for close to 1 month at –20°C, so that it is possible to manufacture shaped foil preforms while the strip is still ductile, which can then be either immediately placed in a jig or returned to cold storage. The alloy can also be readily gas-atomized to

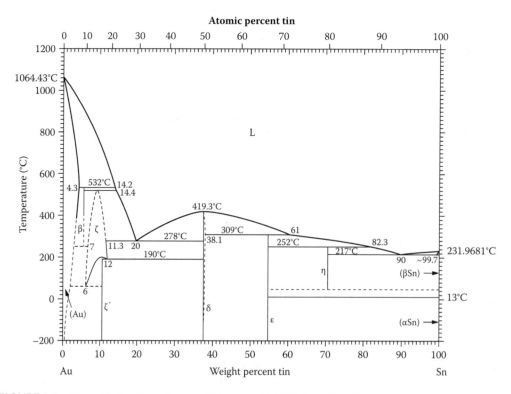

FIGURE 8.7   The gold–tin phase diagram (Courtesy of ASM International).

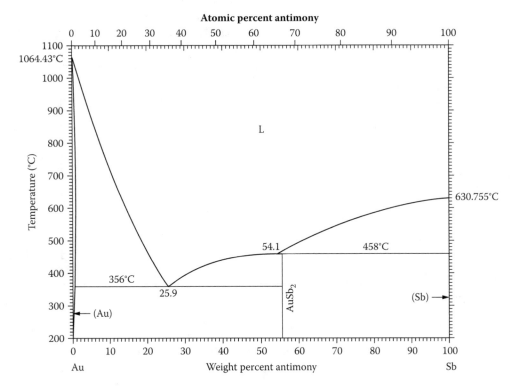

FIGURE 8.8   The gold–antimony phase diagram (Courtesy of ASM International).

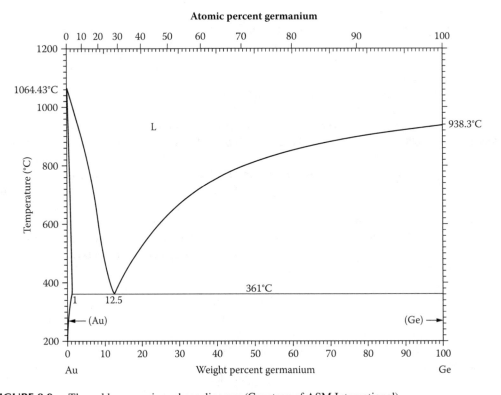

FIGURE 8.9   The gold–germanium phase diagram (Courtesy of ASM International).

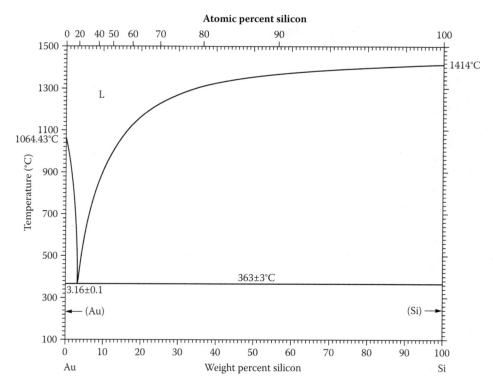

FIGURE 8.10 The gold–silicon phase diagram (Courtesy of ASM International).

form spherical powder and, because it is relatively inert, will survive for long periods in an organic binder medium (optionally containing a flux) without degradation. For this reason, the solder alloy is often used in the form of paste, provided the components being joined are compatible with the chemicals involved.

An alternative method of introducing the Au-20Sn solder into an assembly is to selectively coat the joint area with a thick layer of gold, overlaid with a thinner layer of tin in the thickness ratio of 2Au:1Sn. If electroplated, the layers must be deposited in this order owing to their respective electrochemical potentials. By subjecting the tin to a light acid etch and then immediately applying an outer layer of gold by evaporation or immersion plating, the pre-deposited solder is suitable for use without flux and can be endowed with a shelf-life of several months. However, using this strategy it is difficult to realize solder deposits of more than a few microns because of the need to heat the joints above 420°C in order to destabilize the layer of AuSn intermetallic that will otherwise form an interfacial barrier layer between the two metals. A recent advance in pulse-plating technology now enables co-electroplating of gold–tin solder directly to the surfaces of metal components. Pulse plating extracts a controlled ratio of gold-to-tin atoms from a cyanide-free, weakly acidic solution and so offers the prospect of highly reproducible deposits over large areas to any desired thickness [13,14].

Of the gold bearing solders, only the Au-20Sn alloy has a significant degree of fluidity when molten. In electronic applications, such as die attach of gold metallized chips and the hermetic sealing of ceramic semiconductor packages, the use of fluxes to promote spreading is not usually permitted and joining must be carried out under a protective atmosphere.

The gold–tin eutectic solder is often used for soldering components with thick gold metallizations. There are some difficulties with this, notably the dissolution of gold increases the melting point and prevents good spreading. Other wettable, but insoluble, barrier metallizations for this solder are therefore desirable, but options are limited. Copper, nickel, chromium, and nichrome are wholly unsuitable as they all dissolve readily in the Au-20Sn alloy. Palladium is readily wetted by molten gold–tin solder while it is not highly soluble in it. However, when such joints are aging in the solid state, Kirkendall voids form at the interface between the residual palladium and the Pd_3Sn_2 interfacial intermetallic compound, which can give rise to weak joints [15]. The solubility of platinum in Au-20Sn solder is strongly temperature dependent. At normal soldering temperatures with short process cycles, a thin layer of platinum (200 nm) provides a readily wettable and stable barrier layer [16].

A higher melting point filler metal than the gold–tin eutectic has been developed for joining to thick-film metallizations [17]. It is an alloy of gold–tin with silver and copper additions, substituting for some of the gold. The exact composition and melting range have not been disclosed, but the recommended process temperature is 400°C, compared with 350°C for the Au-20Sn binary alloy. This implies the melting range is probably somewhere around 300–350°C. This modified alloy offers improved ductility and wettability.

Gold–Silicon and Gold–Germanium

On account of their closely similar characteristics as solders, these alloys will be considered together.

Gold–silicon alloys are primarily used as foil preforms for bonding silicon semiconductor chips to gold-metallized pads in ceramic packages. It is particularly employed for this purpose in semiconductor manufacture involving sealed ceramic packages, where the Au-20Sn solder is used for the sealing operation. The gold–silicon and gold–tin solders are used in the initial joining operations of a step soldering sequence, which ends with the soldering of the packages containing the chips onto printed circuit boards, using tin–lead or tin–silver–copper (lead-free) solder at a process temperature in the region of 240°C. However, for this type of application, ceramic packaging is being superseded by inorganic and polymer solutions.

The alloy compositions used as gold–silicon solders are slightly gold-rich with respect to the eutectic composition, generally at or close to Au-2wt% Si; see Figure 8.10. This is deliberate because the eutectic alloy is too hard and brittle even to hot-roll to foil. By making the alloy gold-rich the proportion of the ductile gold phase in the microstructure is sufficient to improve the mechanical properties to tolerable limits. Rapid solidification is unable to produce ductile foil because rapid cooling of molten gold–silicon alloys results in the formation of a large volume fraction of (metastable) gold–silicon intermetallic compounds, principally Au_3Si, which render such foil too brittle to handle [18]. An alternative method of applying the solder to silicon components is simply to coat the back surface of the silicon die with a thin layer of gold, applied by a vapor deposition technique. On heating the gold-metallized silicon to above 363°C, the resulting interdiffusion between the gold and silicon generates liquid Au-Si filler alloy *in situ*.

The gold–silicon solder is highly viscous when molten, in contrast with most other eutectic solders. This characteristic is a direct consequence of the low temperature of the eutectic transformation, relative to the high melting points of both constituent phases, together with the presence of silaceous dross formed on the surface of the alloy and the absence of suitable fluxes that might enhance wetting. Poor fluidity of the solder increases the risk of inadequately filled joints, which mar the performance and reliability of the product. It is generally recommended that gold–silicon foil is given a light etch in hydrofluoric acid and then used within 30 min. This process step removes silica and silicon from the surface, and greatly improves the wetting and spreading behavior.

When using gold–silicon solders, precautions must be taken to ensure that the initial cooling rate of solidified joints is slow and does not exceed 5°C s^{-1}. If this condition is not satisfied, Au_3Si is formed and its subsequent decomposition with time to gold and silicon can produce cracks within joints, due to the associated volume contraction [19].

A number of alloying additions are known to be capable of promoting solder spreading [5]. For the Au-2Si solder, one of the most effective promoters of spreading is tin. Results of spreading tests

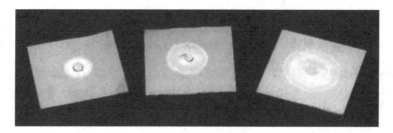

FIGURE 8.11   Spreading behavior of Au-3Si solder containing tin. Left to right, 0%, 4%, 8% tin.

conducted under identical conditions but with different concentrations of tin in the solder showed that increasing the level of tin gave a progressive improvement in solder spread, as illustrated in Figure 8.11. Hardness measurements revealed that this was accompanied by a softening of the alloy, with its hardness decreasing by over 150 HV, a welcome feature because it makes the alloy more amenable to mechanical working into foil and wire for solder preforms. By restricting the concentration of tin to below 8%, the melting temperature of the alloy occurs over a narrow melting range of 356.5–358.0°C. This is closely similar to the solidus temperature of the Au-2Si alloy (363°C) so that switching this binary alloy for one additionally containing tin will therefore not upset subsequent manufacturing steps.

A successful method of promoting wetting of oxidized metal surfaces is to incorporate elements into the filler that are capable of reacting with the oxide to form an adherent bond [20,21]. Research has shown that addition of just 1% of titanium to the Au-2Si and Au-20Sn solders is effective in promoting wetting of bare silicon at soldering temperatures of around 400°C. Concentrations of titanium of up to 2% have been found to have little effect on the melting ranges of the alloys and are actually beneficial in reducing their hardness.

The gold–germanium eutectic composition alloy is sometimes recommended in place of the gold–silicon solder where cost margins are particularly tight. As can be seen from the phase diagram given in Figure 8.9, it offers the benefit of lower gold concentration with little change in melting point. The Au-Ge binary eutectic contains 12.5 wt% Ge compared with 3.2 wt% Si in the case of the Au-Si eutectic alloy. However the price of germanium is much higher than that of silicon, but is considerably less than that of gold. Historically, the price of germanium is about 5% that of gold, although in 2006 it stood at 8% of the gold price, so that the savings in materials cost with respect to gold–silicon is fairly significant. As a solder, gold–germanium is no easier to use than gold–silicon. The Au-12Ge solder reportedly exhibits excellent wetting characteristics on copper, silver, and nickel surfaces [22]. Where it is to be used in conjunction with steel parts, these should be plated with nickel prior to soldering to avoid the formation of embrittling iron germanides.

Gold–Indium

Noneutectic gold–indium alloy of compositions 82Au-18In (melting range 451–485°C) and 75Au-25In (melting range 451–465°C) are offered by several solder suppliers. These alloys are gold rich with respect to the nearest eutectic because of the poor mechanical properties of the eutectic composition alloy (58Au-42In), which is comprised of two brittle intermetallic compounds, AuIn and AuIn$_2$; see Figure 8.5. The demand for gold–indium solders and their inclusion in catalogs stems from the fact that they are the highest melting point solders available. However, they are relatively difficult to use, having inferior wetting and flow characteristics to those of gold–silicon and gold–tin, as indicated in Table 8.3. Also, because there is no suitable flux for these solders, they must be used fluxless, with all the attendant limitations, although the process temperature is sufficiently high that they provide one of the rare examples where a hydrogen-containing atmosphere is effective in reducing surface oxides on a solder.

TABLE 8.3
Contact Angle (in degrees) of Au-12Ge and
Au-18In Solders on Common Metal Substrates

Substrate	Contact angle, degrees	
	Au-12Ge	Au-18In
Copper	4	35
Silver	4	58
Nickel	5	43

The process temperature for gold–germanium was 430°C and gold–indium 550°C, with 5 min hold at peak temperature in a vacuum of 7 mPa (5×10^{-5} torr).

Source: Adapted from Intermediate temperature joining of dissimilar metals, F. M. Hosking, J. J. Stephens, and J. A. Rejent, *Welding Journal*, 1999, April, pp. 127s–136s. (With permission.)

Low Melting Point–High Karat Gold Solders—Options and Limitations

The traditional gold jewelry manufacturing route involves the use of the so-called karat gold solders, which are actually brazing alloys because their working temperatures are above 725°C [23,24]. The high temperatures involved are detrimental to the mechanical strength of high karat gold jewelry because they anneal and soften rapidly when heated above about 450°C. These karat gold brazing alloys are discussed in a later section.

By definition, the temperature of a eutectic transformation is lower than the melting points of the constituent phases. This type of phase transformation therefore offers scope for a large depression of the melting point of gold. Indeed, as we have seen the true "gold solders" that are widely used, namely, gold–tin, gold–germanium, and gold–silicon, belong to this category.

The binary eutectic gold-based alloys, which melt below 450°C, are listed together in Table 8.4 in terms of their composition, caratage, color, and melting point. All of them meet (or exceed) the minimum hallmarking specification for 18 karat gold, because they contain 75% or more of gold. Gold–silicon eutectic solder would also be acceptable for use with 22 karat jewelry, having a gold content that is slightly higher than the 22 karat specification—actually achieving 23 caratage. Furthermore, the alloys listed in Table 8.4 also qualify for joining to gold jewelry due to the relatively high slope of the liquidus phase boundary between the various eutectic points of these alloys and pure gold. In principle, this characteristic should help ensure minimal erosion of jewelry parts by the molten solder and result in good retention of the solder in the joint gap.

TABLE 8.4
Gold–based Eutectic Solders, Some Used for Engineering Applications,
That Are Possible Candidates as Filler Metals for Jewelry Applications

Eutectic Composition	Actual Caratage	Color	Melting Point, °C
Au-3.2Si	23.2	Light yellow	363
Au-12.5Ge	21.0	Pale yellow	361
Au-20.0Sn	19.2	White	278
Au-24.0In	18.2	Grey	458
Au-25.9Sb	17.9	White	360

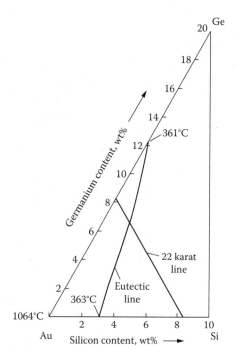

FIGURE 8.12   Liquidus surface of the gold–germanium–silicon system. A continuous eutectic valley links the two binary eutectic points. This eutectic valley crosses the 22 karat fineness at the composition 91.6Au-6.8Ge-1.6Si.

Ternary alloying of these constituents (i.e., of gold with any two of the elements—germanium, indium, antimony, silicon, and tin) mostly results in the formation of intermetallic phases that are not conducive to obtaining good quality soldered joints. The sole exception is the combination of Au-3Si, which is 23 karat, with Au-12.5Ge, which is 21 karat. A continuous eutectic valley extends from the Au-3Si composition to the Au-12.5Ge eutectic point, running in an almost straight line across the gold–germanium–silicon ternary system. This eutectic valley crosses the 22 karat fineness at the composition 91.6Au-6.8Ge-1.6Si, as shown in Figure 8.12.

The microstructure of the gold binary eutectic alloys listed in Table 8.4 is duplex, that is, it comprises a gold-rich phase which, for the gold–germanium–silicon alloy system, is essentially pure gold interspersed with the other constituent phase of the eutectic reaction. In gold–germanium and gold–silicon binary alloys the two constituent phases of the eutectics are virtually the pure metals owing to the very low solubility of germanium and silicon in gold, and vice versa, in the solid state. In the gold–antimony system the second phase is the intermetallic phase $AuSb_2$, which is rich in antimony. All of these base metal phases are silvery or metallic gray in color and will whiten the alloy overall to an extent that is largely determined by their volume proportion. Thus, the Au-3Si alloy contains 81.5 vol% of gold and appears pale gold, while the Au-12.5Ge alloy, with only 72% gold by volume, is even lighter in hue. The Au-26Sb eutectic alloy, which happens to be 18 karat, is completely gray-white in color because its duplex microstructure comprises equal proportions of gold and metallic white phases. The Au-20Sn and Au-24In eutectic alloys have no trace of golden luster whatsoever, because the constituent phases are two grayish intermetallic compounds—Au_5Sn and AuSn, and Au_7In_3, and AuIn, respectively.

What is clear from these considerations is that the condition needed to achieve a rich yellow hue in a eutectic gold alloy containing 75 wt% (18 karat) and therefore color match yellow 18 karat gold jewelry, namely, that the alloy contains a sufficient volume percentage of the yellow gold phase, is not met. This is a largely a consequence of the fact that the elements that enter into eutectic equilibrium with gold have much lower density and therefore constitute a large volume fraction. For this reason, 18 karat gold alloys based on low melting point eutectics predominantly have a base metal color. It is not possible to offset the whitening effect of the base metal by adding the only other metal with a reddish color, namely, copper. Copper reacts with all the base metals that enter into eutectic reaction with gold to form intermetallic phases that are also dull gray or whitish in color. Hence, it is practically impossible to formulate an 18 karat gold solder, which has a melting point below 450°C and, simultaneously, a yellow gold hue [25]. However, there may be greater flexibility with regard to the selection of low melting point gold alloys as solders for white golds.

22 Karat Gold Solders

A suitable solder of 22 karat composition has been identified in the ternary gold, silicon, and germanium alloy system. It takes advantage of the deep eutectic valley that extends through the ternary phase diagram, which intercepts the 22 karat fineness, as mentioned previously and illustrated in Figure 8.12 [26]. In common with the industry standard gold solders used in the electronics industry, foil can be manufactured by hot rolling and wire by hot extrusion. Casting of strip foil by rapid solidification is another possibility. Rapidly solidified foils of karat gold–germanium–silicon alloys are brittle, owing to the formation of the hard, metastable intermetallic $Au_3(Si,Ge)$, which forms when the cooling rate exceeds 5°C sec^{-1} [18]. Ductility can be restored by heat treating the foil at 285°C (~0.9T$_{solidus}$) for 20 min or more, as can be seen from Figure 8.13.

No satisfactory flux has been identified for use with gold–silicon–germanium alloys, or indeed, for the industry-standard gold–silicon and gold–germanium solders. Silicon and germanium react with oxygen when heated to form stable refractory oxides on the free surfaces, which impede wetting by these alloys. Rosin fluxes have been found to be ineffective in either dissolving or disrupting these oxides and stronger acids or halogens corrode the alloys, owing to the large electrochemical potential difference between gold on the one hand and germanium and silicon on the other. Fluxes based on mixtures of salts and hydroxides of the alkali metals show some promise and have the merit that the residues are soluble in water. The fluxing mechanism, in this case, is believed to involve a combination of oxygen exclusion and chemical attack of the refractory oxides. Both the jewelry and the electronics packaging industries would benefit from more research in this area.

Soldering operations with binary gold–silicon and gold–germanium alloys are usually carried out in a nitrogen atmosphere with low oxygen and water vapor content (<5 ppm in total). These and the ternary gold–germanium–silicon alloys can be made more tolerant to the joining environment

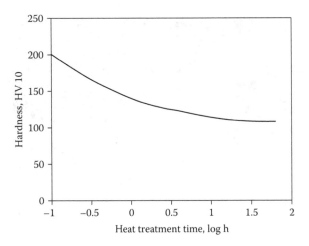

FIGURE 8.13   Progressive recovery of the ductility of strip-cast foils of the 22 karat gold solder 91.6Au-6.8Ge-1.6Si, during heat treatment at 285°C.

by protecting the solder foil or wire with a coating of gold that is impervious to oxygen. An alternative approach, and one that is widely practiced in the electronics industry with the constituent binary solders, is to dip preforms of the filler metal in hydrofluoric acid (then rinse and dry) immediately prior to use. This treatment strips both the oxide and the near-surface nonmetal and, hence, significant reoxidation does not occur until more of the nonmetal has had an opportunity to diffuse through and oxidize at the surface. The shelf-life of solder so prepared is short, typically 30 min at room temperature, but is adequate for handcrafting jewelry.

A solder foil has been successfully developed for use with yellow 22 karat gold jewelry alloys. It comprises a core that is slightly deficient in gold and also the light element, silicon, with respect to the eutectic valley linking the Au-3Si and Au-12.5Ge binary eutectics. It is coated with a gold plating to inhibit formation of dross and thereby facilitate wetting and spreading of the solder. The thinness of the plating means that preforms have a finite shelf-life, although it is several months. The core alloy is of composition 90Au-8Ge-2Si, which has a melting range of 362–382°C. The application of the gold coating of sufficient thickness raises the overall caratage to 22 and reduces the melting range to 362–370°C. In jewelry terminology the two-metal structure of the preforms is referred to as a *double* or *onlay*. The requisite protective atmosphere conditions for use of this solder (<5 ppm combined oxygen and water vapor) can be obtained by using nitrogen gas drawn off a liquid nitrogen tank. The inlet needs to be made leak tight, but the outlet beyond the mouth of the furnace can be left open provided the nitrogen flow rate exceeds 0.5 m/sec. This gas velocity is faster than back-diffusion can occur and, hence, the flowing nitrogen maintains low oxygen and water vapor levels in the furnace. Because there is good heat transfer from the heating elements via the nitrogen at near-atmospheric pressure, the soldering cycle on jewelry items can be accomplished rapidly. Although the furnacing requirements are modest, they are more sophisticated and expensive in terms of capital expenditure than a simple set of torches—which means that the main area of application would be for jewelry manufacture, where it would offer sufficient profit margins.

The joint quality obtained using this plated solder foil at a process temperature of 425°C is excellent, as demonstrated by the T-joint shown in Figure 8.14. The spread and filleting of the solder are comparable to that obtained using conventional high-temperature 22 karat braze with flux, but the low-temperature solder offers the advantages of not requiring post-process cleaning to remove flux residues and not noticeably softening the gold jewelry items. As made, the joints appear whitish with respect to the 22 karat yellow gold jewelry, but the color match is readily restored by a modest temperature heat treatment, at 285°C, maintained for at least 120 min, in a shroud of nitrogen. The

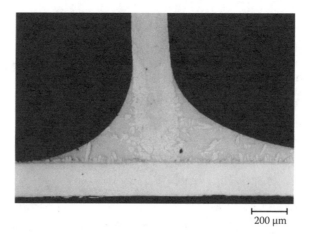

200 μm

FIGURE 8.14 Metallographic cross-section of a T-joint made to 22 karat gold jewelry using a true gold solder (i.e., melting point <450°C).

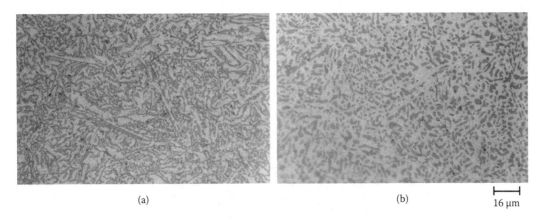

(a)   (b) |—————|
                                                                              16 μm

FIGURE 8.15   Microstructure of the 92.5Au-6Ge-1.5Si alloy. (a) Before heat treatment at 285°C, showing dendritic form of the germanium–silicon precipitates. (b) After heat treatment at 285°C, showing spheroidal form of the germanium–silicon precipitates.

yellowing effect of the heat treatment correlates well with the joint microstructure. On heating, the morphology of the germanium–silicon precipitates throughout the solder, changing from dendritic to spheroidal, so that the same proportion of this phase accounts for a smaller proportion of the area of the free surface of the alloy than it does prior to the heat treatment. In consequence, the yellow gold matrix becomes dominant and the overall color of the alloy favorably alters accordingly (see Figures 8.15a and 8.15b).

The mechanical integrity of joints made to 22 karat gold substrates has been assessed in lap shear and peel resistance tests. Failure always occurred in the 22 karat gold substrate rather than through or adjacent to the joint. This result should not be taken as evidence that the joint is stronger than the parent materials. Such a presumption neglects the role of stress concentration in this style of joint. The important metric is that joint shear strengths typically exceed 210 MPa, with good peel resistance and fracture toughness and are therefore adequate for jewelry applications. Likewise, corrosion tests, designed to assess the susceptibility of joints to degradation from skin acids and household chemicals, have not revealed any deficiencies in the joints.

The different alloying elements used in gold solders and brazes have implications for recycling that need to be considered. Many base metals can cause embrittlement in gold alloys, even at low concentrations, and the treatment of scrap would have to recognize this problem and implement suitable solutions, analogous to the handling of aluminum scrap.

Diffusion Soldering of Karat Gold

The gold–tin alloy system has provided the basis for diffusion soldering process for joining items of high-karat gold jewelry at or below 450°C, which is by convention the upper temperature limit for soldering.

Diffusion soldering is a hybrid joining process (sometimes also referred to as transient liquid-phase joining, or TLP) of soldering with diffusion bonding [11]. It combines the good joint filling, fillet formation, and tolerance to surface preparation (features of conventional soldering) with wide flexibility with regard to service temperature and metallurgical simplicity (features of diffusion bonding). The process uses a thin layer (a few microns) of molten filler to initially fill the joint; during the heating stage the filler metal diffuses into the material of the mating components to form solid phases, raising the remelting temperature of the joint. At this stage, isothermal solidification occurs and further reaction proceeds by solid-state diffusion until the process cycle is complete. Because liquid metal (here molten tin) is generated in the joint, the applied pressure needed is much less than those required for normal diffusion bonding, and are typically in the range of 0.5 to 1 MPa.

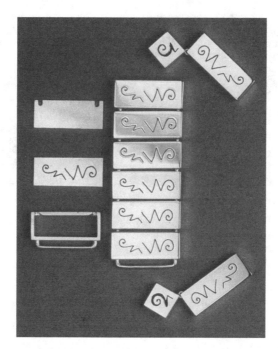

FIGURE 8.16   Parts of an 18 karat gold bracelet and matching earring set assembled using the gold–tin diffusion soldering process at 450°C. The unusually low process temperature enables the face plates to retain much of their work-hardened strength and thereby accept a particularly high surface polish. Each segment of the bracelet measures approximately 8 mm wide.

In trials, it has been found that a tin coating 3 to 4 μm thick is generally sufficient to ensure complete filling of joints between high-karat gold components and the formation of small edge fillets. Provided that the peak process temperature exceeds 419°C, the melting point of the AuSn intermetallic compound, the tin will transform initially to the high gold intermetallic compound Au_5Sn and on continued heating, to gold solid solution. The Au_5Sn compound contains approximately 90 wt% gold and so meets the 18 karat requirement of the jewelry item. Prolonged heating is undesirable as it results in softening of the gold assembly, as reflected by the grain growth and also in Kirkendall voiding in the centerline of the joint. A time of 1 h at 450°C under a compressive loading of 1 MPa was found to be an acceptable compromise for the processing conditions [27]. Figure 8.16 shows a bracelet and matching earring set that was assembled by this method and exhibited at the World Jewelry Trade Fair held in Basel in 1992.

GOLD–BASED BRAZING ALLOYS

Gold is occasionally used as a braze in pure metal form. It has the advantages of resistance to oxidation and tarnishing and being readily worked to foil, provided that the materials being joined are able to withstand being raised in temperature above the melting point of gold (1064°C) without detriment. In this connection, it has been shown to be a suitable braze for the industrially useful $Ni_{50}Ti_{50}$ memory shape alloy as joints of gold do not degrade its essential properties [28].

Lower melting point gold-based brazing alloys may be classified into two groups, according to the application. There are the karat gold brazes, which are mostly used in the jewelry trade, and others that were developed to address the needs of other application sectors, such as the electronics, nuclear power, and aerospace industries [29]. These types of gold-based brazes will be discussed in turn.

Karat Gold Brazing Alloys

Conventional karat gold brazes (often referred to by the misnomer "gold solders") have been tailored to provide goldsmiths with a wide choice of alloys. In addition to generating joints that are mechanically durable (in terms of their strength, ductility, and wear resistance), they must satisfy two special criteria specific to their jewelry application, namely, have the same or higher gold content, or caratage, than the components, and match them in color. The surface texture of the solidified filler metal and its resistance to corrosion must also not differ greatly from that of the joined parts, so that any joints remain essentially invisible to the naked eye.

The fineness/caratage requirement arises from the hallmarking regulations for jewelry. However, an exception is made in the United Kingdom for 22 karat jewelry, where 18 karat gold brazes are allowed. Hallmarking is an independent audit system to ensure caratage conformance that was introduced in the United Kingdom about 700 years ago to protect consumers against unscrupulous traders. In many countries similar hallmarking regulations apply, but in some, including Italy, Germany, and the United States, self-marking by manufacturers is the norm.

The liquidus surface of the ternary gold–silver–copper phase diagram is shown in Figure 8.17. In consequence of the fact that copper and silver enter into a eutectic reaction, a valley in the liquidus surface extends from the binary silver–copper alloy toward the gold-rich end of the ternary phase diagram. This liquidus depression has provided the basis for gold brazing alloys and, traditionally, compositions of brazes for high-karat gold were chosen to lie in the vicinity of the eutectic valley, such that their liquidus lay below the solidus temperature of the jewelry alloy of matching caratage (see Table 8.5). Brazes in current use contain other additions to further lower the liquidus temperature, and to suitably adjust their color to match the jewelry alloys, as discussed below.

Tables 8.6 and 8.7 list a representative range of colored and white karat gold brazing alloys, respectively. The silver, copper, cadmium, zinc, nickel, indium, and gallium additions serve to

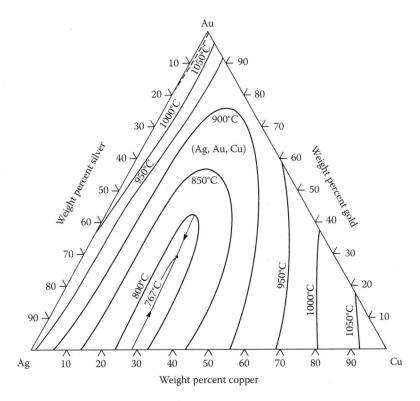

FIGURE 8.17   Liquidus surface of the Ag-Au-Cu ternary system (Courtesy of ASM International).

TABLE 8.5
Selected Gold–Silver-Copper Karat Alloys, Showing the Effect of Varying the Copper-to-Silver Ratio on Their Liquidus Temperatures

Karat Designation	Gold	Silver	Copper	Cu:Ag Ratio	Liquidus Temperature °C
22	91.6	6.3	2.1	1:3	1024
	91.6	4.2	4.2	1:1	971
	91.6	2.1	6.3	3:1	954
18	75.0	21.4	3.6	1:6	976
	75.0	17.0	8.0	1:2	934
	75.0	12.5	12.5	1:1	882
	75.0	8.0	17.0	2:1	882
	75.0	3.6	21.4	6:1	881
9	37.5	53.5	9.0	1:6	905
	37.5	41.5	21.0	1:2	800
	37.5	31.25	31.25	1:1	825
	37.5	21.0	41.5	2:1	875
	37.5	9.0	53.5	6:1	915

Source: After Gold Usage, W. S. Rapson and T. Groenewald, 1978, Academic Press, London, 1978, 80–145. (With permission.)

adjust both the melting temperatures and, more importantly for jewelry applications, the caratage and color of the alloys. Yellow, red, and white brazes are available commercially to match the color of corresponding jewelry alloys. Consideration of Tables 8.6 and 8.7 reveals that colored gold brazes are modified gold–silver–copper jewelry alloys, while white gold brazes, like their counterpart jewelry alloys, contain silver, nickel, zinc, copper, indium, and tin additions, or are based on gold–silver–palladium with possibly copper, zinc, and nickel additions [30]. For each given karat range there is an adjustment of the ratio of the white metal constituents on the one hand and copper on the other, to achieve the correct color balance together with the melting temperature and suitable combination of physical properties. It is possible to isolate a few characteristic features of these brazes. For example, brazes with a higher silver-to-copper ratio tend to be more fluid and able to penetrate narrow joint clearances, because the liquidus temperature rises as the proportion of silver is increased with respect to copper, with other constituents remaining unchanged [31]. This trend can be observed in Table 8.5. Conversely, brazes with a low silver-to-copper ratio are better suited to applications where gaps have to be bridged. As might be expected, the more silver-rich brazes are whiter in color, while the copper-rich filler metals tend toward red. The low silver-to-copper ratio brazes are also usually harder and more receptive to aging treatments for hardening the gold alloy by solid solution strengthening, due to larger atom size differences. Silver atoms are slightly larger than those of gold, whereas copper atoms are 14% smaller.

Zinc is one of the most important base metal constituents of gold brazes, being particularly effective in lowering the liquidus temperature and melting range. However, its volatility, whitening (or bleaching) effect, and detrimental influence on alloy ductility and malleability, as well as on the wetting characteristics of the braze, impose limits on the levels of zinc that can be tolerated. In particular, if zinc is present above about 5%, evolution of vapor becomes significant, resulting in porosity in the brazed joints.

Until recently, many of the gold brazing alloys contained cadmium as an alloying element, which offered the advantages of acting as a melting point depressant and, simultaneously, imparting

TABLE 8.6
Composition of Representative Colored Karat Gold Brazes and Their Melting Ranges

Karat Designation	Gold	Silver	Copper	Zinc	Other	Grade	Color	Melting Range, °C
22	95.0				5.0 Ga		Yellow	415–810
	91.8	2.4	2.0	1.0	2.8 In	Easy	Yellow	850–895
	91.8	3.0	2.6	1.0	1.6 In	Medium	Yellow	895–900
	91.8	4.2	3.1	1.0		Hard	Yellow	940–960
	91.6	0.4	3.0	5.0			Yellow	865–880
	91.6			8.4			Yellow	754–796
21	88.0	6.0			6.0 Ga		Yellow	550–790
	87.5		4.5	4.0	4.0 Sn	Easy	Yellow	662–813
	87.5		5.5	4.8	2.2 In	Medium	Yellow	751–840
	87.5	4.0	3.5	5.0		Hard	Yellow	834–897
18	75.0	12.0	8.0		5.0 Cd		Yellow	826–887
	75.0	9.0	6.0		10.0 Cd		Yellow	776–843
	75.0	5.0	9.3	6.7	4.0 In	Easy	Yellow	726–750
	75.0	6.0	10.0	7.0	2.0 In	Medium	Yellow	756–781
	75.0	6.0	11.0	8.0		Hard	Yellow	797–804
	75.0	5.3	12.2	6.5	1.0 In	Hard	Yellow	792–829
	75.0	6.1	11.0	7.9			Red	805–810
14	58.5	25.0	12.5		4.0 Cd		Yellow	788–840
	58.5	8.8	22.7		10.0 Cd		Yellow	751–780
	58.3	14.4	13.0	11.7	2.5 In	Easy	Yellow	685–728
	58.3	17.5	15.7	6.0	2.5 Sn	Medium	Yellow	757–774
	58.3	20.0	18.2	3.5		Hard	Yellow	795–807
10	41.7	27.1	20.9	5.3	2.5 In + 2.5 Sn	Easy	Yellow	680–730
	41.7	29.4	22.2	4.2	2.5 Sn	Medium	Yellow	743–763
	41.7	33.2	23.9	1.2	2.5 Sn	Hard	Yellow	777–795
9	37.5	31.9	18.1	8.12	3.12 In + 1.25 Ga	Extra easy	Yellow	637–702
	37.5	29.4	19.4	10.6	2.5 In + 0.6 Ga	Easy	Yellow	658–721
	37.5	36.3	18.2	8.0		Medium	Yellow	735–755
	37.5	29.8	27.5	5.2		Hard	Yellow	755–795
	37.5	26.1	27.4	9.0			Red	685–790

favorable flow and wetting characteristics of the molten braze, without strongly whitening the hue. However, the toxicity of the fume of this element is now acknowledged in legal restrictions on its use in brazes and solders in many countries, and it has led to its replacement by tin, indium, and gallium [24,32]. Of these three base metals, gallium has the least bleaching effect [31]. The substitution of these elements for cadmium generally results in an increase in the liquidus temperature and a widening of the melting range. Although gold brazes containing cadmium have been withdrawn from general use in most Western countries, they are still used extensively in many countries, including several major centers of jewelry production in Asia, where such restrictive legislation has not yet been introduced.

The cadmium-free jewelry brazes generally exhibit lower yield strength and enhanced ductility, which makes them more amenable to drawing to fine wire when in the form of a filler metal as well as to further mechanical working operations of a part-fabricated jewelry item. The designation

TABLE 8.7
Composition of Representative White Karat Gold Brazes and Their Melting Ranges

Karat Designation	Gold	Silver	Copper	Zinc	Nickel	Other	Grade	Melting Range, °C
20	83.0			6.7	10.0			855–885
19	80.0			8.0	12.0			782–871
18	75.0		6.0	13.5	5.5		Easy	802–826
	75.0		9.0	7.0	9.0		Hard	843–870
	75.0		6.5	6.5	12.0		Easy	803–834
	75.0		1.0	7.5	16.5		Hard	888–902
14	58.33	22.0	4.42	12.0	1.25	2.0 In	Easy	695–716
	58.33	26.0	3.67	9.0	3.0		Hard	755–805
	58.33	15.75	5.0	15.9	5.0		Easy	707–729
	58.33	15.75	11.0	9.2	5.0		Hard	800–833
10	41.67	35.0	13.5	5.83		1.0 In + 3.0 Sn	Easy	715–745
	41.67	42.0	9.83	3.0		3.5 Sn	Hard	770–808
	41.67	28.1	14.1	6.13	10.0	2.5 Sn	Easy	763–784
	41.67	30.13	15.1	1.1	12.0		Hard	800–832
9	37.5	33.4	23.1			3.0 In + 3.0 Sn		725–735

Source: Courtesy of Cookson Precious Metals Ltd.; and reprinted from *Handbook on Soldering and Other Joining Techniques in Gold Jewellery Manufacture*, M. F. Grimwade, World Gold Council, London, 2002. (With permission.)

ranges "easy," "medium," and "hard" denote the magnitude of the temperature difference between the braze liquidus and the karat jewelry alloy solidus. The larger this value, which is defined as *utility variance*, the easier the braze is to work with, on account of the greater temperature tolerance and hence these designations. This system of designation has significance for jewelry manufacture where multiple brazing operations need to be carried out sequentially. The operator will begin with the hard braze and progress down through medium, easy, and even extra easy to avoid remelting previously made joints. When carrying out repair operations, the easy or extra easy grade of brazing alloy is selected for the same reason. However, considerable care must be taken in applying this designation scheme because different gold solder manufacturers define their easy, medium, and hard alloys differently [33]. Thus, for example, one manufacturer's 14 karat yellow easy gold alloy may have somewhat different liquidus and solidus temperatures to that of another. There is the need to quantitatively define these designations and for a standard system to be adopted by gold solder suppliers.

The fluxes used with both silver and gold jewelry brazes are standard formulations, used also in industrial brazing. The most common fluxes used with gold brazing alloys are based on sodium tetraborate ($Na_2B_4O_7 \cdot 10H_2O$), commonly known as borax, which is fluid above 760°C, applied in the form of a paste.

The majority of jewelry brazing is still carried out by gold- and silversmiths using torches and often by hand [32]. The braze is frequently applied in the form of wire, thin strip or coupons or slugs ("paillons") cut from strip, but increasing use is being made of braze paste [34]. Some jewelry manufacture is highly automated, like much other industrial assembly. A good example is gold chain production, in which the links are formed in one process and then brazed and finished on fully automatic machines at many tens of links per second. In former times, chain was

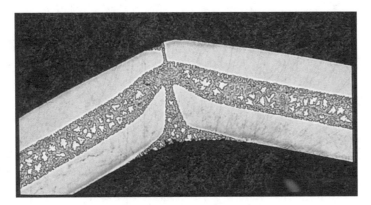

FIGURE 8.18   Section through a link in a brazed karat gold chain, in which the chain stock comprises braze-cored wire (Courtesy of Cookson Precious Metals Ltd., and reprinted from *Handbook on Soldering and Other Joining Techniques in Gold Jewellery Manufacture,* M. F. Grimwade, 2002, World Gold Council, London, UK).

handcrafted, and premium quality items are still made this way today. Chain represents the oldest forms of jewelry, and foxtail-style chain dating from 2500 BCE has been excavated from the city of Ur. It remains one of the most popular styles, although well over 200 other varieties of chain are available presently.

Gold chain is manufactured by machine and then soldered in a separate operation. It is formed from stock wire, which may be a homogeneous precious alloy or cored with a lower melting-point braze. The gaps in the links of chain are closed by brazing. Where homogeneous wire is used, the chain is immersed in a powder of a self-fluxing silver-based brazing alloy containing copper, phosphorus, and small additions of zinc and tin. After passing through this mixture, the chain is rubbed or tumbled in talc to remove excess braze powder adhering to exposed surfaces, leaving it only in the gaps of the links. The brazing operation is then performed by passing the prepared chain through a belt furnace. The stock gold wire is slightly overcarated so that even when aggregating the gold–free braze, on average, chain meets the caratage requirement. Chain made from the cored wire needs only to be formed into links and heated. A cross-section through the joint in a chain link made using braze-cored wire is shown in Figure 8.18 [35]. The most recent technology used in chain-making involves *in situ* welding of chain by microplasma torch or laser, which dispenses with the need for brazes. This new route is also faster and completely avoids color matching problems.

Gold braze pastes are blended mixtures of the karat brazing alloy, in the form of a fine powder combined with an organic binder which may or may not contain a flux depending on whether torch heating or furnace heating with a protective atmosphere is to be used. As is normal practice with braze and solder pastes, the material is dispensed from plastic syringes using a hollow needle of appropriate size. The dosage is dispensed by depressing the plunger. In simple handcrafted jewelry manufacture this may be done by hand, and in more highly controlled operations an electric actuator is used to direct compressed air pneumatically onto the plunger for a predetermined time. Braze paste is available in all caratages up to and including 22 karat, and in the standard colored grades.

Although braze pastes are more expensive than wire and strip, they may provide an overall cost saving, through:

- Shortening the brazing operation and increasing throughput
- Deskilling of the brazing operation, so that joint quality is less sensitive to craft skills of the operators
- Providing a more exact and reproducible dosage of braze to joints
- Reduction in braze scrap, which is expensive

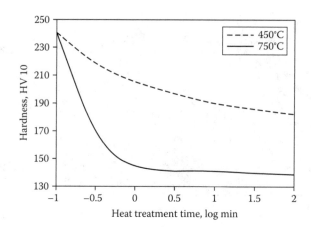

FIGURE 8.19   Hardness of 18 karat yellow gold, type 750Y-3 (75Au-12.5Ag-12.5Cu) in the cold rolled condition as a function of heat treatment time at 450°C and 750°C. Most of the strength generated by cold work is removed by heating for more than a few seconds to typical brazing temperatures.

- Offering better consistency of brazed joints, leading to higher yields
- Enabling a more precise placement of the braze, as compared with paillons

The high working temperatures required for gold brazes have two principal drawbacks. Firstly, the elevated process temperature required for karat gold brazes adds to the complexity or, more specifically, the skill required to use them. Precise and accurate temperature control is necessary because the working temperatures are very close to the melting points of jewelry alloys. For example, the commercially available, cadmium-free, 22 karat brazes have liquidus temperatures that are at most 100°C below the solidus of the corresponding jewelry alloy. Secondly, brazing is inherently detrimental to the mechanical robustness of jewelry. This is because pure gold and high karat gold alloys anneal and soften when heated above about 450°C, as mentioned earlier. Figure 8.19 shows how the hardness of 18 karat yellow gold (type 750Y-3, i.e., 75Au-12.5Ag-12.5Cu) in the cold-rolled condition drops in the course of (furnace) heat treatment at 750°C, as compared with the much smaller change that occurs at 450°C. Most of the strength generated by cold work is removed on heating for more than a few seconds to typical brazing temperatures. This also applies to electroformed (gold–silver alloy) jewelry where the high hardness of the electroform is lost on brazing the end fittings.

Currently this second limitation is circumvented, with varying degrees of success, by making the heating cycle of the joining operation extremely rapid and designing the fabrication sequence such that additional cold working and precipitation/aging treatments can be performed after the joining process have been completed. Nevertheless, the jewelry industry could benefit from the availability of lower melting point brazes for 18 and 22 karat gold items. A recent development toward achieving this objective has been described with the gold-based solders above.

Gold–bearing Brazing Alloys for Industrial Applications

A range of gold–bearing brazing alloys has been developed in response to technological demand, particularly from the electronics, nuclear power, and aerospace industries. Some examples are listed in Table 8.8. These brazing alloys are particularly suited for use in corrosion-resistant assemblies endowed with enhanced mechanical properties and which require joining alloys with matching properties. It is possible to enhance the oxidation resistance of the silver-bearing and nickel-bearing brazing alloy families by adding small percentages of certain elements such as aluminum, nickel, chromium, and manganese. The improved oxidation resistance stems from the ability of these elements to form relatively stable and inert oxide films. However, in chemically aggressive

TABLE 8.8
Industrial Gold Brazing Alloys

Composition, wt%					Melting Range, °C	AWS Designation
Au	Ag	Cu	Ni	Pd		
92.0				8.0	1180–1230	
82.5		16.5	2.0	2.5	899	
82.0			18.0		955	BAu-4
80.0		20.0			910	BAu-2
75.0				25.0	1375–1400	
70.0			22.0	8.0	1005–1045	BAu-6
62.5		37.5			930–940	
60.0	20.0	20.0			845–855	
50.0			25.0	25.0	1120–1125	
37.5		62.5			990–1015	BAu-1
35.0		62.0	3.0		975–1030	BAu-3
30.0			36.0	34.0	1135–1165	BAu-5
29.0	45.0	26.0			767	
20.0		80.0			1020–1040	

environments these modified alloys are susceptible to corrosion. By contrast, noble metals—and especially gold and platinum—are, by their very nature, largely inert chemically and therefore capable of surviving in harsh environments.

The gold-based brazing alloys can be divided into three principal families: gold–copper, gold–nickel, and gold–palladium. The gold content of some of these alloys is less than 50%. Nevertheless, by convention they are classed as gold brazes because the presence of significant proportions of the precious metal constituent. The same applies to brazes containing the platinum-group metals.

The gold–bearing alloys designed for industrial use are superior to base-alloy brazes in the following respects:

- Improved resistance to oxidation
- Good corrosion resistance in most chemical environments
- Enhanced mechanical properties of joints at elevated temperatures
- Relatively low degree of erosion of component metal surfaces during the joining operation
- Produce much smoother brazed fillets, which is of paramount importance in aero jet engines for aerodynamic efficiency
- Require only mild fluxes, on account of the nobility of these alloys

The three principal families of gold–bearing alloy brazes will be considered briefly in turn.

Gold–Copper

The gold–copper binary alloy system, which also provides the basis for many of the brazes used in jewelry, is characterized by a continuous solid solution between the two constituent metals with a liquidus minimum at the Au-20Cu composition, where there is a single melting point (910°C). The gold–copper phase diagram is shown in Figure 8.20. All other alloy compositions in this system have narrow melting ranges, typically less than 20°C. This means that gold–copper brazes possess excellent fluidity and readily form good fillets. Alloys within the range of approximately 40 to 90% gold undergo ordering transformations at low temperature that produce a hardening effect. However, they are sufficiently

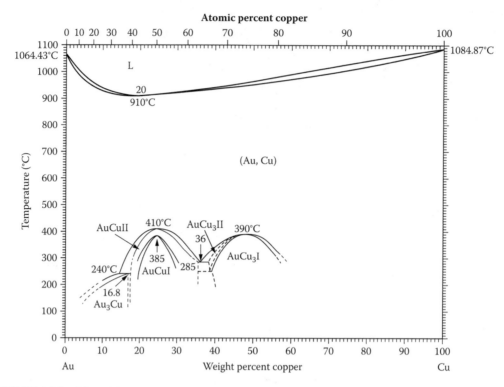

FIGURE 8.20   The gold–copper phase diagram (Courtesy of ASM International).

ductile to be mechanically worked to foil and wire, suitable for making preforms. Gold–copper alloys will readily wet a range of base metals, including many refractory metals.

The addition of nickel to gold–copper brazes helps to improve their ductility; a typical composition is Au-16.5Cu-2Ni, which has a melting point of 899°C. The nickel in the filler metal considerably improves its resistance to creep at elevated temperature (250–750°C) [36]. Adding silver to copper–gold binary alloys effects a significant reduction in the melting point, to 767°C at the composition 45Ag-29Au-26Cu; see Figure 8.17. However, for applications where superior corrosion resistance is required, the gold concentration needs to be kept above 60%, which increases the melting point to about 850°C.

The elevated temperature properties (in particular, strength and corrosion resistance) can be further enhanced by introducing alloying elements such as chromium, manganese, molybdenum, and palladium, exemplified by the alloy Au-34Ni-4Cr-1.5Fe-1.5Mo. The improved mechanical properties at elevated temperature relate to the duplex microstructure, which consists of gold-rich and nickel-rich phases [37]. Gold-based filler metals containing vanadium, such as the Au-15.5Ni-3V-0.7Mo alloy, will wet alumina ceramics, particularly commercial purity grades that contain a glassy phase [38]. Molybdenum is known to increase the ductility of gold-based filler metals and is one reason for its inclusion in both of the above formulations. These additions also promote wetting of refractory materials, notably graphite and carbides. Another example is the Au-34Cu-16Mn-10Ni-10Pd alloy, which was developed for brazing components in the Space Shuttle main engine at about 1000°C, in response to the need for joints with superior oxidation resistance.

Gold–Nickel

Gold forms a continuous series of solid solutions with nickel in a similar manner to the gold–copper alloys, except that the melting range tends to be wider. The liquidus–solidus minimum occurs at the Au-18Ni composition and 955°C. The gold–nickel phase diagram is shown in Figure 8.21.

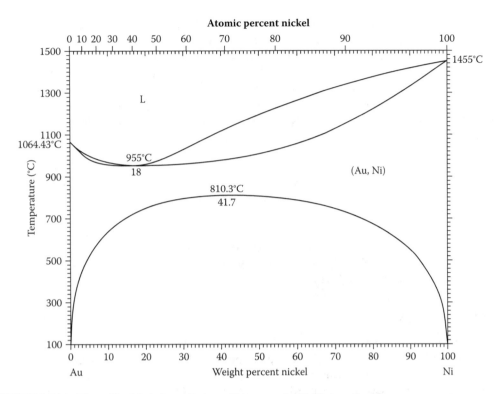

FIGURE 8.21   The gold–nickel phase diagram (Courtesy of ASM International).

A nickel content of 35% is generally the maximum used for brazing alloys, owing to the considerable widening of the melting range as the proportion of nickel is increased. The gold–nickel brazes possess many of the advantages of the gold–copper alloys, but provide the additional benefit of superior resistance to oxidation and improved wetting. Extra constituents enable the proportion of costly gold in the alloy to be significantly reduced without sacrificing essential properties. Thus, the Au-22Cu-8.9Ni-1Cr-0.1B alloy, containing 68% gold (melting range 960–980°C), was developed in the United Kingdom as a cheaper equivalent to the minimum melting point Au-18Ni composition [39]. This substitution provides an instructive example of the considerations that frequently apply in the development of new filler compositions. First, target properties are defined. In this case, the new alloy was required to

- Readily wet heat-resistant steels, in particular martensitic stainless steels such as Jethete M-152 (Fe-12Cr-2.5Ni-1.7Mo-0.3V-0.12C), and produce sound joints to components of these materials
- Completely melt below 1000°C in order to be compatible with the special steels, which are heat treated (the beneficial effects of the heat treatment are lost above this temperature)
- Be free of volatile constituents to enable the braze to be used in (fluxless) vacuum joining operations
- Confer high strength and oxidation resistance to the joints
- Be sufficiently ductile so as to be mechanically workable to foil and wire for preforms

The alloy was designed such that copper substituted for some of the gold and the ensuing loss of oxidation resistance was compensated for by the chromium addition. The deterioration of the wetting characteristics due to the presence of the chromium was made good by the introduction of a small fraction of boron. Finally, the relative proportions of the constituents were then adjusted to minimize the melting range, while optimizing the mechanical properties.

A similar challenge was met in the United States with brazes having a much reduced gold content. These are quinary alloys that contain around 20 wt% gold and 65 wt% nickel, with the balance made up of chromium, iron, and silicon [40]. They melt in the fairly narrow range of 943–960°C and were developed for the assembly of space shuttle engines as alternatives to conventional nickel-based brazes and conferred with an enhanced ductility.

Gold–Palladium

The addition of palladium to the gold–copper, gold–nickel, and gold–copper–nickel alloys improves their resistance to oxidation at elevated temperatures, while boosting the strength of the alloys. These alloys are therefore used mostly for joining superalloy and refractory metal components that need to serve in relatively aggressive environments such as exist in modern jet engines. Commercially available brazes of this type have melting temperatures that reach approximately 1200°C. All these alloys are classified simply as gold–palladium alloys, drawing attention to the two precious metals in this series. Some examples are included in Table 8.8.

ACKNOWLEDGMENTS

ASM International is gratefully acknowledged for permitting the authors to use material from their publications *Principles of Soldering* (2004) and *Principles of Brazing* (2005) in this chapter, including illustrations. Charles Moosbrugger and Ann Britton of ASM International are specifically thanked for their assistance in this regard. Chris Corti of the World Gold Council is also thanked for providing helpful feedback.

REFERENCES

1. *The Solid Phase Welding of Metals*, R. F. Tylecote, Edward Arnold, 1968.
2. Diffusion bonding. Part 1, R. F. Tylecote, *Welding and Metal Fabrication*, R. F. Tylecote, 1967, **35**, 483–489.
3. Diffusion bonding of gold, G. Humpston and S. J. Baker, *Gold Bulletin*, 1999, **31**, 131–132.
4. Explosive joining of precious metals, A. Blatter and D. A. Peguiron, *Gold Bulletin*, 1988, **31**, 93–98.
5. Solder spread: A criterion for evaluation of soldering, G. Humpston, and D. M. Jacobson, *Gold Bulletin*, 1990, **23**, 83–95.
6. Soldering to gold plating, W. B. Harding and H. B. Pressly, *Proc. 50th Annual Conference of the American Electroplaters Society*, 1963, 90–106.
7. Thermal analysis of the AuSn-Pb quasibinary section, G. Humpston and B. L. Davies, *Metal Science*, 1988, **18**, 329–331.
8. Constitution of the AuSn-Pb-Sn partial ternary system, G. Humpston, and B. L. Davies, *Materials Science and Technology*, 1985, **19**, 433–441.
9. Constitution of the Au-AuSn-Pb partial ternary system, G. Humpston and D. S. Evans, *Materials Science and Technology*, 1987, **3**, 621–627.
10. Effects of gold on the properties of solders, R. N. Wild, IBM Federal Systems Division, Owego, Report Number 67-825-2157, 1968.
11. *Principles of Soldering*, G. Humpston and D. M. Jacobson, ASM International, Materials Park, OH, 2004.
12. Dynamical X-ray diffraction study on the phase transformation in rapidly quenched Au71Sn29 alloy, N. Mattern, *Proc. Conf. Advanced Methods of X-ray and Neutron Structural Analysis of Materials*, 1989, 73–76.
13. Pulsed electrodeposition of the eutectic Au/Sn solder for optoelectronic packaging, B. Djurfors and D. G. Ivey, *Journal of Electronic Materials*, 2001, **30**, 1249–1254.
14. Microstructural study of co-electroplated Au/Sn alloys, W. Sun, and D.G. Ivey, *Journal of Materials Science*, 2001, **36**, 757–766.

15. Investigations of Au-Sn alloys on different end metallisations, Anhock, S, *et al.*, Proc. 22nd IEEE/CPMT International Electronics Manufacturing Technology Symposium, Berlin 27–29 April, 1988, 156–165.

16. Preferential reaction and stability of the AuSn/Pt system: Metallisation structure for flip-chip, O. Wada, and T. Kumai, *Applied Physics Letters*, 1991, **58**, 908–910.

17. Low temperature braze system 5087D braze alloy composition, Technical Information Sheet, DuPont Electronics, September 1988.

18. The room temperature dissociation of Au_3Si in hypoeutectic Au-Si alloys, A. A. Johnson and D. N. Johnson, *Materials Science and Engineering*, 1983, **61**, 231–235.

19. Surface cracking in gold-silicon alloys, A. A. Johnson and D. N. Johnson, *Solid State Electronics*, **27**, 1107–1109.

20. Reactive brazing of alumina to metals, R. M. Crispin and M. G. Nicholas, *Brazing and Soldering*, 1984, **6**, 37–39.

21. Wetting of tin-based active solder on sialon ceramic, A. Xian and Z. Si, *Journal of Materials Science Letters*, 1991, **10**, 1315–1317.

22. Intermediate temperature joining of dissimilar metals, F. M. Hosking, J. J. Stephens and J. A. Rejent, *Welding Journal*, 1999, April, 127s–136s.

23. *Gold Useage*, W. S. Rapson and T. Groenewald, Academic Press, London, 1978, 80–145.

24. Cadmium-free gold solder alloys, G. Normandeau, *Gold Technology*, 1996, **18**, 20–24.

25. Do 18 karat gold solders exist? G. Humpston and D. M. Jacobson, *Gold Bulletin*, 1994, **27**, 110–116.

26. A low melting point solder for 22 karat yellow gold, D. M. Jacobson and S. P. S. Sangha, *Gold Bulletin*, 1996, **29**, 3–9.

27. Diffusion soldering: a new low temperature process for joining karat gold jewellery, G. Humpston, D. M. Jacobson and S. P. S. Sangha, *Gold Bulletin*, 1983, **16**, 90–104.

28. Infrared brazing of $Ti_{50}Ni_{50}$ shape memory alloy using gold based brazing alloys, H. Shuie and S. K. Wu, *Gold Bulletin*, 2006, **39**, 200–204.

29. *Principles of Brazing*, G. Humpston and D. M. Jacobson, ASM International, Materials Park, OH, 1985.

30. White Golds: A review of commercial material characteristics and alloy design alternatives, G. Normandeau, *Gold Bulletin*, 1992, **25**, 94–103.

31. Development of 21 karat cadmium-free gold solders, D. Ott, *Gold Technology*, 1996, **19**, 2–6.

32. *Handbook on Soldering and Other Joining Techniques in Gold Jewellery Manufacture*, M. F. Grimwade, World Gold Council, London, 2002.

33. Standardizing the designation of karated gold solders, G. Todd, Proceedings of the Nineteenth Santa Fe Symposium on Jewelry Manufacturing Technology, Sante Fe, NM, 2005, pp. 17.

34. Gold solder pastes, H. H. Hilderbrand, *Gold Technology*, 1993, **9**, 8–12.

35. The 12th Santa Fe Symposium on Jewellery Manufacture, S. Grice, in M. F. Grimwade, *Gold Technology*, 1999, **25**, 21.

36. Elevated temperature creep and fracture properties of the 62Cu-35Au-3Ni braze alloy, J. J. Stephens and F. A. Greulich, *Metals and Materials Transactions A*, 1995, **26**, 1471–1482.

37. Design of high temperature brazing alloys for ceramic-metal joints, S. Kang and H. J. Kim, *Welding Journal*, 1995, **74**, 289s–295s.

38. Microstructural and mechanical characterization of actively brazed alumina tensile specimens, F. M. Hosking, C. H. Cadden, N. Y. C. Yang, S. J. Glass, J. J. Stephens, P. T. Vianco and C. A. Walker, *Welding Journal*, 2000, **81**, 222s–230s.

39. Design and strength of brazed joints, M. H. Sloboda, *Welding and Metal Fabrication*, 1961, **6**, 291–296.

40. Gold-containing brazing filler metals, A. M. Tasker, *Gold Bulletin*, 1983, **16**, 111–113.

9 Jewelry Manufacturing Technology

Christopher W. Corti

CONTENTS

Introduction 191
Current Manufacture of Jewelry 192
Drivers for Change in Jewelry Manufacturing 193
Jewelry Manufacturing Technologies 193
    Lost Wax (Investment) Casting 194
        Technology Trends 196
    Cold Forming Technology: Stamping, Blanking, Coining, and Cold Forging 198
        Technology Trends 199
    Chain-making 199
    Hollow-ware 200
    Electroforming 200
    Soldering and Other Joining Techniques 201
        Trends 203
    Finishing (Polishing and Texturing) 203
    Manufacture of Rings 204
        Trends 204
    Technologies Adapted from Other Industries 205
        Computer-aided Technologies 205
        Cable, Weaving, and Knitting of Wires 206
        Use of Lasers 207
        Robotics 208
        Powder Metallurgy Processes 210
        Computer Modeling of Processes 212
References 212

INTRODUCTION

Gold jewelry and other decorative artifacts have been produced by man since the early Bronze Age. Traditionally, they have been made by skilled goldsmiths using craft skills and simple hand tools. Indeed, it can be said that humans learned the craft of metalworking on native gold and copper and this has led, via the Bronze Age, Iron Age, and our more recent Steel Age, to the modern electronics-based society in which we live today.

Gold was one of the first metals discovered, and it was prized for its rareness, beauty, and value. It has a rich yellow color, one of only two colored metals along with copper, and has been used for adornment in all cultures throughout history. Its other attributes that relate to jewelry usage are that it is very malleable (the most ductile of all metals) and does not corrode or tarnish. It is easily melted in furnaces or by gas torch and can be polished to a high luster.

Many of the modern techniques used in the manufacture of jewelry and today's sophisticated engineered products stem from the early ages of gold. So it is useful to remind ourselves of gold's history [1]. In terms of metallurgy and manufacturing technologies, gold has played its part: Lost wax (or investment) casting was developed in the Middle East over 6000 years ago. Man learned to alloy metals around 4000 years ago by adding copper to native, impure gold and then to refine impure gold around 2500 years ago. Techniques such as soldering, brazing, and welding are often attributed to the Etruscans who developed the art of granulation in about the seventh century BC, but soldered and brazed joints have been found from much earlier times in Egypt (ca. 1350 BC) and in Ur, Mesopotamia (ca. 3200 BC). Powder metallurgy was also developed around alluvial gold in early times too.

Gold was one of the first metals to be used as coinage, back in the seventh century BC by King Alyattes II of Lydia in Asia Minor, where a pattern was stamped into the surface, and this, in turn, led to the need to develop analytical methods, i.e., assaying, to test for purity as man soon learned to debase coinage. This ability to defraud by debasing precious metal, in turn, led to the need for hallmarking of jewelry about 700 years ago in London. This was undertaken by the craft guild, now the Worshipful Company of Goldsmiths, and is probably the United Kingdom's oldest consumer protection legislation. The term *hallmark* stems from the fact that stamping marks on jewelry was performed at the "hall," i.e., Goldsmiths Hall, now home to the London Assay Office. Today, in many countries of the world, all precious metal jewelry must be assayed and hallmarked at an independent Assay Office.

The first medical application for gold was probably in dentistry. Gold wire was used to bind teeth in position 5000 years ago and bridges were first made in gold by the Romans and Etruscans over 2700 years ago. These serve to demonstrate further man's ability to manipulate and shape gold in those early ages.

CURRENT MANUFACTURE OF JEWELRY

Today, much jewelry is still made in traditional workshops around the world by expert goldsmiths using manual metalworking skills but utilizing modern hand tools with limited use of machines. This craft approach leads to inconsistent product quality, a relative lack of design innovation, and is labor intensive. It is limited by available craft skills and is a relatively inefficient and time-consuming production technique.

However, increasingly, jewelry is being made by mass production methods using modern machines and equipment in jewelry factories. Machine-based technology gives a more consistent product quality, higher productivity, and faster production. It does not need specialized craft skills and labor costs are lower, a factor of economic significance in the high labor cost countries of the developed markets of the West especially. Its use enables companies to be more cost competitive. Additionally, such jewelry is more consistent in—and made to higher standards of—quality, meeting the needs of the customer, made compliant with the current standards of fineness and meeting the regulations on health and safety and environmental pollution.

There is a view by many in and outside the jewelry industry that jewelry manufacturing has not changed much over the centuries. Those involved in the jewelry industry know that there is considerable change going on. For example, there is a major shift in the center of gravity of manufacture from the West to the East. Much local manufacture is being lost in the West, with manufacturers outsourcing to or importing from countries of the East that offer low costs. There is also a demand for higher quality in terms of design and manufacture. With regard to customer's disposable income, the jewelry industry is under increasing pressure from other luxury goods. It is a time of much change. The jewelry manufacturing industry's response to such pressure is reflected by changes in the manufacturing technology that it uses. This chapter will attempt to summarize some of these trends and developments.

This chapter is an overview of the main technologies used in gold jewelry manufacture (but will not cover the setting of gemstones). There is relatively little published literature, some of which is dated (e.g., books on investment casting), and it is not covered comprehensively in this chapter. Rather, the main, more modern sources are referenced, particularly those published in *Gold Technology* journal, the Santa Fe Symposia proceedings, and the handbooks published by the World Gold Council.

DRIVERS FOR CHANGE IN JEWELRY MANUFACTURING

With increasing internationalization, cost competitiveness is the major driver for change in the industry to meet tougher competition from overseas producers, but there are also other important factors: Innovative jewelry design is increasingly being demanded by the consumer, and this requires use of new decorative effects and materials that depend on technology. Product development and speed to market are also increasingly important: He who gets to market first reaps the largest rewards.

As well as cost pressure, there are other important factors: First, health, safety, and pollution of the environment are subject to increasingly tighter regulation in most countries. There is also an increasing pressure in export and some domestic markets to meet the tight controls imposed by hallmarking and similar regulations as hallmarking becomes mandatory in more countries. This impacts on quality control and hence in the technologies used in manufacture.

All these drivers for change lean heavily on the implementation of improved technology. For jewelry manufacturers to prosper, the optimum use of technology and best practice is essential. In this context, quality, in its wider sense, is increasingly the focus of product differentiation in the market. Attainment of quality standards also relies on improved technology and Best Practice, the latter reflected in the implementation of more sophisticated quality assurance and production systems that are not covered in this chapter.

JEWELRY MANUFACTURING TECHNOLOGIES

As well as improvements in jewelry alloys (see Chapter 7, "The Metallurgy of Gold"), there are a number of modern production technologies that dominate gold jewelry manufacturing:

1. Lost wax (or investment) casting
2. Stamping (and other cold forming technologies)
3. Chain-making
4. Hollow-ware
5. Electroforming
6. Soldering and other joining techniques (welding)
7. Finishing (polishing and texturing)
8. Manufacture of rings

There are also some newer technologies entering into the sector, including some that are rapidly gaining acceptance in the industry. Many are adaptations of engineering technologies that have been tailored to the needs of jewelry manufacturing, for example:

1. Computer-aided design (CAD)
2. Computer-aided manufacturing (CAM), including rapid prototyping (RP) and rapid manufacturing (RM)
3. Manufacture of cable, knitting/weaving of wires
4. Use of lasers
5. Robotics
6. Powder processes—powder metallurgy, metal injection molding (MIM), and laser sintering
7. Computer modeling of processes

On the jewelry materials side, there has been considerable development of alloys tailored to the requirements of specific manufacturing routes and end-market needs, such as alloys for investment casting, microalloyed high karat gold, and alloys designed for special color effects such as black, brown, and blue gold produced by oxidation.

LOST WAX (INVESTMENT) CASTING

One of the earliest metal technologies developed was the lost wax casting process, developed for copper and gold in the Middle East at least by the fourth millennium BC. The earliest known example in gold was made in Ur, Mesopotamia in 2600 BC [2]. Today, the lost wax (or investment) casting process is the most widely used process for the mass manufacture of gold jewelry. However, it has gone through considerable evolution, particularly in recent years, with the development of more sophisticated equipment and materials from those used in the dental industry from which it was adapted in the mid-twentieth century. The history of the modern development of casting machines can be found in Gainsbury [3].

Lost wax casting of karat golds is a complex, near-net shape process consisting of many steps [4,5]:

1. A master model is made in a hard alloy such as nickel silver (a copper–nickel–zinc alloy) or often in silver.
2. A rubber mold is made by surrounding the master model with sheet rubber in a mold frame. It is then placed in a heated press and vulcanized. On cooling, it is cut with a scalpel into two halves, thus releasing the master model.
3. The rubber mold is used to make multiple copies of the master model in wax by use of a wax injector which injects molten wax, often under a vacuum to remove air from the mold, into the mold cavity. On cooling, the mold is separated and the wax model is removed to give an exact copy of the original master model.
4. The waxes are assembled into a "tree" around a central feeder or sprue fixed to a rubber base. This tree is cleaned of dust, and placed in a metal cylinder, known as a *flask*. To make the refractory mold, special investment powder (comprising silica powder plus a binder, typically gypsum, i.e., calcium sulfate) is mixed with water to form a slurry and poured into the flask around the wax tree. Thus the wax tree is "invested," i.e., coated with the refractory mold material. It is placed under a low vacuum to remove air bubbles and allowed to set and harden to form the refractory mold.
5. The flask is then inverted and the wax removed by melting in steam or in air in a furnace (the burnout oven). The investment mold is then carefully heated in the burnout oven in set stages to the maximum burnout temperature of 750°C for several hours and then cooled down to the temperature required for casting (typically in the range 450–650°C).
6. The hot flask is placed in a casting machine. The gold metal or alloy is melted in a crucible and then cast into the investment mold. It is allowed to cool and solidify and is then quenched hot into water which helps to break off the investment mold material to leave the cast gold tree (Figure 9.1). The castings are cut off from the tree, assembled into the jewelry pieces, and polished.

To obtain good quality castings in high yield requires that all the steps in the process are performed properly, and much R&D has been done over the years to get a better understanding of each process step and its impact on the casting quality. A bibliography of the relevant literature up to 2002 can be found in the *Handbook of Investment Casting* [5]. Defects in castings can be a major quality problem with investment casting, particularly porosity. Our understanding of the nature of casting defects and their causes has been greatly enhanced in recent years, mainly through studies at FEM in Germany, and this has been comprehensively summarized in the *Handbook on Casting*

FIGURE 9.1 **(See color insert following page 212.)** Casting tree after removal of investment.

and Other Defects by Ott [6]. This is an excellent reference of such defects—their characteristics, causes, and control.

Porosity can be due to solidification shrinkage or to gas dissolved in the molten alloy. Shrinkage porosity is due to premature solidification in the feed sprue or thinner sections of the casting before the heavy sections of the casting have solidified, thus preventing feeding of more molten metal to these regions. An example is shown in Figure 9.2 where the characteristic shape (outlining the metal dendrites) can be seen. In contrast, gas porosity arises from a chemical reaction between the mold material (gypsum, calcium sulfate) and the molten metal or use of unclean scrap in the melt charge. This reaction, which is accelerated by a reducing atmosphere, generates sulfur dioxide gas which is soluble in the melt but is ejected on solidification. It is usually finer in scale, round in shape, and may be found either concentrated near to the surface or totally throughout the cross-section. An example is shown in Figure 9.3. Many defects such as gas porosity, sandy surfaces, fins, and watermarks can be attributed to poor investment mold practice and poor temperature control in burnout.

There are two types of casting machine in use: the older, original technology is the centrifugal casting machine and the more modern, preferred is the static vacuum-assist machine [7]. In centrifugal casting, the crucible containing the molten metal and the mold (flask) are placed in tandem on a horizontal rotating arm, the rotation of which accelerates the molten metal from the crucible into the mold. In contrast, in the static machine, filling of the mold from the bottom-pouring crucible is achieved by gravity, shown schematically in Figure 9.4. Melting can be done under vacuum or under an inert gas atmosphere and the hot flask is placed below the crucible in a second chamber that can be evacuated to remove air prior to casting; on pouring the melt, a gas overpressure can be applied to assist mold filling.

The static casting machines allow more control over casting parameters and are preferred for quality and productivity reasons [5,7]. Use of advanced computer control technology in the static

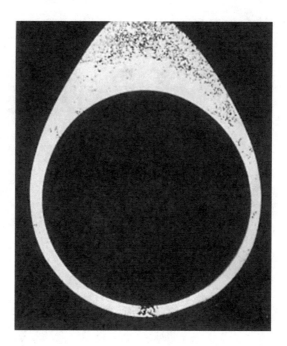

FIGURE 9.2   Karat gold ring showing shrinkage porosity (cross-section).

casting machine reduces operator error. The advantages of both types of machine are summarized in Table 9.1. There is a wide range of casting machines suitable for casting gold of various capacities and features, particularly the static machines, on the market from a large number of manufacturers.

Technology Trends

The biggest shift in the lost wax casting of gold (and silver) jewelry over the last 20 years has been from the use of the old centrifugal casting technique to the newer static, vacuum-assisted technology. The static, vacuum-assisted technology enables better casting quality, with reduced levels of defects, to be routinely achieved through its inherently better process control. Importantly in a mass production situation, it allows higher productivity through the use of larger casting flasks [5]. Use of induction heating in melting allows fast production cycles.

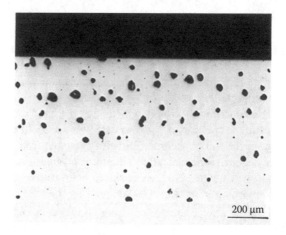

FIGURE 9.3   Gas porosity in cast karat gold (cross-section).

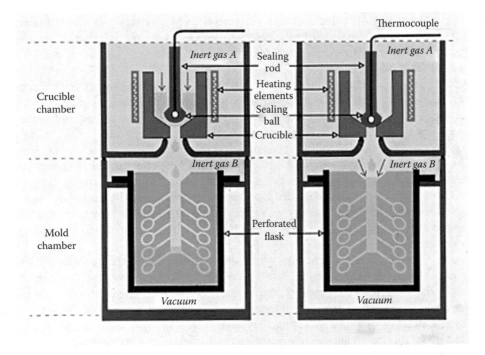

FIGURE 9.4 Schematic: melting and casting in a static, vacuum assist, casting machine (Courtesy of Neutec USA).

The most advanced machines have moved away from use of computer-assisted programmable technology to computer-controlled, self-programming technology. This reduces the risk of operator error through use of artificial intelligence (AI) computer technology borrowed from the space industry. The necessity for vacuum in the casting chamber prior to casting with the subsequent use of gas overpressure has been questioned more recently with the development of machines incorporating

TABLE 9.1
Comparison between Centrifugal Casting and Static Casting for Gold

Centrifugal Casting	Static Casting
Basic programming	Sophisticated programming, up to self-programming
It is possible to have an inert atmosphere, but only in a few models	It is easy to have a controlled atmosphere. It is also possible to have different atmosphere composition in crucible and flask chambers
Relatively small flasks	Larger flasks [h> 200 mm (≈ 8 in)]
Max. charge weight ~800 g	Max. charge weight even >1500 g
High pouring turbulence	Lower pouring turbulence (with a correct feed system)
Risk of investment mold erosion, because of high metal flow and pressure	No risk of investment mold erosion
Feed system not critical	Critical feed system (it should be suitably designed)
Relatively low productivity (8–10 casts/h)	Higher productivity (20 casts/h in the most sophisticated machines, using larger flasks)
Relatively lower capital cost	High capital cost (for the more sophisticated machines)

Source: From V. Faccenda and D. Ott, 2003. *Handbook of Investment Casting*, London: World Gold Council.

Flowlogic™ technology [8], which features pressure differential casting, in which a pressurized (neutral) atmosphere is introduced to the melting chamber. A recent development has been the incorporation of vibration to the flask which, it is claimed, improves mold filling and refines the grain size of the casting [9].

Coupled with this trend to static, vacuum-assisted casting has been the continual improvement in the various ancillary equipment and consumable materials used in the complete process, tailored to meet specific needs such as investment powders for stones-in-place casting and the available equipment range to suit production volume [5,10]. Again, much of this improvement aims to remove operator error and process variability and to reduce defect levels. There are now better waxes and more sophisticated intelligent wax injectors, improved rubbers (natural and synthetic, including silicones) for molds as well as powders for investment molds. Many problems in casting can be attributed to defective waxes and degraded investment powders or poor burnout of the flask, which leads to weak molds. Improved burnout ovens (fan-assisted, rotating hearths, etc.) are available with better, more uniform temperature control [5,10]. These improvements meet the demand for better quality and higher productivity.

An ongoing problem has been knowing what happens during the burnout process and during the casting process in terms of temperature distribution and its evolution over time. Some *in situ* measurements using high speed thermometry, carried out at Stuller Inc., have thrown considerable light on these processes. These show the large temperature variations that can exist within the mold in both burnout and during casting; in casting, the measurements indicate some measure of turbulence in the metal as it fills the mold as well as variations in solidification time, depending on wax model position on the tree [11–13].

Another significant trend in lost wax casting is the increasing use of casting gemstones in place, where the stones are set into the waxes prior to investing and casting. This has considerable economic benefit as it removes the need to hand set individual stones in the traditional, time-consuming way by expert stone setters. It is seen as primarily a process for the cheap, mass market, but it is increasingly being taken up at the high end of the market [14]. Many gemstones including diamonds can be cast in place, but there is much skill and know-how involved. Special attention needs to be given to wax design and tolerances [15–17], and limits are placed on burnout temperatures of the flask (typically 630°C maximum). Special investment powder formulations are designed to protect the gemstones during burnout [18,19].

The latest development in investment casting is the use of computer modeling of the process to optimize casting quality and yield [20–22]. This is European Union (EU)-funded work conducted at FEM in Germany, together with European industrial partners, for both silver and gold casting. The model predicts metal flow and mold filling, the temperature distribution, and its evolution over time, including time to solidification. The model enables the effects of feed sprue size (diameter and length), position, and wax orientation to be determined and the probability of casting defects— shrinkage and gas porosity—to be minimized. The modeling has been experimentally checked against actual casting studies and found to be very consistent. For large-scale casters in particular, this is an important tool that saves much trial and error experimentation and its use in modern casting practice is predicted to grow.

Cold Forming Technology: Stamping, Blanking, Coining, and Cold Forging

Stamping is a commonly used cold-forming process to mass produce hollow lightweight jewelry such as Creole earrings, brooches, lockets, and pendants as well as jewelry findings such as clasps and components for some designs of chain. It is, perhaps, the earliest machine technology used to produce precious metal jewelry, stemming from work developed by early pioneers in the industrial revolution of the eighteenth century in the United Kingdom such as Matthew Boulton (many of his original stamping dies can be seen at the Birmingham Assay Office, UK). It was superseded as the major machine production technology by investment casting in the mid to late twentieth century.

In the stamping process, a thin sheet or strip of the karat (spelled 'carat' in the UK) gold is placed between a series of matching halves of hardened steel dies and progressively punched and cut out (blanked) to form the required three-dimensional shapes. The strip size is important in terms of optimum material utilization because of the material and fabrication costs. A proper nesting or orientation of the parts on the strip can help this considerably. The temper or hardness of the material is also important in blanking, and harder material may give a better quality blank with less distortion and clinging to the tooling as well as reduced burring [23].

With regard to tooling design, the material thickness, temper (hardness), and alloy composition determine the force needed to carry out the deformation required and also influence the clearance between punch and plate in blanking. The choice of tool steel for the dies depends on the design and application of the tool. Typically, dies are made by CAD/CAM techniques and wire electrical discharge machining. These give more accuracy and faster production than traditional manual methods. Manufacture of the steel dies is an expensive operation, thus stamping is economically viable only when high numbers of pieces are being produced.

Coins and medals are also made by this technique, first by blanking out the basic shape, and then stamping (coining) on both sides to imprint the design in one or more stages. The surface finish (polish) of the embossed pattern depends on the quality (polish) of the initial sheet material and of the die surfaces. Many jewelry findings are also made by cold forging wire or thin strip between a matching pair of tool steel dies [24]. Klotz has reviewed modern stamping and cold forging technologies for the production of jewelry findings [23–24] and the relevant working characteristics of karat golds [25]. Some of the defects that can arise in the cold forming of findings have been reviewed by Klotz and Grice [26]. The metallurgical basis of sheet metal forming has been reviewed by Grimwade [27].

In the older stamping technology, parts were stamped out in a sequence in separate steel die sets, each step deforming the component further toward the desired end shape. Typically, hand-operated or steam-powered screw presses or fly presses would be used. Today, complex "progression" die sets are used in which the strip moves in stages through a single die and emerges as pieces in the final shape. Modern hydraulic presses are used, although some hand-operated presses are still to be found in small workshops.

Generally, matching pairs of stamped pieces are soldered together to form the completed (hollow) jewelry piece and then polished. Soldering can be done with conventional solder in wire form or with solder paste but use of solder flush strip for one-half of a pair of stamped pieces is preferred in some countries [28]. This uses a sheet of the karat gold material being stamped with a thin layer of karat gold solder on one side (inside) prior to stamping. The two halves are then jigged together, e.g., on a carbon board, and passed through a belt furnace for soldering; the solder layer melts and gives a good soldered joint [29].

Technology Trends

With the higher prices of precious metals, there has been a noticeable increase in some markets for lighter, thin-walled hollow jewelry to meet price limits without compromising on volume. In stamping, wall thicknesses can be reduced to around 0.10 mm, and the use of progression tooling with up to 20 stamping operations in one set of tools enables faster and cheaper production at up to 1 stroke per second.

CHAIN-MAKING

Chain is the only really continuous production technology used in the jewelry industry. It can be mass produced on specialized chain machines using round, oval, or square wire in a variety of patterns. The production of gold wire is described in Taimsalu [30] and the manufacture and types of chain in a series of articles in Taimsalu [31] and Walters [32]. The wire is fed continuously into the machine, wrapped around shaped steel formers to create the link, cut and the next link is interwoven as the wire is fed in again and the process cycle repeated.

Traditionally, the links of the chain are soldered after chain-making, using a powder soldering technique [33,34]. This method involves impregnating the link gaps with a solder alloy powder containing silver, copper, and phosphorus with small additions of zinc and tin. The powder mix is dispersed in a carrier such as castor oil and an organic solvent. The chain lengths are passed through this mixture and then rubbed or tumbled with talc to remove excess mixture from the surfaces. Heating in a belt furnace under a controlled atmosphere drives off the oil, melts the solder, and joins the links. It will be noted that gold is not present in the solder alloy and therefore the wire used to make the chain has to be overkarated, typically by 4 parts per thousand, to compensate and ensure that the soldered chain will have the required karatage to meet hallmarking requirements on fineness.

Alternatively, soldered-cored wire (where the solder alloy is of the required karatage) is used to make the chain, which is also postsoldered in a belt furnace. On more modern machines, instead of postsoldering, the solid wire links are automatically welded by laser or microplasma welding torch *in situ* on the machine [35].

The metallurgical condition of the wire, as well as its dimensional tolerance, is important as it needs to have sufficient strength to cope with the tensile stresses involved in the process, but it also needs adequate ductility to minimize the risk of fracture. This aspect has been discussed in Agarwal [36], while testing of chain for service performance has been discussed in Agarwal [37].

Some chain designs, such as herringbone, require a flattened profile and this is achieved by rolling the chain in a rolling mill. It then needs some secondary operations to make it flexible [32d]. Decoration can be added by diamond machining to produce bright cut facets. Use of frozen water (ice) to secure the chain in position on large drums for machining is now practiced and enables better control of quality [32d].

More complex chains can be made by laser welding single chains together or from stamped components that are bent and interlocked. It is also possible to make hollow, lightweight chain by use of a small, thin tube around a base metal core, of defined cross-section profile, often containing a solder wire [38]. After processing, the base metal core is leached out in acid. Chain, solid and hollow, can be made in all karatages (8–24 kt) and in the full color range of karat golds.

Hollow-ware

Hollow bangles, earrings, and other jewelry can be produced by various techniques in which karat gold strip is formed into tubes by drawing it through steel dies or through a train of several rolls in a rolling mill. The tubes so formed can be welded or soldered seamless tubes or more simply seamed tubes (with the seam hidden on the inside of a bangle). These can be made as plain round tubes, with patterned or textured surfaces, or made with twists, flutes, and other profiles, which can then be shaped into circles and other shapes by wrapping them around dies or formers. Often a base metal core is used around which the tube is formed and supported [38]. This has to be removed by chemical dissolution at the end of the process. Such cores have a specific profile cross-section and may contain a solder wire to enable postsoldering of the seam.

Generally, end fittings or findings may be soldered on to give a final piece of jewelry. Surface decoration may be added by polishing or texturing (e.g., by sand blasting) and may include diamond faceting to produce a decorative pattern of bright facets.

This technique can also be applied to the manufacture of hollow, lightweight chain and balls. The production of hollow gold jewelry, including chain, has been reviewed by Raw [38].

Electroforming

Electroforming of complex three-dimensional lightweight shapes is a technique that is growing in popularity. It is a technique that offers unique possibilities in jewelry design [39,40]. Some companies specialize in electroformed designs. It is basically a process in which pure or karat gold is electroplated onto a shaped former or mandrel [41–46]. The latter can be a low-melting-point metal

FIGURE 9.5   (See color insert following page 212.) Examples of electroformed 18 karat gold jewelry (Courtesy of Umicore Galvanotechnik and Carla Corporation).

[43] but more recent technology developments now enable electroforming onto a wax model (coated with silver paint to make it electrically conducting), such as those produced for lost wax casting [44]. Control of both consistency and uniformity of thickness as well as karatage/fineness is very important in a mass production situation where 50–75 pieces may be electroformed simultaneously in the bath. This is best achieved by computer control of the plating bath [46]. Generally electroformed articles will be around 100–150 μm thick and even up to 250 μm thick for large items, the final thickness being a compromise between cost and mechanical strength. The wax or metal former is removed at the end of the process by melting or chemical dissolution via small holes in the piece. End fittings may be soldered to yield the final piece of jewelry. Some examples of electroformed jewelry are shown in Figure 9.5.

There are two electroforming technologies available. The earlier development produces a deep yellow gold–copper–cadmium alloy deposit and the other, more recent development a paler yellow gold–silver alloy, both in the range of 8 to 18 karats. Note that use of cadmium-containing materials is no longer allowed in the European Union countries due to health concerns. Electroforming of red and white gold is not possible, although 22kt gold is now claimed possible on a commercial scale [47]. Electroforming of pure, 24 karat gold is also carried out and is particularly popular in the Far East. When plating is done at high deposition rates, a textured (dendritic) surface is produced that does not require further finishing. All commercial electroforming baths are based on the use of gold potassium cyanide (GPC) as the main gold plating component, with other additions. Electroforming produces thicker deposits than those in conventional electroplating, so baths are designed to produce low-stress deposits.

Soldering and Other Joining Techniques

Soldering (strictly, it is brazing as it is performed at temperatures >450°C; see Chapter 8, "Gold in Metal Joining") continues to be a very important process in jewelry manufacture for joining various component parts to make the finished piece. It involves heating the pieces to be joined and applying the solder alloy, which melts and runs along the joint gap by capillary action. Heating is typically by gas torch although oil lamps and blowpipes are still to be found in some traditional workshops in India and the Far East [33]. For mass production, soldering can be done in a belt furnace, usually under an inert or reducing atmosphere, obviating the need to flux the joint.

Solder alloys must fulfill a number of requirements for successful joint production [48]:

- The solder alloy has a lower melting range than that of the parent metal or metals being joined, i.e., the liquidus of the solder is below the solidus temperature of the karat gold alloy used to make the parts to be soldered.
- There is no melting of the parent metal surfaces during soldering but some diffusion will occur at the liquid–solid interface to produce a strong bond on completion of the process.
- The solder alloy must be chemically and metallurgically compatible with the parent metal. For example, the formation of brittle intermetallic compounds at the interface is undesirable.
- The solder should have good fluidity when molten.
- The solder alloy should have the same karatage as that of the parent gold alloy parts being joined (but there are exceptions, such as use of 18 karat solders for 22 karat items, which is allowed in some countries).
- The solder alloy should have a good color match if possible with the parent karat gold so that solder lines are less visible.

The solder is usually applied to the joint in the form of wire or strip that can be touched against the joint when it is sufficiently hot such that the solder instantly melts and runs into the gap. Alternatively, the solder may be applied as small cut pieces of thin sheet (known as *paillons*) by placing them over the joint seam or occasionally in the gap prior to heating. Preferably, gold solder paste is used and has a number of advantages in terms of faster production, lower skills necessary, no necessity for the separate application of flux, less rework and cleaning up post soldering, fewer defects, and reduced costs [48,49].

In mass production processing, it may be more convenient to solder jewelry items using a continuous belt furnace with a protective gas atmosphere to prevent oxidation. In this situation it is not usually necessary to use a flux. However, the part surfaces must be clean and grease, dirt, and oxide free, and the solder has to be placed in position before the parts are placed on the furnace belt. In this case, the solder will either be in the form of paillons or solder paste (a mixture of solder powder and a binder with or without flux) or present as solder flush sheet and solder filled parts. It is necessary to hold the pieces in fixture jigs in most cases [48]. Furnace brazing has the added advantage that the pieces are uniformly heated and that the temperature is controlled throughout the process.

Karat gold solders for jewelry application are discussed in Chapter 8 and so will not be discussed in detail here. The liquidus temperature of the solder alloy should be at least 20°C below the solidus of the karat gold alloy being soldered to enable a skilled craftsman to make the joint without melting the parent metal surfaces. For many years, the alloying metals most commonly used to depress the liquidus, impart good wetting characteristics, and give good molten flow and capillary action have been zinc (Zn) and cadmium (Cd). Unfortunately, there is a serious toxicity problem with cadmium. It has a low melting point of 321°C, boils at 767°C, and has a high vapor pressure. This means that it readily forms a vapor that reacts with air to form poisonous cadmium oxide fume. Exposure to this fume can cause long-term health problems to workers in the jewelry industry including scrap refiners. Although good ventilation and exhaust systems should always be in place in a workshop, escape of cadmium into the atmosphere causes environmental pollution and can get into the food chain. Many countries have banned the use of cadmium in solder alloys or placed severe restrictions on its use in the workshop. The subject of cadmium toxicity in a jewelry context has been discussed elsewhere [48–50].

Much development work has been done during the last 2–3 decades to produce solder alloys, including pastes, of all karatages up to 22 kt that are cadmium free. The alloying additions commonly used to depress the liquidus are zinc, tin, and indium; also, gallium, germanium, and silicon are possible. A series of solder alloy compositions in a particular karatage can be designed by controlling the amount and type of alloying addition so that different melting ranges are obtained.

TABLE 9.2
Composition and Melting Range Data for 18 kt Solder Alloys

Type	%Au	%Ag	%Cu	%Zn	%Sn	%In	%Cd	Liquidus °C	Solidus °C
18 kt easy	75.0	5.0	9.3	6.7	—	4.0	—	750	726
18 kt medium	75.0	6.0	10.0	7.0	—	2.0	—	781	765
18 kt hard	75.0	6.0	11.0	8.0	—	—	—	804	797
18 kt medium	75.0	2.8	11.2	9.0	—	—	2.0	788	747
18 kt hard	75.0	-	15.0	1.8	—	—	8.2	822	793
18 kt casting alloy	Typical yellow gold composition							875	855

Table 9.2 shows the compositions and melting ranges of three 18 kt cadmium-free solder alloys and two containing cadmium compared with the solidification range for a typical 18 kt yellow gold casting alloy.

The "easy" grade solder has a liquidus temperature which is 105°C below the solidus temperature of the casting alloy. This is followed by the "medium" grade solder and then the "hard" grade solder with liquidus temperatures 74°C and 46°C, respectively, below the solidus of the casting alloy. This variation is extremely useful for step soldering operations on assemblies with a number of joints. The first joints are made using the hard solder. Later joints are made with the medium and easy solders at lower temperatures, which avoids remelting the earlier joints. The definitions of these designations in terms of melting temperatures vary from manufacturer to manufacturer, and standardization of the designation of solder grades has been advocated [51], although to date not progressed by the industry. There are a range of cadmium-free solders at all karatages up to 22 karat available for colored and white golds [48,52–55].

Trends

While soldering remains the dominant joining technique, use of laser welding is an alternative approach that is finding increased use, especially in chain manufacture and also for repair work, where heating by the soldering torch may damage gemstones [48]. Laser welding has an advantage in that there are no karatage conformance or color matching problems, as found with solders. Tungsten inert gas (TIG) welding also finds application. Electrical resistance welding is also used, particularly for tack welding prior to soldering.

A diffusion bonding technique is used for bonding the layers of multicolored rings, which may be karat golds of different colors or combinations of karat gold and other precious metals such as platinum and silver as well as stainless steel. The precisely machined component parts are held together under pressure and heated, usually by induction heating, under an inert atmosphere to a high temperature for a period of time to effect a diffusion bond [56]. This technique is also used for bonding layers in the Mokume Gane technique for patterned jewelry [57].

FINISHING (POLISHING AND TEXTURING)

Generally, with few exceptions, all jewelry needs to be polished to a high luster at the end of the manufacturing process. Traditionally, this is done on motorized polishing wheels impregnated with polishing compounds such as jewelers rouge. This is labor intensive, with high gold loss (the gold polished away) and consistency of quality cannot be controlled. In mass production and in the more progressive workshops, use is made of modern mechanized polishing methods. These involve tumbling the jewelry mixed in grinding or polishing media. These are generally abrasive shapes (cones, pyramids) in which the abrasive particles are bound together as sintered ceramic or supported in plastic binders, or with polishing compounds in natural media such as wood chips or nut shells.

Also, polishing can be achieved by burnishing (hammering) the surface with hard materials such as steel pins and cones or porcelain balls. Generally, polishing is achieved in a sequence of grinding and polishing steps, either wet or dry, each of which leads to a further reduction in surface roughness and is aimed to achieve the final degree of polish in the most efficient way. The basic principles of the process are described in Faccenda [58] and Faccenda and Corti [59]. Typical equipment includes the original basic tumbler (rotating barrel), which has been described by Dreher [60,61], and the more advanced and faster techniques such as vibratory polishers, centrifugal disc machines (turbo), centrifugal planetary barrels, and magnetic burnishers [59,62–68].

Comparisons of these techniques have been reported in Dreher [69] and Goodrich [70]. More recent innovations include the drag finisher for rings and similar products, where the rings are attached to fixtures that rotate in the static media, thus dragging the jewelry through the media. Electropolishing is also practiced as part of a polishing procedure [58,71]. The scientific basis of grinding and polishing can be found in Faccenda [58] and Faccenda and Corti [59], along with details of the various polishing machines and their operation. The magnetic barrel, which contains steel pins, is a burnishing machine and is popular for removing residual investment mold from castings.

Other surface textures such as satin or matt finishes can be produced by various techniques such as sandblasting, wire brushing, acid etching, and vibratory engravers (pens) while bright facets and engraving can be obtained by engine turning, diamond machining and laser engraving. Chemical brightening methods such as acid treatment or "bombing" in a cyanide-hydrogen peroxide mixture are also practiced in some regions but are increasingly being discouraged due to health and safety concerns. Such treatments can enhance the color to the deep yellow of pure gold by preferentially dissolving away the base metal and silver from the surface layer as well as giving a good finish.

MANUFACTURE OF RINGS

Most plain rings are manufactured from wrought material, either from sheet or strip or from solid bar or tube. More complex shapes can be machined from solid bar stock but more frequently are made by investment casting.

Thin plain rings can be made by simply bending a strip into a circle and soldering it together. Better quality, thicker rings are traditionally made by cutting (blanking) flat washers from sheet material and then inverting these in stages over a conical die into a ring shape and ring rolling to final size [72,73]. This process results in considerable scrap from unused sheet material, with final material usage of less than 30%, and it is a time-consuming process (up to 10 days).

Other approaches include machining from solid bar stock or from tube on computer numerically controlled (CNC) lathes or multiaxis machining centers, increasingly direct from CAD files. This enables more complex patterns to be produced, and diamond finishing on a lathe yields a highly polished surface. Tube is typically made by extrusion or by forming from strip by controlled bending in rolling mills, and is usually welded to give a seamless product.

Trends

For more complex designs, the use of multiaxis machining centers, controlled directly from CAD files, is seeing increased use for mass production [74]. In addition, their use can enable customization of generic designs for specific customers. Diffusion bonding of multicolored rings is a technology also becoming more widespread (see section on soldering and other joining techniques).

For plain wedding bands in karat golds, a powder metallurgy technique, *press and sinter,* has been developed that offers a number of advantages over the traditional technique described above. This uses water-atomized karat gold powders that are automatically fed into steel dies and pressed in a hydraulic press. The green parts are then sintered at high temperature, 780°C, under a controlled atmosphere and cold forged to a reduced height to yield high density parts that are then mechanically worked by ring rolling to the correct size. Yield, in terms of reduced scrap, is very high and production is faster, both factors leading to reduced inventories and costs. Technically, the product is metallurgically superior, with a finer grain size and greater ductility which, in turn, enables a larger range of sized rings to be produced from one basic ring blank. Yields from the powder of >85% are achieved, thus reducing the amount of metal needed in production and, coupled with faster production cycles of about 5 days, leads to lower financing costs of metal inventory. This patented process, developed by Engelhard in the United Kingdom, has been used in mass production since 1998 [72,73].

TECHNOLOGIES ADAPTED FROM OTHER INDUSTRIES

As discussed earlier, a number of modern technologies used in other engineering industries are being adapted to the needs of the jewelry industry, some of which are gaining rapid acceptance.

Computer-aided Technologies

The use of CAD (Computer-Aided Design) for the design of jewelry has now found wide acceptance and there are several software systems developed and used in the industry [75–78]. These tend to be based on creating shapes rather than working to engineering parameters [75]. Increasingly, CAD is being used for innovative design and for faster new product development, which is at the heart of all successful jewelry businesses [75,76,79,80].

The traditional design approach involves designing in two dimensions with pencil and paper. This must then be translated into a physical three-dimensional model in wax or metal, which is time consuming and prone to error in reproduction. The modern approach to design involves three-dimensional virtual drafting of designs on computers—CAD. This can be coupled electronically with CAM (Computer Aided Manufacturing) or, more specifically, with RP (Rapid Prototyping) technology [77–82]. The computer files produced by CAD are used to make a master model (often wax or resin but also metal) by CAM or by RP—also known more specifically as solid freeform fabrication (SFF)—technologies. CAM is a material removal technique in which a solid piece is machined away to the desired shape. RP technology is a material additive technique in which the three-dimensional model is built up layer by layer using thermosetting or photosetting polymeric materials. Such models can be used as the master model to produce a rubber mold from which conventional waxes can be produced for investment casting [77,82].

This approach has a number of advantages:

- CAD provides an easier and more flexible design capability.
- From a business perspective, it speeds product development and hence speed to market.
- From a production perspective, jewelry can be designed in one place and the design transmitted in seconds to a production site anywhere around the world. The design skills do not have to be located with the production centers.
- The use of CAD also enables initial designs to be adjusted easily and quickly to meet product criteria or to improve appearance, without the need for making several physical model prototypes. The old trial-and-error approach is unnecessary.
- The use of CAD and RP/SFF gives designers more design freedom to create innovative designs.
- Furthermore, CAD allows matching pieces in a set or variations in size to be easily designed. The opposing, mirror image design of a pair of earrings is a good example of the benefits of CAD.
- When coupled with RP/SFF equipment, models in wax or metal can be easily produced to form the master model of the design or used directly in manufacture.

The advent of RP/SFF technology, adapted to meet the needs of the jewelry industry, is having a substantial impact. It has also developed into a RM (Rapid Manufacturing) approach, for example, by use of direct metal laser sintering (DMLS) of powders (see later) [83,84].

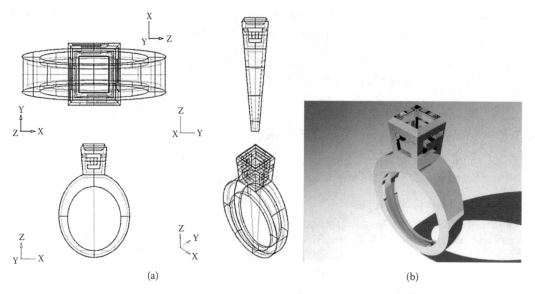

(a)   (b)

FIGURE 9.6   Typical CAD design.

Examples of the use of CAD and RP/SFF models are shown in Figures 9.6 and 9.7. Their use is described, for example, in the training materials, CDs and DVDs, produced by The Goldsmiths' Company [85,86].

Cable, Weaving, and Knitting of Wires

Cable-making technology using metal wire is an old technology used for ships and bridges at the large scale and electrical cables at the smaller scale. This latter technology has been adapted into the jewelry industry for precious metal where fine cables are used as an alternative to chain for necklaces, often to support pendants. Small-scale cable machines using about five to nine spools of wire are commercially produced. Knitting and weaving of wires is similarly finding increased usage, usually at the craft end of the industry. Some examples are shown in Figures 9.8 and 9.9 [88].

FIGURE 9.7   Making a rapid prototype (RP) resin model (Courtesy of Grant MacDonald) .

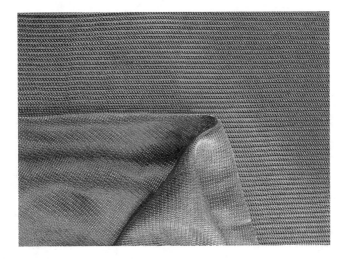

FIGURE 9.8 Knitting of wire (Courtesy of W Böhm).

Use of Lasers

There are an increasing number of lasers (generally of the Nd:YAG type) on the market designed specifically for use in jewelry manufacturing for tasks such as welding, cutting, marking, and engraving. They offer a number of advantages and can give a fast payback on capital investment. They are easy to use and allow precise control and positioning of work pieces. Their application in jewelry is manifold [35,87,89–91]:

- Repair of manufacturing defects such as casting porosity.
- Manufacture of jewelry, where several component parts are laser-welded to form the complete piece. Upmarket gem-set rings may be made from several component castings, for example, as this enables all surfaces, including internal ones, to be polished and the ring gem-set even prior to assembly.
- Repair of broken or worn jewelry. An advantage here is the ability to weld close to gemstones without damage, obviating removal of the stones prior to repair and then resetting afterward.
- Soldering and welding in manufacture and repair.

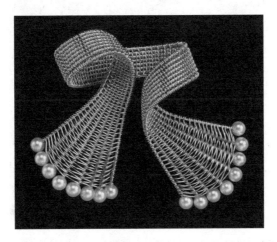

FIGURE 9.9 **(See color insert following page 212.)** Weaving of wires (Courtesy of Barbara Berk Design).

FIGURE 9.10   Laser welding in gold jewellery production. (From Laser applications in gold jewelery production. S. Valenti, *Gold Technology*, No, 34, Spring 2002, 14–20. With permission.)

- Marking and hallmarking of jewelry, both for security and hallmarking requirements.
- Decoration—laser engraving and patterning. This can be aided by CAD and performed on complex three-dimensional objects.

Some examples are shown in Figures 9.10 to 9.12. Laser hallmarking (Figure 9.13) was developed in the United Kingdom in 1998 [92] and has found rapid application as it does not damage (bruise) the item being marked, a problem with conventional stamping, especially on delicate items. The safe use of lasers in jewelry manufacturing has been recently reviewed [93].

A new technique, not yet commercialized for karat gold, is laser forming where the laser beam is rastered over the sheet surface in a controlled pattern, inducing thermal stresses that cause it to self-deform in a controlled way [94,95]. This technique also presents unique design opportunities. Another newer use of lasers is in DMLS, a method of SFF or RM, mentioned previously and described in the later section, Powder Metallurgy Processes.

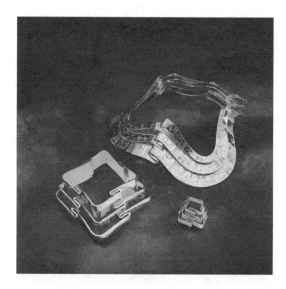

FIGURE 9.11   Laser engraved gold jewelry (Prize winner in Gold Virtuosi design competition. With permission of World Gold Council.)

FIGURE 9.12 Laser cutting in gold jewellery production. (From Laser applications in gold jewelry production. S. Valenti, *Gold Technology*, No. 34, Spring 2002, 14–20. With permission.)

Robotics

The use of robots in engineering applications is well established, e.g., for welding car bodies. Their use in jewelry manufacture is still relatively rare; however, there are one or two jewelry companies in the United States who are using robots to automatically grind and polish jewelry to a high standard, with improvements in quality and consistency and speed of production. Menon [96] has described their use for grinding and polishing rings. Robots have also been exhibited by a machine manufacturer for automatic operation of wax injection, to produce the waxes for investment casting from rubber molds on wax injectors, suggesting another potential application. Control of injection conditions, and hence consistency in wax quality and weight, is much more easily attained and a production schedule involving multiple rubber molds can be programmed for the robot, giving the necessary flexibility in production.

In other engineering sectors, robots are used where there are repetitive tasks and where health and safety considerations present a hazard to operators. There are similar applications within the jewelry production sector, for example, in the transfer of molten metal and flasks or molds to the casting machine, in the weighing and mixing of investment powders, and in quenching the molds after casting, which present opportunities for the use of robots.

Powder Metallurgy Processes

As a near-net shape technique for manufacturing engineering parts from powders in the automotive and aerospace industries, powder metallurgy processing is well established. It is surprising that it has not found more application in the jewelry and watch industries where the high cost of precious metals and the low material utilization of the conventional technologies should make it attractive [97–99]. However, the manufacture of gold and platinum plain wedding rings from water-atomized powders by a press and sinter technique, developed by Engelhard in the United Kingdom [72,73], is now a commercial process although still limited in its application (Figure 9.14). It is described in more detail earlier (Manufacture of Rings). It is faster, reduces costs with little scrap, and gives a technically superior product. However, the cost of the tooling makes this an economic technique only for high production quantities, and that is a difficulty in an industry where production runs of a model design tend to be relatively short.

For more complex pieces MIM (Metal Injection Molding), in which the metal powders are combined in a (wax) binder and injected into a mold cavity to form the desired shape, which is then sintered to high

FIGURE 9.13 Laser engraved Hallmark (Courtesy of Birmingham Assay Office).

FIGURE 9.14 Gold wedding rings from powder—tomorrow's technology today (The steps in the process). (From Gold wedding rings from powder tomorrow's technology today. P. Raw, *Gold Technology*, No. 27, November 1999, 2–8. With permission.)

density, Figure 9.15, is the obvious progression of the technology [98–102] and work in the industry is underway to develop its application in both Europe and the United States [101]. However, the high tooling costs make this a technique for high production runs. The precious metal clays (PMCs) developed by Mitsubishi Materials Corporation are MIM-type materials targeted at the designer/craft end of the industry [103] but do have some potential application in machine-based mass production.

The making of jewelry directly from powders has many advantages, not the least of which is the elimination of manufacturing scrap and faster production, which reduces the interest charges on the cost of the metal. It is predicted that further advances in the MIM technology to produce more complex shapes will be implemented in the industry in the near future, particularly where long production runs can justify the tooling costs, as Strauss has described [101].

More recently, the use of a RM technology in which layers of powder are sintered together, layer by layer to build a complex three-dimensional piece of jewelry (Figure 9.16), is finding interest in our industry [83,101], as is the related direct metal laser melting (DMLM) technique [104]. These allow new freedoms in jewelry design and are technologies that will find application in niche areas, for example, in complex shapes such as custom-designed jewelry and linked chains and bracelets that can be made in one process.

Computer Modeling of Processes

The use of computer modeling to optimize process parameters, increase efficiency, and reduce development time is well established in the engineering industries. However, it has only recently found application in the jewelry industry for investment casting, as described in the first section,

FIGURE 9.15 Jewelry and components made by powder metallurgy. (From J. T. Strauss, 2008, Powder metallurgy in jewelry manufacturing: Status report and discussion. In *Proceedings of the Santa Fe Symposium on Jewelry Manufacturing Technology*, ed. E. Bell, 295–306. Albuquerque, USA: Met-Chem Research. With permission.)

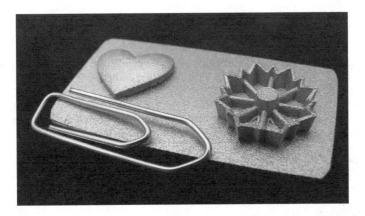

FIGURE 9.16   Parts produced by Direct Metal Laser Sintering (DMLS) (Courtesy of EOS Ltd) .

a process particularly prone to poor yields, inconsistent quality, and high defect rates. Its use in other jewelry manufacturing technologies, particularly stamping and cold forging, could also be beneficial.

REFERENCES

Note that articles from *Gold Bulletin* and *Gold Technology* can be downloaded free at www.goldbulletin.org and www.utilisegold.com, respectively. Articles from *Aurum* magazine may be obtained from the World Gold Council: industry@gold.org. Proceedings of the Santa Fe Symposia can be obtained from www.santafesymposium.org.

1. *Gold Bulletin*, www.goldbulletin.org, archive and index under "Archeology and history"; articles on the history of gold technologies therein.
2. The long history of lost wax casting, L. B. Hunt, *Gold Bulletin*, 1980, **13(2)**, 63–79.
3. P. E. Gainsbury, Jewellery investment casting. In *Investment Casting*, ed. P. R. Beeley and R. F. Smart, 1995, 408–440, London: The Institute of Materials.
4. Back to basics: Investment casting, Part I, C. W. Corti, *Gold Technology*, 2000, No. 28, 27–32.
5. V. Faccenda and D. Ott, 2003. *Handbook of Investment Casting*, London: World Gold Council.
6. D. Ott, 1997. *Handbook on Casting and Other Defects*, London: World Gold Council.
7. Investment casting: Centrifugal or static vacuum assist?, V. Faccenda, *Gold Technology*, 1998, No. 23, 21–26.
8. Going with the flow, S. Wade, *AJM Magazine*, December 2004, 14–16.
9. P. Hofmann, 2006. Vibration technology for the solidification process of investment casting. In *Proc. 3rd Jewellery Technology Forum*, May 2006, 189–199. Vicenza, Italy: Legor Group.
10. V. Faccenda and M. Condo, 2001. Investment casting: An integrated process. In *Proceedings of the Santa Fe Symposium on Jewelry Manufacturing Technology*, ed. E. Bell, 97–119. Albuquerque, NM: Met-Chem Research.
11. S. Aithal, D. Busby, and J. McCloskey, 2002. Evaluation of mold burnout by temperature measurement and weight loss techniques. In *Proceedings of the Santa Fe Symposium on Jewelry Manufacturing Technology*, ed. E. Bell, 1–24. Albuquerque, NM: Met-Chem Research.
12. P. Dubois, S. Aithal, and J. McCloskey, 2002. Temperature measurement in mold cavities during vacuum-assisted, static pouring of 14kt yellow gold. In *Proceedings of the Santa Fe Symposium on Jewelry Manufacturing Technology*, ed. E. Bell, 131–156. Albuquerque, NM: Met-Chem Research.
13. J. McCloskey, P. Dubois, and S. Aithal, 2002. Temperature measurements in 14 kt yellow gold during counter-gravity pouring of investment casting molds. In *Proceedings of the Santa Fe Symposium on Jewelry Manufacturing Technology*, ed. E. Bell, 353–374. Albuquerque, NM: Met-Chem Research.
14. H. Schuster, 2008. Stone-in-place casting for high-end jewelry. In *Proceedings of the Santa Fe Symposium on Jewelry Manufacturing Technology*, ed. E. Bell, 283–294. Albuquerque, NM: Met-Chem Research.
15. A. Menon and J. Martin, 1996. Casting gemstones in place. In *Proceedings of the Santa Fe Symposium on Jewelry Manufacturing Technology*, ed. D. Schneller, 69–81. Albuquerque, NM: Met-Chem Research.

16. H. Schuster, 1999. Stone casting with invisible setting. In *Proceedings of the Santa Fe Symposium on Jewelry Manufacturing Technology*, ed. D. Schneller, 369–378. Albuquerque, NM: Met-Chem Research.

17. H. Schuster, 2000. Problems, causes and their solutions on stone-in-place casting process: Latest developments. In *Proceedings of the Santa Fe Symposium on Jewelry Manufacturing Technology*, ed. E. Bell, 315–321. Albuquerque, NM: Met-Chem Research.

18. Advances in investment casting materials, C. J. Cart, *Gold Technology*, No. 23, 1998, 18–20.

19. I. McKeer, 2004. Stone-in-place casting: the investment perspective. In *Proceedings of the Santa Fe Symposium on Jewelry Manufacturing Technology*, ed. E. Bell, 293–314. Albuquerque, NM: Met-Chem Research.

20. J. Fischer-Bühner, 2005. Computer simulation of investment casting. In *Proceedings of the Santa Fe Symposium on Jewelry Manufacturing Technology, 2006*, ed. E. Bell, 193–216. Albuquerque, NM: Met-Chem Research; and *Proc 3rd Jewellery Technology Forum*, May 2006, 249–268. Vicenza, Italy: Legor Group.

21. M. Actis Grande, 2007. Computer simulation of the investment casting process: Widening of the filling step. In *Proceedings of the Santa Fe Symposium on Jewelry Manufacturing Technology*, ed. E. Bell, 1–18. Albuquerque, NM: Met-Chem Research.

22. J. Fischer-Bühner, 2007. Advances in the prevention of investment casting defects by computer simulation. In *Proceedings of the Santa Fe Symposium on Jewelry Manufacturing Technology*, ed. E. Bell, 149–172. Albuquerque, NM: Met-Chem Research.

23. Production of gold findings by stamping. F. Klotz, *Gold Technology*, No. 33, 2001, 13–16.

24. Cold forging of karat gold findings. F. Klotz, *Gold Technology*, No. 35, 2002, 11–17.

25. F. Klotz, 2003. Cold forging karat gold findings. In *Proceedings of the Santa Fe Symposium on Jewelry Manufacturing Technology*, ed. E. Bell, 151–168. Albuquerque, NM: Met-Chem Research and F. Klotz, 2007. A comparison of nickel white gold, palladium white gold, 950 platinum-ruthenium and 950 palladium in the manufacture of findings. In *Proceedings of the Santa Fe Symposium on Jewelry Manufacturing Technology*, ed. E. Bell, 123–268. Albuquerque, NM: Met-Chem Research.

26. Live and let die (struck). F. Klotz and S. Grice, *Gold Technology*, No. 36, 2002, 16–22.

27. M. Grimwade, 1992. Die press stamping and sheet metal forming operations. Santa Fe Symposium on Jewelry Manufacturing Technology, Albuquerque, NM. (Proceedings unpublished to date, but see report in *Gold Technology*, No. 8, November 1992, 2–8.)

28. G Normandeau, 1991. Bi-metal solder flush products and their use for high productivity manufacture. In *Proceedings of the Santa Fe Symposium on Jewelry Manufacturing Technology*, ed. D. Schneller, 149–170. Albuquerque, NM: Met-Chem Research.

29. M. Grimwade, 2002. *Handbook on Soldering and Other Joining Techniques,* London: World Gold Council.

30. The production of karat gold chain wire. P. Taimsalu, *Aurum*, No. 14, 1983, 49–55.

31. Manufacturing karat gold machine chain, P. Taimsalu, *Aurum,* No. 15, 1983, 43–46.

32a. Gold chains and mesh – I, J. Walters, *Aurum*, No. 31, Autumn 1987, 47–57.

32b. Gold chains and mesh – II, J. Walters, *Aurum*, No. 32, Winter 1987, 36–47.

32c. Gold chains and mesh – III, J. Walters, *Aurum*, No. 34, Summer 1988, 40–49.

32d. Gold chains and mesh – IV, J. Walters, *Aurum*, No. 36, Winter 1988, 72–77.

33. Joining of gold jewelry from Bombay, India to Valencia, Spain. H. H. Hildebrand, *Gold Technology*, No. 14, November 1994, 22–29.

34. A. M. Reti and P. A. Fossaluzza, 1991. Precious metal chain. In *Proceedings of the Santa Fe Symposium on Jewelry Manufacturing Technology*, ed. D. Schneller, 287–307. Albuquerque, NM: Met-Chem Research.

35. Overview: Joining technology in chain manufacture. A. Canaglia, *Gold Technology*, No. 17, October 1995, 20–25.

36. Metallurgical aspects of chain manufacture. D. P. Agarwal, *Gold Technology*, No. 17, October 1995, 16–18.

37. Evaluation of strength and quality of chains. D. P. Agarwal, *Gold Technology,* No. 24, September 1998, 2–5; *Proceedings of the Santa Fe Symposium on Jewelry Manufacturing Technology*, ed. D. Schneller, 1995, 23–268. Albuquerque, NM: Met-Chem Research.

38. Hollow karat gold jewelry from strip and tube. P. Raw, *Gold Technology*, No. 35, Summer 2002, 3–10.

39. The shape of the future – recent developments in electroforming, C. W. Corti. *AJM Magazine*, Issue 3/03, March 2003, 51–59.

40. Electroforming for innovative design. S. Müller and B. Biagi, *Gold Technology*, No. 27, November 1999, 41–46.

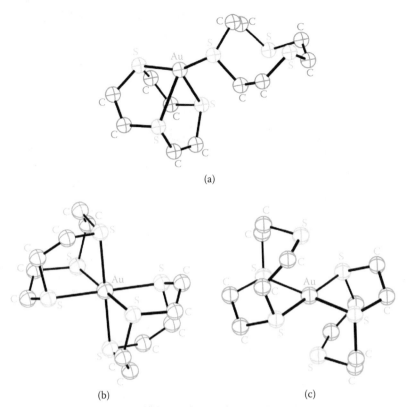

(a)

(b)   (c)

COLOR FIGURE 3.1   The structures of [Au(1,4,7-trithiacyclononane)$_2$]$^{+/2+/3+}$ showing the structural variations in different oxidation states. (a) Monocation Au^{I}, (b) DicationAu^{II} (c) Trication Au^{III}. For clarity hydrogen atoms are omitted. [Data from Bis(1,4,7-trithiacyclononane)gold dication: a paramagnetic, mononuclear AuII complex, A. J. Blake, J. A. Greig, A. J. Holder, T. I. Hyde, A. Taylor, and M. Schröder, *Angew. Chem. Int. Ed.* 1990, **29**, 197; and Gold thioether chemistry – synthesis, structure, and redox interconversion of [Au(1,4,7-trithiacyclononane)$_2$]$^{+/2+/3+}$, A. J. Blake, R. O. Gould, J. A. Greig, A. J. Holder, T. I. Hyde, and M. Schröder, *J. Chem. Soc., Chem. Commun.*, 1989, 876.]

COLOR FIGURE 3.2   Photograph of a thin gold film that has been briefly immersed into alcohol solution of 4-pyridinethiol. Dissolution of the gold film begins from the corners and edges. (From Oxidation of elemental gold in alcohol solutions, M. T. Risnen, M. Kemell, M. Leskel, and T. Repo, *Inorg. Chem.*, 2007, **46**, 3251. With permission.)

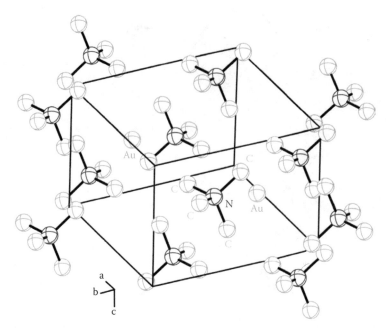

COLOR FIGURE 3.3 The crystal structure of $(Me_4N)Au^{-I}$. For clarity hydrogen atoms are not shown. (Data from Synthesis and crystal structure determination of tetramethylammonium auride, P. D. C. Dietzel and M. Jansen, *Chem. Commun.*, 2001, 2208.)

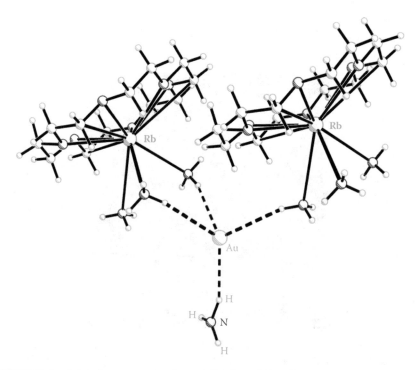

COLOR FIGURE 3.4 A drawing comparing the coordination of Rb^+ and Au^- and emphasizing the hydrogen bonding of ammonia molecules to the auride ion in $[Cs([18]crown-6)(NH_3)_3]Au^{-I} \cdot NH_3$. (Data from $[Rb([18] crown-6)(NH_3)_3]Au \cdot NH_3$: gold as acceptor in $N\text{-}H \cdots Au^-$ hydrogen bonds, H. Nuss and M. Jansen, *Angew. Chem. Int. Ed.*, 2006, **45**, 4369.)

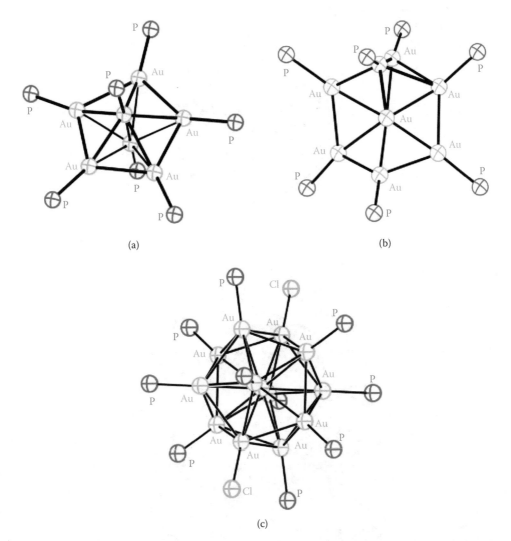

COLOR FIGURE 3.17 The structures of cluster cations. (a) [Au$_7$(PPh$_3$)$_7$]$^+$. (From Intercluster compounds consisting of gold clusters and fullerides: [Au$_7$(PPh$_3$)$_7$]C$_{60}$ •THF and [Au$_8$(PPh$_3$)$_8$](C$_{60}$)$_2$, M. Schulz-Dobrick and M. Jansen, *Angew. Chem. Int. Ed.*, 2008, **47**, 2256.) (b) [Au$_8$(PPh$_3$)$_7$]$^{2+}$.(From Reactions of cationic gold clusters with Lewis bases. Preparation and X-ray structure investigation of [Au$_8$(PPh$_3$)$_7$(NO$_3$)2.2CH$_2$Cl$_2$ and Au$_6$(PPh$_3$)$_4$[Co(CO)$_4$]$_2$, J. W. A. Van der Velden, J. J. Bour, W. P. Bosman, and J. H. Noordik, *Inorg. Chem.*, 1983, **22**, 1983.) (c) [Au$_{13}$(PMe$_2$Ph)$_{10}$Cl$_2$]$^{3+}$. (From Synthesis and X-ray structural characterization of the centered icosahedral gold cluster compound [Au$_{13}$(PMe$_2$Ph)$_{10}$Cl$_2$](PF$_2$)$_3$; the realization of a theoretical prediction, C. E. Briant, B. R. C. Theobald, J. W. White, L. K. Bell, and D. M. P. Mingos, *Chem. Commun.*, 1981, **5**, 201.) For clarity, phenyl rings were omitted.

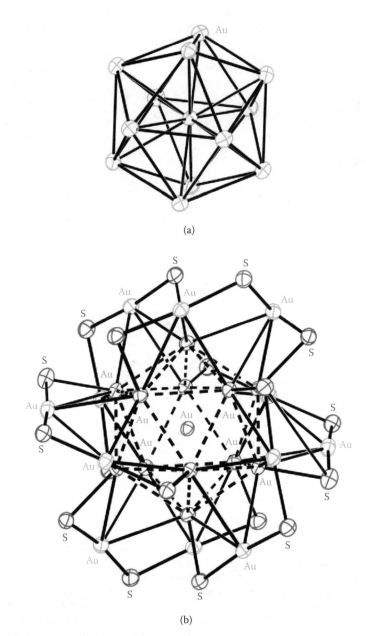

(a)

(b)

COLOR FIGURE 3.18 The structure of the $[Au_{25}(SCHCHPh)_{18}]^+$. (a) the central gold atom and the icosahedral array of 12 gold atoms at the core; (b) the entire cation without the 18 CH_2CH_2Ph groups. In (b) the 13 core gold atoms are shown in green and bonds to the central gold atom are omitted for clarity. (Data from Crystal structure of the gold nanoparticle $[N(C_8H_{17})_4][Au_{25}(SCH_2CH_2Ph)_{18}]$, M. W. Heaven, A. Dass, P. S. White, K. M. Holt, and R. W. Murray, *J. Am. Chem. Soc.*, 2008, **130**, 3754.)

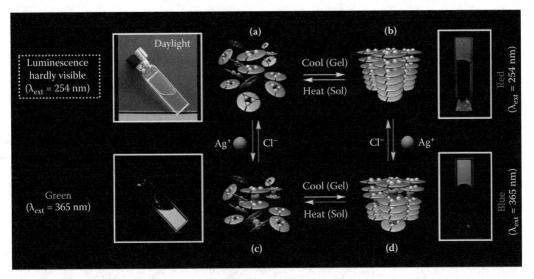

COLOR FIGURE 5.9 Luminescence profiles of the trinuclear gold(I) complex in hexane. Pictures and schematic self-assembling structures of (a) sol, (b) gel, (c) sol containing AgOTf (0.01 equiv.), and (d) gel containing AgOTf (0.01 equiv.). (Reproduced from Phosphorescent organogels via "metallophilic" interactions for reversible RGB-color switching, A. Kishimura, T. Yamashita, T. Aida, *J. Am. Chem. Soc.* 2005, **127**, 179. With permission.)

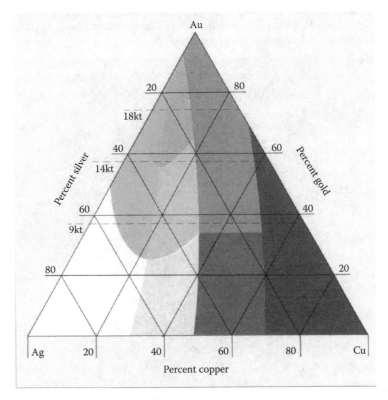

COLOR FIGURE 7.21 Color triangle indicating color variations in the ternary Au-Cu-Ag system [33]. (From A plain man's guide to alloy phase diagrams: their use in jewellery manufacture – Part 2, M. Grimwade, *Gold Technology*, 2000, **30**, 8. With permission.)

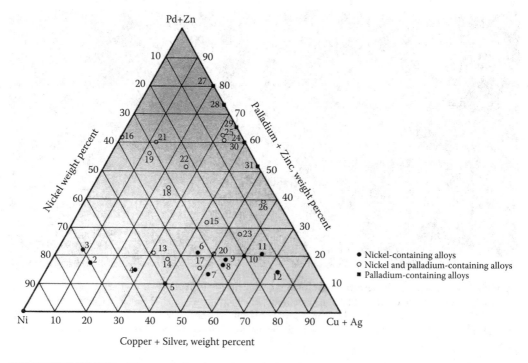

COLOR FIGURE 7.22 Diagram indicating color variations for 18ct white Au depending on Ni, Pd+Zn, and Cu+Ag contents [46]. (From 18ct white gold jewellery alloys: Some aspects of their metallurgy, C. P. Susz and M. Linker, *Gold Bulletin*, 1980, **13(1)**, 15. With permission.)

COLOR FIGURE 9.1 Casting tree after removal of investment.

COLOR FIGURE 9.5 Examples of electroformed 18 karat gold jewelry (Courtesy of Umicore Galvanotechnik and Carla Corporation).

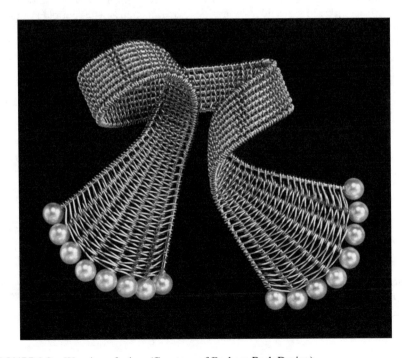

COLOR FIGURE 9.9 Weaving of wires (Courtesy of Barbara Berk Design).

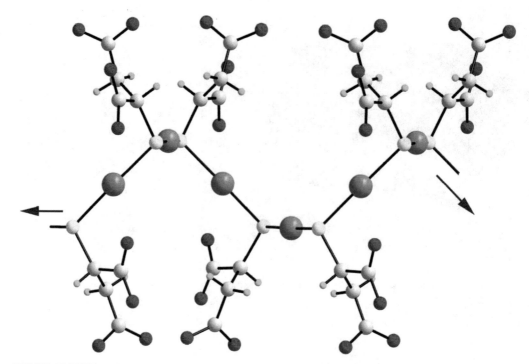

COLOR FIGURE 10.2 Representation of the supramolecular aggregation of the aurothiomalate anion, $[\{Au(SCH(CO_2)CH_2CO_2\}_2]^{4-}$, in the solid-state leading to helical strands with fourfold symmetry, as determined by X-ray crystallography. Color code: gold, orange; sulfur, yellow; oxygen, red; carbon, gray; and hydrogen, green.

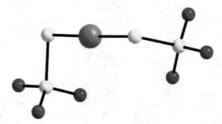

COLOR FIGURE 10.3 Representation of the molecular structure of the aurothiosulfate anion, $[Au(SSO_3)]^{3-}$, as determined by X-ray crystallography. Color code: gold, orange; sulfur, yellow; oxygen, red.

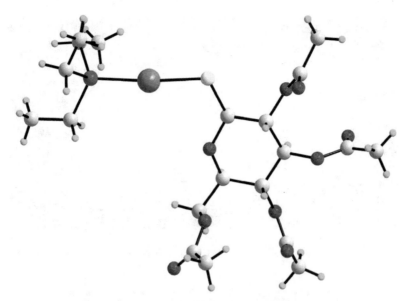

COLOR FIGURE 10.4 Representation of the molecular structure of triethylphosphinegold(I) tetraacetylthioglucose, Et₃PAuSATg, as determined by X-ray crystallography. Color code: gold, orange; sulfur, yellow; phosphorus, pink; oxygen, red; carbon, gray; and hydrogen, green

COLOR FIGURE 13.1 Gold bonding wires wound on spools, ready for use in the electronics industry.

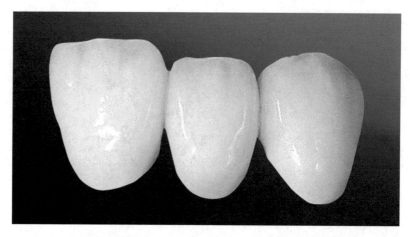

COLOR FIGURE 14.11 Goldtech Bio2000 three unit bridge. (From The Argen Corporation. With permission.)

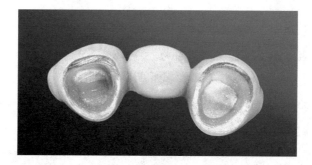

COLOR FIGURE 14.12 Obverse of Goldtech Bio2000 three unit bridge. (From The Argen Corporation. With permission.)

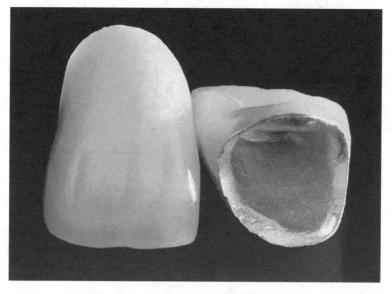

COLOR FIGURE 14.13 Crowns made from electroplated gold. (From Wieland Dental + Technik GmbH & Co. With permission.)

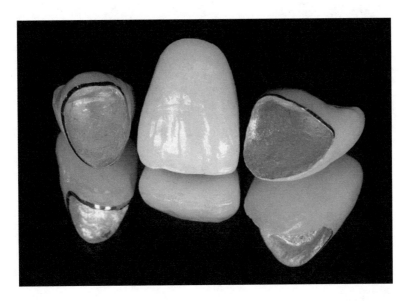

COLOR FIGURE 14.16 Crowns fabricated from CAPTEK®. (From Precious Chemicals, Ltd. With permission.)

COLOR FIGURE 14.17 Sintered crowns fabricated by 3D printing. (Image courtesy of imagen. With permission.)

COLOR FIGURE 14.18 Crowns infiltrated with gold ready for porcelain. (From imagen. With permission.)

(a)　　　　　　　　　　　　　　　　(b)

COLOR FIGURE 15.1 (a) Depicts a bowl having bright gold bands carried out by hand gilding for the handles, and machine banding where the gold is aligned with the color. (b) Depicts a combination of burnish and raised matt gold effects with multicolored enamel decoration. Application is by a combination of decal and machine banding.

COLOR FIGURE 15.2 An ornately decorated flute glass containing bright-banded gold and raised matte gold on the thick filmed area.

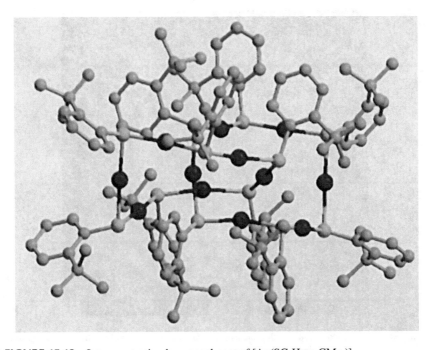

COLOR FIGURE 15.12 Interpenetrating hexagonal core of $[Au(SC_6H_4\text{-}o\text{-}CMe_3)]_{12}$.

COLOR FIGURE 15.14 Ink-jet printed gold oleylamine nanoparticle films printing onto paper at room temperature.

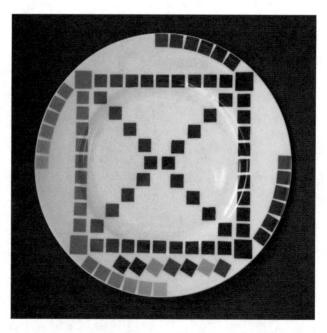

COLOR FIGURE 15.16 Display plate depicting the differing shades of Purple of Cassius available to a decorator. The light shades on the left have low concentrations of gold, 0.5–1 wt%, whereas the darker shades have 2–2.5 wt% gold present. The color hue can be varied by addition of silver during the preparation. The silver will alloy with gold and cause a red shift in the colored appearance.

COLOR FIGURE 15.17 Sèvres No. 601 developed at the *Manufacture de Sèvres*.

COLOR FIGURE 15.18 Gold colloids (5–35 nm in size) deposited onto three metal oxides with differing dielectric constants; red (on silica) dielectric constant = 4.5; purple (on alumina) dielectric constant = 9–11.5; blue (on titania), dielectric constant = 110. The color shifts from red to blue with increasing dielectric constant.

COLOR FIGURE 15.23 A colored glass film prepared from the comb-shaped copolymer stabilized gold nanoparticles and combined with an oxide sol gel to produce a durable low temperature "type" Purple of Cassius.

COLOR FIGURE 15.25   The head of a silver statuette of a youth with a gilded headdress from the Treasure of the Oxus (fifth century BC). It is not clear from the picture, but the seam around the edge of the head-dress where the two gold sheets come together has been burnished (shinier area).

COLOR FIGURE 15.26   Funerary papyrus of Neforronpet (fourteenth century BC), who is depicted with his wife Hunro. The papyrus was originally decorated with gold leaf on the head-bands, collars, armlets, and ankles of the two figures, but most is unfortunately now lost.

41. Production of gold jewelry by electroforming. M. Perrone, *Aurum*, No. 10, 1982, 31–33.
42. Electroforming of karat gold alloys – I. General principles. G. Desthomas, *Aurum*, No. 14, 1983, 19–26; repeated in *Gold Technology*, No. 4, May 1991, 2–9.
43. Electroforming of karat gold alloys – II. Jewelry production. G. Desthomas, *Aurum*, No. 15, 1983, 17–21.
44. Electroforming in gold jewelry production. F. Simon, *Gold Technology*, No. 4, May 1991, 10–15.
45. Electroforming of gold alloys – the Artform process. G. Desthomas, *Gold Technology*, No. 16, July 1995, 4–15.
46. Recent developments in the field of electroforming, a production process for hallmarkable hollow jewelry. F. Simon, *Gold Technology*, No. 16, July 1995, 22–29.
47. Pino Aliprandini, Exhibited at Basel Fair, 2007.
48. M. F. Grimwade, 2002. *Handbook on Soldering and Other Joining Techniques*, London: World Gold Council.
49. Health, safety and environmental pollution in gold jewelry manufacture. M. F. Grimwade, *Gold Technology*, No. 18, April 1996, 4–10.
50. J. Bellows, 1988, Work related hazards in the jewelry industry, 89–103; C. Freeman and C. Jones, 1989, Cadmium toxicity and health effects among jewellers, 131–160; P. Pryor, 1990, Personal protection equipment versus engineering/ventilation controls, 187–207. In *Proceedings of the Santa Fe Symposium on Jewelry Manufacturing Technology*, ed. D. Schneller. Albuquerque, NM: Met-Chem Research.
51. G. Todd, 2005. Standardizing the designation of karated gold solders. In *Proceedings of the Santa Fe Symposium on Jewelry Manufacturing Technology*, ed. E. Bell, 471–487. Albuquerque, NM: Met-Chem Research.
52. G. Normandeau, 1989. Cadmium-free gold brazing alloys. In *Proceedings of the Santa Fe Symposium on Jewelry Manufacturing Technology*, ed. D. Schneller, 179–209. Albuquerque, NM: Met-Chem Research.
53. G. Normandeau, 1990. Cadmium free gold solders: An update: Indium toxicity and potential workplace exposures. In *Proceedings of the Santa Fe Symposium on Jewelry Manufacturing Technology*, ed. D. Schneller, 239–292. Albuquerque, NM: Met-Chem Research.
54. Cadmium-free gold solder alloys. G. Normandeau, *Gold Technology*, No. 18, April 1996, 20–24.
55. Development of 21 karat cadmium-free gold solders. D. Ott, *Gold Technology*, No. 19, July 1996, 2–6.
56. K. Weisner, 2005. Sintering technology for jewelry and multicolor rings. In *Proceedings of the Santa Fe Symposium on Jewelry Manufacturing Technology*, ed. E. Bell, 501–519. Albuquerque, NM: Met-Chem Research.
57. J. Binnion, 2004. Non-traditional Mokume Gane materials: Diffusion bonding of iron to precious metals. In *Proceedings of the Santa Fe Symposium on Jewelry Manufacturing Technology*, ed. E. Bell, 87–104. Albuquerque, NM: Met-Chem Research.
58. V. Faccenda, 1999. *Handbook on Finishing*, London: World Gold Council.
59. Polishing: The basic principles. V. Faccenda and C. W. Corti, *Gold Technology*, No. 26, July 1999, 11–15.
60. Bulk finishing of gold jewellery – I. Wet tumbling: Basic principles. M. Dreher and P. Taimsulu, *Aurum*, No. 9, 1982, 34–39; also in *Gold Technology*, No. 26, July 1999, 2–6.
61. Bulk finishing of gold jewellery – II. Tumbling in practice. M. Dreher and P. Taimsulu, *Aurum*, No. 10, 1982, 46–51.
62. Bulk finishing of gold jewellery – III. Centrifugal tumbling. M. Dreher and P. Taimsulu, *Aurum*, No. 13, 1983, 45–48.
63. Getting gold to glitter or all that glitters is gold – the mass finishing of jewellery. S. Alviti, *Gold Technology*, No. 26, July 1999, 11–15.
64. Polishing up on finishing: Finishing jewelry by machine. M. Moser, *Gold Technology*, No. 31, Spring 2001, 29–34.
65. S. Alviti, 2002. The effects of burnishing on the surface of cast gold and silver jewelry. In *Proceedings of the Santa Fe Symposium on Jewelry Manufacturing Technology*, ed. E. Bell, 25–38. Albuquerque, NM: Met-Chem Research.
66. S. Alviti, 2003. The effects of burnishing on the surface of cast gold and silver jewelry: Phase II. In *Proceedings of the Santa Fe Symposium on Jewelry Manufacturing Technology*, ed. E. Bell, 1–14. Albuquerque, NM: Met-Chem Research.
67. M. Moser, 2003. How to get that 'ready to sell' finish from your disc-finishing machine. In *Proceedings of the Santa Fe Symposium on Jewelry Manufacturing Technology*, ed. E. Bell, 267–288. Albuquerque, NM: Met-Chem Research.
68. M. Moser, 2006. Important parameters that influence the result in mass-finishing. In *Proceedings of the Santa Fe Symposium on Jewelry Manufacturing Technology*, ed. E. Bell, 377–386. Albuquerque, NM: Met-Chem Research.

69. M. Dreher, 1991. Comparative study on metal finishing techniques on standard sample of cast gold. In *Proceedings of the Santa Fe Symposium on Jewelry Manufacturing Technology*, ed. D. Schneller, 27–44. Albuquerque, NM: Met-Chem Research.

70. D. Goodrich, 1994. Practical mass finishing techniques and applications for jewelry. Santa Fe Symposium on Jewelry Manufacturing Technology, Albuquerque, NM: Met-Chem Research. (Proceedings not yet published but see report by M. Grimwade in *Gold Technology*, No. 15, April 1995, 16–27.)

71. Back to basics: Electroplating and electropolishing. C. W. Corti, *Gold Technology*, No. 35, Summer 2002, 19–26.

72. Gold wedding rings from powder – tomorrow's technology today. P. Raw, *Gold Technology*, No. 27, November 1999, 2–8.

73. P. Raw, 2000. Mass production of gold and platinum wedding rings using powder metallurgy. In *Proceedings of the Santa Fe Symposium on Jewelry Manufacturing Technology*, ed. E. Bell, 251–270. Albuquerque, NM: Met-Chem Research.

74. Investing in technology: B & N Rings Ltd, C. W. Corti, *Technical Bulletin (of the Goldsmiths' Company, London)*, No. 5, April 2007, 10–11.

75. The role of CAD/CAM in the modern jewelry business. L. C. Molinari, M. C. Megazzini, and E. Bemporad, *Gold Technology*, No. 23, April 1998, 3–7.

76. Design opportunities through production technology. W. Böhm, *Gold Technology*, No. 23, April 1998, 8–11.

77. S. Adler and T. Fryé, 2005. The revolution of CAD/CAM in the casting of fine jewelry. In *Proceedings of the Santa Fe Symposium on Jewelry Manufacturing Technology*, ed. E. Bell, 1–24. Albuquerque, NM: Met-Chem Research.

78. S. Patrick, 2005. CAD software for jewelry design: A comprehensive survey. In *Proceedings of the Santa Fe Symposium on Jewelry Manufacturing Technology*, ed. E. Bell, 409–422. Albuquerque, NM: Met-Chem Research.

79. CAD-CAM technology: transforming the goldsmith's workshop. M. G. Malagoli, *Gold Technology*, No. 34, Spring 2002, 31–35.

80. Rapid prototyping: Application to gold jewelry production. L. C. Molinari, M. C. Megazzini, and A. Ungarelli, *Gold Technology*, No. 20, November 1996, 10–16.

81. S. Adler, 2008. Reverse engineering of complex geometries utilizing X-ray tomography for CAD/CAM applications. In *Proceedings of the Santa Fe Symposium on Jewelry Manufacturing Technology*, ed. E. Bell, 1–15. Albuquerque, NM: Met-Chem Research.

82. A. Nooten-Boom II, 2005. CAD/CAM for models and molds with CNC. In *Proceedings of the Santa Fe Symposium on Jewelry Manufacturing Technology*, ed. E. Bell, 365–398. Albuquerque, NM: Met-Chem Research.

83. T. Norlén, 2006. Jewelry manufacturing using selective laser technique. In *Proceedings of the Jewelry Technology Forum*, June 2006, 25–35. Vicenza, Italy: Legor Group.

84. J.T. Strauss, 2008. Powder metallurgy in jewelry manufacturing: Status report and discussion. In *Proceedings of the Santa Fe Symposium on Jewelry Manufacturing Technology*, ed. E. Bell, 295–306. Albuquerque, NM: Met-Chem Research.

85. J. Robertson, April 2005. An Introduction to Computer Aided Design (CAD) for the Jewelry & Silversmithing Industries. Technical Report, The Goldsmiths' Company, London (available in print or CD-ROM).

86. F. Cooper and A.-M. Carey, April 2004. Rapid Prototyping Applications for the Jewelry & Silversmithing Industries. Technical Report, The Goldsmiths' Company, London (available in print or CD-ROM); and An Introduction to Rapid Prototyping – From CAD to Casting. DVD No. 4 in Technology and Training Masterclass series, 2008, The Goldsmiths' Company, London.

87. New laser process technologies for optimised gold jewellery manufacture. C. Esposito, R. Faes, M. L. Vitobello van der Schoot, *Gold Technology*, No. 20, November 1996, 30–34.

88. B. Berk, 2004. Textile techniques in metal: A designer's perspective on sheet and wire. In *Proceedings of the Santa Fe Symposium on Jewelry Manufacturing Technology*, ed. E. Bell, 55–86. Albuquerque, NM: Met-Chem Research; *ibid*, 2006. Part II. In *Proceedings of the Santa Fe Symposium on Jewelry Manufacturing Technology*, ed. E. Bell, 55–100. Albuquerque, NM: Met-Chem Research.

89. Laser applications in gold jewellery production. S. Valenti, *Gold Technology*, No. 34, Spring 2002, 14–20.

90. D. A. Goodrich, 1997. Nd:YAG laser engraving of jewelry. In *Proceedings of the Santa Fe Symposium on Jewelry Manufacturing Technology*, ed. D. Schneller, 313–338. Albuquerque, NM: Met-Chem Research.

91. C. Volpe and R. D. Lanam, 1998. Utilization of lasers for the joining of gold and platinum for jewelry. In *Proceedings of the Santa Fe Symposium on Jewelry Manufacturing Technology*, ed. D. Schneller, 97–141. Albuquerque, NM: Met-Chem Research.

92. Breakthrough in laser marking for precious metals. Anon., *Gold Technology*, No. 24, September 1998, 10–11.

93. J. Jones, 2007. Laser safety in jewelry manufacturing. In *Proceedings of the Santa Fe Symposium on Jewelry Manufacturing Technology*, ed. E. Bell, 229–230. Albuquerque, NM: Met-Chem Research.

94. S. Silve and H. Zhao, 2004. Laser forming as a method of producing designed objects. In *Proceedings of the Santa Fe Symposium on Jewelry Manufacturing Technology*, ed. E. Bell, 401–434. Albuquerque, NM: Met-Chem Research.

95. S. Silve and H. Zhao, 2005. Laser bending silver. In *Proceedings of the Santa Fe Symposium on Jewelry Manufacturing Technology*, ed. E. Bell, 423–469. Albuquerque, NM: Met-Chem Research.

96. A. J. Menon, 2003. Areas of research and developments for the jewelry industry. In *Proceedings of the Santa Fe Symposium on Jewelry Manufacturing Technology*, ed. E. Bell, 227–242. Albuquerque, NM: Met-Chem Research.

97. J. T. Strauss, 1997. Powder metallurgy (P/M) applications in jewelry manufacturing. In *Proceedings of the Santa Fe Symposium on Jewelry Manufacturing Technology*, ed. D. Schneller, 105–131. Albuquerque, NM: Met-Chem Research.

98. Metal injection molding (MIM) for gold jewelry production. J. T. Strauss, *Gold Technology*, No. 20, November 1996, 17–29.

99. J. T. Strauss, 2003. P/M (Powder metallurgy) in jewelry manufacturing: Current status, new developments and future projections. In *Proceedings of the Santa Fe Symposium on Jewelry Manufacturing Technology*, ed. E. Bell, 387–412. Albuquerque, NM: Met-Chem Research.

100. K. Wiesner, 2003. Metal injection molding (MIM) technology with 18 ct gold – a feasibility study. In *Proceedings of the Santa Fe Symposium on Jewelry Manufacturing Technology*, ed. E. Bell, 443–462. Albuquerque, NM: Met-Chem Research.

101. J. T. Strauss, 2008. Powder metallurgy in jewelry manufacturing: Status report and discussion. In *Proceedings of the Santa Fe Symposium on Jewelry Manufacturing Technology*, ed. E. Bell, 295–306. Albuquerque, NM: Met-Chem Research.

102. Uses of gold in jewelry. E. Drost and J. Hausselt, *Interdisciplinary Science Reviews*, 17(3), September 1992, 271–280.

103. Design opportunities through innovative materials. S. Takahasi, N. Uchiyama, and A. Nishio, *Gold Technology*, No. 23, April 1998, 12–17.

104. Rapid prototyping, Part I. The current 'state of the art' and future developments for rapid prototyping, F. Cooper, *Technical Bulletin (of the Goldsmiths' Company, London)*, No. 7, April 2008, 10–11; see also Part II, No. 8, September 2008, 4–9.

10 Biomedical Applications of Gold and Gold Compounds

Elizabeth A. Pacheco, Edward R. T. Tiekink,
and Michael W. Whitehouse

CONTENTS

Introduction..217
Chrysotherapy—The Use of Gold In Medicine..218
Pharmacokinetics of Gold Drugs..219
Mechanisms of Action(s) of Gold Drugs..221
Aurocyanide [Dicyanoaurate(I)]—an Important Metabolite..221
Other Gold Metabolites...222
Antitumor Properties of Gold(I) and Gold(III) Compounds..223
Some Other Promising Therapeutic Uses for Gold..226
    Nanoparticles...226
    Malaria...226
    Hiv (Aids Virus)...227
    Asthma...227
Conclusions and Future Prospects..227
References..228

INTRODUCTION

Since antiquity, gold has been prized not only for its beauty but also for medicinal purposes. The therapeutic use of gold in China and other ancient civilizations has been traced back to around 2500 BC [1]. *Swarna bhasma* (gold ash) is a remedy containing pure gold used by ayurvedic physicians in India since ancient times to treat an array of disorders and diseases including loss of memory, infertility, bronchial asthma, rheumatoid arthritis, and diabetes mellitus [2]. Gold and silver coatings on medications made by Taoist philosophers of ancient China circa 600 BC were used to help enhance the quality of their potions [1]. The Roman naturalist Pliny the Elder and Greek physician Dioscorides both recorded the medicinal use of gold in the first century AD [1].

The modern medicinal use of gold, termed *chrysotherapy*, developed from the pioneering work of the German microbiologist and Nobel laureate Robert Koch. Koch's report that gold cyanide exhibited antibacterial effects against the tubercle bacillus *in vitro* initiated studies for the treatment of both human and bovine forms of tuberculosis with gold derivatives over the next 40 years [1]. An assumption by the French physician Jacques Forestier, namely, that rheumatoid arthritis (RA) was an infectious disease analogous to tuberculosis, led to clinical studies showing that gold compounds did indeed have antiarthritic activity *in vivo*. While there is still inconclusive evidence concerning the causes of RA, there is no doubt that the introduction of gold compounds to treat arthritis was an important milestone in therapeutic medicine. Recently, there has been a growing interest in gold compounds being used as antitumor agents as well as their possibly of possessing antiviral activity.

CHRYSOTHERAPY—THE USE OF GOLD IN MEDICINE

Two classes of chrysotherapeutic agents are currently used to treat rheumatoid arthritis. Gold(I) thiolates were the first class of drugs used in this context. They are generally polymeric, charged, water-soluble [3], and usually administered intramuscularly owing to their poor absorption from the gut. Examples include sodium aurothiomalate with various trade names (e.g., Myochrisine®, Myocrisin®, and Tauredon®), aurothioglucose (e.g., Aureotan®, Solganol®, and Auromyose®), sodium aurothiosulfate (e.g., Sanochrysin® and Fosforcrisolo®) and sodium aurothiopropanol sulfonate (e.g., Allochrysine®); their chemical formulae are depicted in Figure 10.1. Crystals have not been obtained for all of these compounds, precluding detailed X-ray crystallographic analysis, so their real or precise solid-state structures may be more complex than otherwise assumed, as in Figure 10.1.

A crystal structure is available for the aurothiomalate anion in the salt $[Na_2Cs(H)][\{Au(SCH(CO_2)CH_2CO_2\}_2]$, shown in Figure 10.2 [4]. While the cation does not correspond to that in the medicinal formulation, the observed polymeric structure found for the anion conforms to expectation. The structure has a gold–sulfur backbone comprising two interpenetrating spirals that each has fourfold helical symmetry. A racemic mixture is formed, with R-thiomalate forming right-handed helices and S-thiomalate forming left-handed helices. In the structure there are two crystallographically distinct gold atoms with equivalent Au-S distances of 229 pm [4]. By contrast, the bond angle of one of the gold atoms is essentially linear, i.e., 179°, while some distortion is evident about the second gold atom having an S-Au-S angle of 169°. The Au-S-Au angle is approximately 99° [4].

A crystal structure is also available for the aurothiosulfate anion, $[Au(SSO_3)_2]^3$[5], as illustrated in Figure 10.3. Its monomeric configuration again illustrates the tendency of gold to exist in linear coordination geometries. Here, the sulfur atoms are monodentate in contrast to the bidentate bridging mode seen in the structure of aurothiomalate [4], which must occur as there is a 1:1 ratio between gold and sulfur by contrast to the aurothiosulfate anion which has a 1:2 gold to (coordinating) sulfur ratio.

At present, the second class of chrysotherapeutic agents comprises only one member, triethylphosphinegold(I) tetraacetylthioglucose (Et$_3$PAuSATg), commonly known as auranofin and marketed as, for example, Aktil®, Crisinor®, Crisofin®, or Ridaura®. Unlike the first class of gold thiolates, auranofin is monomeric, neutral, and administered orally due to its lipophilicity [3]. Its

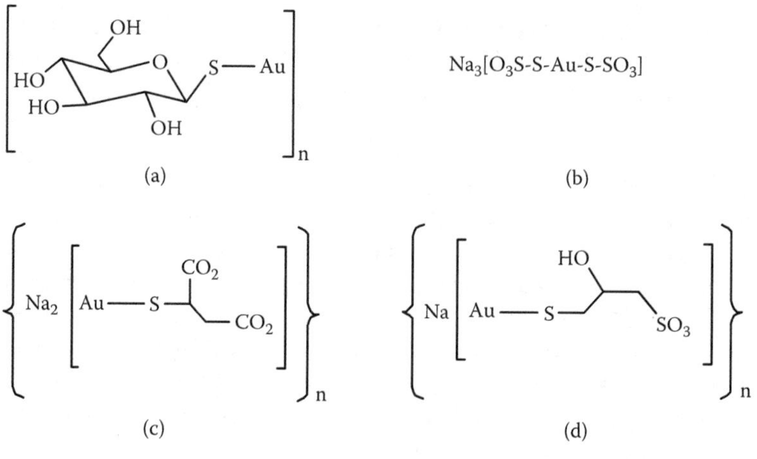

(a)

$Na_3[O_3S\text{-}S\text{-}Au\text{-}S\text{-}SO_3]$

(b)

(c)

(d)

FIGURE 10.1 Chemical structures of gold thiolates used in the treatment of rheumatoid arthritis: (a) aurothioglucose, (b) sodium aurothiosulfate, (c) sodium aurothiomalate, and (d) sodium aurothiopropanol sulfonate.

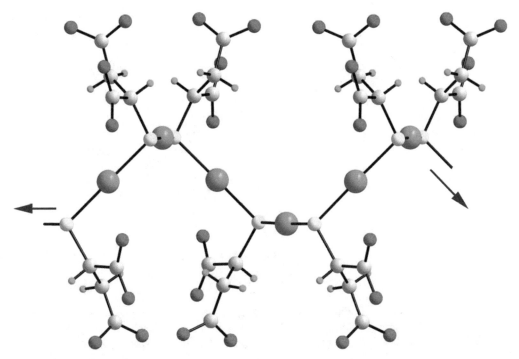

FIGURE 10.2 (See color insert following page 212.) Representation of the supramolecular aggregation of the aurothiomalate anion, $[\{Au(SCH(CO_2)CH_2CO_2)_2\}_2]^{4-}$, in the solid-state leading to helical strands with fourfold symmetry, as determined by X-ray crystallography. Color code: gold, orange; sulfur, yellow; oxygen, red; carbon, gray; and hydrogen, green.

crystal structure has been determined and reveals a two-coordinate gold(I) compound containing the anion derived from 2,3,4,6-tetra-O-acetyl-β-1D-thiolglucose and triethylphosphine as ligands [6] (see Figure 10.4). Although auranofin has a linear coordination geometry similar to the first class of gold therapeutic agents, its geometry is defined by sulfur and phosphorous atoms rather than two sulfur atoms. The Au-S bond distance is 229 pm while the Au-P bond distance is shorter at 226 pm. The S-Au-P bond angle of 174° is slightly distorted due to steric effects [6].

PHARMACOKINETICS OF GOLD DRUGS

Due to the chemical differences between the binary gold thiolates and auranofin, as well as their different modes of administration, they demonstrate varying modes of biodistribution. Because the actual structures in solution of many of the injectable gold thiolates are still largely unknown,

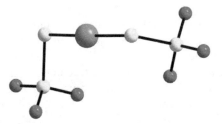

FIGURE 10.3   (See color insert following page 212.) Representation of the molecular structure of the aurothiosulfate anion, $[Au(SSO_3)]^{3-}$, as determined by X-ray crystallography. Color code: gold, orange; sulfur, yellow; oxygen, red.

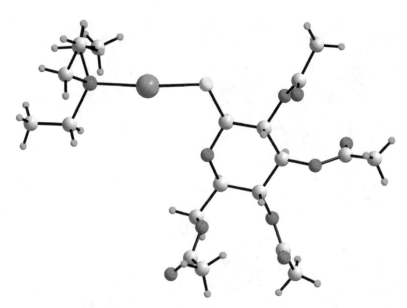

FIGURE 10.4 (See color insert following page 212.) Representation of the molecular structure of triethylphosphinegold(I) tetraacetylthioglucose, Et₃PAuSATg, as determined by X-ray crystallography. Color code: gold, orange; sulfur, yellow; phosphorus, pink; oxygen, red; carbon, gray; and hydrogen, green.

pharmacokinetic studies rely heavily on the measurement of elemental gold. These injectable thiolates are usually administered as solutions or suspensions containing 50 mg of gold intramuscularly once a week to be delivered into and via the vascular system [3]. Over 95% of administered gold is bound to albumin. The remainder of the gold is bound to the macroglobulins [7,8]. Nearly 75% of gold is excreted via the kidney while the remainder is excreted in the faeces. There is wide distribution throughout the reticulo-endothelial system after the administration of gold compounds via injection. In particular, gold is found in the bone marrow, lymph nodes, phagocytic cells of the liver, and the synovium, the tissue overlaying and nourishing diarthrodal joints [9].

Auranofin is usually given daily in capsules containing 3 or 6 mg gold and delivered via the gastrointestinal tract [3]. Like the injectable gold therapies, auranofin also binds primarily to albumin [10,11]. Following the administration of [195]Au-labeled auranofin, approximately 25% of the administered dose was detected in the plasma within 1–2 h, implying that only small amounts of gold are actually absorbed and retained by the body [12,13]. The half-life of the gold in plasma is around 15–25 days. The gold is almost entirely eliminated from the body in 55–80 days [14]. Gold from orally ingested auranofin is excreted much more quickly than the injectable gold thiolates. Consequently, radiolabeling studies show only around 1% of [195]Au from auranofin is still detectable in the body after 180 days whereas after the same period of time, up to 30% of [195]Au from gold sodium thiomalate is still detectable [15].

Due to their differences in retention and distribution, injectable gold thiolates and auranofin elicit different therapeutic responses and side effects. Injectable gold treatments induce significant therapeutic benefits after 3–6 months of treatment, including reducing the pain and swelling of inflamed joints, and preventing further bone and cartilage destruction. Adverse side-effects from treatment may be manifested almost immediately and can sometimes persist for a few months after the last administration. They include metallic taste, oral ulcers, kidney and bone marrow problems, dermatitis, discoloration of the skin, and diarrhea [1,3]. Side effects of auranofin include diarrhea, skin rash, and gastrointestinal problems. In comparative trials [16], injectable gold sodium thiomalate was more effective than auranofin but more patients (perhaps as many as 30%) may also abandon treatment due to these side effects.

MECHANISMS OF ACTION(S) OF GOLD DRUGS

Rheumatoid arthritis is not a well-understood disease: it cannot be cured, only treated. Many different drugs, exploiting differing mechanistic pathways, are used to treat the disease. Just as the structures of many of the gold drugs are still unknown, the mechanism(s) of action of therapeutic gold is not well understood. Nevertheless, 30% or more of gold-treated patients may achieve a state of remission [17].

After treatment with an injectable gold thiolate, most of the gold is bound to the protein serum albumin. Evidently, the administered gold drug undergoes a substitution reaction whereby one of the thiolate ligands (e.g., thiomalate) is displaced by an albumin molecule. Further transformation, again by ligand exchange, may involve the substitution of the second thiomalate for another albumin molecule. As a result, the gold atom becomes surrounded by protein and this gold–protein complex can then be transported to the therapeutic target, a site of inflammation. Once delivered to this site, the gold may be extracted from the protein carrier(s) by cyanide ions, a metabolite of thiocyanate that can be generated naturally within sites of inflammation. The anionic species $[Au(CN)_2^-]$, formed locally, may be the actual active metabolite [3].

The *in vivo* transformation(s) of auranofin follows a similar pattern. The gold–sulfur bond of auranofin is easily cleaved with the gold then binding to albumin. Next, another available thiol such as albumin or glutathione will coordinate to the gold. Accompanying this process is the reduction of a coordinated albumin molecule with concomitant oxidation of triethylphosphine (Et_3P) to triethylphosphineoxide ($Et_3P=O$), which is released. Now, the gold is surrounded by two bioligands (reduced albumin and albumin or glutathione) and may be carried to the site of inflammation where it may undergo further ligand exchange, e.g., with cyanide, as described previously [18,19].

While both classes of gold drugs undergo facile ligand exchange, gold monitored as the metal is generally retained *in vivo* much longer than the original dosed drug. When a study was carried out giving dogs triply-radiolabeled auranofin, i.e., $(Et_3^{31}P)^{195}Au(^{35}SATg)$, it was found that each component of the drug was rapidly distributed throughout the body but suffered a different metabolic fate. The half-lives for excretion of the ^{31}P and ^{35}S were 8 and 6 h, respectively, while the half-life for gold excretion was 20 days [17,20]. When the same radiolabeled auranofin was also added to whole blood *in vitro*, bond cleavage occurred with rapid ligand exchanges occurring for each component of auranofin [17]. Thus, within 20 min, gold was primarily transferred into the red blood cells and the acetylthioglucose was mainly bound in the serum, while the phosphine was distributed among the red blood cells, low-molecular-weight species, and, to a lesser extent, extracellular proteins [17].

Because the gold drugs used in chrysotherapy undergo such rapid ligand exchange *in vivo* and each component of the administered drug has a different metabolic fate, they should probably be considered *prodrugs*. The gold-bound ligands present in the original drug, being rapidly displaced *in vivo* by natural ligands (protein and low-molecular-weight thiols), are really no more than "carriers" for introducing the gold into the body. As the phosphine and acetylthioglucose ligands of auranofin are easily displaced, the cells and tissues are not exposed for extended periods to the dosed drug. Therefore, the protein-bound metabolites become the dominant species *in vitro*, to serve as a circulating reservoir from which other, more bioreactive gold metabolites might be engendered.

Because all the current antiarthritic gold drugs undergo facile ligand exchange with the same endogenous ligands (thiols, cyanide, and possibly others), both classes of drugs appear to have similar modes of action. Common metabolites, including metallic gold Au^0 and aurocyanide, $[Au(CN)_2^-]$, are formed from both water-soluble gold thiolates and water-insoluble auranofin.

AUROCYANIDE [DICYANOAURATE(I)]—AN IMPORTANT METABOLITE

Robert Koch first reported the bacteriostatic properties of aurocyanide, $[Au(CN)_2^-]$ [21]. This is actually a common metabolite of both oligomeric gold(I) thiolates and auranofin, present in the serum and excreted in the urine of patient receiving chrysotherapy [22]. In a study of patients receiving aurothiomalate treatment, it was found that gold was taken up by the red blood cells of smokers at a much higher level than by nonsmokers [18]. The reason was found to be that the hydrogen cyanide (HCN) present in inhaled cigarette smoke reacts with serum-bound gold to form $[Au(CN)^-_2]$, and thereby allows gold to be taken into red blood cells in the form of this low-molecular-weight "metabolite."

However, cyanide can also be locally generated at a site of inflammation such as within the joints of an arthritic patient. Here, any circulating thiocyanate can be converted into cyanide by hypochlorite, generated during the oxidative burst of activated leukocytes, attracted to the inflamed joint by various chemotactic factors. In these instances, the neutrophil enzyme myeloperoxidase converts gold thiomalate to aurocyanide via the oxidation of thiocyanate [23]. Aurocyanide, now produced locally within that site as an active metabolite, can be taken up by these cells to limit the extent of the oxidative burst and associated tissue damage, caused by reactive oxygen species (derived from inflammatory cells). Aurocyanide may be formed wherever myeloperoxidase is functional and thiocyanate is available. If there is an excess of thiols in these cells, thiolysis of the aurocyanide can occur liberating cyanide as a secondary endotoxin.

A major therapeutic effect of aurocyanide is that it inhibits the oxidative burst of polymorphonuclear leukocytes and the proliferation of lymphocytes *in vitro* [18]. These cells are active participants in the development and maintenance of inflammatory processes of rheumatoid arthritis. The fact that *in vitro* aurothiomalate actually weakly stimulates superoxide production by leukocytes, but, in the presence of thiocyanate, inhibits superoxide production indicates that aurocyanide is in fact an active metabolite of chrysotherapy [24]. An *in vivo* study further confirmed the intrinsic pharmaco-activity of aurocyanide by observing rats being treated with aurothiomalate. When treated with aurothiomalate alone, the development of collagen-induced arthritis in the rats was actually stimulated! However, when the rats were also given a low dose of NaSCN in their drinking water (in addition to their gold therapy) the aurothiomalate treatment significantly inhibited the development of collagen-induced arthritis [18].

OTHER GOLD METABOLITES

In addition to facilitating the conversion of a gold(I) thiolate to aurocyanide, myeloperoxidase in (or secreted from) polymorphonuclear leukocytes may also oxidize Au(I) to Au(III) [20]. It is now believed that Au(III) generated *in vitro* may be the cause of some of the side effects of treatment with monovalent gold drugs. In an attempt to understand the chemistry of Au(III)-biomolecule interactions, the oxidation of the amino acid glycine by Au(III) was investigated. Extensive deamination of glycine was observed with the formation of glyoxylic acid, NH^{4+}, formic acid, CO_2, and metallic gold [25,26].

It was concluded by Abraham and Himmel [27] that colloidal metallic gold (Au^0) might be a safer and more effective alternative to currently used gold thiolates for treating rheumatoid arthritis. They postulated that metallic gold may be the active "ingredient" in chrysotherapy because (1) gold thiolates can cause a high incidence of side effects, and (2) metallic gold (Au^0) was generated *in vivo*.

A small clinical trial tested the proposal that metallic gold might be an effective treatment for rheumatoid arthritis. Ten arthritis patients with erosive bone disease who were not responding to previous treatment were given 30 mg colloidal Au^0 orally daily for 24 weeks, then weekly for 4 weeks, and then finally monthly for 5 months [27]. There was no evidence of toxicity in the patients. However, there was a decrease in tenderness and swelling of joints after the first week that continued through the entire period of treatment.

More recently, a laboratory study established three forms of experimental arthritis (mycobacterial-induced arthritis, collagen-induced arthritis, and pristane-induced arthritis) in rats and then treated each group with either nanosized Au^0 or aurothiomalate [28]. Both gold preparations were administered subcutaneously. The development of all three forms of induced arthritis was suppressed by

Au^0, while aurothiomalate suppressed only the development of mycobacterial adjuvant. The Au^0 preparation was approximately 1000 times more potent than aurothiomalate (per gold content).

ANTITUMOR PROPERTIES OF GOLD(I) AND GOLD(III) COMPOUNDS

While the study of gold compounds for arthritis has been pursued for many decades, it is only since the 1980s that there has been ever-increasing interest in gold compounds for use as antitumor agents. One driving force behind this new interest is the current prominence of the inorganic drug cisplatin, *cis*-diamminedichloroplatinum(II) [29]. As cisplatin has been proven to be an effective oncolytic treatment, especially for head, neck, ovarian, and testicular cancers, it is logical that other metal-containing compounds should be investigated for efficacy [30]. Gold(III) compounds are of particular interest because gold(III) is isoelectronic with platinum(II) and both species form similar coordination compounds. Further, when a long-term (retrospective) study was conducted into the mortality rates of chrysotherapy patients, it was found that the patients' incidences of cancer were no higher, or even actually lower, than for patients not on chrysotherapy [31]. This may suggest a link between gold and anticancer activity. Furthermore, several known anticancer drugs, such as cyclophosphamide and 6-mercaptopurine, have also been used successfully as antiarthritic agents when other anti-inflammatory drugs have failed [32,33].

Both classes of antiarthritic gold drugs have been tested for potential antitumor activity. Early in the screening process it was observed that, generally, gold(I) thiolates were inactive. That the presence of a phosphine ligand is important for activity is shown by the data in Table 10.1. Much less activity was observed for the compounds without a phosphine ligand such as the gold(I) thiolates, as well as those with carbon or nitrogen donors [34]. The most active gold compounds included those with the P-Au-S arrangement, thus auranofin analogs seem to hold significant promise for the development of new and effective antitumor agents [35]. A clear difference between the two classes of gold drugs is that the lipophilic phosphinegold compounds can readily enter cells while the more hydrophilic oligomeric thiolates cannot. Consequently, the oligomeric thiolates do not accumulate in cells [36].

A comparative study with 6-mercaptopurine (6-MP) and Ph_3PAuCl demonstrates the gain in activity with the P-Au-S coordination arrangement. Individually, the thiol ligand and Ph_3PAuCl were potent. However, when combined together in a new gold compound, $Ph_3PAu(6-MP)$, a significantly more potent species was obtained [37].

Pharmacological studies of gold(I) compounds have most recently focused on complexes containing new ligands that may enhance cytostatic activity and be particularly effective against cisplatin-resistant cell lines. Gold(I) compounds with 1,2-bis(diphenylphosphine)ethane (dppe) and

---

TABLE 10.1
Cytotoxicity and Antitumor Activity of Selected Gold Thiolates and Auranofin Analogs

	B16 Melanoma IC_{50} (µM)	P388 Leukemia ILS_{max} (%)[a]
$[AuSH]_n$	166	15
$Au(SCH_2CH_2OH)$	140	24
$(C_5H_5N)AuCl$	125	35
$(CH_3CH_2)_3PAuSH$	2	68
$(CH_3CH_2)_3PAuCl$	1	36
$(CH_3CH_2)_3PAuCN$	0.4	68

[a] *Values* greater than 40 signify a reduction in the tumor at the termination of the trial.

Source: From Correlation of the in vitro cytotoxic and in vivo antitumor activities of gold(I) coordination complexes, C. K. Mirabelli, R. K. Johnson, D. T. Hill, et al., *J. Med. Chem.*, 1986, **29**, 218. (With permission.)

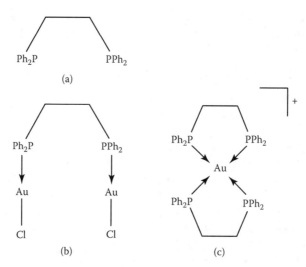

FIGURE 10.5 Chemical structures of (a) (diphenylphosphino)ethane (dppe), (b) [dppe(AuCl)$_2$], and (c) [Au(dppe)$_2$$^+$].

1,2-bis(dipyridylphosphine)ethane ligands, where a phenyl group has been substituted by a pyridyl substituent, have shown antitumor activity, especially among cisplatin-resistant cell lines; see Figure 10.5 for chemical structures. However, in contrast to cisplatin, DNA does not appear to be the principal target of these gold compounds. Instead, their cytotoxicity may derive from their ability to alter mitochondrial function and inhibit protein synthesis [38,39]. A study of gold-phosphines with various phenyl or pyridyl ligands revealed that their IC$_{50}$ values were dependent on the lipophilicity of the drug and therefore its likely uptake by the targeted tumor cells [40].

The compound [(AuCl)$_2$(dppe)] has shown activity against the M5076 reticulum cell sarcoma, subcutaneous mammary adenocarcinoma, and P388 leukemia, including a subline resistant to cisplatin [38]; see Figure 10.5 for chemical structure. Crucially, [(AuCl)$_2$(dppe)] undergoes ring closure when reacted with (further) dppe to form the tetrahedral compound [Au(dppe)$_2$]$^+$, so this latter cation may in fact be the active species. Though [Au(dppe)$_2$]$^+$ is stable in the presence of serum proteins, thiols, and disulfides, it is cytotoxic to cells lines, generating DNA-protein cross-links, and causing DNA strand breaks in cells. It preferentially inhibits protein synthesis relative to DNA and RNA synthesis [41].

Gold(III) compounds have also shown promise as potential antitumor agents [42]. The motivation for studying gold(III) compounds derives from their similarity to cisplatin because both gold(III) and platinum(II), as in cisplatin, generally form four-coordinate compounds with a square-planar geometry. However one clear difference between these species is that gold(III) is easily reduced, particularly in the reducing environment of mammalian tissues *in vivo*. However, through the careful choice of ligands including those with nitrogen or oxygen donors, the gold(III) oxidation state can be stabilized. In particular, the incorporation of multidentate ligands such as polyamines, terpyridine, and cyclam can also help stabilize gold(III) for therapeutic delivery; see Figure 10.6 for chemical structures.

When a series of Au(III) compounds was prepared incorporating the 2-dimethylaminomethylphenyl (damp) ligand, their antitumor activity suggested that proteins may be the main target of several of the investigated gold(III) compounds instead of DNA [43]; see Figure 10.6 for chemical structure of damp. Gold(III)-damp compounds show a preference for S-donor ligands such as glutathione and cysteine but only limited reactivity with nucleosides and their bases [43].

A number of gold(III) compounds have been prepared by Messori et al., including [Au(en)$_2$]Cl$_3$, [Au(dien)Cl]Cl$_2$, [Au(cyclam)](ClO$_4$)$_2$Cl, [Au(terpy)Cl]Cl$_2$, and [Au(phen)Cl$_2$]Cl [44]; see Figure 10.6 for chemical structures of the nitrogen-donor ligands. The compounds were stable under biological

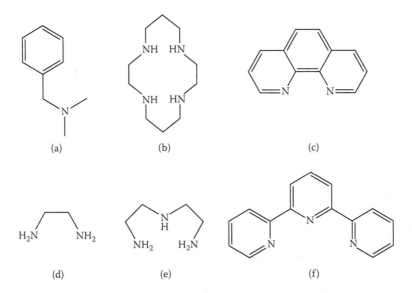

FIGURE 10.6 Chemical structures of (a) 2-dimethylaminomethylphenyl (damp), (b) 1,4,8,11-tetraazacy-clotetradecane (cyclam), (c) phenanthroline (phen), (d) 1,2-ethylenediamine (en), (e) diethylenetriamine (dien), and (f) terpyridine (terpy).

conditions and cytotoxic against the ovarian cell line A2780. The IC_{50} range of these compounds was 0.20–10 μM. By contrast, the free ligands ethylenediamine (en), diethylenetriamine (dien), and cyclam have IC_{50} values in excess of 100 μM, indicating the importance of its gold content for promoting cytotoxicity. By contrast, the uncoordinated phenanthroline (phen) and terpyridine (terpy) ligands were more cytotoxic than their gold(III) compounds. All the gold(III) compounds were shown to interact with DNA and extensively alter its solution behavior. So, in these compounds, DNA may represent the primary target as it is in therapy with most platinum antitumor compounds [26].

Other gold ligands have been studied with particular interest in compounds formed with dithiocarbamates (S_2CNR_2) and diphosphines. Both gold(I)- and gold(III)-dithiocarbamates are cytotoxic to tumors [45–48]. *In vitro* cytotoxicity screening indicated that some gold(III) dithiocarbamates are more cytotoxic than cisplatin by up to four orders of magnitude [47,48]. Compounds with diphosphine are more stable than auranofin due to reduced ligand exchange in the presence of serum proteins, thiols, and disulfides. Another advantage of investigating gold–diphosphine compounds is the opportunity to selectively substitute the R groups of the diphosphine ligands and therefore enhance or tailor the lipophilicity or hydrophilicity of the compounds [41], so there may be some selectivity for targeting the tumors rather than the host's normal tissues.

The mechanism(s) by which gold(III) compounds exert their cytotoxicity is often considered to resemble the cytotoxic action of platinum(II) compounds because of the similarities in the geometry of Au(III) and Pt(II) compounds. This simple physical analogy disregards the chief chemical differences between them, namely, that Au(III) is a powerful oxidant but Pt(II) is not. There are a few gold(III) compounds that appear to directly damage DNA, just like platinum(II) drugs. However, for most gold(III) compounds, DNA is not the primary target. Instead, gold(III) compounds may exert their cytotoxic effects by damaging mitochondrial functions through modification of essential specific proteins like thioredoxin reductase [41].

Gold(I) N-heterocyclic carbenes (NHC) constitute another class of compounds gaining prominence for their potential as antitumor agents, there being similarities between NHC ligands and phosphines ligands in terms of their metal coordination chemistry [49]. Consequently gold(I)-NHC compounds are being studied as alternatives to gold(I) phosphines, particularly to gain a greater understanding of the role that phosphine ligands may play in conferring antitumor activity [50,51]. Gold(I)-NHC compounds also show antimitochondrial activity [41]. NHC ligands are relatively easy to synthesize from imidazolium salts and form neutral or cationic gold compounds. With appropriately chosen ligands NHC compounds can now be synthesized to have variable hydrophilic and lipophilic properties [50,51].

SOME OTHER PROMISING THERAPEUTIC USES FOR GOLD

NANOPARTICLES

Recently, nanoparticles have become of great interest for their unique properties, especially those containing gold [52] (see Chapter 16). Reduction of water-soluble gold(III) compounds such as $HAuCl_4$ readily generates metallic gold (Au^0) nanoparticles [53]. Methods have been developed for attaching amino acids, peptides, proteins, etc. onto these nanoparticles to stabilize them, prevent their aggregation, and even render them water soluble. Currently, there is much interest in using nanoparticles as scaffolds for drug delivery [54,55].

Many conventional drugs, especially those given orally or intravenously, become widely dispersed throughout the body and do not localize specifically at the target of the therapy [54]. Therapies using nanoparticles may have more success targeting specific sites of interest, such as diseased tissues enriched with pathogenic cell populations (tumors, immunoinflammatory cells, etc.). This prospect may be fulfilled through either (1) passive targeting by controlling the nanoparticle size to ensure selective trapping at the site of interest, or (2) active targeting by attaching tissues-specific antibodies or ligands to the nanoparticle that have sufficient affinity to direct their homing to the desired target. Both targeting methods have limitations. Passive targeting is limited by its reliance on size-specific filtration to selectively deliver the nanoparticles. Nonspecific tissue-binding and/or activation of (unwanted) immune responses are some pitfalls of active targeting. When gold nanoparticles are selectively retained at a target site, it is possible to "activate" them through the absorption of radiation of an appropriate wavelength [55]. The radiation will either generate heat at the localized site (for tissue naturation) or facilitate local release of a chemical toxin prebound to the particle.

MALARIA

Over 400 million people throughout Africa, India, and South America are infected with the malaria parasite *Plasmodium falciparum* [56], so there is an urgent need to develop new drugs to combat present and evolving drug-resistant strains of this parasite. One stratagem has been to attach organic drugs to metal-containing fragments which can enhance the activity of the original drug [3,57]. Two examples include using chloroquine to treat malaria and clotrimazole for Chagas disease; see Figure 10.7 for chemical structures. Both Au(I) and Au(III) compounds containing chloroquine display *in vitro* activity against both chloroquine-sensitive *and* chloroquine-resistant strains of *Plasmodium falciparum* [56].

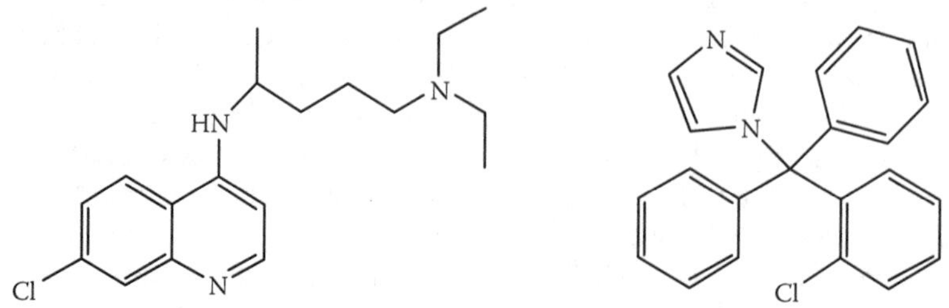

(a)   (b)

FIGURE 10.7   Chemical structures of (a) chloroquine and (b) clotrimazole.

HIV (AIDS VIRUS)

Gold compounds show some promise as antiviral agents, particularly for their potential anti-HIV activity in the context of new or supplementary treatments for the AIDS pandemic. The enzyme reverse transcriptase (RTase) converts viral RNA into DNA in the host cell. Aurothioglucose (AuSTg) has been reported as an inhibitor of RTase [58]. In addition it shows activity in cell-free extracts, but is unable to actually enter cells where RTase is active [59]. A related compound [Au(STm)$_2^-$] can be generated *in situ* (STm represents thiomalate). It appears to inhibit the infection of MT-4 cells by HIV strain HL4-3 without inhibiting the RTase activity of the virions. The target site has been identified as Cys-532 on gp160, a glycoprotein of the viral envelope [59]. However, oligomeric aurothiomalate, aurothioglucose, and auranofin were inactive.

H9 cells are a line of T-lymphocytes susceptible to HIV infection. Aurocyanide can be taken up by H9 cells and hinders HIV proliferation within these cells even at concentrations as low as 20 ppb. Concentrations as low as these are well tolerated by arthritic patients. Thus, aurocyanide might be a useful adjunct to other existing HIV treatments [60].

Metallic gold nanoparticles are similar in size to proteins. If they also carry multiple protein-binding ligands, these nanoparticles might be able to disrupt protein–protein interactions involved in disease pathogenesis, e.g., initial viral-binding to lymphocyte surface proteins. Many biologically inactive small molecules have been conjugated to gold nanoparticles. For example, when 2.0-nm-diameter gold nanoparticles were first coated with mercaptobenzoic acid and then conjugated with a derivative of a known CCR5 antagonist (SDC-1721), this otherwise inactive small molecule now became an active drug, i.e., by conjugation to the gold nanoparticles that inhibited HIV fusion to human T cells [61].

ASTHMA

Some gold drugs can inhibit molecular events underlying asthma, including IgE-dependent histamine release and neutrophil chemotaxis [62]. Two studies using oral gold and one using injectable gold examined their therapeutic effects in asthma patients [63]. In all three studies there was a reduction in requirements for oral corticosteroids (acting as anti-inflammatory agents), a useful attribute for sparing some of the adverse consequences of steroid therapy. However, there was no objective difference in the measure of lung function. Patients receiving gold therapy did show improvement in their subjective self-assessment and reduction in total serum IgE concentrations.

CONCLUSIONS AND FUTURE PROSPECTS

Biomedical applications for gold-based therapeutics have greatly expanded in recent years. While the mechanism(s) of action for gold drugs in rheumatoid arthritis patients are still not fully understood, even after 80 years of continued use, it is now recognized that many of these therapeutic gold compounds are transformed to common metabolites *in vitro*. It will be interesting to see if gold therapy can now be refined based on the concept that the antiarthritic gold(I) drugs, as currently used, are only *prodrugs* or precursors for generating more active metabolites, including, quite paradoxically, metallic gold itself. The development of these metabolites may lead to more potent drugs while at the same time reduce the total gold "burden" for the patients.

Even more promising is the potential for totally new therapeutic applications of gold compounds. The development of nanoparticles as scaffolds on which to construct new therapeutic entities (gold carrier *plus* specifically designed low-molecular-weight pharmaco-active species) may provide novel bioregulants that more efficiently and more directly target those tissues requiring therapeutic intervention. This strategy may also decrease side effects caused by random distribution and accumulation of the active drugs/toxins throughout other body compartments of the patients.

Both HIV and malaria are diseases in desperate need of new therapeutic developments to combat emerging resistant strains of the causative pathogens. Further study in these areas will be exciting, particularly if these life-threatening diseases can be controlled by new hybrid combinations of gold *plus* more conventional organic drugs.

REFERENCES

1. Clinical pharmacology of gold, W. F. Kean and I. R. L. Kean, *Inflammopharmacology*, 2008, **16**, 107.
2. Evaluation of chemical constituents and free-radical scavenging activity of *Swarna bhasma* (gold ash), an ayurvedic drug, A. Mitra, S. Chakraborty, B. Auddy, et al., *J. Ethnopharm.*, 2002, **80**, 147.
3. Gold compounds in medicine: potential anti-tumour agents, E. R. T. Tiekink, *Gold Bull.*, 2003, **36**, 117.
4. Crystal structure of the antiarthritic drug gold thiomalate (myochrysine): A double-helical geometry in the solid state, R. Bau, *J. Am. Chem. Soc.*, 1998, **120**, 9380.
5. Crystal structure of sodium gold(I) thiosulfate dihydrate $Na_3Au(S_2O_3)_2.2H_2O$, H. Ruben, A. Zalkin, M. O. Faltens, and D. H. Templeton, *Inorg. Chem.*, 1974, **13**, 1836.
6. (2,3,4,6-Tetra-O-acetyl-1-thio-β-D-glucopyranosato-S)(triethylphosphine)gold, $C_{20}H_{34}AuO_9PS$, D. T. Hill and B. M. Sutton, *Cryst. Struct. Comm.*, 1980, **9**, 679.
7. Difference in the pharmacokinetics, protein binding, and cellular distribution of gold when different gold compounds are used, J. D. Herrlinger, *Modern Aspects of Gold Therapy*, eds. M. Schettenkirchner and W. Muller, 1983, 52–57, Karger, Basel.
8. Determination of thiomalate in physiological fluids using high performance liquid chromatography and electrochemical detection, S. R. Rudge, D. Perret, P. L. Drury, et al., *J. Pharm. Bio. Anal.*, 1983, **1**, 205.
9. A biological effect on platelets by the minor component of gold sodium thiomalate – a by-product of heat sterilization and exposure to light, W. F. Kean, C. J. Lock, Y. B. Kassam, et al., *Clin. Exp. Rheum.*, 1984, **2**, 321.
10. The mammalian biochemistry of gold: An inorganic perspective of chrysotherapy, C. F. Shaw, *Inorg. Perspec. Biol. Med.*, 1979, **2**, 287.
11. Bis(L-cysteinato)gold(I): chemical characterization and identification in renal cortical cytoplasm, C. F. Shaw, G. Schmitz, H. O. Thompson, et al., *J. Inorg. Biochem.*, 1979, **10**, 317.
12. Blood gold concentrations in children with juvenile rheumatoid arthritis undergoing long term oral gold, E. H. Giannini, E. J. Brewer, and D. A. Person, *Ann. Rheum. Dis.*, 1984, **43**, 228.
13. Biologic actions and pharmacokinetic studies of auranofin, D. T. Walz, M. J. DiMartino, D. E. Griswold, et al., *Am. J. Med.*, 1983, **75**, 90.
14. Auranofin versus injectable gold. Comparison of pharmacokinetic properties, A. K. Blocka, *Am. J. Med.*, 1983, **75**, 114.
15. Pharmacodynamics of ^{195}Au-labeled aurothiomalate in blood, N. L. Gottlieb, P. M. Smith, and E. M. Smith, *Arthritis Rheum.*, 1974, **17**, 171.
16. Comparison of auranofin, gold sodium thiomalate, and placebo in the treatment of rheumatoid arthritis, J. R. Ward, N. J. Williams, M. J. Egger, et al., *Arthritis Rheum.*, 1983, **26**, 1303.
17. Pharmacokinetics of auranofin in animals, A. P. Intoccia, T. L. Glanagan, D. T. Walz, et al., *Bioinorganic Chemistry of Gold Coordination Compounds,* eds. B. M. Sutton, and R. G. Franz, 1983, 21–33, SK&F, Philadelphia.
18. Aurocyanide, dicyano-aurate(I), a pharmacologically active metabolite of medicinal gold complexes, G. G. Graham, M. W. Whitehouse, and G. R. Bushell, *Inflammopharmacology*, 2008, **16**, 126.
19. Redox chemistry and $[Au(CN)^-_2]$ in the formation of gold metabolites, C. F. Shaw III, S. Schraa, E. Gleichmann, et al., *Metal-Based Drugs*, 1994, **1**, 351.
20. Gold-based therapeutic agents, C. F. Shaw, *Chem. Rev.*, 1999, **99**, 2589.
21. Uber bacteriologische Forschung, R. Koch, *Dtsch. med. Wochenschr.*, 1890, **16**, 756.
22. Dicyanogold(I) is a common human metabolite of different gold drugs, R. C. Elder, Z. Zhao, Y. Zhang, et al., *J. Rheumatol.*, 1993, **20**, 268.
23. Chrysotherapy: A synoptic review, R. Eisler, *Inflammation Res.*, 2003, **52**, 487.
24. The activation of gold complexes by cyanide produced by polymorphonuclear leukocytes. II. Evidence for the formation and biological activity of aurocyanide, G. G. Graham and M. M. Dale, *Biochem. Pharmacol.*, 1990, **39**, 1697.
25. Gold(III)-induced oxidation of glycine, Z. Juan, G. Zijian, J. A. Parkinson, et al. *Chem. Commun.*, 1999, **15**, 1347.

26. Gold coordination complexes as anticancer agents, I. Kostova, *Anti-Cancer Agents Med. Chem.*, 2006, **6**, 19.

27. Management of rheumatoid arthritis: rationale for the use of colloidal metallic gold, G. E. Abraham, and P. B. Himmel, *J. Nutrit. Environ. Med.*, 1997, **7**, 295.

28. Nanogold-pharmaceutics: (i) The use of colloidal gold to treat experimentally-induced arthritis in rat models; (ii) Characterization of the gold in *Swarna bhasma* a microparticulate use in traditional Indian medicine, C. L. Brown, G. Bushell, M. W. Whitehouse, et al., *Gold Bulletin*, 2007, **40**, 245.

29. Platinum binding to DNA: Structural controls and consequences, T. W. Hambley, *J. Chem. Soc.*, 2001, 2711.

30. Anti-cancer potential of gold complexes, E. R. T. Tiekink, *Inflammopharmacology*, 2008, **16**, 138.

31. Cancer in rheumatoid arthritis: a prospective long-term study of mortality, J. F. Fries, D. Block, P. Spitz, and D. M. Mitchell, *Am. J.Med.*, 1985, **78**, 56.

32. Treatment of rheumatoid arthritis with cytostatic drugs, I. Lorenzen, *Ann. Clin. Res.*, 1975, **7**, 195.

33. Role of disease-modifying antirheumatic drugs versus cytotoxic agents in the therapy of rheumatoid arthritis, J. R. Ward, *Am. J. Med.*, 1988, **85**, 39.

34. Correlation of the in vitro cytotoxic and in vivo antitumor activities of gold(I) coordination complexes, C. K. Mirabelli, R. K. Johnson, D. T. Hill, et al., *J. Med. Chem.*, 1986, **29**, 218.

35. Gold derivatives for the treatment of cancer, E. R. T. Tiekink, *Crit. Rev. Oncol. Hematol.*, 2002, **42**, 225.

36. Gold compounds and their application in medicine, E. A. Pacheco, E. R. T. Tiekink, and M. W. Whitehouse, in Gold Chemistry. Applications and Future Directions in Life Sciences, 2009, 283–319. ed. F. Mohr, John Wiley & Sons, in press.

37. Anti-tumor activity, in vitro and in vivo, of some triphenylphosphinegold(I) thionucleobases, L. K. Webster, S. Rainone, E. Horn, and E. R. T. Tiekink, *Metal-Based Drugs*, 1996, **3**, 63.

38. In vivo antitumor activity and in vitro cytotoxic properties of bis[1,2-bis(diphenylphosphino)ethane] gold(I) chloride, S. J. Berners-Price, C. K. Mirabelli, R. K. Johnson, et al., *Cancer Res.*, 1986. **46**, 5486.

39. Role of lipophilicity in determining the cellular uptake and antitumor activity of gold phosphine complexes, M. J. McKeage, S. J. Berners-Price, P. Galletis, et al., *Cancer Chemother. Pharmacol.*, 2000, **46**, 343.

40. Mechanisms of cytotoxicity and antitumor activity of gold(I) phosphine complexes: The possible role of mitochondria, M. J. McKeage, L. Maharaj, and S. J. Berners-Price, *Coord. Chem. Rev.*, 2002, **232**, 127.

41. Targeting the mitochondrial cell death pathway with gold compounds, P. J. Barnard and S. J. Berners-Price, *Coord. Chem. Rev.*, 2007, **251**, 1889.

42. A screening strategy for metal antitumor agents as exemplified by gold(III) complexes, S. P. Fricker, *Metal Based Drugs*, 1999, **6**, 291.

43. Antitumor properties of some 2-(dimethylamino)methyl.phenylgold(III) complexes, R. G. Buckley, A. M. Elsome, S. P. Fricker, et al., *J. Med. Chem.*, 1996, **39**, 5208.

44. Gold(III) complexes as potential antitumor agents: solution chemistry and cytotoxic properties of some selected gold(III) compounds, L. Messori, F. Abbate, G. Marcon, et al., *J. Med. Chem.*, 2000, **43**, 3541.

45. Cytotoxicity profiles for a series of triorganophosphinegold(I) dithiocarbamates and triorganophosphinegold(I) xanthates, D. de Vos, S. Y. Ho, and E. R. T. Tiekink, *Bioinorg. Chem. Appl.*, 2004, **2** 141.

46. A novel anticancer gold(III) dithiocarbamate compound inhibits the activity of a purified 20S proteasome and 26S proteasome in human breast cancer cell cultures and xenografts, V. Milacic, D. Chen, L. Ronconi, et al., *Cancer Res.*, 2006, **66,** 10478.

47. Gold(III) dithiocarbamate derivatives for the treatment of cancer: solution chemistry, DNA binding, and hemolytic properties, L. Ronconi, C. Marzano, P. Zanello, et al., *J. Med. Chem.*, 2006, **49**, 1648.

48. Gold dithiocarbamate derivatives as potential antineoplastic agents: Design, spectroscopic properties, and in vitro antitumor activity, L. Ronconi, L. Giovagnini, C. Marzano, et al., *Inorg. Chem.*, 2005, **44**, 1867.

49. Review of gold(I) N-heterocyclic carbenes, I. J. B. Lin and C. S. Vasam, *Can. J. Chem.*, 2005, **83**, 812.

50. Mitochondrial permeability transition induced by dinuclear gold(I)-carbene complexes: Potential new antimitochondrial antitumor agents, P. J. Barnard, M. V. Bakers, S. J. Berners-Price, and D. A. Day, *J. Inorg. Biochem.*, 2004, **98**, 1642.

51. Synthesis and structural characterization of linear Au(I) N-heterocyclic carbene complexes: New analogs of the Au(I) phosphine drug Auranofin, M. V. Baker, P. J. Barnard, S. J. Berners-Price, et al., *J. Organomet. Chem.*, 2005, **690**, 5625.

52. Gold nanoparticles: assembly, supramolecular chemistry, quantum-size-related properties, and applications toward biology, catalysis, and nanotechnology, M-C. Daniel and D. Astruc, *Chem. Rev.*, 2005, **104**, 293.

53. Therapeutic possibilities of plasmonically heated gold nanoparticles, D. Pissuwan, S. M. Valenzuela, and M. B. Cortie, *Trends Biotech.*, 2006, **24**, 62.

54. Factors determining the efficacy of nuclear delivery of antisense oligonucleotides by gold nanoparticles, Y. Liu and S. Franzen, *Bioconjugate Chem.*, 2008, **19**, 1009.

55. Colloidal metallic gold is not bio-inert, C. L Brown, M. W. Whitehouse, E. R. T. Tiekink, and G. R. Bushell, *Inflammopharmacology*, 2008, **16**, 133.

56. Toward a novel metal-based chemotherapy against tropical diseases. 1. Enhancement of the efficacy of clotrimazole against *Trypanosoma cruzi* by complexation to ruthenium in RuCl$_2$(clotrimazole)$_2$, R. A. Sanchez-Delgado, K. Lazardi, L. Rincon, et al., *J. Med. Chem.*, 1993, **36**, 2041.

57. Toward a novel metal-based chemotherapy against tropical diseases. 7. Synthesis and in vitro antimalarial activity of new gold-chloroquine complexes, M. Navarro, F. Vasquez, R. A. Sanchez-Delgado, et al. *J. Med. Chem.*, 2004, **47**, 5204.

58. Organic thio compounds having a sulfur-gold linkage as specific inhibitors for retrovirus reverse transcriptase, and their application as medicaments, H. Blough, *Abstracts 3rd International Conf. Gold and Silver in Medicine*, 1993, **14**, Manchester, UK.

59. Aurothiolates inhibit HIV-1 infectivity by gold(I) ligand exchange with a component of the virion surface, T. Okada, B. K. Patterson, S. Q. Ye, and M. W. Gurney, *Virology*, 1993, **192**, 631.

60. Transport of the dicyanogold(I) anion, K. Tepperman, Y. Zhang, P. W. Roy, et al., *Metal-Based Drugs*, 1994, **1**, 433.

61. Inhibition of HIV fusion with multivalent gold nanoparticles, M-C. Bowman, T. E. Ballard, C. J. Ackerson, et al., *J. Am. Chem. Soc.*, 2008, **130**, 6896.

62. Alternative agents in asthma, A. J. Frew and M. J. Plummeridge, *J. Allergy Clin. Immunol.*, 2001, **108**, 3.

63. A placebo controlled multicenter study of auranofin in the treatment of corticosteroid-dependent chronic severe asthma, I. L. Bernstein, D. I. Bernstein, J. W. Dubb, et al., *J. Allergy Clin. Immunol.*, 1996, **98**, 317.

11 Gold Electroplating

Antonello Vicenzo and Pietro L. Cavallotti

CONTENTS

Introduction 231
Historical Note 232
Electrodeposition Fundamentals 234
    Deposition Techniques by Electrochemical Methods 234
        Deposition without Use of an External Current Source 234
        Electrodeposition 236
Gold Electroplating Baths and Processes 238
    Thermodynamic Aspects 238
        Gold Complexes 238
    Kinetics of Deposition 241
        Cyanide Electrolyte 241
        Sulfite Electrolyte 243
        Thiosulfate and Mixed Sulfite Thiosulfate Electrolyte 244
    Classification of Gold-Plating Processes 245
        Cyanide-Plating Baths 247
        Noncyanide Baths 253
    Pulse Electrodeposition 254
    Composite Electrodeposition 257
    Electroforming 258
Applications of Electroplated Gold 259
    Decorative Uses 260
    Engineering Uses 264
        Soft Gold 264
        Hard Gold 268
Conclusions 271
References 271

INTRODUCTION

Electroplated gold has been reviewed many times since the 1960s. An extensive review of gold plating appeared in a book by Fischer and Weimer as early as 1964 [1]. A book entirely devoted to *Gold Plating Technology*—bearing, in fact, this very title—appeared in 1974, edited by Reid and Goldie, collecting the works of the greatest authorities in the field at that time [2]. This book remains an irreplaceable reference to understand the development of gold plating and to appreciate, in particular, the enormous impact the birth and rapid growth of the electronics industry had on gold-plating technology starting in the 1950s. The name and reputation of those who contributed to this book still make it an exemplary publication among the technical manuals devoted to electroplating. A chapter on gold electroplating is included in the book by Rapson and Groenewald, *Gold Usage* [3], providing an essential and readable introduction to plating fundamentals, as well as a wealth of information on gold electrolyte formulations, especially on gold alloys plating. This reference, though in part somewhat outdated, may nevertheless be useful to the reader wanting an introduction to the gold-plating industry, as well as to the engineer interested in the fundamentals of gold plating. More recently, an essential, well-constructed review on the electrodeposition of gold focused on plating for electronics appeared in the fourth edition of *Modern Electroplating* [4]. Finally, the special relevance gold electrodeposition has had for electronic applications since the 1950s is evidenced by a conspicuous number of review papers published over the last 30 years or so, among which those published in the *Gold Bulletin* should be mentioned as the most authoritative, for example, Christie and Cameron [5] and Kato and Okinaka [6].

The properties of electroplated gold, as often is the case for electrodeposited metals and alloys, are scattered in a number of technical and scientific publications, the only comprehensive review being the chapter devoted to it in the book of Safranek [7].

Finally, though electroless plating is not treated in the present work, it is only right and fair to mention the research work accomplished by Okinaka on the autocatalytic deposition of gold over a long span of time and presented in detail in a series of review papers [8–10].

HISTORICAL NOTE

The history of gold electrodeposition is linked to the very birth of the electroplating industry. A detailed and delightful account of its intricate development, which brilliantly evokes the excitement raised by the discovery, as well as the long controversy about the priority rights of the invention, was given by Hunt [11]. The historical background of gold plating is also very nicely sketched by Weisberg in his introductory chapter to the book *Gold Plating Technology* [2].

The electrodeposition process first industrialized and commercially developed was, notoriously, the electroforming of copper. However, no doubt the electrodeposition of gold and silver was a strong driver for the nascent electroplating industry back in the 1840s. Before that time, the technique of gold gilding, mainly from mercury amalgam but also in the form of immersion gilding (a rudimentary immersion plating technique from dilute gold chloride solution) was standard practice for gold coating of a variety of items, exclusively for decorative purpose. The widespread use of mercury amalgam gilt with its lethal or disabling effects on workers had raised great concern and was a strong motivation for the search of alternatives to mercury gilding, at least among the scientific community of the time.

The first ever recorded gold-plating experiment by electrodeposition, truly a great event and the seed of a novel industry, was performed by Brugnatelli at the University of Pavia, shortly after and thanks to Volta's discovery of the pile at the beginning of the nineteenth century. Brugnatelli, a chemistry professor in the same university and good friend of Volta, in fact used Volta's pile for electrodepositing metals and, in particular, gold on medals and cups. Brugnatelli's discovery did not have any publicity and was destined to escape notice for many years. The following events that eventually led to the birth of the gold-plating industry are relatively well-documented and culminated in the issuing of a patent in the joint names of George Richards and Henry Elkington in March 1840. A decisive contribution to the patent completion was given by John Wright, a surgeon and a chemist, who should be credited with the idea of using cyanide for gold-plating electrolyte [11].

For several decades on, the practice of gold plating was consigned to jewelers' and artisans' shops, where it lived on silently, becoming of widespread use for decorative purposes. Heavy plating and electroforming were also significantly developed and gained widespread use in the second half of the nineteenth century, but afterward the relevant technique and production capability met a fate of rapid decline and obliviousness [2]. The most important advances in gold-plating technology occurred during the second half of the twentieth century as schematically shown in the diagram of Figure 11.1, a snapshot of its entire history.

In modern times, starting from the second half of the 1940s, the birth and rapid development of the electronic industry totally changed the scenario that had determined the development of gold plating up to that time. The use of gold layers as a contact surface in connectors, switches, and

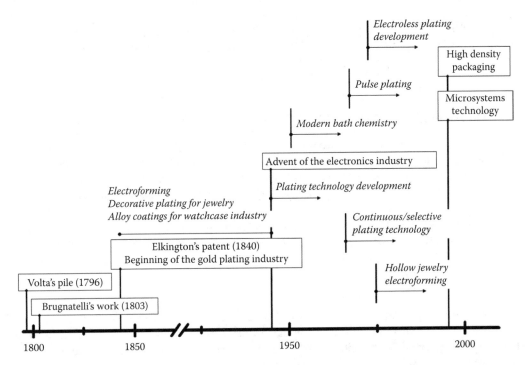

FIGURE 11.1   The evolution of gold plating technology through its main advances over the two centuries of its history.

more specifically in edge contacts for printed circuit boards, imposed new and strict requirements on coating properties, and this was the chief driver of the revolution in gold plating initiated with the development of bright acid gold baths in the late 1950s. This symbiotic connection between electronics manufacture and gold-plating technology was the source of all the most important innovations that occurred in this field over the last half century, with the only notable exception of gold electroforming. The impact of electronic manufacture was manifold, since it imposed not only innovation in the standard bath chemistry of that time, particularly with the formulation of acid cyanide baths and the use of metal cations as hardening additives, but also fostered the search for alternative chemistry, culminating in the development of gold sulfite bath, or alternative processes such as electroless deposition, which has progressed with a steady pace since its official birth at AT&T Bell Laboratories in 1970 [12]. Similarly, the technical development and the industrial implementation of pulse current gold plating received a strong surge from the need of dense, nominally pore-free thin layers to comply with process economics and life-service expectations in contacts manufacture. Finally, the requirement for high throughput process and the ever increasing need of an efficient use of gold was the driving force for the development of high-speed and selective plating technology.

Cyanide-based baths for gold and gold alloy plating—more especially acid cyanide bath—are by far the most used and reliable processes; however, a few drawbacks needing improvement remain, such as hydrogen evolution during plating, incorporation of cyanide in the coatings, relatively high-tensile residual stress, and low compatibility with masking materials used in patterned gold electrodeposition. Moreover, cyanide is a hazardous chemical, demanding special care and compliance with strict regulations, which may somewhat discourage its use, provided that technical specifications can be met by alternative processes.

These are some of the reason why R&D activity in gold plating has never really stopped. Even today, the issuing of patents related to gold and gold alloy plating—either by electrochemical deposition or electroless/immersion plating (the later for pure gold films)—is continuing at an astonishingly good and steady pace. This is the symptom of a lively and developing technology, which is continuously asked to face new and more stringent requirements or to fit the needs of innovative applications, by both a continuous refinement of the technology at all levels of its articulation and by the invention of new process and technology solutions.

ELECTRODEPOSITION FUNDAMENTALS

The following section offers a concise introduction to the fundamentals of electrodeposition. Electroless deposition techniques, though not included in this chapter, are also briefly mentioned for the sake of completeness. A complete and exhaustive treatment of the subject is obviously outside the scope of this chapter, and the interested reader should study the relevant technical and scientific literature.

For a metal electrode M in contact with a solution of its ions M^{z+} an electrochemical equilibrium is established at the surface, corresponding to equal rate of the forward and backward reaction of deposition and dissolution, respectively, according to the reaction:

$$M^{z+} + ze^- \rightleftarrows M$$

This equilibrium condition is characterized by a definite value of the potential for a given set of experimental conditions, i.e., purity of the metal, metal cations concentration, and temperature. Changing the experimental conditions with respect to standard values also changes the equilibrium potential of a given redox couple. For the pure metal (unity thermodynamic activity) in contact with a solution of its ions (for which the molar concentration may be used instead of the activity), the equilibrium potential of the redox couple is given by the Nernst equation:

$$E = E^0 + \frac{RT}{zF} \ln[M^{z+}] = E^0 + \frac{0.0591}{z} \log[M^{z+}]$$

where E^0 is the standard potential of the couple, i.e., the value of the metal potential with reference to the hydrogen electrode in standard conditions of temperature and activity of all the chemical species involved.

At equilibrium, net reaction is not taking place at the surface. By changing the potential of the electrode in the negative or positive direction, i.e., by cathodic or anodic polarization of the electrode, the deposition or the dissolution reaction is made possible at the surface, respectively. It should be noted that the equilibrium electrochemical potential of a redox couple is a fundamental property physically meaningful, i.e., a measurable quantity, only for relatively fast electron transfer reactions, i.e., reactions that requires a low overvoltage in order to occur at a finite measurable rate.

DEPOSITION TECHNIQUES BY ELECTROCHEMICAL METHODS

Electrochemical processes for metal deposition are broadly classified in electrolytic and electroless techniques. The electrolytic method implies the use of an external current source, whereas electroless methods—though relying on an electrochemical mechanism—only require the immersion of the substrate into the plating solution. Electroless methods comprise two essentially different processes, displacement or immersion plating and autocatalytic deposition. In Figure 11.2 the basic functioning of these processes is schematically represented.

Deposition without Use of an External Current Source

Deposition of a metal or an alloy film without recourse to an external current source is accomplished in the process of immersion plating and electroless plating. The term *electroless* is sometimes used

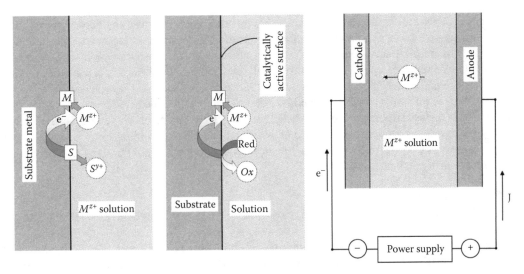

FIGURE 11.2 Schematic representation of the electrochemical processes used for deposition of metals. From left to right: displacement deposition; autocatalytic deposition; electrodeposition.

indiscriminately with reference to both methods of deposition, a habit that by itself would not be censurable except that the expression "electroless plating" was coined by Brenner and Riddell to indicate autocatalytic processes of metal deposition [13].

The process known as immersion plating is the simplest and oldest electrochemical technique for metal deposition. Immersion plating is the result of a displacement reaction in which a more noble metal cation is reduced to the elemental state thanks to the oxidation/dissolution of a more active (less noble) metal. A schematic representation of the process is shown in Figure 11.2: the substrate metal S gets oxidized to the ionic state S^{y+} supplying electrons for the reduction of the more noble metal cation M^{z+} which is reduced to the elemental state. The distinct feature of the immersion plating process is therefore its reliance on a spontaneous reaction. Thus, the feasibility of the process is dictated by the relative nobility of the metals involved as film and substrate material. The relative nobility is determined by the position in the electrochemical series yet can be altered and even reversed with respect to the dictate of the latter by recourse to complex chemistry. An example of reversed nobility is encountered in the immersion tin process used to deposit a thin layer of tin on copper substrate.

The displacement reaction is a self-limited process since it may proceed only as long as the surface of the less noble metal is accessible to the electrolyte. Accordingly, the thickness of immersion plating deposits is always limited, e.g., the maximum thickness of gold thin films by displacement reaction is about 0.2 μm. The main use of immersion gold is as a solderable finish onto nickel-phosphorous or electroplated nickel surfaces. The gold is plated from slightly acidic cyanide electrolytes usually to a final thickness in the range of 0.05–0.1 μm.

Autocatalytic deposition is far more important than immersion processes in the plating industry, and its prominence in the field also explains why it is commonly referred to as electroless deposition *tout court*. The discrimination between electroless or autocatalytic processes and immersion or displacement deposition should be particularly cared about in the case of gold deposition, as emphasized in an early review of the subject [14].

The autocatalytic deposition is basically a heterogeneous redox reaction in which metal cations are reduced by a suitable reducing agent in a controlled way. Schematically, see also Figure 11.2, the deposition reaction of a metal cation M^{z+} is made possible by the oxidation of a reducing agent *Red* present in solution, which supplies ze^- electrons and produces the *Ox* species. Control over the reaction is granted by the catalytic nature of the reaction itself, which is catalyzed by the metal or alloy being deposited. If the substrate material is not catalytically active, the deposition reaction must be triggered by an alternative procedure, typically the formation of active metal clusters at the surface, or may be activated via an initial spontaneous displacement reaction forming active nuclei of the metal to be deposited at the surface of a less noble metal substrate. In principle, as long as the surface maintains its catalytic activity in a stable reactive environment, there is no intrinsic limitation to the thickness of the deposited layer, though in practice such conditions are not readily achievable or industrially significant. A distinctive feature of autocatalytic deposition is the possibility of forming an alloy thanks to the codeposition of an element coming from the molecule of the reducing agent (e.g., P from hypophosphite or B from borohydride), or through the concomitant reduction of different metal ionic species.

Electrodeposition

Electrodeposition, also called electroplating or simply plating, is the process of depositing a metal or an alloy layer from a solution of its ions onto the surface of a bulk material—the substrate—by virtue of the flow of an electric current through an electrochemical cell, i.e., the plating cell. The substrate is usually a metallic material but may be of different nature, even a nonconductive material, provided it is suitably coated with a thin metal layer produced by a different process, e.g., electroless or vapor phase deposition.

Electrodeposition is performed in an electrochemical cell consisting of two metal electrodes and an electrolyte, i.e., an ionic conductor. A simplified scheme of an electroplating cell is presented in Figure 11.2. A direct current source, e.g., a rectifier, supplies current when a potential difference is applied across the cell. The current flow is that of electrons in the external circuit whilst the charge transport in the solution is by means of electrically charged species, the ions, which under the potential difference across the system travel either toward the negative electrode (cathode) if positively charged (cations) or toward the positive electrode (anode) if negatively charged (anions).

The current flow through the electrode–electrolyte interface is made possible by the exchange of electrons between solution species and the electrode surface. This charge transfer process may be from the electrode to the electrolyte or in the opposite direction, the former case resulting in the electrode operating as a cathode, the latter in the electrode operating as an anode. The transfer of electrons toward the solution side of the electrode–electrolyte interphase requires electrons to be fed to the cathode through the external circuit and implies that chemical species in the solution layer close to the cathode surface undergo electrochemical reduction. In other words, a metal ion or complex may be reduced to the metallic state and, under special conditions, form a compact adherent layer on the electrode surface. The transfer of electrons toward the electrode side of the interphase, i.e., to the surface of the anode, entails either the electrochemical oxidation of solution species at the surface of the electrode or the electrochemical oxidation of the electrode material itself. In the electroplating technology, the first type of electrode is called an inert or insoluble (oxygen evolution) anode; the second type is called a soluble anode.

More than one reaction may take place at each electrode and reactions taking place at the anode and at the cathode are independent from each others apart from the compliance of material balance requiring that the number of electrons consumed at the cathode equal the number of electrons liberated at the anode.

The charge flow through the cell and the amount of the substances transformed at the electrodes are related by Faraday's laws. The electrochemical equivalent of gold, that is the mass of the gold reduced or oxidized by the unit charge is calculated as the ratio of the molecular mass of the metal 196.966 and the charge zF required to transform a mole of material where z is the ionic valence and F is the Faraday's constant (the charge of a mole of electrons, 96485 C).

Because the charge required to transform a mole of material depends on the valence of the reactant and product species, it is obvious that the electrochemical equivalent of Au(I) is three times that of Au(III). The charge passed through the cell in a given period of time is conveniently expressed as the product of the cell current (A) and the time (h). The electrochemical equivalent of Au(I) per 1 A h is then calculated as follows:

$$\frac{MM_{Au}}{zF} \cdot 1\ Ah = \frac{196.966}{96485} \cdot 3600 = 7.349\ g$$

The electrochemical equivalent of Au(III) is, therefore, about 2.450 g.

A straightforward and reliable procedure to determine the current efficiency of a plating bath is deduced from the above as the measurement of the weight gain of a coupon plated in the laboratory using a sample of the solution. The latter may be also used to calculate the average thickness of the deposit if the area of the plated part and the density of the plated material are known.

As the potential of the electrode is increased or reduced, other electrochemical reactions may become possible and may therefore occur in competition with each other at the same electrode surface. Since electrochemical reactions are surface reactions, the rate of each reaction is conveniently referred to the unit surface area of the electrode. In fact, the rate of an electrochemical reaction is given as the current density and is expressed in the SI units of A m^{-2}, though multiples and submultiples are widely employed in practice. The kinetic behavior of a given electrochemical system, that is the electrode–electrolyte system, is experimentally characterized by recording the polarization curve at the electrode, i.e., the electrode potential–current density relationship.

As the electrode potential is shifted in the cathodic direction the current density increases steadily starting from the equilibrium or immersion potential of the electrode E_{imm} (region I in Figure 11.3), then apparently tends to a plateau—the diffusion limited current density for the electrode process under consideration—(region II), and finally, as the potential is moved further in the cathodic direction, it starts increasing again since a new electrode process is activated (region III) running in parallel to the previous one.

This is the general behavior that may be expected whenever there are two possible reactions in a given potential range. Most of the time in metal electrodeposition, these two reactions are the reduction of metal cations and the discharge of hydrogen ions, but other cathodic processes are also possible, e.g., the discharge of dissolved oxygen. A similar behavior is to be expected in the electrodeposition of binary alloys when the difference between the equilibrium potential of the two metal redox couples is remarkable. The typical shape of the polarization curve as briefly discussed

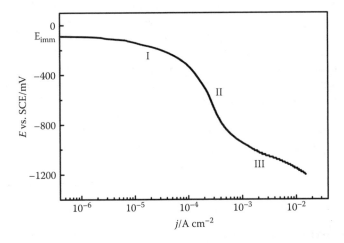

FIGURE 11.3 A typical linear polarization curve in a E vs. log j diagram showing the main features of the kinetic behavior of electrode processes. The curve was recorded at a gold rotating disk electrode (600 rpm) in an acid gold cyanide bath containing cobalt at pH 4.5 and 35°C; for gold deposition the electrode should be polarized at potential in region III.

earlier is easily understood in view of the different elemental steps involved in the macrokinetic picture of the electrodeposition process, that is, convective-diffusive transport of the species reacting at the electrode and electron transfer reaction. These elemental processes may be thought to be confined to a thin layer of electrolyte close to the surface, the diffusion layer, and to the surface itself, respectively. The rate determining step of the overall process changes according to the reaction rate. Depending primarily on the concentration of the electroactive species, e.g., the metal cation, there is a shift from a surface kinetics (region I of the curve) to a diffusion-controlled regime (region II) as the value of the cathodic electrode polarization increases.

The technique of electrodeposition is far more complex than this simplified picture may lead readers to think. A plating cell is basically an electrochemical reactor whose performance must be optimized especially with respect to the current density distribution at the electrodes, which translates into the distribution of plated thickness in the case of metal electrodeposition. The degree of uniformity of the plated thickness over the object being plated is a fundamental feature of a plating process and is referred to as the *macrothrowing power*. The collection of basic design rules and relevant plating measures for improving throwing power forms the "spell book" of the electroplater. However, service requirements dictating the design of components and articles may occasionally pose more than a challenge to the ingenuity of an experienced electroplater. A layman's guide to the basics of metal distribution in electroplating is described in a paper by Silman [15]. It is readily realized that thickness distribution in gold plating has an immediate economic implication.

The distribution at the microscale, that is, at the scale of surface imperfections such as roughness, crevices, and scratches, is as much an important process feature, since it determines the ability of a plating process to reduce the surface roughness of the substrate and to fill surface depressions, thus producing a leveling effect on the plated surface.

GOLD ELECTROPLATING BATHS AND PROCESSES

The high value of the standard potential means that simple ionic gold species are not stable in aqueous solution. Gold is, in fact, invariably deposited from complex electrolytes, and since the earliest time of gold plating the gold complex most commonly used in plating baths has been the cyanide Au(I) complex. Over recent decades, gold-plating processes based on alternative chemistry, namely, the Au(III) cyanide complex and the Au(I) sulfite complex, reached significant commercial use in specific applications; nonetheless the Au(I) cyanide chemistry remains the dominant base formulation for gold-plating baths. There are literally hundreds of formulations in use, a proliferation partly caused by commercial policy and competition, as well as empirical development of alloy plating baths over a long span of time. On the other hand, it is equally true that selecting a plating bath for gold may not be a common engineering task whenever tight specifications must be met, so that the plethora of electrolyte formulations is also the result of a continuous improvement work toward electrodeposits featuring enhanced and well-balanced properties.

The high and variable price of gold is the reason why gold-plating solutions, like those of precious metals in general, are relatively dilute electrolytes, compared to the concentration common to most of other plating processes. This economic constraint has a direct bearing on processing since the low concentration of the metal influences, to a large extent, the global formulation and conduction of the bath, and is responsible for a generally increased sensitivity to the change of almost any control parameter.

THERMODYNAMIC ASPECTS

Gold Complexes

Gold exists in oxidized form with two valences: Au(I) and Au(III). Other oxidation states are possible, yet the chemistry of a plating bath is concerned only with the +1 and +3 oxidation states.

TABLE 11.1
Standard Potential of Gold Redox Couples

Redox couple	Half Reaction	Standard Potential	Ref.
Au(I)/Au	$Au^+ + e = Au$	1.83 V	[16]
		1.71	[17]
		1.695	[16]
Au(III)/Au	$Au^{3+} + 3e = Au$	1.52 V	[16]
Au(III)/Au(I)	$Au^{3+} + 2e = Au^+$	1.36	[16]
		1.40	[18]

The Au^+ and Au^{3+} free gold cations are inherently unstable in aqueous solutions, nevertheless they can be regarded as forming very weak complexes in water, $Au(H_2O)_2^+$ and $Au(H_2O)_4^{3+}$, which easily undergo ligand exchange reactions in the presence of complexing species stronger than water itself. Because of the instability of the ionic state in aqueous solution, the equilibrium potential of the redox couples Au(I)/Au and Au(III)/Au is not liable to direct measurement and is in fact derived by thermochemical calculations.

Calculated values of the standard potential of the redox couples Au(I)/Au, Au(III)/Au, and Au(III)/Au(I) taken from different bibliographic sources [16–18] are reported in Table 11.1.

The exact value of the standard potential of the Au(I)/Au and Au(III)/Au couples is of little practical significance; what is important is the instability of the ionic state in aqueous solutions and the need to stabilize the gold ionic species through complexation.

Gold complexes, with a few exceptions, are either 2-coordinated with the two ligand molecules bonded to the central Au^+ ion in linear configuration or 4-coordinated with the ligand molecules bonded to the central Au^{3+} ion in a square-planar configuration. Complex stability in aqueous solution is assessed by determining the equilibrium stability constant of the complex formation reaction. The formation of an n-coordinated complex of a z-valent metal cation M^{z+} with a ligand species L^{y-} occurs by the equilibrium reaction:

$$M^{z+} + nL^{y-} \rightleftharpoons M(L)_n^{z-ny}$$

for which a β_n equilibrium constant is defined as follows:

$$\beta_n = \frac{[M(L)_n^{z-ny}]}{[M^{z+}][L^{y-}]^n}$$

where the brackets are used to denote activity of the species enclosed.

The value of the equilibrium constant provides a measure of the stability of the complex and it is therefore also referred to as the stability constant of the complex. The stability constants of Au(I) and Au(III) complexes relevant to gold plating and related processes are reported in Table 11.2 along with the values of the standard potential for the reduction of gold from the same complexes. The data included in the table were collected from the compilation of Senanayake [19] (stability constants) and from Bard et al. [16] (standard or formal potentials).

The chemistry of gold-plating baths is largely associated with the Au(I) cyanide complex $Au(CN)_2^-$. The high value of the stability constant of the complex accounts for its wide and enduring acceptance for the formulation of gold electrolytes intended either for electrodeposition, displacement, or electroless plating. Moreover, the Au(I) cyanide complex is stable in a wide pH range down to about 3.0 and such a large pH stability field is key to its exploitation in a countless number

TABLE 11.2
Stability Constant and Standard or Formal Potential for Gold(I)
Complexes

Complex	log β_n	Half-reaction	E^0/V
$Au(CN)_2^-$	38.3	$Au(CN)_2^- + e = Au + 2CN^-$	−0.595
$Au(SO_3)_2^{3-}$	26.8	$Au(SO_3)_2^{3-} + e = Au + 2SO_3^{2-}$	—
$Au(S_2O_3)_2^{3-}$	26.1	$Au(S_2O_3)_2^{3-} + e = Au + 2S_2O_3^{2-}$	0.153
$Au(CS(NH_2)_2)_2^+$	23.3	$Au(CS(NH_2)_2)_2^+ + e = Au + 2CS(NH_2)_2$	0.380
$Au(SCN)_2^-$	17.2	$Au(SCN)_2^- + e = Au + 2SCN^-$	0.662
$AuCl_2^-$	9.71	$AuCl_2^- + e = Au + 2Cl^-$	1.154

Source: Based on data from Gold leaching in noncyanide lixiviant systems: Critical issues
on fundamentals and applications, G. Senanayake, *Minerals Engineering*, 2004,
17, 785; and *Standard Potentials in Aqueous Solutions*, A. J. Bard, R. Parsons, and
J. Jordan, Marcel Dekker, New York, 1985.

of formulations for gold and gold alloys electrodeposition. Decomposition of the complex occurs below pH 3.0 with precipitation of insoluble AuCN and release of cyanide:

$$Au(CN)_2^- \rightleftarrows AuCN(s) + CN^-$$

Plating baths based on the Au(I) cyanide complex and buffered at pH in the neutral or acidic range do not contain any free cyanide. In fact, free cyanide tends to separate as gaseous hydrogen cyanide, a lethal poison, at pH below about 8.0 [20]:

$$H^+ + CN^- \rightleftarrows HCN(g)$$

Therefore, the lower limit of the pH range for alkaline bath containing free cyanide should be fixed at above 8.0, whereas the low pH limit for Au(I) cyanide acid bath is about 3.0. The recommended values are actually higher than that, usually not less than 9.0 and 3.5, respectively. The Au(III) cyanide complex is used in formulating special purpose baths for plating pure gold or color gold. The stability constant of the Au(III) cyanaurate complex $Au(CN)_4^-$ is reported as $\sim 10^{56}$ [3] from which a standard reduction potential of about 0.41 V can be calculated. The cyanaurate complex is stable to very low pH values and in fact the pH of trivalent plating solutions is usually in the highly acidic range.

Apart from cyanide complexes, by no means irreplaceable in the gold-plating industry, the Au(I) sulfite and thiosulfate complexes entered into use dating from the 1970s, though with quite a different record since then. In fact, the sulfite process has gained a good acceptance for specific applications whereas the actual exploitation of the thiosulfate complex, which is invariably encountered in a mixed thiosulfate-sulfite complex system, is still not in any significant commercial use either as an electroless or electroplating process.

The lack of reliable data for the sulfite complex of Au(I) is noteworthy. As recently noted by Green [21], a low value of the stability constant of about 10^{10} is often quoted for the sulfite complex, e.g., [22], especially in the literature on gold plating. However, a much higher value was reported by Webster [23]—which is included in Table 11.2—and in the compilation of stability constants of gold complexes in Rapson and Groenewald [3]. Interestingly, the speciation analysis of the gold sulfite electrolyte performed by Green and Roy suggested that the aforementioned low value of the stability constant of $Au(SO_3)_2^{3-}$ is unrealistic [24].

Similarly, there are no reliable data on the standard reduction potential for the gold sulfite complex, for which values in the range 0–0.4 were reported [25]. However, assuming for the stability constant the previously quoted value of 26.8 [23], the standard reduction potential of the sulfite complex can be calculated as 0.11 V, which is in excellent agreement with the formal potential of the sulfite complex at a mercury amalgam electrode, namely, 0.116 V, as measured by Baltrûnas et al. [26].

The gold thiosulfate complex $Au(S_2O_3)_2^{3-}$ has a fairly high stability constant (see Table 11.2), and a wide pH range of thermodynamic stability. Despite these characteristics, a gold thiosulfate electrolyte has never been successfully developed because of the limited stability of the ligand. In fact, at neutral or acid pH, thiosulfate is unstable towards decomposition in colloidal sulfur and sulfite [27]:

$$3S_2O_3^{2-} + 3H^+ \rightarrow 3S + 3HSO_3^-$$

while in alkaline solution, sulfide ions may be formed by disproportionation of the thiosulfate ion, according to the following reaction:

$$S_2O_3^{2-} + 2OH^- \rightarrow SO_4^{2-} + H_2O + S^{2-}$$

The sulfide formed in the preceding reaction can be oxidized to sulfur on a gold surface at potential in the range 0–0.4 V [28] forming an inhibiting adsorbed layer which is stable in the potential range between −0.6 and +0.4 V [29]. The instability of thiosulfate in such a wide range of conditions rules out its use as a ligand in electroplating baths. On the other hand, a mixed thiosulfate-sulfite bath, as first proposed by Osaka et al. [30], shows good stability and is currently being developed for pure gold electrodeposition.

KINETICS OF DEPOSITION

Most of the kinetic studies conducted in gold electrodeposition were concerned with the electrolytes most widely used for gold plating, i.e., acid or neutral Au-cyanide baths and the Au-sulfite bath. In the following account, the main results emerging from these studies are presented.

Cyanide Electrolyte

The high stability of the Au(I) cyanide complex reflects in an extremely small, actually vanishing, concentration of the metal ions. It is therefore reasonable to assume that the reduction or deposition of the metal at the cathode occurs from the complex ions, i.e., without formation of an intermediate simple ionic species. The direct reduction from the complex is generally considered as the most likely mechanism also for the deposition from cyanide complexes of other metal cations. In other words, the complex species in solution is the species undergoing electron transfer at the surface, and if any chemical modification of its structure takes place, this must be thought of as a surface reaction not involving complete decomposition of the complex itself and thus excluding the formation of the free cation.

The kinetics of gold deposition was studied in alkaline cyanide electrolytes containing an excess of free cyanide and from phosphate or citrate buffered electrolytes at pH in the neutral and acid range, respectively.

According to Maja [31] deposition of gold from alkaline cyanide solution in the presence of excess cyanide occurs through a two-step mechanism of adsorption (reaction (11.1)) and discharge (reaction (11.2)) of the complex $Au(CN)_2^-$ irrespective of the overpotential,

$$Au(CN)_2^- \rightarrow Au(CN)_2^-\big|_{ads} \tag{11.1}$$

$$Au(CN)_2^-\big|_{ads} + e^- \rightarrow Au + 2CN^- \tag{11.2}$$

Maja's conclusion was confirmed by the work of Cheh and Sard [32], who derived the polarization curve by galvanostatic measurements at a gold rotating disk electrode in 0.0052 M KAu(CN)$_2$ and 0.54 M KCN solution at pH 10.2 and 60°C. The exchange current density for the discharge of the complex was calculated from the slope of the polarization curve at zero overpotential and from the measured value of the cathodic limiting current density as 0.082 A dm^{-2}. This early picture of the discharge mechanism of the Au(I) cyanide complex from alkaline solution was revisited in the light of the work of Harrison and Thompson [33], who found the reduction to be first order in free cyanide concentration, a result later confirmed also by Beltowska-Brzezinska et al. [34]. To account for this observation, a reaction mechanism was postulated in which the species undergoing reduction is AuCN as an adsorbed intermediate in equilibrium with the bulk species Au(CN)$_2^-$. The reduction reaction is therefore described by a two-step sequence consisting of the adsorption equilibrium (reaction (11.3)) and electron transfer step (reaction (11.4), rate determining):

$$\text{Au(CN)}_2^- \rightarrow \text{AuCN}\big|_{\text{ads}} + \text{CN}^- \tag{11.3}$$

$$\text{AuCN}\big|_{\text{ads}} + e^- \rightarrow \text{Au} + \text{CN}^- \tag{11.4}$$

MacArthur [35] drew similar conclusions from his work on the deposition of gold in a 0.0052 M KAu(CN)$_2$ and 0.846 M KCN solution at pH 12.2 and 61°C, though differentiating between a low and high field discharge mechanism. Based on a voltammetry study of the system, MacArthur inferred that two different reaction paths were conceivable for the Au(I) cyanide complex: reduction through the adsorbed intermediate AuCN$_{\text{ads}}$, at low overpotential, and direct transfer between the complex in solution and the metal atom at larger overpotential.

MacArthur [35] as well as Cheh and Sard [32] also studied the kinetics of the Au(I) cyanide complex reduction from phosphate and citrate buffered solutions at pH 7.5 and 5.0, respectively.

According to Cheh and Sard [32] the polarization measurements in the citrate solution were irreproducible, and the limiting current density not measurable because of the overlapping of the hydrogen evolution reaction, thus precluding the derivation of kinetic parameters. Similarly in a phosphate electrolyte, the equilibrium potential and consequently the exchange current density could not be determined. MacArthur [35] failed to obtain strong evidence from voltammetric analysis that the two reaction paths proposed for alkaline electrolyte were also operative in neutral and acid solutions, though this lack of evidence did not discourage him from stating that the same deposition mechanism should hold for the different electrolytes. In this respect, however, it should be noted that the impossibility of determining the equilibrium potential was a clear indication that the reduction of the Au(I) cyanide complex suffered from a stronger kinetic hindrance in neutral or acidic electrolyte compared to alkaline electrolyte.

These early works on the kinetics of gold electrodeposition were primarily a contribution to the understanding of some fundamental aspects of the process. However, they were also instrumental to clarifying the issue of carbon codeposition in gold deposits from cyanide electrolyte [36,37], an issue that was rapidly emerging as a serious reliability threat in the years around 1970.

The need for an unifying treatment of the kinetics of soft and hard gold deposition was raised by Eisenmann [38], who performed a detailed experimental study of the cathodic reduction of the dicyanoaurate complex from both a commercial cobalt-hardened gold bath and a lead-doped soft gold bath, at pH 4 and 7, respectively, using both steady-state and transient techniques in combination with a rotating disk electrode. Eisenmann committed himself to merging all the experimental results into an unifying picture through an elaborate discussion of the results, though he could not actually explain in a thoroughly convincing manner some characteristics of the different systems, particularly when discussing the effects of adsorption on the crystallization behavior. Nevertheless, his conclusion was that one common mechanism could be assumed for the deposition of both hard and soft gold from dycianoaurate, involving adsorption equilibria preceding and following the rate determining electron transfer step. This was essentially the same reaction path already postulated for the reduction of the Au(I) cyanide complex from alkaline solution in the presence of free cyanide [33] as well as in neutral or acid solution at low overpotential [35], as outlined by reactions (11.3) and (11.4).

Experimental evidence supporting the adsorbed intermediate mechanism emerged from several investigations, in particular the work of Reinheimer [39], Okinaka and Nakahara [40,41] and Angerer and Ibl [42]. This mechanism was definitely confirmed in the light of further and direct evidence of the implication of an AuCN species in the discharge process, namely, the observation of AuCN inclusion in a gold layer electroplated from a citrate buffered bath at pH 4.7 and 40°C, as reported by Holmbom and Jacobson [43], and the imaging by *in situ* scanning tunneling microscopy of AuCN ad-layers formed from dicyanoaurate anion $Au(CN)_2^-$ on an Au(111) surface in $KAu(CN)_2$ solution [44].

The unifying picture proposed by Eisenmann, though not substantially challenged by following investigations, needed to be refined in order to unveil the mechanistic details which could explain the striking difference in the microstructure of soft and hard gold layers. In this respect, a fundamental contribution came from the study of heavy metal cations adsorption at a gold electrode and its relationship with the electrocrystallization of soft gold layers [45], a peculiar aspect of the kinetics of the gold cyanide complex discharge that was further investigated by Bindra et al. [46]. In the later work, the influence of trace heavy metal ions –Co(II), Ni(II), Tl(I), and Pb(II)– on the kinetics and microstructure of gold deposits from a phosphate buffered gold cyanide electrolyte at pH 6.5 and 65°C was studied. Convincing evidence of two distinct kinetic mechanisms was presented, namely, a direct discharge mechanism from the solution phase species $Au(CN)_2^-$ for soft gold and an adsorption mediated discharge mechanism from the intermediate $AuCN_{ads}$ for hard gold deposition. The type of active mechanism was recognized as directly determined by the base metal ions additive: the underpotential deposition of Pb^{2+} and Tl^+ ions on the gold surface prevents the formation of an adsorbed AuCN layer thus permitting the direct discharge of the complex (soft gold deposition); on the other hand, the inability of Ni^{2+} and Co^{2+} ions to be underpotentially deposited on the gold surface, results in the formation of an AuCN ad-layer hindering the direct discharge of the gold complex (hard gold deposition). In this case, the kinetic behavior is further complicated by the superposition of the hydrogen evolution reaction triggered by the reduction of the Ni^{2+} and Co^{2+} ions at potential more positive to that of the dicyanoaurate complex.

Sulfite Electrolyte

The kinetic behavior in the electrodeposition of gold from sulfite solution was studied by Derivaz et al. [47]. Polarization curves at a gold coated copper electrode were recorded in sodium sulfate and sodium sulfite electrolytes at pH in the range 6.5–9.5 in the absence and in the presence of the Au(I)-sulfite complex. The polarization curves obtained in the base electrolyte showed a cathodic wave starting at a potential about −0.850 mV SCE and, at more cathodic potential, a diffusion limited current density plateau. The value of the limiting current density was found to rise with increasing sulfite concentration and lowering pH of the electrolyte. The authors realized that the observed cathodic electrochemical behavior was related to the chemical stability of the sulfite ion. In fact, the sulfite ion is in equilibrium with the bisulfite ion according to:

$$SO_3^{2-} + H^+ \rightleftarrows HSO_3^-$$

The pK of the equilibrium sulfite/bisulfite, i.e., the pK_2 of sulfurous acid, is 7.2 from data reported in Wagman et al. [48]. The equilibrium concentration of the two ionic species can be readily calculated showing that the bisulfite ion is increasingly present in sulfite solution as the pH drops below about 9.0. This observation could explain the observed increase of the limiting current density with lowering pH and eventually allowed to ascribe the wave in the polarization curve to the cathodic reduction of bisulfite.

The same study [47] also inferred from the analysis of voltammetric curves that the electrochemical reduction of bisulfite leads to the formation of the dithionite ion according to the redox reaction:

$$2HSO_3^- + 2e^- \rightarrow S_2O_4^{2-} + 2OH^- \tag{11.5}$$

The standard potential of the HSO_3^- / $S_2O_4^{2-}$ couple was calculated by Chao as 0.100 NHE [49].

The dithionite is supposed to be stable only in acidic medium [50], which explains the inability of Derivaz et al. [47] of quantitatively characterizing the electrochemical conversion of the bisulfite to dithionite ion.

Further studies suggested a more complex picture of the kinetics of gold deposition from $[Au(SO_3)_2]^{3-}$ solution. Different and possibly overlapping mechanisms should be taken into account. Critical parameters in deciding the type of mechanism are the pH of the electrolyte and the overvoltage.

In alkaline solution (at pH > 8.5) and at low overvoltage (cathode potential higher than about –0.8 V NHE), a two-step process was proposed involving decomposition of the complex followed by reduction of an adsorbed ad-ion [51], i.e.,

$$[Au(SO_3)_2]^{3-} \rightleftarrows Au_{ads}^+ + 2SO_3^{2-} \tag{11.6}$$

$$Au_{ads}^+ + e^- \rightarrow Au \tag{11.7}$$

while, at cathode potential lower than about –0.8 V NHE, the deposition of gold is supposed to occur by direct reduction of the complex [47,51]

$$[Au(SO_3)_2]^{3-} + e^- \rightarrow Au + 2SO_3^{2-}$$

In neutral or weakly alkaline solutions, at pH below about 8.0, it was proposed that gold deposition occurs also through the chemical reduction of the Au(I) complex by the cathodically formed dithionite species (reaction (11.5)) [47]:

$$2[Au(SO_3)_2]^{3-} + S_2O_4^{2-} + 2OH^- \rightarrow 2Au + 2HSO_3^- + 4SO_3^{2-}$$

The reduction of the gold sulfite complex by dithionite is also responsible for the instability of the sulfite electrolyte at pH below about 9.0, a well-known drawback of this type of bath. The role of dithionite in the decomposition of the sulfite complex with the formation of colloidal gold has been recently discussed by Green [21, 24].

An interesting contribution to the understanding of gold reduction from sulfite electrolyte is to be found in a combined voltammetry and spectroscopy study of gold deposition from a slightly acidic pH 6.0 electrolyte [52]. The interpretation of the observed voltammetric behavior was based on the assumption that the reduction of the sulfite complex followed a chemical-electrochemical mechanism, as previously proposed by Horkans and Romankiw [51] for alkaline solution (reactions (11.6) and (11.7)). In addition, two different regimes were postulated, namely inhibited and uninhibited deposition, occurring at low and high overpotential, respectively. The regime of inhibited deposition was thought to depend on the intervening surface passivation caused by reduction of active sulfite ion to elemental sulfur. Based on the results of *in situ* Raman spectroscopy, the authors concluded that the two different regimes correspond to the adsorption and prevailing coverage at the gold surface of different species: namely, at low overpotential, the adsorption of active sulfite via the S-atom and, at high overpotential, the adsorption of the Au(I) sulfite complex via O-coordination.

Thiosulfate and Mixed Sulfite Thiosulfate Electrolyte

The kinetic behavior of the gold thiosulfate electrolyte has not been studied in much detail because of the experimental difficulties arising from the inherent instability of the system as discussed earlier. Similarly, the characterization of the kinetic mechanism of deposition from the mixed sulfite-thiosulfate bath is far from being satisfactory. This gap is partly justified because during the initial development of the mixed electrolyte the main interest was focused on the issue of bath stability and deposit hardness in view of its use for soft gold plating.

A detailed study of the electrochemical reduction of the gold thiosulfate complex was performed by Sullivan and Kohl [53]. Kinetic parameters for the reduction of the gold thiosulfate complex derived from polarization curves at a rotating disk electrode were reported, suggesting that the reduction of the thiosulfate complex occurs through an intermediate species structurally similar to the reactant in the pH range 4.0–6.4 with a faster kinetics compared to the Au(I) cyanide complex.

Early studies on the electroless deposition of gold from a sulfite-thiosulfate bath suggested that the active species was either the thiosulfate complex [54] or a mixed sulfite-thiosulfate complex [55]. In another study [53], the influence of sulfite addition on the cathodic reduction of the thiosulfate complex was shown to cause a negative shift of the reduction potential which was tentatively explained as being due to the formation of a mixed complex or to surface inhibition through sulfite adsorption at the electrode. This polarization effect was confirmed in a later study [56].

Osaka et al. [57] were only indirectly concerned with the mechanism of deposition when studying the mechanism of sulfur inclusion in gold electrodeposits from the mixed electrolyte. Two different routes for the discharge of the complex were envisaged, depending on thiosulfate concentration: the formation of an adsorbed intermediate followed by the rate determining electron transfer step at low thiosulfate concentration, according to the following reaction scheme:

$$2[Au(S_2O_3)_2]^{3-} \rightleftarrows (Au_2S_2O_3)_{ads} + 3S_2O_3^{2-}$$

$$(Au_2S_2O_3)_{ads} + 2e^- \rightarrow 2Au + S_2O_3^{2-}$$

and direct discharge of the complex at high concentration, according to the following reaction,

$$[Au(S_2O_3)_2]^{3-} + e^- \rightarrow Au + 2S_2O_3^{2-}$$

Indirect evidence for the proposed mechanism was found in the adsorption behavior of the thiosulfate anion and the $Au(S_2O_3)_2^-$ complex at gold surfaces [57,58]. Similar conclusions about the nature of the gold complex and its role as electroactive species in the mixed electrolyte were reached in a voltammetric study of the thiosulfate–sulfite bath at a rotating disk electrode [59].

The proposed mechanism briefly outlined above does not attribute any role to the sulfite ion. The later species is assumed to act as a purely chemical stabilizer in the mixed electrolyte and not to take part in the surface reaction nor to interfere with the formation of the gold deposit. This assumption does not appear to be acceptable in view of the expected strong interaction between the sulfite ion and the gold surface [52] and cannot be reconciled with the aforementioned polarization effect caused by sulfite addition to thiosulfate electrolyte [53,56]. Besides, the postulated unimportance of the sulfite ion is also questioned by the results of the speciation model of gold in the mixed sulfite-thiosulfate bath developed by Green and Roy [24] and in the light of experimental evidence confirming the formation of mixed-ligand species in this system [60]. In summary, a detailed kinetic study of the mixed thiosulfate–sulfite electrolyte has yet to be performed; this may be the only way to gather relevant information and possibly conclusive evidence about the very nature of the electroactive species and its discharge reaction mechanism.

CLASSIFICATION OF GOLD-PLATING PROCESSES

The current status of gold plating in decorative and industrial applications has been shaped over recent decades by a number of diverse requirements, concerned either with plating technology or deposit properties, resulting in the development of a variety of processes. A comprehensive and systematic classification is probably neither feasible nor useful. Nevertheless, the whole spectrum

TABLE 11.3
Classification of Gold-Plating Processes According to Electrolyte pH and Type of Deposits

		Deposition Mechanism			Gold Complex	
		Electrolytic	Electroless	Displacement	Cyanide	Noncyanide
Electrolyte pH	Acid	×		×	×	
	Neutral	×	×	×	×	×
	Alkaline	×	×		×	×
Type of deposit	Hard gold	×			×	
	Soft gold	×	×	×	×	×
	Color gilding (flash deposition)	×			×	
	Gold alloy	×			×	×
	Electroformed gold	×			×	×
	Electroformed gold alloy	×			×	×

of gold-plating processes can be easily arranged based on a few classification criteria, namely, the electrolyte type and pH, the deposition mechanism and the type of deposit obtained, whenever this is key to discriminate the final use. A gold-plating processes categorization according to these set of criteria is proposed in Table 11.3.

Gold is plated either with high purity, in excess of 99.9%, or as a pure gold, typically hard gold deposits, with Au content in the range from 99.0 to 99.7% (excluding carbon, nitrogen, and potassium, i.e., the common impurities of gold deposits from cyanide electrolytes), or as an alloy in a wide range of composition. The thickness of gold and gold alloy deposits may vary from as low as 50 nm, for purely decorative purpose, to a few hundreds of micrometers for jewelry electroforming.

As already noted above, gold electrodeposition processes can be schematically classified according to the key characteristics of the bath chemistry, i.e., type of ligand and pH of the electrolyte:

- Cyanide alkaline baths are used almost exclusively for decorative purposes, for the electrodeposition of alloys in a range of different colors, or for electroforming of Au-Ag and Au-Cu-Cd alloys.
- Cyanide acidic baths containing hardening metal additives (typically Co or Ni, but also Fe or In) are primarily used for engineering purposes, but also in decorative applications to produce hard, wear resistant high karat (Au 96–98%) colored gold deposits.
- Cyanide neutral baths are largely employed for the electrodeposition of soft gold layers serving as a bonding surface.
- Sulfite alkaline baths are mainly used in engineering applications to produce layers of pure gold with a range of properties, in particular hardness and ductility; they are also used to deposit alloy coatings otherwise not achievable by cyanide based baths, in the electroforming of prosthetic dentistry and, to a minor extent, of decorative and jewelry items.

For cyanide-based baths, the electrolyte pH is the most important process parameter and its value broadly defines other characteristics of the plating bath and actually designates the field of applications of a class of processes. The nature of the deposition mechanism, i.e., electroplating or electroless deposition, is important not only for obvious fundamental reasons but also in relation to the final application. The most important classification key is eventually the type of deposit, as this is directly linked to the final use and often also closely associated with the electrolyte type and its basic operative parameters, namely, the pH. The composition of the plating solution changes according to the equipment used for deposition.

For the barrel-plating application, where deposition is accomplished at low current density, the concentration of gold is maintained at the low end of the range commonly utilized for gold electrolytes in order to reduce precious metal losses due to drag-out. Conversely, electrolytes especially formulated for high-speed plating machines, e.g., reel-to-reel plating systems, are characterized by the highest value of gold concentration since a much higher current density is used. The concentration of the gold complex in electrolytes used for rack plating and semiconductor plating applications cover the intermediate range between those above.

Cyanide-Plating Baths

Gold cyanide baths are formulated either with the Au(I) or the Au(III) cyanide complex. The Au(I) cyanide baths are by far the largest class of electrolytes, Au(III) baths being used especially for deposition on difficult to plate materials.

There are basically three types of Au(I)-cyanide plating baths according to the pH range: acid baths (pH in the range 3.5–5.5); neutral baths with pH in the range 5.5–8.2; and alkaline baths (pH > 9.0). The free-cyanide content is the other major compositional parameter which differentiates the type of electrolyte. Alkaline baths contain significant concentration of free-cyanide, while neutral and acidic baths are exempt from free cyanide. The gold compound used for cyanide baths is the double salt potassium gold cyanide, $KAu(CN)_2$, with a gold content of 68.2%. The double potassium salt is preferred to the double sodium salt because of its higher solubility. The latter can be used where the solubility is not an issue, e.g., for dilute baths used in barrel plating, though it is hardly used. The potassium cation may be codeposited in the form of salt as an inclusion in the gold layer depending on the type of bath and operative conditions.

Alkaline Electrolytes

Pure Gold Electrolytes Alkaline baths are used exclusively for decorative purposes, for the deposition of thin layers of gold in color finishing of jewelry items and other decorative objects. The composition of typical alkaline cyanide baths used for the electrodeposition of gold is reported in Table 11.4 [2,3,61,62].

TABLE 11.4

Examples of Alkaline Cyanide Baths Composition and Operating Conditions for the Electrodeposition of Matte or Bright Gold (Concentration in g L^{-1})

	High Purity Matte Gold	Rack	Barrel	Bright Au(Ag)	Bright
$KAu(CN)_2$	6–20	12	6	8–20	12
$KAg(CN)_2$	—	—	—	0.6–1.2	0.8
KCN	15–30	20	30	60–100	80
K_2HPO_4	15–30	20	30	—	—
K_2CO_3	20–30	20	30	—	(20)
pH	11.0–12.0	11.0–11.5	11.0–11.5	11.0–12.0	12
T/°C	50–60	50–60	55–65	20–30	25
c.d./A dm^{-2}	0.1–1.0	0.1–0.5	0.1–0.5	0.1–1.0	0.3–0.8

Source: From *Gold Plating Technology*, ed. F. H. Reid and W. Goldie, Electrochemical Publications, Ltd., Ayr, Scotland, 1974; Gold plating, A. M. Weisberg, *Metal Finish.*, 2007, **105(10)**, 205; Electroplated golds, in *Gold Usage*, W. S. Rapson and T. Groenewald, Academic Press, London, 1978, pp.196–269; and Gold in modern electroplating, 2nd Ed., ed. F. A. Lowenheim, John Wiley & Sons, 1963, pp. 207–223.

Excellent throwing power, high tolerance to metal contaminants and to changes in operating conditions are the main characteristics of alkaline cyanide baths. However, the alkaline baths are definitely inferior to the other classes of cyanide baths as far as hardness and sliding wear behavior of deposits is concerned. In addition, the presence of free cyanide poses safety and environmental issues that must be duly managed. These are the basic reasons why alkaline cyanide baths have been largely replaced by other processes in industrial applications. The pH value of alkaline cyanide baths is usually in the range 9–11, in the presence of substantial concentration of free cyanide. The gold concentration is usually kept at 5–10 g L^{-1}, while the KCN concentration may vary in a wider range depending on the type of plating process. Baths for barrel plating feature lower metal concentration (Au 4 g L^{-1} or less) and a higher free cyanide metal ratio compared to standard rack formulations.

Free cyanide from KCN acts as a stabilizer of the gold cyanide complex, helps in preventing metal impurities in the bath from depositing, is essential to anode dissolution when a gold anode is used, and improves throwing power by both increasing bath conductivity and deposition overpotential. The concentration of potassium cyanide usually increases with electrolyte aging because of gold replenishment with KAu(CN)$_2$ and the addition of potassium hydroxide to adjust the pH, where insoluble anodes are used.

The primary function of potassium carbonate is that of a conductivity salt and by this action it also contributes to throwing power, although a high concentration, found in aged baths, may be deleterious to thickness distribution and deposit appearance. The concentration of carbonate increases with electrolyte life due to cyanide oxidation and the carbonation of the solution with the absorption of carbon dioxide from the air. Potassium phosphate is basically used as a pH buffer, particularly to prevent local pH drop at the anode. Potassium hydroxide serves various functions in the bath, such as the improvement of throwing power and inhibition of the anodic oxidation of cyanide. Potassium hydroxide may also be necessary to impart solubility to either inorganic or organic addition agents.

Standard operating conditions of temperature (T) and current density (c.d.) are 50–60°C and 0.1–0.5 A dm^{-2}, respectively. Higher T or c.d. may be used for alloy plating. In any case, the optimum c.d. value depends on the gold concentration and electrolyte stirring. Accordingly, current efficiency of alkaline cyanide baths can be very high, usually in excess of 90%, provided that appropriate conditions of gold concentration and electrolyte stirring are maintained. The operating temperature in barrel plating may be slightly higher compared to rack-plating applications, while current density is approximately from one-third to half that of rack plating. In any case, the operative temperature of alkaline cyanide baths should be controlled below about 70°C, since at higher temperature hydrolysis of cyanide becomes important with build-up of breakdown products. Degradation of cyanide also occurs by anodic oxidation but hydrolysis is more important in hot baths.

Alkaline baths suffer from organic contamination while being relatively tolerant of inorganic contaminants (metal cations), depending on free cyanide content [63,64]. Periodic carbon treatment is mandatory in order to maintain organic contamination at a low acceptable level. This should be performed in an auxiliary tank for best results (high level of removal and low loss of precious metal). Continuous filtration through a carbon packed filter is current practice.

Alkaline cyanide baths may be used to obtain either matt or bright deposits. Gold deposits from alkaline cyanide baths in the absence of brighteners or grain refiners have a columnar structure and a matt appearance. Gold from alkaline cyanide baths is usually brightened by addition of metals such as Sn, Sb, and Ag. Low alloyed Ag deposits (Ag content less than 1%) are particularly attractive for the ease of operation and control of the baths and because of room temperature operation.

Gold Alloy Electrolytes   The development of gold alloy plating began soon after the establishment of the gold-plating industry around 1850, basically for the achievement of different color finishes [11]. Like the long and slow development of gold plating, gold alloy plating remained (for almost a century) a craftsmanship art lacking any connection with science and engineering.

The electrodeposition of gold alloys for decorative color finishing was reviewed by Rapson and Groenewald [3]. The classical book of Brenner [65] remains a valuable reference for alloy plating of any type and should be referenced by the reader interested in a wider treatment of the subject. A paper by Arrowsmith [66] is a useful short introduction to understanding the origin of the color of gold alloys and the basic mechanism behind the possible change of color especially in connection with electroplated gold.

Gold alloy electrodeposition has many diverse uses, primarily for decorative applications (the most important being thick coatings and electroforming) with only occasional industrial uses [67]. Today, plating bath suppliers offer processes easily tuneable for the deposition of either thin or thick gold alloy layers in a range of colors. Silver-brightened baths can be formulated for the deposition of a range of alloy compositions, with gold content as low as 10 karat, and can be used to build up thickness at lower cost on items that are then given a finish layer, to provide the final color and properties, e.g., wear resistance.

Flash gold alloy deposition with thickness typically in the range 0.1–0.2 μm is used for decorative purposes and is traditionally performed on an empirical basis, the final color shade more than the composition being the primary consideration [61]. Alkaline cyanide solutions buffered with phosphate are used for flash deposition of color gold finishes. Typically the gold concentration is kept as low as viable, consistent with the use for thin film deposition, with the twofold objectives of reducing metal inventory and bath losses through drag-out. A low gold content also implies a limited concentration of free cyanide, which in turn is a desired characteristic in view of the necessity to operate color gold baths at high temperature to obtain bright deposits.

Decorative gold flash baths can be formulated for any gold shade, for the achievement of the desired color designation according to the international standard [68], by addition of so-called metallic brighteners as the cyanide complex of the relevant metal. A variety of different shades and colors can be produced by alloying with metals such as silver, copper, nickel, tin, or zinc. Table 11.5 reports some example formulations of color gold baths for flash deposition.

These formulations may cover most general requirements and describe the range of possibilities rather than a manual for operation. All operating parameters should be carefully adjusted and a good operating practice is essential to the achievement of the desired results.

The surface finish of the base material will influence the final appearance of the color finish and this should be a concern, particularly when flash coloring items with different surface texture.

TABLE 11.5
Examples of Alkaline Cyanide Baths for Flash Color Gold Deposition (Concentration in g L^{-1})

	24ct Yellow	White	Green	Pink	Rose
$KAu(CN)_2$	2	0.4	2	0.82	6
KCN	7.5	15	7.5	4	4
K_2HPO_4 - Dibasic potassium phosphate	15–20	15–20	15–20	15–20	
KOH/NaOH	—	—	—		15
K_2CO_3	—	—	—		30
$K_2Ni(CN)_4$	—	1.1	—	0.2	1
$KCu(CN)_2$	—	—	—	2.7	2.7
$KAg(CN)_2$	—	—	0.25		
T/°C	60–70	60–70	50–70	60–70	60–80
c.d./A dm^{-2}	1–4	3–6	1–3	3–4	2–5

Source: From Electroplated golds, in *Gold Usage*, W. S. Rapson and T. Groenewald, Academic Press, London, 1978, pp.196–269; and Gold plating, A. M. Weisberg, *Metal Finish.*, 2007, **105(10)**, 205.

The color of the base material has an influence on the final appearance although this issue is easily resolved by deposition of a minimum thickness of the gold layer.

Current density, temperature, and concentration of free cyanide are the most important operating parameters having the greatest effect on the final result. As a general guide, with the exception of silver brightened solutions, decreasing current density or temperature will favour gold deposition, while higher than recommended values of either current density or temperature give an initially richer color but make control of the final shade increasingly difficult. When copper is used as an alloying element, free cyanide concentration should be carefully adjusted to enable reproducible results. As a general remark, copper deposition, i.e., pink and red shades, are favored by a low free cyanide concentration. Free cyanide content for a given color finish is kept as low as possible in order to limit its degradation in the hot baths, though a higher concentration, in principle, allows for an easier control of the color.

The gold concentration in flash deposition electrolytes may be slightly reduced for barrel plating applications, while increasing the concentration of free cyanide and/or of dipotassium phosphate. Barrel plating operating conditions allow for deposition rates that are obviously lower compared to rack application and a flash deposit is obtained in a few minutes rather than in a few tenths of seconds as it is in the case of rack plating. Alkaline gold and gold alloy baths are usually operated with inert anodes of stainless steel [61,62].

Neutral Electrolytes

Neutral baths for pure gold plating, i.e., gold cyanide baths with pH in the range 5.5–8.2 and obviously without free cyanide, had probably their first massive use in the industry for barrel plating of electronic components thanks to their excellent throwing power, higher tolerance against metal impurities and lower susceptibility to bipolar effects compared to acid electrolyte conditions [69,70].

Neutral baths are used especially for semiconductors and printed circuit-board applications, whenever the compatibility with masking polymer materials and photoresist is not an issue. Basically, the preference for neutral baths in these applications is dictated by the high purity of the gold layer, greater than 99.9%, the high plating speed, high current efficiency and the range of layer properties achievable. A variety of formulations is commercially available for barrel, rack, and reel-to-reel plating applications, the latter either for conventional direct current or pulse current deposition.

The basic formulation of the electrolyte simply consists of $KAu(CN)_2$ and a phosphate buffer in the neutral to slightly acidic range of pH. Operating conditions of current density and temperature are selected based on the gold concentration and depend upon the desired type of deposit. Either soft or hard gold deposits of high purity can be obtained from such electrolytes either by simply changing the operative parameters (without any special additive) or by minimal variation of the composition with metal or organic additives. In particular, hard gold deposits can be obtained without addition of hardening metal additives, as first described by Reinheimer [39]. This is the reason why neutral baths for the deposition of hard gold became known as additive free hard gold (AFHG) baths [71]. AFHG baths are not exclusive to the class of neutral electrolytes; acid formulations are also possible for low speed operation. Current efficiency is usually in the range 75–90% and it is mainly influenced by temperature, gold concentration, and flow conditions. In this respect, high-speed AFHG baths may operate at a current efficiency in excess of 85%, either at 25°C and current density values in the medium to high range (i.e., 10–15 A dm^{-2}), or at 45°C and a current density higher than 15 A dm^{-2}.

The electrodeposition of soft gold is traditionally obtained in the presence of metal additives, such as Pb(II) and Tl(I), which are known as depolarizers for the reduction of the dicyanoaurate gold complex [45]. However, recourse to metal depolarizers is currently discouraged because of their adverse effect on wire bonding performance [72,73] and for the concern raised by their toxicity. Alternatively, soft gold electrodeposits can be obtained by a judicious choice of the operating parameters, including gold concentration, electrolyte pH, temperature and current density, with

TABLE 11.6

Basic Formulations of Neutral and Acid Baths for the Electrodeposition of Soft Gold and Additive-Free Hard Gold (AFHG) (Concentration in g L^{-1})

	Soft Gold [2,61]		AFHG [71]		
	Rack or Barrel		High Speed	Low Speed	High Speed
$KAu(CN)_2$	10–30	8–20	15–30	40	44–59
KH_2PO_4	60	—	—	100	100
$K_3C_6H_5O_7$	60	70	90	—	—
pH (KOH or H_3PO_4)	5.5–8.0	6.0–8.0	4.5–5.5	4.3–4.5	6.5–7.5
T/°C	60	70	60–70	25	40
c.d./A dm^{-2}	0.1–1.5	0.1–0.5	10–40	1–2	20–30
Current efficiency/%	90	90	95	75–95	85–95

electrolyte aging being duly managed by proprietary chemistry. Soft gold deposition is primarily favored by a relatively low current density—the recommended c.d. range depending on the gold concentration and on flow conditions in the plating equipment—and high temperature in the range from 60 to 75°C. Accordingly, fine gold electrolytes formulations are commercially available either for rack, barrel, or high-speed applications. Example formulations of neutral baths for the electrode-position of soft gold and AFHG deposits are reported in Table 11.6 [2,61,71]. Acid formulations are also included for completeness.

An important processing issue in gold plating from neutral as well as acid baths is the choice of the anode and its interaction with the electrolyte. Anodes in neutral baths can be platinized titanium, dimensional stable mixed metal oxide electrodes, or even soluble anodes in high speed operations with close anode to cathode spacing. Two types of degradation processes were shown to occur with conventional platinized titanium anodes in high-speed gold plating applications: the oxidation of the cyanoaurate (I) complex to the cyanoaurate (III) species [74]; and, in citrate buffered baths, the oxidative degradation of the citrate [39]. The accumulation of the trivalent gold species leads to a loss of current efficiency since a higher charge is required for its reduction at the cathode [75]. A soluble gold anode or a RuO_2-TiO_2 type anode were shown to be significantly less active toward the oxidation of the cyanoaurate (I) species, in fact a much lower saturation concentration of the Au(III) cyanide complex was observed; the gold anode was particularly effective in this respect [75]. However, the use of a soluble gold anode requires regular maintenance operations and entails a changing current density distribution over the service life of the anode. Therefore it is not usually regarded as a feasible solution. The mixed-metal oxide anode with RuO_2 as the active layer turned out to be unstable in high-speed deposition conditions, showing a limited service life; anodes with an IrO_2 catalytic layer performed much better [76].

Acid Electrolytes

The development of the first acid gold bath based on $KAu(CN)_2$ dates back to the late 1950s [77] after the discovery that potassium dicyanoaurate was stable in acid solutions down to pH 3 [2]. The great advantage of acid gold electrolytes was the possibility to deposit bright gold layers and in fact they were originally intended for jewelry plating. The fast-growing electronic industry was, however, the natural destination of this class of electrolytes, particularly due to their higher compat-ibility with printed circuit board manufacturing and the much improved wear resistance of deposits compared to the standard, at that time, alkaline formulation of gold-plating baths.

Commercial acid baths are usually based on a citrate buffer and feature a pH value about 4, but various formulations based either on citrate or phosphate buffer are available with pH in the range 3.5–5.5. The pH of the bath is adjusted by addition of potassium hydroxide and citric or phosphoric acid.

The codeposition of alloying elements is readily obtained from acid baths resulting in an impressive change of the properties of deposits, in particular the hardness. Hard bright gold deposits are usually produced from acid baths containing small additions of cobalt or nickel; baths including small additions of other hardening metals, such as indium, are also available.

Current efficiency is a critical process parameter in hard gold deposition and it is affected by many variables: primarily the concentration of Au(I), the concentration of Co(II), or other metal additive, the pH and the temperature, the flow conditions and the type and concentration of impurities. An additional variable is the concentration of Au(III) [75]. The cyanoaurate Au(III) complex may slowly build-up as a result of anodic oxidation of the cyanide Au(I) complex and has an obvious effect on current efficiency, as already discussed earlier with reference to neutral baths. Control of the oxidation of the Au(I) complex is also the reason why acid Au(I) cyanide baths may include a mild reducing agent in their formulation. For most electrolytes and operating conditions the current efficiency is in the range of 30–40%. Current efficiency values about or below 30% are observed depositing from pH 3.5 baths at gold concentration below about 6 g L^{-1}. In such conditions, current efficiency is relatively insensitive to change in current density and electrolyte stirring.

Cobalt (nickel) content in the gold deposits is a complex function of deposition current density and electrolyte flow at fixed electrolyte pH and deposition temperature; besides, the extent of codeposition also depends on the nature of the ligand by which the metal cation is stabilized in solution. Cobalt content can be as low as 0.1% at high current density and moderate electrolyte flow, increasing up to about 0.5% at low current density and under strong stirring. An important influence on the codeposition of the hardening metal is that of the electrolyte pH; at pH above about 5.0 the codeposition of either cobalt or nickel is practically suppressed [78–80]. The deposition temperature was found to affect Co and Ni codeposition in an opposite way; the Co and Ni content decreasing and increasing, respectively, as the deposition temperature increases [80]. In that study, Raub et al. observed that the change of the cobalt content with increasing temperature was correlated with the decreasing trend of carbon codeposition, whilst the opposite was found for Ni hardened gold deposits, suggesting that an increasing share of the nickel was actually forming a solid solution with gold. However, the influence of temperature on cobalt codeposition in hard gold deposits was discounted in a later study [81].

Example formulations of acid baths for the electrodeposition of hard gold deposits are reported in Table 11.7 [2,61].

Alkaline cyanide electrolytes have been in use for years in color gold flash deposition, but they are currently being increasingly substituted by weakly acidic baths. Weakly acidic baths are also largely used for the electrodeposition of thick gold alloy deposits in a range of

---

TABLE 11.7

Example Formulations of Acid Gold Baths for the Electrodeposition of Co (Ni) Hardened Gold (Concentration in g L^{-1})

	Rack or Barrel	Co-HG Low Speed	Co-HG High Speed	High-Speed Continuous Plating
KAu(CN)$_2$	8	20	22	8–16
C$_6$H$_8$O$_7$	40	10	75	—
K$_3$C$_6$H$_5$O$_7$	40	—	110	90
Co(II) or Ni (II)	0.2–0.5	0.1	0.075	0.7
pH	3.8–5	4.0	4.2	3.8–4.3
T/°C	20–30	30	65	30–50
c.d./A dm^{-2}	0.5–1.0	1	10–40	10–40
Current efficiency/%	30–40	40	40	—

compositions. In particular, acid gold plating baths are formulated for the deposition of karat colors, that is gold alloy deposits with a color matching a lower caratage than actually assayed. These baths can be formulated for the deposition of various shades of gold from pale 0-1 N to warm 3 N yellow [61], usually at pH in the mildly acidic range, from 4.0 to 5.0, and with addition of standard metal brighteners Co (yellow gold) or Ni (color gold).

Trivalent Gold Cyanide Electrolytes

Acid Au(III) cyanide baths were introduced for industrial applications in the late 1970s; see Wilkinson for a short account of its development [22], and have seen an increasing use since then. The main limits of Au(III) cyanide baths are the corrosiveness, the low rate of deposition, and the comparatively inadequate wear resistance when used as a contact finish, compared to hard gold from acid Au(I)-cyanide electrolytes. The acid Au(III) bath are especially recommended for deposition onto difficult to plate materials (because of passivity) such as stainless steel and chromium. They are currently also formulated for decorative plating (rack and brush) in a range of color shades and for deposition of thick, ductile coatings.

Noncyanide Baths

The gold(I) sulfite complex has been known since the 1840s [2] but its successful exploitation in the electroplating industry dates from relatively recent times. The first commercial electrolytes based on the sulfite complex were developed in the 1960s, in the wake of the issuing of Smith [82] and Shoushanian's [83] patents. Gold sulfite solutions were known to be exposed to gradual degradation with electrolysis time. Visual evidence of the decomposition of the complex is shown in the change of the solution from colorless to light purple, followed eventually by precipitation of gold. The degradation of the electrolyte was ascribed to the disproportionation of free Au(I) ions, formed as an intermediate species in the discharge process [84]. This interpretation was then disproved by evidence of the active role of dithionite in the degradation of sulfite electrolyte [24].

The low stability of the electrolyte was the reason why the bath was originally formulated with an alkaline pH and with the addition of chelating agents, as well as recommended for operation at low current density. Notwithstanding the open issue of stability, the quality of gold deposits from the alkaline sulfite electrolyte could be substantially improved thanks to the use of additives as grain refiners (As, Se, Te, Sb) in concentrations ranging from 1 to 400 ppm or the addition of metal compounds, especially cadmium and lead, as brighteners and hardening additives [82,83].

The starting point for the subsequent development in the chemistry of the sulfite bath was the work of Smagunova et al. [85], resulting in the formulation of more stable amino-sulfite baths. The amino-sulfite formulation was further studied and modified for alloy deposition, e.g., Au-Cu [86]. Further developments in the patent literature revealed the intention of challenging the extreme limit of stability of the sulfite complex. Sulfite electrolytes at pH as low as 4.0 were discussed in a patent by Morrissey [87], though a pH of about 6.0 should be taken as a safe acceptable lower limit since in more acidic conditions excessive evolution of sulfur dioxide occurs [88]. Operation of the sulfite bath at pH in the range 6.0–7.0 has the distinct advantage of promoting sulfite decomposition and the volatilization of sulfur dioxide, thus reducing the accumulation of sulfite caused by consumption and replenishment of the gold complex.

Sulfite baths utilize a gold sulfite compound as the source of gold, usually potassium or sodium gold sulfite $X_3Au(SO_3)_2$ (X=Na or K), in the presence of an excess of sodium or potassium sulfite at pH in the range 8.5–10. The potassium form is the sulfite compound, which is commonly used in bath formulation, though ammonium gold sulfite baths are also available and show improved stability in the slightly acidic-neutral pH range ~6.0–8.0 [89].

The base formulation of most sulfite based electrolytes for pure gold electrodeposition comprises the gold sulfite compound at concentration 12–16 g L^{-1} (about 8–10 g L^{-1} as Au) sodium sulfite in the range 50–80 g L^{-1}, and additional components such as a buffer and a chelating compound. Most commercial formulations are operated at pH 9.5–10.0 and temperature in the range 50–60°C, though slightly acidic or neutral baths are also available. The operating current density depends on gold concentration and flow conditions or cathode movement. For a standard 10 g L^{-1} gold concentration, a current density of 0.3 A dm^{-2} is recommended; a linear change of operating c.d. with the gold concentration may be reasonably assumed as the latter is increased. For the best stability, commercially available sulfite baths contain proprietary stabilizing additives and are preferably operated at alkaline pH.

Sulfite electrolytes for colored gold alloys are also in use for decorative deposition thanks to their excellent distribution and leveling properties and also for their superior ductility compared to gold alloy deposits from acid cyanide electrolytes giving similar shades and colors [90]. Alloy plating from sulfite electrolytes comprises binary and ternary compositions, such as Au-Cu, Au-Pd, and Au-Cu-Pd—already reported by Rapson [3] and later incorporated into proprietary processes [91–93]. The later is an alloy finish widely used in the electroplating of spectacle frames. Example formulations of neutral and alkaline sulfite baths for the electrodeposition of gold alloy deposits are reported in Table 11.8.

After many years of controversial reputation, pure gold sulfite baths have found growing acceptance particularly in a few specialized industrial applications, for example the electrodeposition of soft gold for interconnection and assembly in semiconductor manufacture. This is thanks to their greater compatibility with standard resist materials used in photolithographic patterning processes compared to the standard cyanide electrolyte for soft gold deposition. In fact, even trace amounts of cyanide were reported to damage a standard photoresist for through-mask plating [94]. Problems may still arise with alkaline soluble positive photoresists, if an alkaline bath is used, slowly releasing contaminants in the plating solutions as a consequence of a mild attack. This does not compromise resist adhesion or structural and shape integrity of the pattern, but leads to the build-up of impurities in the bath that may adversely affect deposit properties, primarily surface roughness and residual stress. These issues prompted the search for a bath chemistry free from cyanide, alternative to the standard sulfite bath. A thiosulfate-sulfite mixed ligand electrolyte was first described by Osaka et al. [30]. This type of bath was developed especially for microelectronic packaging applications; the electrodeposition of Au bumps for tape automated bonding (TAB) of integrated circuits. Mixed sulfite-thiosulfate baths are operated at temperatures of 55–60°C, c.d. in the range from 0.3 to 0.7 A dm^{-2} with electrolyte stirring. The electrolyte compositions so far proposed in the literature are reported in Table 11.9.

The concentration of thiosulfate in the electrolyte was shown to be the primary factor determining the sulfur content of deposits and consequently their residual stress and mechanical properties [57]. The addition of thallium as a grain refiner was shown to reduce the deposits residual stress to almost zero, at least for a definite set of operative conditions: a temperature of 50°C and c.d. within the range 0.4–0.6 A dm^{-2} [30]. This electrolyte was found to be adequate for interconnection purposes even in the absence of additives [95].

The issue of electrolyte ageing requires further investigation, though according to Liew et al. [96,97] the process performed well on large-scale operation under industrial conditions.

PULSE ELECTRODEPOSITION

Pulse plating is the electrodeposition of metals by a periodic variation of the applied current. In pulse plating three parameters can be varied independently: the pulse current density, the pulse time, and the off-time. Theoretical aspects of pulse plating were discussed by Ibl [98].

The main advantages of the pulse current deposition technique is the increased nucleation rate caused by the higher instantaneous current density and the interphase relaxation during the off-time period. Nucleation intensity has an obvious effect on a deposit's grain size and microstructure and both these factors may significantly impact on the physical and mechanical properties of electrodeposited metals.

TABLE 11.8
Example Formulations and Essential Operating Parameters of Sulfite Plating Baths for Electrodeposition of Pure Gold and Gold Alloy

Type of Electrolyte	Composition/%	$[M^{z+}]$/g L^{-1}	pH	T/°C	μm min^{-1}	c.d./A dm^{-2}
Neutral	Binary and ternary alloys					
	Au-Cu 33	Au 8	6.6	60	0.25–0.30	1.0
		Cu 5				
	Au-Cu3-Pd5	Au 8	8.0	55–60	0.25–0.30	0.5
		Cu 0.1				
		Pd 1.8				
Alkaline	Pure gold, binary and	Au 99.9	Au 10	9.0–10	50–65	0.20–0.30	0.3–0.5
	ternary alloys	Au-Cd1	Au 10 Cd 1	9.2	50–60	0.30	0.5
		Au-Cu24-Cd1	Au 4 Cu 10 Cd 0.1	9.7	55–60	0.25	0.5

TABLE 11.9
Mixed Sulfite–Thiosulfate Bath for Soft Gold Electrodeposition (Concentration in mol l^{-1} unless Otherwise Noted)

Composition and Operative Conditions	Ref. [30]	Ref. [30]	Ref. [95]
$NaAuCl_4$	0.06	0.06	0.05
$Na_2S_2O_3\ 5H_2O$	1.1–1.4	0.42	0.42
Na_2SO_3	1.1–1.4	0.42	0.42
Na_2HPO_4	0.30	0.30	—
Tl(I) (added as Tl_2SO_4)/ppm	—	5	—
pH	6.0	6.0	7.4
T/°C	60	60	55
c.d./A dm^{-2}	0.5	0.5	0.35–0.7

The study of pulse current electrodeposition of gold was initiated by the need to improve the quality of gold layers, particularly with respect to porosity [99] and electrical resistance [100], and more generally by the aim of achieving the same or better properties of direct current plated films at lower thickness [101]. Later, pulse plating received renewed momentum in view of further advantages concerning the possible reduction of residual stress [102,103] and the implementation of gold electrodeposition in through-mask plating applications [51,103]. The main and most important effects of pulsed current plating on the microstructure and physical properties of gold deposits had already clearly emerged when a review was published by Raub and Knödler on this subject [102].

Most studies on pulse electrodeposition of gold concentrated on standard acid electrolytes utilizing Fe-group metal ions as hardening additives, particularly Co. In these electrolytes, pulse plating was shown to have quite strong effects [102]: in particular, carbon content was substantially decreased by pulse plating while Co or Ni content was shown to increase as a function of the off-time. Correspondingly, drastic changes in physical properties and microstructure were observed. Au(Co) deposits produced at 1 A dm^{-2} showed an average crystal size of 25 nm, a high density of nanopores, with size in the range 0.5–5 nm, and correspondingly low density, about 17.0 g cm^{-3}; pulse plated deposits (10 ms on-time, 100 ms off-time) had grain size in the range of 100 nm and were found to be almost exempt from inclusions, showing density of 19.2 g cm^{-3} [104]. The concentration of codeposited cobalt was reported to increase slightly, from about 0.2 to 0.25%, and a small drop in hardness was also noticed, together with increased ductility. Pulse plating influence on composition and microstructure of Co or Ni hardened gold deposits was particularly strong when operating at a 10% duty cycle (10 ms on-time, 100 ms off-time). Not surprisingly, the wear behavior of pulse-plated gold was adversely affected by these compositional changes, resulting in a much higher susceptibility to adhesive transfer and wear in sliding abrasion tests [104]. Pulse current deposits showed transition from an elastic to a plastic deformation wear regime at a lower load, compared to direct current deposits. This behavior was related to higher elastic modulus—that is, lower maximum elastic deformation—and to lower thermal stability of the pulse current deposits compared to direct current coatings. In contrast, direct current and pulse current (0.9 ms on-time, 9 ms off-time) gold deposits from a pH 5.5 gold cyanide bath in the absence of hardening additives, did not display any remarkable property modification. The only noticeable effect was a slightly lower grain size and a more uniform microstructure of the pulse plated gold [105]. These results consistently suggest that the main influence of pulse plating is through the removal of adsorbed (Co or Ni) species during the off-time or through preventing their formation during the cathodic pulse.

The impact of pulse electrodeposition on morphology and especially on residual porosity in gold films has been recognized in the technical literature [106]. The effect of pulse deposition on the residual porosity of metal electrodeposits was considered as one of the most appealing peculiarities of the pulse plating technique, deserving both a detailed analysis and an attempt at modeling its beneficial influence [107]. Reduction in porosity is likely to be the result of both increased nucleation intensity and desorption of hydrogen during off-time. Hydrogen evolution during cathodic deposition of metals is generally recognized as an engendering factor in porosity of electrodeposits [108] and its role in causing nanopores in Co and Ni hardened gold deposits from acidic baths is supported by several studies [40,109–111]. Similarly, gold deposits from room temperature baths containing no hardening additives showed a high density of nonmetallic inclusions, probably gas bubbles, in sizes up to 15 nm [40]. Possible effects of hydrogen entrapment on the properties of electrodeposited gold were found in studying thermally induced defects and structural damage of pure gold electrodeposits [112] and electroformed Au-Cu [112,113] and Au-Cu-Cd alloys [114].

The implementation of pulse-plating processes for gold deposition in industrial applications seems to be of limited importance. In particular, pulse plating can be used with high-speed formulations of neutral and acid baths, but it is not reported as a widely used practice. On the other hand, gold deposition by pulse current is essential to some specialized applications, such as the fabrication of X-ray masks—utilized in the LIGA process (LIGA is the German acronym for Lithographie, Galvanoformung, Abformung)—having a gold absorber layer deposited by pulse plating from a sulfite electrolyte for an effective control of the residual stress level in the electroformed gold [115].

Composite Electrodeposition

Composite coatings are obtained by electrodeposition from a plating bath in which the particles to be codeposited are dispersed in suspension. Particle size may range from a few micrometers to less than 100 nm, in which case the term nanocomposite coating is also used.

Many factors are involved in the electrocodeposition process [116,117]: the electrolyte formulation, including particle concentration (bath load), the operative conditions, the type and intensity of electrolyte stirring, and the kinetics of deposition. Not surprisingly, process development and control are not straightforward and the industrial implementation of electrocodeposition processes is a challenging task.

Research work conducted over many years [118–121] has demonstrate that through composite electrodeposition, significant improvements in the properties of the pure metal matrix can be achieved, including hardness, wear resistance, and corrosion behavior. In particular, the incorporation of ceramic or other hard particles is an effective way to improve coating hardness and wear resistance. Hardness increase can be explained according to the Orowan mechanism of dispersion hardening [122], as long as particle size is less than 1 µm [119]. This increase depends on the interparticle distance, i.e., on particle size and volume fraction of the hard phase.

Research in the field of gold or gold alloy matrix composites, as coatings or electroformed parts, started in the 1980s to provide industry with coatings of improved mechanical properties and thermal stability [123]. The processes studied in the frame of academic research activities are not particularly numerous and most of them were aimed at the preparation of wear resistant coatings, with low and stable contact resistance for connectors application: Au\Al$_2$O$_3$ composite [124–126]; Au\TiN [127]; Au\diamond [127]; Au(Co) and Au(Ni)\PTFE [128–130]. Composite plating was also studied for the development of electroforming processes or thick coating deposition, more especially of the following composites: Au\B$_4$C [121,131,132], Au-Cu\B$_4$C [133–136]; Au-Cu-Cd\B$_4$C [131], Au\UDD and Au\CNT composite, that is ultra dispersed diamond (UDD) or carbon nanotubes (CNT) codeposition in pure gold matrix [137,138].

The findings emerging from these studies confirmed that electrocodeposition is a promising solution for the issue of mechanical strength and thermal stability of electroplated gold and gold alloys. The extent of codeposition was shown to depend on the nature of the dispersed phase and could reach value as high as 40% by volume (for Au\BC$_4$ and Au-Cu\BC$_4$ composites) corresponding to a weight fraction of about 8%, a 2-karat reduction with respect to pure Au.

Thermal stability, a material issue especially relevant to gold alloy deposits such as Au-Cu-Cd and Au-Cu that are nanocrystalline and of metastable phase structure in the as-deposited state [139–141], was shown to be remarkably improved by codeposition of ceramic particles.

More recently, the introduction of submicron and nanoscale particles for codeposition in electrode-posited layers has given impetus to new research efforts, after demonstrating that equal or even better properties of the matrix metal can be achieved at lower thicknesses and with a low mass fraction of incorporated particles [142–144]. A distinct advantage of the codeposition of nanometer sized particles is the possibility to reduce the undesired effects of particles codeposition on appearance, surface texture and roughness of deposits, provided that the nanoparticles are effectively dispersed in the layer. This in turn would allow the exploitation of electrocodeposition processes in microsystems manufacturing, the type of application obviously outside the scope of conventional codeposition processes. However, this remains a field largely unexplored: the literature about the codeposition of nanoparticles in metal matrices is limited, and in particular results about Au matrix nanocomposite are very scarce. A wear testing study on Au-Cu-Cd\Al$_2$O$_3$ (50 nm particle size) composites is possibly the first published experimental evidence demonstrating the benefits of nanoscale ceramic particles codeposited in an electrodeposited gold alloy matrix [125]. Significant increases in microhardness and/or wear resistance were demonstrated for Au\Al$_2$O$_3$ nanocomposites from a neutral cyanide bath [145] and for a Au\ UDD nanocomposite from either standard cyanide [146] or sulfite electrolyte [137,138]. Codeposition of carbon nanotubes in a pure gold matrix from a pH 7.5 sulfite electrolyte, resulting in carbon content in the range 0.05–0.1%, was shown to produce a remarkable improvement in wear performance [138]. Though promising, these results appear isolated and preliminary.

Finally, it must be stressed that the most critical aspect of composite plating with respect to its industrial implementation is the achievement of a homogenous dispersion of particles, an essential requisite for properties improvement. Therefore, a primary requirement for process development is a careful adjustment of hydrodynamic conditions. The issues to be faced in nanocomposite electrodeposition are even more complicated and will certainly call for much work both at the process design stage and during experimental development. The most important aspects are listed as follows: the chemical nature, size and shape of nanoparticles, a reasonable cost (at the present time not suitable for conventional electroplating processes, but maybe acceptable in precious metal plating), the availability of nanoparticles with tight size distribution and, in case of agglomeration, the possibility of breaking up agglomerates by simple treatments. The stabilization of suspended particles in the plating bath and the definition, by systematic investigation, of the optimized parameters of codeposition are also important.

Electroforming

Electroforming is basically the deposition of a metallic material onto a substrate, called a mandrel or mould, which has the shape of the final product. After deposition to the final thickness, the electroformed object is separated from the mandrel by a suitable procedure in the form of a free standing object, thus allowing reuse of the mandrel itself for reproduction of duplicates of the same object. Alternatively, as in jewelry electroforming, either a low melting point alloy or wax mandrel can be used, made in multiple copies by wax casting in a rubber master model. The mandrel is then removed after electroforming by melting and any residues dissolved out by chemical cleaning procedures, thus leaving a hollow electroformed object.

Gold and gold alloy electroforming is used in different manufacturing processes: for jewelry making in alternative to the traditional procedure of microcasting [147,148]; in the production of a variety of decorative objects; in prosthetic dentistry to form the substructure for porcelain inlays and crowns [149,150]; and in other applications such as the production of pure gold sputtering targets [67] and prostheses for medical purposes [151].

The unparalleled advantage of electroforming in jewelry manufacturing is the capability of producing hollow objects that are thinner, lighter, and more elaborate in design compared to those produced by traditional casting techniques. The effective exploitation of this potential resulted in the development of highly specialized technology for gold alloy electroforming requiring dedicated computer controlled equipment. The need to develop a system-technology integrated process is one of the reasons why gold alloys electroforming has been barely cultivated in the academic lab, though a few exceptions do exist and should be noted [152–155].

Alloy electroforming processes currently used in the jewelry industry are for production of 24 karat gold, 14 and 18 karat Au-Cu-Cd [148], and a range of Au-Ag [147,156] alloys. Proprietary processes comprise a complete production system, including the electroforming module, the automatic control unit, the dewaxing equipment, and other facilities. Bath composition and pH and operating parameters (basically, temperature and current density) change according to the desired composition of the alloy. The need to conform to hallmarking/fineness standards, without depositing excess gold is an important consideration in commercial systems.

The color of Au-Cu-Cd and Au-Ag electroformed alloys is usually in the range of pale yellow to yellow or yellowish-white, respectively, though the former type of alloy can also be obtained with a 3-4 or 4 N color, i.e., a pinkish to pink color, by increasing the copper content. The final color of the electroformed object can be restored by pure gold flash deposition. Notoriously, red (pink) and white gold alloys are unavailable as commercial electroforming processes. Pure gold electroforming is similar to conventional electroplating as far as processes and technology are concerned. However, optimized electrolyte composition and operating conditions, more stringent (and preferably automatic control of plating bath composition) are all important features in the implementation of a reliable process and are critical to the achievement of adequate strength and low residual stress in the electroformed gold. Both cyanide and sulfite based electrolytes are available for fine gold electroforming, operating in the slightly acidic or neutral pH range [147,157]. The plating rate is mostly in the range of 0.2–0.3 μm min^{-1} depending as usual on the electrolyte gold content and stirring.

Pure Au electroforming from sulfite baths has spread over the years to the field of restorative dentistry. The adoption of this electrolyte and process in the decorative field is limited because of the relatively low mechanical strength of the electrodeposited gold, with microhardness usually in the range 100–120 HV, in the absence of metal brighteners, which though effective as hardeners, have the drawback of inducing an ordinarily unacceptable high level of residual stress.

Electroforming is today a robust and competitive technology with a few technical issues deserving due care, alongside economic evaluation of the process. First, electroforming can be very time consuming, although is not labour intensive, provided that automatic process control is used. Model design must comply with some limitations such as avoiding narrow recesses, sharp angles, and abrupt or large changes in either cross section or wall thickness. In addition, fine surface details are bound to weaken or disappear as thickness builds up and should be avoided. Residual stress in the electroformed parts may be unacceptably high, causing deformation after mandrel removal; preparation and (if permanent) maintenance of the mandrels should be performed with great care.

APPLICATIONS OF ELECTROPLATED GOLD

The use of gold and gold alloy deposits is widespread in the electrical and electronic industries as well as in decorative applications. Electrodeposition is the leading technology in gold plating and the processes based on cyanide electrolytes are the industry standard.

As already discussed (see Table 11.3), there are many different commercial plating processes for electrodeposited gold and its alloys. The variety of processes has arisen from the need to meet a large and diverse range of application requirements, but it is also the result of commercial competition. This variety of formulations is especially prevalent in the electrodeposition of gold for decorative applications, while there is a greater degree of uniformity in the plating processes designed for functional uses, primarily for deposition on electric contacts, connectors, and for interconnection purposes in semiconductor devices, in photonics and the sensor industry.

A summary of gold electroplating processes is schematically presented in Table 11.10. The main characteristics of the different plating process classes are outlined, e.g., pH working range, typical thickness range, throwing power, and the availability of special formulations for different plating applications (rack, barrel, high-speed selective plating, brush plating). Finally, the most important uses are indicated.

The wide range of processes and operating conditions used in gold electroplating is mirrored by a large properties database, scattered in countless publications, technical reports, manufactures' datasheets, and the like. There is no obvious way to systematically arrange and present such a large amount of data, which should be critically examined whenever reliable information is required for the preliminary selection of a plating process and coating material suitable for the final application. Gold electrodeposits are especially characterized by a few crucial material properties, namely, the density, which is related to composition, purity and microstructure of deposits; the hardness, which is closely associated with different types of gold deposit, as defined by the process class and the material purity; and residual stress, a thin film and coating characteristic which is of special relevance to advanced gold plating applications in microelectronic packaging and microfabrication. Of course, there are other important properties for gold and gold alloy deposits, particularly ductility and contact resistance, and other material characteristics relevant to wear and corrosion behavior which are of paramount technical importance. However, it is beyond doubt that composition (purity) and hardness are the reference material properties usually specified in the engineering uses of gold electrodeposits.

Table 11.11 presents a collection of such data for different types of gold and gold alloy electrodeposits, with the objective of providing the basic information relating to material properties and its correspondence to the process classes already introduced in Table 11.10.

Decorative Uses

Gold and gold alloy deposits, either as single layer or duplex coatings, have been in use over many years for the application of a variety of decorative finishes on a wide range of products, including jewelry items and watches (straps and cases), writing instruments, eyeglass frames, plumbing fixtures and other decorative objects including personal accessories and fashion articles (e.g., buckles, buttons, cuff links). Precious metal finishes for decorative purpose must fulfil two fundamental requirements: aesthetic appeal and protective properties, namely, tarnish and wear resistance. Color and appearance are the subject of specifications that need to be preserved over the product life as the most important perception of quality. Hardness and scratch-resistance properties are also of paramount importance in achieving a reasonable product lifetime.

Depending on the application, the target market and product quality, thin films or thick coatings are applied with thickness in the range from a fraction of a micron to several microns, providing microhardness values up to 350–400 HV, as well as different colors and color shades. The protection of the substrate material is guaranteed primarily by a suitable underlayer or underplating (nickel, palladium, palladium-nickel, or bronze) and by a judicious choice of gold layer thickness, depending on the final use of the plated object. In Europe, items for jewelry use fall under the EU Nickel Directive regulations, which limit the nickel release from surfaces in contact with the skin; use of nickel underlayers in not desirable. Thin films are by far the most popular finish for gold-plated articles. Thick coatings are used only for high-quality products in watchmaking, jewelry manufacturing, and for other personal objects, whenever cost is not a limiting factor and outstanding tarnishing resistance and wear performance are mandatory requirements.

The watch industry is probably the most important application of gold plating in the decorative field. Watchcases and straps for the high-quality market segment may first be plated with a gold alloy coating (12 to 18 karat) ranging in thickness between 5 and 30 μm and finished with a high caratage gold layer. Gold plated pieces of jewelry of high quality such as rings, bracelets, and necklaces may be plated with high karat hard layers (thickness up to several micrometers), while many other pieces of jewelry are usually flash gold plated in strike color gold baths applying a thickness as low as 0.2 μm.

TABLE 11.10
Summary of Gold-Electroplating Processes; Essential Characteristics, Plating Applications, and Uses

Type of Electrolyte	pH Range	Type of Deposit	Main Features	Thickness Range and Plating Applications	Uses
			Cyanide		
Alkaline	9.0–13.0	High purity	Usually pH 10.0–11.0; Very good throwing power; Low sensitiveness to metal contaminants; Matt or bright	0.05–0.10 μm (decorative); 0.5–5.0 μm (industrial); Rack and barrel plating	Various decorative items, primarily flash gold custom jewelry; Uncommon in industrial use
		Alloy	Usually pH 10.0–12.0; Good throwing power; Formulation according to desired gold shade, thickness, and application (rack and barrel); Special formulation for electroforming	Decorative: usually 0.05–0.10 μm; Occasionally decorative-functional and industrial: 1–10 μm; Electroforming: >100 μm	Flash deposition for jewelry color finishes; Underlayer in thick duplex coating; Hollow jewelry electroforming
Neutral	5.5–8.2	High purity	Very good throwing power; Little or no attack on photoresist materials	Large thickness range; Thick bright deposits possible; Rack, barrel, high-speed selective plating	Industrial applications for: Semiconductors; Printed circuit boards
Acid	3.5–5.5	99.0–99.3%	Hard gold by codeposition of Co, Ni, In, Fe; Usually mildly acidic at pH about 4; Throwing power from poor to good	Usually 0.5–5 μm; Rack, barrel, high-speed selective plating; brush plating	Industrial: plating of contact surfaces in contacts, switches, connectors, circuit breakers, PCB, etc.; Decorative, either flash color gold or thick coatings
		High karat gold alloy	Karat color deposits (i.e., deposits matching a karat color but actually assaying higher Karatage); Usually pH 4.0–5.0	Wide thickness range; Thick deposits possible; Rack, barrel plating	Decorative flash deposition; Thick deposits for decorative-functional applications

(Continued)

TABLE 11.10 (CONTINUED)
Summary of Gold-Electroplating Processes; Essential Characteristics, Plating Applications, and Uses

Type of Electrolyte	pH Range	Type of Deposit	Main Features	Thickness Range and Plating Applications	Uses
Au(III) electrolyte	<2.0 – usually highly acidic	Pure and alloyed	May be color matched in varying Au shades	Usually thin layers According to formulation may be suitable to plate thick coating Rack and brush plating	Strike layer on difficult to plate substrate, e.g., stainless steel Decorative and industrial
			Noncyanide		
Sulfite based	6.0–10.5	High purity	Exceptional throwing power Neutral bath for ductile smooth deposit Wide thickness range Bright deposit also at high thickness	Industrial and decorative-functional Rack, barrel; can be formulated for brush application Pure gold electroforming	Semiconductors Microfabrication Pure gold electroforming (particularly prosthetic dentistry)
		Alloy	Exceptional throwing power Alkaline or slightly acidic: Cu, Ni, or Cd alloyed gold; ternary alloys (e.g., Au-Cu-Pd)	Depending on formulation, large thickness range form flash up to several μm thick deposits.	Decorative (flash color gold deposits) and decorative-functional, e.g., custom jewelry, eyeglass frames, etc.)

TABLE 11.11
Density, Hardness, and Stress Data for a Range of Gold and Gold Alloy Electrodeposits

Type of Electrolyte and Deposit	Composition/%	Density/g cm⁻³	Microhardness/ HV^(a)	Internal Stress/MPa	Ref.	Notes
Cyanide						
Alkaline pure gold	>99.9Au	18.9–19.2	60–90	–7 to –14	[7]	pH 10, 65°C
Alkaline gold alloys	75Au25Ag	17.1	270–290	–	[158]	
	66Au34Ag	13.5	190	–	[7]	
	75Au14Cu11Cd	17.2	380–420	–	[158]	
	75Au18Cu7Cd	15.0	385	–	[160]	
Neutral pure soft gold	>99.9Au	19.2	70–100	10–50	[90]	
		19.1	45–100	0	[176]	
Neutral hard gold	99.9Au	17.4	180 KHN (25 g)	–	[71]	AFHG
		19.2	130–200	–20	[195]	Organic hardener
Acid hard gold	99.0–99.6Au	17.8 (0.1%Co)	110–180	118–195	[90]	
		17.5 (0.15%Co)	130–180	100	[195]	
		17.3 (0.2%Co)	–	–	[71]	
		–	180–200 (10 g)	–	[185]	0.3%Co
Noncyanide						
Alkaline sulfite	>99.9Au	19.2	160–190	30–70	[90]	As(III) brightened
			125 (25 g)	–	[168]	pH 9, 47°C, As(III)
			66 (1 g)	<10	[94]	pH 8, 60°C
Neutral sulfite				>40	[174]	pH 6.5, 35°C, no As(III)
				–32 to –40		same with As(III) 25 ppm
				–5		same with pulse plating
Neutral baths gold alloys	66Au34Cu	15.9	300–350	9	[158]	From manufacture's data sheet
	92Au3Cu5Pd	17.0	280–310	–	–	
Mixed sulfite-thiosulfate	>99.9Au	–	80–100 (5 g)	–	[57]	pH 6, 60°C
		–	70–93^(b)	40–70	[95]	pH 7.4, 55°C

(a) 20 g load unless otherwise noted.

(b) Nanoindentation hardness measured at maximum load of 1 mN.

Color finishing using flash gold deposits ranging from pale yellow to orange red is quite popular, and according to the application, especially to the severity of wear and corrosion factors, different operating sequence and layering schemes are followed. The gold deposit may be a duplex coating, such as for watchcases, or a single layer. The thickness is increased to meet special conditions of corrosion or wear (e.g., eyeglass frame parts worn close to the skin). Usually, flash gold-plated items are top-coated with transparent lacquer to improve wear endurance of thin deposits.

As a general remark, while the wear and tarnish resistance of gold alloy deposits is satisfactory for most current decorative applications, the corrosion performance is usually inadequate for engineering uses, a partial exception being 18 karat Au-Cu-Cd alloy coatings [158,159] (typical composition: Cu 18% and Cd 7% by weight) [160]. Operating limitations for alloy electrodeposition processes, e.g., not being suited to high-speed plating operation, are another important factor discouraging their use for industrial purposes.

ENGINEERING USES

The most important use of electroplated gold in engineering applications is for electrical connection, either as a contact surface or as a bonding termination. Electroplated gold is also used in a number of minor specialized applications, again in the field of microelectronics and optoelectronics technology, serving multiple functions, from electric interconnection to sensing and environmental endurance, as well as in space and vacuum technology and in the chemical industry.

Gold is the ultimate choice for electrical connections, due to a unique combination of properties: high electrical and thermal conductivity, low electric contact resistance and good bondability (if high purity gold); long-term resistance to tarnish, very low susceptibility to electrochemical migration and high wear resistance if hardened by codeposition of metal additives. In particular, a contact finish material should satisfy a number of diverse and strict technical requirements, including:

- Protection against corrosion and high resistance to oxidation
- Wear resistance, especially against fretting (and in this respect perform well in a high vibration environment)
- Adequate thermal stability

An additional advantage of gold as a contact material is that gold does not catalyze polymerization of atmospheric organic contaminants thereby forming an insulating layer, a well-known reliability threat for contact surfaces of palladium and its alloys [161]. Finally, gold electroplating is a cost effective, industrially robust technology. In particular, gold can be deposited as a thin coating and selectively by processes amenable to a high level of automation. These unequalled properties of gold, explain why plated gold has been the preferred finish for interconnection systems in demanding applications for years. As an electroplated contact material for contacts, connectors, relays, and switches in the manufacture of aerospace, automotive, telecom and information technology electronics, gold is essential whenever safety is a basic design requirement. For the same reasons, gold is also recommended for highly miniaturized electrical contacts with low contact forces [162].

Electroplated gold used in the electronics industry is broadly classified as either soft gold or hard gold [163]. Soft gold is primarily used as a surface finish for wire bonding and soldering of surface mounted components. Hard gold is used as a contact finish on electrical connectors and contacts requiring a wear resistant and low contact resistance surface layer.

Soft Gold

High purity and low hardness are the key properties for soft gold deposits. The high purity obviously results in a range of physical and mechanical properties that have immediate relevance to specific industrial uses, i.e., an electrical and thermal conductivity close to that of bulk pure gold, excellent bondability, low hardness, and high ductility. In addition, soft gold may be obtained by electrodeposition with quite a low level of residual stress.

Soft gold is widely used in semiconductor, optoelectronic, and microsystem applications, mostly as an interconnection material and bonding surface [21]. The leading application is in semiconductor manufacturing, either for chip-package bonding or for on-chip metallization. High purity gold is traditionally used as a solderable surface finish in the conventional wire bonding assembly process, for joining pads of a semiconductor device to pads of a carrier substrate; it is also used in the form of electrodeposited bumps in more advanced interconnection technologies, which are increasingly used in the manufacture of small-sized electronic devices and in telecommunication applications, such as TAB or thermosonic flip-chip bonding [163]. Other processes in the same manufacturing area still make use of electroplated pure gold either as a readily solderable surface or for electrical connection, for example, the fabrication of microwave integrated circuits [164] and metal electrodes for high density memory devices [165]. Finally, high purity electrodeposited gold is also increasingly used in microfabrication technologies either for device interconnection or to serve as a functional element [21].

Soft gold plating is traditionally performed from neutral or slightly acidic gold cyanide baths at high temperature, basically the same bath that when operated at about room temperature yields so-called additive free hard gold deposits [39]. The high purity (Au >99.9%) of soft gold deposits from standard cyanide baths translates into a very low content of embedded carbon, as low as 0.001% [39], and results in a relatively coarse columnar microstructure with grain size typically in the range of 1–2 μm [4]. The carbon content, which is mainly caused by the incorporation of AuCN [41], is actually the key factor determining the hardness of pure gold electrodeposits as demonstrated by the work of Reinheimer [39] and discussed most convincingly by Okinaka [58]. The relationship between hardness and carbon content is demonstrated by the plot shown in Figure 11.4, where all the data reported by Reinheimer [39] are displayed.

These data refer to pure gold deposits plated from additive free phosphate and citrate buffered electrolytes operated in different conditions of pH and current density and either at 65°C (hot bath) or 25°C (cold bath). Nonetheless, the impact of carbon contamination on hardness and on the deformation behavior of gold deposits is apparent. Soft gold deposits with hardness in the range 60–80 KHN show a very low carbon content (C < 10 ppm), corresponding to the shaded area in the graph. Three further regions can be recognized in the graph according to carbon content and hardness: in

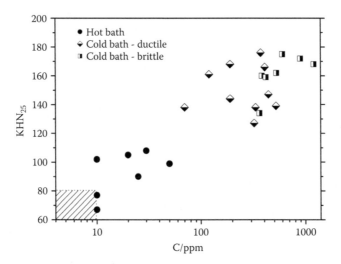

FIGURE 11.4 Hardness versus carbon content in gold films plated from additive free phosphate and citrate buffered Au(I)-cyanide solutions, either at 65°C (hot bath) or 25°C (cold bath). (Data from Carbon in Gold Electrodeposits, H. A. Reinheimer, *J. Electrochem. Soc.*, 1974, **121**, 490.)

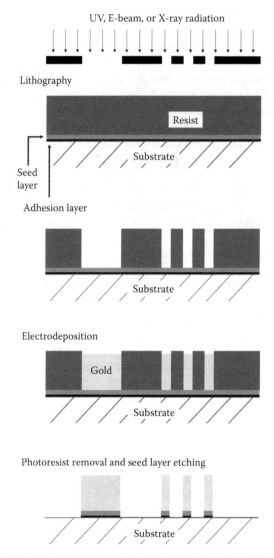

FIGURE 11.5   Schematic outline of process steps in through-mask electrodeposition.

the 10–50 ppm range, the hardness stays relatively low at values in the range from 80 to 120 KHN; deposits with carbon content up to about 350 ppm show distinctly higher hardness, from 120 to 170 KHN, and are ductile; with a further increase of the carbon content (>350 ppm), the hardness remains close to the upper limit of the later range and deposits are brittle.

For most of the applications mentioned above, the through-mask plating technique [166] is routinely used for the electrodeposition of patterned gold. This process consists in electroplating through a patterned photoresist onto a substrate typically covered with a seed layer, i.e., a thin metal film serving as the conductive base for plating, and an underlying adhesion layer of a refractory metal, both usually produced by physical vapor deposition techniques. After electrodeposition and photoresist removal, the seed layer and the adhesion layer are both removed by wet etching. A schematic representation of process steps in through-mask plating technology is shown in Figure 11.5.

The applicability of through-mask plating technology is limited by the requirement that the masking material must be compatible with the electrolyte used for plating. Conventional photoresist materials typically restrict the use of through-mask plating to acidic or weakly alkaline (and preferably to cyanide free) plating baths. Besides, the use of gold cyanide baths in these applications is discouraged because of safety and disposal concerns. Essentially, notwithstanding property requirements of the plated metal, the resist compatibility issue is the reason why cyanide free baths are preferred for through-mask plating and micro-electroforming applications. The gold sulfite bath is typically employed for this purpose because of its superior compatibility with photoresists compared to neutral cyanide baths. The sulfite electrolyte effectively prevents underplating, that is resist lifting and deposition under the resist resulting in defective filling and reproduction of the pattern. In addition, its excellent throwing power and microdistribution are additional advantages over the cyanide bath, making it especially suitable for through-mask electrodeposition processes. Finally, gold deposits from the sulfite bath can be suitably tailored to meet the rigid set of criteria specified for such demanding applications as TAB interconnection [167,168] and LIGA processes [169].

Gold bumps for TAB assembly must meet severe criteria, including shape and size uniformity, ductility, and low hardness. Since a large number of bumps must be joined for each chip, coplanarity to the base and deformability of the gold bumps are essential requirements, in order to accommodate inevitable variation in thickness. Solderability and low electrical resistance are additional requirements that can be effectively met thanks to the high purity of gold electrodeposits from standard sulfite baths. The as-plated hardness of gold from a pH 8.5 sulfite bath operated at 40°C was reported to be as low as 80 KHN in the presence of Tl(I) as brightener, without any detriment to bondability [170].

The fabrication of X-ray blocking masks used in the patterning of polymethylemethacralate PMMA photoresist in the LIGA process [115,171] is another special application of gold electrodeposition from a sulfite bath. The patterned PMMA is used as a mould to form high aspect ratio microstructures. The gold mask is produced by through-mask electroplating onto a suitable substrate, such as a Si wafer, from a sulfite electrolyte. As already mentioned earlier, the preference for the sulfite bath is motivated first by its improved compatibility with the photoresist material used in through-mask electrodeposition, though a further important aspect in this particular application is the achievement of an effective control of residual stress in the plated gold [172]. A low level of residual stress is in fact a key requirement for X-ray masks, both during processing and in operation, to guarantee planarity of the electroformed structure. In this respect, strict control of Au(I) concentration in the sulfite bath is an essential prerequisite [173]. Residual stress may be effectively controlled and almost suppressed by using suitable stress-relief additives, typically Tl(I) and As(III) compound, as well as by pulse plating. According to Kelly et al. [174] gold deposits from a 15–20 g L^{-1} Au(I) sulfite solution at pH 6.5 showed tensile stress in the range of 50–60 MPa when obtained at c.d. 0.2–0.4 A dm^{-2}; upon addition of As(III) (from As$_2$O$_3$) residual stress became compressive, ~30–40 MPa and could be further reduced to close to zero (slightly compressive), by pulse plating. These changes were correlated with microstructural modification of the layer, noting that the formation of porosity, twins and dislocations occurred extensively in deposits from simple baths, while upon addition of As(III) 25 ppm all these microstructural defects were remarkably reduced as revealed by transmission electron microscopy investigations. Crystal size did not change appreciably, while microhardness decreased from 114 (base bath) to 87 HV (As(III) 25 ppm) and to 77 HV (same but with pulse current) [174]. It should be stressed that the results reported in that work refer to peculiar conditions, particularly the pH value of 6.5 and a deposition temperature of 35°C. The addition of As(III) as As$_2$O$_3$ to alkaline sulfite baths may produce quite different effects depending on the electrolyte pH and operating conditions, acting as a brightener and hardener additive.

As noted earlier, gold cyanide or gold sulfite electrolytes are the standard baths for soft gold electrodeposition. However, the search for alternative chemistry has been continuously pursued over the last few years in connection with expanding applications in the field of microfabrication technologies, with the need for still more robust control of the electrodeposition process. The electrolyte-photoresist compatibility is also the primary concern when using a conventional masking material in contact with a standard mildly alkaline sulfite bath, because of the slow dissolution of the photoresist during continuous operation. The achievement of both low hardness and a low level of residual stress in the plated gold is the other fundamental issue.

Currently, a gold thiosulfate-sulfite solution is regarded as the most promising alternative to conventional soft gold-plating baths, particularly for interconnection purposes [95,175]. In this mixed ligand system the hardness was found to increase with the sulfur content of the deposits, while the sulfur content decreases with the total concentration of the sulfur-bearing components of the bath, thiosulfate and sulfite, or in the presence of thallium Tl(I) [30]. Hardness values in the range of 80–90 HV were obtained at a sulfur content of about 50 ppm, for a total ligand concentration in excess of 2 M. Further studies showed that the primary source of sulfur was the thiosulfate and that sulfur was not codeposited in elemental form but most likely as the $Au_2S_2O_3$ compound, through its formation as an adsorbed intermediate in the discharge of the complex [57]. Using a slightly modified formulation of the mixed ligand system [95], see Table 11.9 for composition, almost full compliance with the requirements for electrodeposited gold bumps was reported, i.e., photoresist compatibility, thickness uniformity, hardness in the range of 70–90 HV and good performance in both thermocompression and wire bonding. The relationship between sulfur content and hardness of gold electroplated from the mixed ligand system was discussed in detail by Okinaka [58].

Hard Gold

Excellent corrosion resistance and high electrical conductivity make gold an ideal contact surface. However, it is a relatively soft metal, prone to galling or cold welding. The inherent tendency of pure soft gold to adhesive wear is the main disadvantage of a gold coated surface for sliding contacts and connectors, where hardness and wear resistance are key properties for achieving reasonable service life in compliance with reliability standards. In this respect, it should be stressed that high hardness is not by itself a guarantee of good wear performance, which is a rather complex function of many different variables, including material properties, coating system architecture, and environmental conditions.

Alloying and its outstanding hardening effect is not a straightforward solution for improving wear performance, because the high hardness of a gold alloy deposit, coupled to low ductility, can make the layer brittle and unsuitable for use in contacts and connectors. Moreover, the tarnish resistance of alloyed gold is generally reduced compared to pure gold and as a result the contact resistance may be unacceptably affected.

The need to preserve low contact resistance and to achieve the demanding wear performance requirements needed from a gold contact finish led to the development of hard gold electrolytes, i.e., gold-plating baths containing metal ions as hardening and brightening additives, usually Co(II) or Ni(II) species [2], and later to the introduction of additive free hard gold plating [71]. Hard gold layers containing small amounts of Co or Ni, as low as 0.1% and less than 0.4% by weight, show hardness in the range of 140–200 HV [176], but levels as high as 250 HV (Co-HG) [7] or 200 HV (Ni-HG) [177] can be obtained, depending on electrolyte pH and deposition current density (though at the price of a complete loss of ductility). The hardness of AFHG deposits is comparable or slightly lower than that of metal hardened gold, yet preserving a relatively high ductility [71]. The later characteristic is most likely responsible for the poor performance of AFHG layers in dry sliding wear [178].

Today, either cobalt or nickel-hardened gold layers (Co-HG and Ni-HG, respectively) are ubiquitous in electrical and electronic equipment, as contact finishes for high-reliability separable connectors, switches, printed circuit board contacts and other applications where hardness and resistance to mechanical wear are basic requirements. The hard gold finish is plated along with a nickel underlayer on a copper or copper alloy substrate [179] which is a connector element, a contact spring, a pin-socket contact, or the edge of a printed circuit board. The nickel underlayer, usually of thickness in the range 1.25–2.5 μm, has multiple functions [180,181]; working as a diffusion barrier and a protective coating for the substrate [182], imparting stiffness and providing mechanical support to the contact layer [183,184]. The latter is a basic requirement, since hard gold layers plated on a substrate or underlayer unable to provide adequate support would eventually fail under the applied contact forces by brittle fracture. On the other hand, the reduced compliance of the nickel underplate/gold finish system may be deleterious to contact resistance stability because insulating surface films that may form upon exposure to high temperature are less likely to fracture under contact load [180].

The thickness of gold contact finishes may be specified in a relatively large range from 0.5 to about 5.0 μm, balancing service life expectations and minimal usage of precious metal. For the less critical applications the gold thickness is usually specified at 0.5–1.2 μm, while for the most demanding applications the gold thickness is specified in the high range up to 5 μm [162].

The hardening effect observed in hard gold deposits is predominantly determined by grain size refinement, as shown by Lo et al. for Co-HG [185], and in part by lattice strain [186,187]. If only the low range (below 0.4%) of metal hardener concentration is considered, the same hardening mechanism may be assumed for Co-HG and Ni-HG, since both deposits consist of extremely small crystallites with size in the range from 20 to 30 nm [40,41]. AFHG deposits show grain size in a wider range from 25 to 75 nm [40]. The determining influence of the grain size on the strength of hard gold is readily appreciated if the above values are compared to the standard grain size of soft gold deposits which is in the range of 1–2 μm [2].

The microstructural characteristics of hard gold, i.e., an extremely small grain size and high lattice strain, resulting in generally high tensile residual stress in the deposits, are both related to the incorporation of impurities. In fact, hard gold deposits, namely, Co or Ni-HG as well as AFHG (though the purity of the later is comparatively much higher), are characterized by a very high density of nonmetallic inclusions, about $\sim 10^{17}$ cm^{-3} [40], mostly in the range of 2–5.5 nm [40,41,188]. Such a high concentration of nonmetallic inclusions accounts for the large fraction of codeposited impurity elements found in metal hardened gold deposits, as revealed in early studies [36,78,189], including carbon, nitrogen, hydrogen, oxygen, and potassium as well as the metal hardening additive Co or Ni. On the other hand, the main contaminant in AFHG deposits is carbon, up to 0.1% [39], nonmetallic inclusions being AuCN nanocrystals [43,41]. Examples of impurity contents of Co-HG, Ni-HG, and AFHG deposits are reported in Table 11.12.

The chemical identification of inclusions has been the focus of a number of investigations over a time span of some 20 years, i.e., the time lapse between Munier's paper on "polymer" codeposition in hard gold [36] and Nakahara's paper unveiling the chemical nature of inclusions in Co-HG [188]. In this respect, it is worth noting that the only chemical compound directly identified as an inclusion in both AFGH [43] and Co-HG [188] deposits is AuCN.

An important characteristic of Co-HG is that the metal is incorporated in the deposit in two different chemical states: as metallic cobalt in solid solution with Au and as a cobalt complex compound [190,191]. According to Cohen et al., less than 70% of the Co is in the metallic state, i.e., substitutionally dissolved in the Au matrix, the remainder being present in the form of a cobalt hexacyanate $CoCN_6^{3-}$ complex [192]. The cobalt complex was identified as the $K_3Co(CN)_6$ compound from the

TABLE 11.12
Typical Composition (wt%) of Metal Hardened Gold Co or Ni-HG and Additive-free Hard Gold AFHG Deposits

Type of Gold Deposit	[80]	Co-HG [163],[193]	[71]	Ni-HG [80]	AFHG [71]
Co or Ni	0.20	0.20	—	0.11	—
C	0.205	0.28	0.136	0.16	0.055
H	0.050	0.040	0.034	—	0.010
O	0.043	0.093	—	—	—
N	0.136	0.17	0.12	—	0.03
K	0.25	0.28	0.275	0.15	0.22

analysis of the residue formed upon dissolution of Co-HG in mercury [190], though the possibility of nonmetallic cobalt being included in the divalent form, i.e., as the complex $K_4Co(CN)_6$, cannot be excluded based on these results [58]. The Co-hexacyanate compound and AuCN may account for only a fraction of the total carbon content of Co-HG; in fact, the nitrogen to carbon atomic ratio in Co-HG deposits is significantly lower than unity, about 0.5–0.6 according to Raub et al. [80], an indication that additional carbon containing inclusions exist in Co-HG. The issue of the chemical nature of inclusions in hard gold is therefore still not completely resolved [193], a gap in the characterization of hard gold that reflects in the continuing uncertainty remaining in understanding its behavior as a contact finish.

Optimized wear and friction behavior was shown to depend critically on the amount of co-deposited cobalt and the form in which it is incorporated, as a metal or as a complex. The presence of complex cobalt was recognized as a necessary condition for wear resistant Co-HG coatings. According to De Doncker and Vanhumbeeck [194] a nonmetallic Co content in excess of 0.08% combined with a metallic Co amount between 0.06 and 0.20% showed excellent wear resistance and low friction coefficient. Based on the data presented by De Doncker and Vanhumbeeck as well as on the work of Whitlaw et al. [195], the ratio of complex Co to metallic Co appeared as the most sensitive parameter for predicting the wear behavior of hard gold deposits. This was confirmed in a work by Celis et al. [196], where it was demonstrated that wear resistant Co-HG deposits were characterized by a ratio of complex Co to metallic Co greater than unity. The reasons for the high wear resistance of Co-HG are still unclear; low ductility (elongation less than 0.4%) [178] or high internal stress (>100 MPa) [196] were in turn indicated as relevant factors for the achievement of a suitable balance of properties, providing for high wear resistance in Co-HG and, most especially, for prevention of galling or adhesive wear [193].

The type and amount of impurities codeposited in Ni-HG is not significantly different from those discovered in Co-HG. Therefore, based on the role of inclusions in affecting the wear resistance of Co-HG, the wear behavior of Ni-HG is expected to be comparable to that of Co-HG [58]. A few published studies on the subject confirm this expectation [177,197,198]. As already noted earlier, the wear resistance of AFHG is significantly lower than metal hardened gold deposits: according to Antler [178], AFHG gold has similar wear performance to soft gold.

In an increasing number of applications, thermal stability of the contact resistance is a key characteristic for reliability of contacts and connectors as either the environmental conditions or localized heating in electrical contacts cause exposure to high temperatures. The impressive increase in hardness and the positive enhancement of wear resistance of metal hardened gold, in particular of Co-HG (both consequences of the impact of Co codeposition on microstructure and purity of the deposits), are, however, achieved at the price of a measured decrease in thermal stability of the contact resistance compared to pure gold surfaces [199]. On the other hand the microstructure of Co-HG displays a remarkable thermal stability with no appreciable softening after a short (~1 h) exposure to 250°C (see e.g., Harris et al. [200]). As the annealing temperature exceeds 300°C, in addition to substantial softening, the surface may suffer from severe degradation due to decomposition of codeposited cyanide complex [200,201].

The contact resistance of Co-HG deposits with a 0.1–0.2% Co content is nearly stable up to ageing time in the range from 400 to 500 h at 125°C [180,194,202,203], while it increases rapidly at temperatures in excess of 150°C. Moreover, the contact resistance stability degrades rapidly with raising Co content in the layer [180,194]. Cobalt surface segregation and oxidation was eventually recognized as the root cause of the unacceptable increase in contact resistance of Co-HG. Different cobalt compounds were detected at the surface depending on the temperature of air annealing, namely, cobalt oxide CoO at 150°C and cobalt hydroxides at 200°C [204]. In an ESCA (electron spectroscopy for chemical analysis) study of the surface films on Co-HG electrodeposits, only CoO was observed below 175°C, while in the temperature range above 175°C the formation of potassium oxide K_2O was inferred from ESCA and quantitative AES (Auger electron spectroscopy) [205]. In that study, the formation of K_2O was related to decomposition of potassium gold cyanide inclusions.

Ni-HG is known to display a significantly more stable contact resistance compared to Co-HG on exposure to high temperatures [202] and may be acceptable for contacts in connectors at 150°C or even at 200°C, depending on the expected service life [180]. However, the thermal stability of contact resistance of Ni-HG is still inadequate for some applications, especially in electronic and electrical components for the automotive industry. In recent years, environmental exposure conditions in automotive applications have increased in severity, e.g., engine compartment connectors should be designed to operate continuously at 150°C. Au flashed palladium was shown to be a reliable alternative to hard gold deposits in such demanding applications [193]. Palladium and palladium alloys definitely display better contact resistance stability compared to hard gold [206]. However, palladium contact surfaces may degrade rapidly in aggressive environments, namely, in the presence of high concentrations of chloride, and are notoriously prone to fretting and frictional polymerization [207,208]. Antler demonstrated that a gold flash on electrodeposited palladium was a viable solution to reduce friction and increase wear resistance of the contact surface thanks to its effectiveness as a solid lubricant [209]. The gold overplate is typically a hard gold flash 0.05–0.1 μm thick, over a pure palladium layer with a thickness of 0.5–1.5 μm, in turn over a nickel underplate.

CONCLUSIONS

The strong link between the electronic industry and the science and technology of the electrochemical deposition of gold has been the key factor stimulating innovation in this field over the last half century. New expectations and unresolved issues still call for innovative solutions, as a recent paper emphasizes [175]. In this respect, the focus of current research and development work is on the use of gold deposits as a contact surface, still the most important use of gold films in electronics. The trend toward miniaturization and the ever-increasing severity of the environmental or operating conditions, as well as cost reduction issues, are maintaining a strong interest in improved gold contacts as well as in alternative finishes. Wear resistance of Co and Ni hardened gold deposits is the result of several factors, including deposit's microstructure and morphology, hardness, residual stress and ductility. There is no straightforward route to optimized wear resistance based on the selection of operating parameters for deposition. The development of electrolytes and processes for hard gold deposition for connectors and PCB contacts has been largely empirical and the relative importance of the different factors involved in determining the film properties is still not completely understood.

Similarly, recent progress in the development of gold electroless processes is fostered by the requirements of new advances in microelectronic interconnections technology. Though still having relatively small commercial relevance, electroless plating of gold is expected to play an important role in the future, under the challenge of an evolution in electronic packaging design. Mixed immersion-autocatalytic processes are currently proposed as a substitute for pure immersion gold deposition, with the objective of drastically reducing the risk of etching the Ni-P underlayer, the source of a serious reliability hazard in the microelectronic industry [210]. The interest in electroless gold processes will receive further impetus from the continuous trend toward miniaturization and the strict requirements on selectivity and uniformity of contacts plating.

In summary, there is no doubt that the future of gold plating will continue to depend primarily on the vitality of its connection with the electronic industry. Parallel to and partially overlapping with this long-standing connection, the emergence and development of new technologies, primarily in the fields of renewable energy, sensing, and biomedical applications, will be an important driver of the continuous development of gold-plating technology.

REFERENCES

1. *Precious Metal Plating*, J. Fischer, D. E. Weimer, Robert Draper, Teddington, 1964.
2. *Gold Plating Technology*, ed. F. H. Reid and W. Goldie, Electrochemical Publications, Ltd., Ayr, Scotland, 1974.

3. Electroplated golds, in *Gold Usage*, W. S. Rapson and T. Groenewald, Academic Press, London, 1978, pp.196–269.
4. Electrodeposition of gold, P. A. Kohl, in *Modern Electroplating*, 4th Edition, eds. M. Schlesinger and M. Paunovic, John Wiley & Sons, New York, 2000, pp. 201–226.
5. Gold electrodeposition within the electronics industry, I. R. Christie and B. P. Cameron, *Gold Bull.*, 1994, **27**, 12.
6. Some recent developments in non-cyanide gold plating for electronics applications, M. Kato and Y. Okinaka, *Gold Bull.*, 2004, **37**, 37.
7. Gold and Gold Alloys, in *Properties of Electrodeposited Metals and Alloys*, William H. Safranek, Elsevier, 1974; American Electroplaters & Surface Finishers S; 2nd Edition, 1986, pp. 141–194.
8. Electroless Plating of Gold and Gold Alloys Y. Okinaka, in *Electroless Plating: Fundamentals and Applications*, eds. G. O. Mallory and J. B. Hajdu, pp. 410–420, AESF, Orlando, FL, 1990.
9. Electroless deposition processes: fundamentals and applications. Y. Okinaka and T. Osaka, in *Advances in Electrochemical Science and Engineering*, Vol. 3, eds. H. Gerischer and C. W. Tobias, VCH, Weinheim, 1994.
10. Electroless Deposition of Gold, Y. Okinaka and M. Kato, in *Modern Electroplating*, 4th Edition, eds. M. Schlesinger and M. Paunovic, John Wiley & Sons, New York, 2000, pp. 705–727.
11. The early history of gold plating, L. B. Hunt, *Gold Bull.*, 1973, **6**, 16.
12. Electroless gold deposition using borohydride or dimethylamine borane as reducing agent, Y. Okinaka, *Plating*, 1970, **57**, 914.
13. Nickel plating on steel by chemical reduction, A. Brenner and G. E. Riddell, *J. Res. National Bureau Standards,* 1946, **37**, 31; Deposition of nickel and cobalt by chemical reduction, A. Brenner and G. E. Riddell, *J. Res. National Bureau Standards*, 1947, **39**, 385.
14. Electroless solutions, Y. Okinaka, in *Gold Plating Technology,* eds. F. H. Reid and W. Goldie, Electrochemical Publications, Ltd., Ayr, Scotland, 1974, pp. 82–102.
15. Designing for gold plating, H. Silman, *Gold Bull.*, 1976, **9**(2), 38.
16. *Standard Potentials in Aqueous Solutions*, A. J. Bard, R. Parsons, and J. Jordan, Marcel Dekker, New York, 1985.
17. Gold, G. M. Schmid and M. E. Curley-Fiorino, in *Encyclopaedia of Electrochemistry of Elements*, Vol. 4, ed. A.J. Bard, Marcel Dekker, New York, 1975.
18. *Atlas d'équilibres électrochimiques*, M. Pourbaix, Gauthier-Villars, Paris, 1963; English translation: *Atlas of Electrochemical Equilibria in Aqueous Solutions*, Pergamon Press, Oxford, 1966.
19. Gold leaching in non-cyanide lixiviant systems: critical issues on fundamentals and applications, G. Senanayake, *Minerals Engineering*, 2004, **17**, 785.
20. Cyanide oxidation at nickel anodes – Part I: Thermodynamics of $CN-H_2O$ and $Ni-CN-H_2O$ systems at 298 K, G. H. Kelsall, *J. Electrochem. Soc.*, 1991, **138**, 108.
21. Gold electrodeposition for microelectronic, optoelectronic, and microsystem applications, T. A. Green, *Gold Bull.*, 2007, **40**, 105.
22. Understanding gold plating, P. Wilkinson, *Gold Bull.*, 1986, **19**(3), 75.
23. The solubility of gold and silver in the system $Au-Ag-S-O_2-H_2O$ at 25°C and 1 atm, J. G. Webster, *Geochim. Cosmochim. Acta*, 1986, **50**, 1837.
24. Speciation analysis of Au(I) electroplating baths containing sulfite and thiosulfate, T. A. Green and S. Roy, *J. Electrochem. Soc.,* 2006, **153**, C157.
25. The Anodic Behaviour of Gold – Part I, M. J. Nicol, *Gold Bull.*, 1980, **13**(2), 46.
26. Electrochemical gold deposition from sulfite solution: application for subsequent polyaniline layer formation, G. Baltrūnas, A. Valiūnienė, J. Vienožinskis, E. Gaidamauskas, T. Jankauskas, and Ž. Margarian, *J. Appl. Electrochem.* 2008, **38**, 1519.
27. *Advanced Inorganic Chemistry — A Comprehensive Text*, F. A. Cotton and G. Wilkinson, 4th Edition, John Wiley & Sons, New York, 1980.
28. An investigation of the deposition and reactions of sulphur on gold electrodes, I. C. Hamilton and R. Woods, *J. Appl. Electrochem.*, 1983, **13**, 783.
29. Electrochemical properties of sulfur adsorbed on gold electrodes, D. G. Wierse, M. M. Lohrengel, and J. W. Schultze, *J. Electroanal. Chem.*, 1978, **92**, 121.
30. Electrodeposition of soft gold from a thiosulfate-sulfite bath for electronics application, T. Osaka, A. Kodera, T. Misato, T. Homma, Y. Okinaka, and O. Yoshioka, *J. Electrochem. Soc.*, 1997, **144**, 3462.
31. Il comportamento elettrochimico dell'oro in bagni cianidrici, M. Maja, *Atti Accad. Sci. Torino: Classe Sci. Fis. Mat. Nat.*, 1965, **99**, 1111.

32. Electrochemical and structural aspects of gold electrodeposition from dilute solutions by direct current, H. Y. Cheh I and R. Sard, *J. Electrochem. Soc.*, 1971, **118**, 1737.

33. The reduction of gold cyanide complexes, J. A. Harrison and J. Thompson, *J. Electroanal. Chem.*, 1972, **53**, 113.

34. Gold cyanide ion electroreduction on gold electrode, M. Bełtowska-Brzezinska, E. Dutkiewicz, and W. Ławicki, *J. Electroanal. Chem.*, 1979, **99**, 341.

35. A study of gold reduction and oxidation in aqueous solutions, D. M. MacArthur, *J. Electrochem. Soc.*, 1972, **119**, 672.

36. Polymer codeposited with gold during electroplating, G. B. Munier, *Plating*, 1969, **56**, 1151.

37. Inkorporation Kohlenstoffhaltiger Fremdstoffe bei der Galvanishen Abscheidung von Gold und Goldlegierungsueberzuegen in Sauren und Alkalischen Cyanidbaedern, E. Raub, Ch. J. Raub, A. Knödler, and H. P. Wiehl, *Werkstoffe und Korrosion*, 1972, **23**, 643.

38. Kinetics of the electrochemical reduction of dicyanoaurate, E. T. Eisenmann, *J. Electrochem. Soc.*, 1978, **125**, 717.

39. Carbon in gold electrodeposits, H. A. Reinheimer, *J. Electrochem. Soc.*, 1974, **121**, 490.

40. Structure of electroplated hard gold observed by transmission electron microscopy, Y. Okinaka and S. Nakahara, *J. Electrochem. Soc.*, 1976, **123**, 1284.

41. Inclusions in electroplated additive-free hard gold, S. Nakahara and Y. Okinaka, *J. Electrochem. Soc.*, 1981, **128**, 284.

42. On the electrodeposition of hard gold, H. Angerer and N. Ibl, *J. Appl. Electrochem.* 1979, **9**, 219.

43. Incorporation of gold cyanide in electrodeposited gold, G. Holmbom and B. E. Jacobson, *J. Electrochem. Soc.*, 1988, **135**, 788.

44. Electrochemical scanning tunneling microscopy and ultrahigh-vacuum investigation of gold cyanide adlayers on Au(111) formed in aqueous solution, T. Sawaguchi, T. Yamada, Y. Okinaka, and K. Itaya, *J. Phys. Chem.*, 1995, **99**, 14149.

45. Electrodeposition of gold, J. D. E. McIntyre and W. F. Peck, Jr., *J. Electrochem. Soc.*, 1976, **123**, 1800.

46. The effect of base metal ions on the electrochemical and structural characteristics of electrodeposited gold films, P. Bindra, D. Light, P. Freudenthal, and D. Smith, *J. Electrochem. Soc.*, 1989, **136**, 3616.

47. Étude du placage électrolytique de l'or en milieu sulfitique, J.-P. Derivaz, A. Resin, and S. Losi, *Surf. Technol.*, 1977, **5**, 369.

48. NBS tables of chemical thermodynamic properties, D. D. Wagman, W. H. Evans, V. B. Parker, R. H. Schumm, I. Halow, S. M. Bailey, K. L. Chuney, and R. L. Nuttal, *J. Phys. Chem. Ref. Data*, 1982, **11**, Suppl. 2.

49. The sulfite/dithionite couple: Its standard potential and Pourbaix diagram, M. S. Chao, *J. Electrochem. Soc.*, 1986, **133**, 954.

50. The electrochemical reaction of sulphur-oxygen compounds – Part 1. A review of literature on the electrochemical properties of sulphur/sulphur-oxygen compounds, *Electrochim. Acta*, 1992, **37**, 2775.

51. Pulsed potentiostatic deposition of gold from solutions of the Au(I) sulfite complex, J. Horkans and L. T. Romankiw, *J. Electrochem. Soc.*, 1977, **124**, 1499.

52. An electrochemical and in-situ Raman investigation of the electrodeposition of gold from a sulphite electrolyte, A. Fanigliulo and B. Bozzini, *Trans. Inst. Met. Finish.*, 2002, **80**(4), 132.

53. Electrochemical study of the gold thiosulphate reduction, A. M. Sullivan and P. A. Kohl, *J. Electrochem. Soc.*, 1997, **144**, 1686.

54. Electrochemical behaviour of electroless gold plating with ascorbic acid as a reducing agent, M. Kato, K. Niikura, S. Hoshino, and I. Ohono, *J. Surf. Finish. Soc. Jpn.*, 1991, **42**, 729.

55. Stable non-cyanide electroless gold plating which is applicable to manufacturing of fine pattern wiring boards, T. Inoue, S. Ando, H. Okudaira, J. Ushio, A. Tomizawa, H. Takehara, T. Shimazaki, H. Yamamoto, and H. Yokono, in *Proc. 45th IEEE Electron. Comp. Technol. Conf.*, 1995, 1059–1067.

56. Substrate (Ni)-catalyzed electroless gold deposition from a noncyanide bath containing thiosulfate and sulfite I. Reaction mechanism, M. Kato, J. Sato, H. Otani, T. Homma, Y. Okinaka, T. Osaka, and O. Yoshioka, *J. Electrochem. Soc.*, 2002, **149**, C164.

57. Mechanism of sulfur inclusion in soft gold electrodeposited from the thiosulfate-sulfite bath, T. Osaka, M. Kato, J. Sato, K. Yoshizawa, T. Homma, Y. Okinaka, and O. Yoshioka, *J. Electrochem. Soc.*, 2001, **148**, C659.

58. Significance of inclusions in electroplated gold films for electronics applications, Y. Okinaka, *Gold Bull.*, 2000, **33**, 117.

59. Characterisation of a thiosulphate–sulphite gold electrodeposition process, M. J-Liew, S. Sobri, and S. Roy, *Electrochim. Acta*, 2005, **51**, 877.

60. Interaction of gold(I) with thiosulphate-sulfite mixed ligand systems, W. N. Perera, G. Senanayake, and M. J. Nicol, *Inorg. Chim. Acta*, 2005, **358**, 2183.
61. Gold plating, A. M. Weisberg, *Metal Finish.*, 2007, **105**(10), 205.
62. *Gold in Modern Electroplating*, 2nd Edition, ed. F. A. Lowenheim, John Wiley & Sons, New York, 1963, pp. 207–223.
63. Chemical behavior of the components of the KCN/KAu(CN)$_2$ electroplating system, H. G. Silver, *J. Electrochem. Soc.*, 1969, **116**, 26C.
64. Analysis of gold foils and electroplating solutions for metallic and anionic impurities, H. G. Silver, *J. Electrochem. Soc.*, 1969, **116**, 741.
65. *Electrodeposition of Alloys*, Vol. II, A. Brenner, Academic Press, New York, 1963, pp. 494–541.
66. The colour of electroplated golds, D. J. Arrowsmith, *Gold Bull.*, 1986, **19**, 117.
67. Gold electroforms and heavy electrodeposits — decorative and industrial applications, D. Withey, *Gold Bull.*, 1983, **16**, 70.
68. Colours of gold alloys — definition, range of colours and designation, International Standard ISO 8654, 1987.
69. Metal distribution in barrel plating. further studies, F. I. Nobel and D. W. Thomson, *Plating*, 1970, **57**, 469.
70. Gold plating in the electronics industry, F. H. Reid, *Gold Bull.*, 1973, **6**, 77.
71. Additive-free hard gold plating for electronics applications, Part I and Part II, F. B. Koch, Y. Okinaka, C. Wolowodiuk, and D. R. Blessington, *Plat. Surf. Finish.*, 1980, **67**(6), 50; 1980, **67**(7), 43.
72. Effects of gold plating additives on semiconductor wire bonding, D. W. Endicott, H. K. James, and F. Nobel, *Plat. Surf. Finish.*, 1981, **68**(11), 58.
73. Effects of grain refiners in gold deposits on aluminum wire-bond reliability, S. Wakabayashi, A. Murata, and N. Wakobauashi, *Plat. Surf. Finish.*, 1982, **69**(8), 63.
74. New analytical methods for cobalt containing hard gold plating solutions, Y. Okinaka and C. Wolowodiuk, *Plat. Surf. Finish.*, 1979, **66**(9), 50.
75. Cyanoaurate(III) formation and its effect on current efficiency in gold plating, Y. Okinaka and C. Wolowodiuk, *J. Electrochem. Soc.*, 1981, **128**, 288.
76. High speed gold plating: Anodic bath degradation and search for stable low polarization anodes, C. G. Smith and Y. Okinaka, *J. Electrochem. Soc.*, 1983, **130**, 2149.
77. Electroplating Bright Gold, E. C. Rinker and R. Duva, U.S. Patent 2,905,601 (1959).
78. The insertion of foreign substance into gold-cobalt and gold-nickel electrodeposits from acid baths, A Knödler, *Metalloberfläche – Angewandte Elektrochemie*, 1974, **28**, 465.
79. Gold plating from the acid cyanide system: some aspects of the effect of plating parameters on codeposition, L. Holt, *Trans. Inst. Met. Finish.*, 1973, **51**, 134.
80. The properties of gold electrodeposits containing carbonaceous material, Ch.J. Raub, A. Knödler, and J. Lendvay, *Plat. Surf. Finish.*, 1976, **63**, 35.
81. High temperature gold deposition from acid cyanide baths, Ch.J. Raub, H. R. Khan, and M. Baumgärtner, *Gold Bull.*, 1986, **19**(2), 70.
82. Gold Plating Bath and Process, P. T. Smith, U.S. Patent 3,057,789 (1962); U.S. Patent 3,666,640 (1972).
83. Gold Plating Bath and Process, H. H. Shoushanian, U.S. Patent 3,475,292 (1969).
84. Gold plating using disulfiteaurate complex, H. Homma and Y. Kagaya, *J. Electrochem. Soc.*, 1993, **140**, L135.
85. N. A. Smagunova, J. P. Gavrilova, and A. K. Yudina, U.S.S.R. Patent 231,991 (1968).
86. Dépôt électrolytique d'alliages or-cuivre en milieu non cyanuré, S. A. Losi, F. L. Zuntini, and A. R. Meyer, *Electrodepos. Surf. Treatment*, 1973, **1**, 3.
87. Non-cyanide Electroplating Solution for Gold or Alloys Thereof, R. J. Morrissey, U.S. Patent 5,277,790 (1994).
88. A versatile non-cyanide gold plating system, R. J. Morrissey, *Plat. Surf. Finish.*, 1993, **80**(4), 75.
89. Aqueous Solution of Monovalent Gold and Ammonium Sulfite Complex, Process for the Preparation Thereof and Electrolytic Bath Obtained Therefrom for the Plating of Gold or Gold Alloys, P. Laude, E. Marka, and Z. Morrens, U.S. Patent 4,192,723 (1980).
90. A comparison of cyanide and sulphite gold plating processes, D. R. Mason and A. Blair, *Trans. Inst. Met. Finish.*, 1977, **55**, 141.
91. Electrodeposition of Gold-Palladium Alloys, P. Stevens, U.S. Patent 4,048,023 (1977).
92. Composition for the Electroplating of Gold, H. Middleton and P. C. Hydes, U.S. Patent 4,199,416 (1980).
93. Electrodeposition of Gold Alloys, P. Wilkinson, U.S. Patent 4,366,035 (1982).

94. Fabrication of gold bumps using gold sulfite plating, H. Honma and K. Hagiwara, *J. Electrochem. Soc.*, 1995, **142**, 81.

95. Electrodeposition of gold from a thiosulfate-sulfite bath for microlectronic applications, T. A. Green, M.-J. Liew, and S. Roy, *J. Electrochem. Soc.*, 2003, **150**, C104.

96. Development of a non-toxic electrolyte for soft gold electrodeposition: An overview of work at University of Newcastle upon Tyne, M. J. Liew, S. Roy, and K. Scott, *Green Chemistry*, 2003, **5**, 376.

97. Characterisation of a thiosulphate-sulphite gold electrodeposition process, M. J. Liew, S. Sobri, S. Roy, *Electrochim. Acta*, 2005, **51**, 877.

98. Some theoretical aspects of pulse electrolysis, N. Ibl, *Surface Technol.*, 1980, **10**, 81.

99. Effect of deposition method on porosity in gold films, D. L. Rehrig, *Plating*, 1974, **61**, 43.

100. Der Einfluss von Stromimpulsen auf die Physikalischen Eigenschaften von Elektrolytisch Abgeschiedenen Gold-Kobaltschichten (The influence of pulse current on the physical properties of electrodeposited gold cobalt Coatings), F. H. Reid, *Metalloberflaeche - Angew. Elektrochemie*, 1976, **30**, 453

101. Design factors in pulse plating, A . J. Avila and M. J. Brown, *Plating*, 1970, **57**, 1105.

102. The electrodeposition of gold by pulse plating, Ch. J. Raub and A. Knödler, *Gold Bull.*, 1977, **10**(2), 38.

103. Characterization of gold layers selectively plated by a pulsed current, F. Gerhard, *Thin Solid Films*, 1989, **169**, 105.

104. Effect of pulsed current plating on structure and properties of gold-cobalt electrodeposits, W. Fluehmann, F. H. Reid, P.-A. Maüsli, and S. G. Steinemann, *Plat. Sur. Finish.*, 1980, **67**(6), 62.

105. The properties of gold deposits produced by DC, pulse and asymmetric AC plating, J. W. Dini and H. R. Johnson, *Gold Bull.*, 1980, **13**(1), 31.

106. Pulse plating, G. Devaraj, S. Guruviah, and S. K. Seshadri, *Mater. Chem. Phys.*, 1990, **25**, 439.

107. The effect of pulsating potential electrolysis on the porosity of metal deposits, K. I. Popov, D. N. Keca, D. A. Draskovic, and B. I. Vuksanovic, *J. Appl. Electrochem.*, 1976, **6**, 155.

108. Hydrogen in electrodeposits, Ch. J. Raub, *Plat. Sur. Finish.*, 1993, **80**(9), 30.

109. Annealing behavior of electrodeposited gold containing entrapments, K. C. Joshi and R. C. Sanwald, *J. Electr. Mat.*, 1973, **2**, 533.

110. Structure and properties of a gold 0.4wt% nickel electrodeposit, P. S. Willcox and J. R. Cady, *Plating*, 1974, **61**, 1117.

111. Gold-cobalt electrodeposits—microstructure and surface topography, S. J. Harris and E. C. Darby, *Gold Bull.*, 1976, **9**, 81.

112. Hydrogen incorporation and embrittlement of electroformed Au, Cu and Au-Cu, B. Bozzini,G. Giovannelli, S. Natali, B. Breveglieri, P. L. Cavallotti and G. Signorelli, *Eng. Fail. Anal.*, 1998, **6**, 83.

113. Chelating agents effects on hydrogen incorporation in the electrodeposition of Au-Cu, B. Bozzini and P. L. Cavallotti, in *Hydrogen at Surfaces and Interfaces*, ed. G. Jerkiewicz, J. M. Feliu, B. N. Popov, ECS Proceedings Vol. 2000-16, 174–184, The Electrochemical Society, Pennington 2000.

114. Electrodeposition and characterization of Au-Cu-Cd alloys, B. Bozzini and P. L. Cavallotti, *J. Appl. Electrochem.*, 2001, **31**, 897.

115. Low-stress gold electroplating for X-ray masks, W. Chu, M. L. Schattenburg, and H. I. Smith, *Microelectron. Eng.*, 1992, **17**, 223.

116. Electrochemical codeposition of inert particles in a metallic matrix, A. Hovestad and L. J. J. Janssen, *J. Appl. Electrochem.* 1995, **40**, 519.

117. Review of electrocodeposition, J. L. Stojak, J. Fransaer, and J. B. Talbot, in *Advances in Electrochemical Science and Engineering*, Vol.7, eds. R. C. Alkire and D. M. Kolb, Wiley VCH, 2002, 193–223.

118. Electrolytic and electroless composite coatings, J. P. Celis and J. R. Roos, *Rev. Coat. Corr.*, 1982, **5**, 1.

119. Electrocomposites and their benefits, V. P. Greco, *Plat. Sur. Finish.*, 1989, **76**(7), 62.

120. Electroforming of metal matrix composite: dispersoid grain size dependence of thermostructural and mechanical properties, M. Verelst, J. P. Bonino, and A. Rousset, *Mater. Sci. Eng. A*, 1991, **135**, 51.

121. Mechanical properties of ACD and ECD particulate metal-matrix composite thin films, B. Bozzini, G. Giovannelli, M. Boniardi, and P. L. Cavallotti, *Composites Sci. Technol.*, 1999, **59**, 1579.

122. Mechanical properties of multiphase alloys, J. L. Strudel, in *Physical Metallurgy*, Vol. 3, 4th Edition, eds. R.W. Cahn and P. Haasen, Cambridge University Press, 1996, 2114–2206.

123. Electrodeposited gold composite coatings, C. Larson, *Gold Bull.*, 1984, **17**(3), 86.

124. Electrochemical aspects of the codeposition of gold and copper with inert particles, C. Buelens, J. P. Celis, and J. R. Ross, *J. Appl. Electrochem.*, 1983, **13**, 541.

125. Electrochemical aspects of the plating of gold and composite gold-alumina, C. Buelens, J. P. Celis, and J. R. Roos, *Trans. Inst. Met. Finish.*, 1985, **63**, 6.

126. M. De Bonte, J. P. Celis, and J. R. Roos, in *Precious Metals — Modern Technologies and Application*, EAST Report 1991, Eugen G. Leuze Verlag, Saulgau/Württ, 1992, p. 81.

127. Advanced materials by electrochemical techniques, A. Zielonka and H. Fauser, *Z. Physik. Chem.*, 1999, **208**, 195.

128. Determination of the incorporation rate of PTFE particles in Au-Co-PTFE composite coatings by infrared reflection absorption spectroscopy, Z. Serhal, J. Morvan, P. Berçot, M. Rezrazi, and J. Pagetti, *Sur. Coat. Technol.*, 2001, **145**, 233.

129. Quantitative characterisation of Au-Co/PTFE composite coatings by differential enthalpic analysis, Z. Serhal, P. Berçot, J. Morvan, M. Rezrazi, and J. Pagetti, *Sur. Coat. Technol.*, 2002, **150**, 290.

130. Au–PTFE composite coatings elaborated under ultrasonic stirring, M. Rezrazi, M. L. Doche, P. Berçot, and J.Y. Hihn, *Surf. Coat. Technol.*, 2005, **192**, 124.

131. Mechanical properties of ACD and ECD particulate metal-matrix composite thin films, B. Bozzini, G. Giovannelli, M. Boniardi, and P. L. Cavallotti, *Comp. Sci. Technol.* 1999, **59**, 1579.

132. Morphology evolution of Au/B$_4$C electrodeposited composites, B. Bozzini and P. L. Cavallotti, *Prakt. Metallogr.*, 2001, **38**, 88.

133. An investigation into the electrodeposition of Au-Cu matrix particulate composites, B. Bozzini, G. Giovannelli, and P. L. Cavallotti, *J. Appl. Electrochem.*, 1999, **29**, 687.

134. An investigation into the electrodeposition of AuCu-matrix particulate composites. Part II: Baths not containing free cyanide, B. Bozzini, G. Giovannelli and P. L. Cavallotti, *J. Appl. Electrochem.*, 2000, **30**, 591.

135. Hydrodynamic problems related to the electrodeposition of AuCu/B$_4$C composites, *Electrochim. Acta*, 2000, **45**, 3431.

136. Electrokinetic behavior of gold alloy and composite plating baths, B. Bozzini, P. L. Cavallotti, and G. Giovannelli, *Met. Finish.*, 2002, **100**(4) 50.

137. Electrodeposition of Au/nanosized diamond composite coatings, P. Cojocaru, A. Vicenzo, and P. L. Cavallotti, *J. Solid State Electrochem.*, 2005, **9**, 850.

138. Properties of ECD gold composite with nanostructured carbon-based materials, P. Cojocaru, M. Santoro, A. Vicenzo, and P. L. Cavallotti, *ECS Trans.*, 2007, **3**(30), 15.

139. Electrodeposited Au-Cu alloys: Structural analysis, thermal stability measurements and mechanical characterization, L. Battezzati, M. Baricco, S. Barbero, A. Zambon, I. Calliari, A. Variola, B. Bozzini, P. L. Cavallotti, G. Giovannelli, and S. Natali, *La metallurgia italiana* 93(2) (2001) 17. (In Italian)

140. Microstructure and thermal stability of nanocrystalline electrodeposited Au-Cu alloys, L. Battezzati, M. Baricco, M. Belotti, and V. Brunella, *Mater. Sci. Forum*, 2001, **360–362**, 253.

141. Metastable structures in electrodeposited AuCu, B. Bozzini, G. Giovannelli, and S. Natali, *Scripta Mat.*, 2000, **43**, 877.

142. Electrodeposition and sliding wear resistance of nickel composite coatings containing micron and submicron SiC particles, I. Garcia, J. Fransaer, and J. P. Celis, *Surf. Coat. Technol.*, 2001, **148**, 171.

143. Mechanical properties of nickel silicon carbide nanocomposites, A. F. Zimmerman, G. Palumbo, K. T. Aust, and U. Erb, *Mater. Sci. Eng. A*, 2002, **328**, 137.

144. Pulse co-electrodeposition of nano Al$_2$O$_3$ whiskers nickel composite coating, N. S. Qu, K. C. Chan, and D. Zhu, *Scripta Mat.*, 2004, **50**, 1131.

145. Investigations on the electrochemical preparation of gold-nanoparticle composites, F. Wünsche, A. Bund, and W. Plieth, *J. Solid. State Electrochem.*, 2004, **8**, 209.

146. Electroplating of gold-nanodiamond composite coatings, E. N. Loubnin, S. M Pimenov, A. Blatter, F. Schwager, and P. Y. Detkov, *New Diam. Front Carbon Technol.*, 1999, **9**, 273.

147. Recent developments of electroforming with precious metals for jewellery, F. Simon, *Trans. Inst. Met. Finish.*, 1997, **75**(3), B56.

148. Electroforming of gold alloys—The ARTFORM process, G. Desthomas, *Gold Technol.*, July 1995, No. 16.

149. Electroforming technology for galvanoceramic restorations, B. S. Vence, *J. Prosthet. Dent.*, 1997, **77**, 444.

150. Gold in dentistry: Alloys, uses and performance, H. Knosp, R. J. Holliday, and C. W. Corti, *Gold Bull.*, 2003, **36**(3), 93.

151. Electroforming, T. Harta and A. Watson, *Met. Finish.*, 2007, **105**, 331.

152. Pulse-plating of Au-Cu-Cd alloys I. Experiments under controlled mass transport conditions, A. Ruffoni and D. Landolt, *Electrochim. Acta*, 1988, **33**, 1273.

153. Pulse-plating of Au-Cu-Cd alloys II. Theoretical modelling of alloy composition, A. Ruffoni and D. Landolt, *Electrochim. Acta*, 1988, **33**, 1281.

154. Hydrodynamic effects in the electrodeposition of Au-Cu-Cd alloys: an experimental and numerical study, B. Bozzini and P. L. Cavallotti, *Trans. Inst. Met. Finish.*, 2000, **78**, 227.

155. Electrodeposition and characterization of Au-Cu-Cd alloys, B. Bozzini and P. L. Cavallotti, *J. Appl. Electrochem.*, 2001, **31**, 897.

156. Recent developments in the field of electroforming, a production process for Hallmarkable hollow jewellery, F. Simon, *Gold Technol.*, July 1995, No. 16.

157. Method for Electroplating Gold, L. Greenspan and H. K. Straschil, US Patent 3,637,473 (1969).

158. Alloy gold deposits: have they any industrial use?, D. R. Mason, A. Blair, and P. Wilkinson, *Trans. Inst. Met. Finish.*, 1974, **52**, 143.

159. Mechanical and tribological characterisation of electrodeposited Au-Cu-Cd, B. Bozzini, A. Fanigliulo, E. Lanzoni, and C. Martini, *Wear*, 2003, **255**, 903.

160. A comparative study of metal distribution from 18-karat and 24-karat gold plating solutions, F. I. Nobel, D. W. Thomson, and W. R. Brasch, *Plating*, 1975, **62**, 462.

161. Organic deposits on precious metal contacts, H. W. Hermance and T. F. Egan, *Bell Labs Tech. J.*, 1958, **37**, 739.

162. *Electrical Contacts: Fundamentals, Applications and Technology*, M. Braunovic, N. K. Myshkin, and V.V. Konchits, CRC Press, Boca Raton, FL, 2006.

163. Some recent topics in gold plating for electronics applications, Y. Okinaka and M. Hoshino, *Gold Bull.*, 1998, **31**, 3.

164. Understanding high-frequency circuits: Electroplating can provide high aspect ratios, dense packing, and accurate pattern reproduction, *Metal Finish.*, 2005, **103**(2), 22.

165. Plating of noble metal electrodes for DRAM and FRAM, P. C. Andricacos, J. H. Comfort, A. Grill, D. E. Kotecki, V. V. Patel, K. L. Saenger, and A. G. Schrott, US Patent 5,789,320, 1998.

166. A path: from electroplating through lithographic masks in electronics to LIGA in MEMS, L. T. Romankiw, *Electrochim. Acta*, 1997, **42**, 2985.

167. High performance gold plating for microdevices, A. Gemmler, W. Keller, H. Richter, and K. Ruess, *Plat. Sur. Finish.*, 1993, **81**(8), 52.

168. Microbump formation by noncyanide gold electroplating, H. Watanabe, S. Hayashi, and H. Homma, *J. Electrochem. Soc.*, 1999, **146**, 574.

169. Experimental study of the microstructure and stress of electroplated gold for microsystem applications, J. J. Kelly, N. Yang, T. Headley, and J. Hachman, *J. Electrochem. Soc.*, 2003, **150**, C445.

170. Soft gold electroplating from a non-cyanide bath for electronic applications, K. Wang, R. Beica, and N. Brown, 2004 IEEE/SEMI Int'l Electronics Manufacturing Technology Symposium, San Jose, CA.

171. Progress in the fabrication of high-aspect-ratio zone plates by soft X-ray lithography, R. Divan, D. C. Mancini, N. Moldovan, B. Lay, L. Assoufid, Q. Leonard, and F. Cerrina, *Proc. Design and Microfabrication of Novel X-Ray Optics*, July 9 2002, Seattle, WA, SPIE - The International Society for Optical Engineering, 2002, **4783**, 82.

172. Stress-controlled X-ray mask absorber using pulse-current plating, T. Ogawa, T. Soga, Y. Maruyama, H. Oizumi, and K. Mochiji, *J. Vac. Sci. Technol. B*, 1992, **10**, 1193.

173. Electrodeposition of low stress gold for X-ray mask, S.-L. Chiu and R. E. Acosta, *J. Vac. Sci. Technol. B*, 1990, **8**, 1589.

174. Experimental study of the microstructure and stress of electroplated gold for microsystem applications, J. J. Kelly, N. Yang, T. Headley, and J. Hachman, *J. Electrochem. Soc.*, 2003, **150**, C445.

175. Development of new electrolytic and electroless gold plating processes for electronics applications, T. Osaka, Y. Okinaka, J. Sasano, and M. Kato, *STAM*, 2006, **7**, 425.

176. Properties of some gold electrodeposits, R. Duva and D. G. Foulke, *Plating*, 1968, **55**, 1056

177. Wear resistance of electroplated nickel-hardened gold, L.-G. Liljestrand, L. Sjögren, L. B. Révay, and B. Asthner, *Compon. Hybr. Manuf. Technol.*, 1985, **CHMT-8**(1), 123.

178. Sliding wear of metallic contacts, M. Antler, *IEEE Trans. Compon. Hybr. Manuf. Technol.*, 1981, **CHMT-4**, 15.

179. Standard specification for electrodeposited coatings of gold for engineering uses, B 488 - 01, ASTM 2006.

180. Gold-plated contacts: Effect of thermal aging on contact resistance, M. Antler, *Plat. Surf. Finish.*, 1998, **85**(12), 85; also in Proceedings of the Forty-Third IEEE Holm Conference on Electrical Contacts, 1997, p.121.

181. Significance of contact finish requirements, R. Sard and R. G. Baker, *Plat. Sur. Finish.*, 1980, **67**(4), 42.

182. Corrosion inhibition and wear protection of gold plated connector contacts, S. J. Krumbein and M. Antler, *IEEE Trans. Parts Mater. Packag.*, 1968, **PMP-4**, 3.

183. Mechanisms of wear of gold electrodeposits, A. J. Solomon and M. Antler, *Plating*, 1970, 57, 812.

184. Wear of gold electrodeposits: effect of substrate and of nickel underplate, M. Antler and M. Drozdowicz, *AT&T Tech. J.*, 1979, **59**, 323.

185. Hardening mechanisms of hard gold, C. C. Lo, J. A. Augis, and M. R. Pinnel, *J. Appl. Physics*, 1979, **50**, 6887.

186. Structure and mechanical properties of gold-cobalt electrodeposits, E. C Derby and S. J. Harris, *Trans. Inst. Met. Finish.*, 1975, **53**, 138.

187. X-ray diffraction studies of electrodeposited and sputtered hard gold, E. T. Eisenmann, *J. Electrochem. Soc.*, 1980, **127**, 1349.

188. Direct observation of inclusions in cobalt-hardened gold electrodeposits, S. Nakahara, *J. Electrochem. Soc.*, 1989, **136**, 451.

189. Structure of polymer codeposited in gold electroplates, M Antler, *Plating*, 1973, **60**, 468.

190. "Polymer" inclusion in cobalt-hardened electroplated gold, Y. Okinaka, F. B. Koch, C. Wolowodiuck, and D.R. Blessington, *J. Electrochem. Soc.*, 1978, **125**, 1745.

191. The chemical state of cobalt in cobalt hardened gold electrodeposits, H. Leidheiser Jr., A. Vertes, M. L. Varsanyi, and I. Czako-Nagy, *J. Electrochem. Soc.*, 1979, **126**, 391.

192. Characterization of cobalt-hardened gold electrodeposits by Møssbauer spectroscopy, R. L. Cohen, F. B. Koch, L. N. Schoenberg, and K. W. West, *J. Electrochem. Soc.*, 1979, **126**, 1608.

193. Electroplated contact materials for connectors and relays, Y. Okinaka, in *New Trends in Electrochemical Technology,* Vol.3, Microelectronic Packaging, eds. M. Datta, T. Osaka, and J.W. Schultze, CRC Press, Boca Raton, FL, 2005.

194. Cobalt in gold electrodeposits, R. De Doncker and J. Vanhumbeeck, *Trans. Inst. Met. Finish.*, 1985, **62**(2), 59.

195. Wear properties of high speed gold electrodeposits, K. J. Whitlaw, J. W. Souter, I. S. Wright, and M. C. Nottingham, *IEEE Trans. Compon. Hybr. Manuf. Technol.*, 1985, **CHMT-8**, 46.

196. Cobalt hardened gold layers for electrical connectors: optimization of wear properties, J. P. Celis, J. R. Roos, W. Van Vooren, and J. Vanhumbeeck, *Trans. Inst. Met. Finish.*, 1989, **67**(3), 70.

197. Tribological properties of gold for electric contacts, *IEEE Trans. Parts Hybr. Packag.*, 1973, **PHP-9**, 4.

198. Friction and wear of electroplated hard gold deposits for connectors, G. L. Horn and W.A. Merl, *IEEE Trans. Parts Hybr. Packag.*, 1974, **PHP-10**, 53.

199. Gold-plated contacts: Effect of heating on reliability, M. Antler, *Plating*, 1970, **57**, 615.

200. The effects of heating on the microstructure and surface topography of gold-cobalt electrodeposits, S. J. Harris, E. C. Darby, K. Bridger, and A. E. Mason, *Trans. Inst. Met. Finish.*, 1976, **54**, 115.

201. Temperature sensitive properties of gold and gold alloy electrodeposits, Ch.J. Raub, H. R. Khan, and J. Lendvay, *Gold Bull.*, 1976, **9**(4), 123.

202. Kontakteigenschaften galvanischer Hartgoldschichten bei erhöhten Temperaturen (Contact properties of electroplated hard gold at elevated temperature), M. Huck, *Metall.*, 1992, **46**(1), 32.

203. A new concept for electroplated electronic contact golds, R. T. Hill and K. J. Whitlaw, *IEEE Trans. Compon. Hybr. Manuf. Technol.*, 1979, **CHMT-2**, 324.

204. Secondary ion mass spectrometric analysis of cobalt-hardened gold electroplate surfaces, R. Schubert, *J. Electrochem. Soc.*, 1981, **128**, 126.

205. High temperature film formation on cobalt-hardened gold, J. H. Thomas III and S. P. Sharma, *J. Electrochem. Soc.*, 1979, **126**, 445.

206. The development and application of palladium contact materials, M. Antler, *Platinum Met. Rev.*, 1987, **31**, 13.

207. Fretting of electric contacts: an investigation of palladium mated to other materials, M. Antler, *Wear*, 1982, **81**, 159.

208. Survey of contact fretting in electrical connectors, M. Antler, *IEEE Trans. Compon. Hybr. Manuf. Technol.*, 1985, **CHMT-8**, 87.

209. Friction and wear of electrodeposited palladium contacts: thin film lubrication with fluids and with gold, *IEEE Trans. Compon. Hybr. Manuf. Technol.*, 1986, **CHMT-9**, 485.

210. Electroless Ni/immersion Au interconnects: Investigation of black pad in wire bonds and solder joints, P. Snugovsky, P. Arrowsmith, and M. Romansky, *J. Electron. Mater.*, 2001, **30**, 1262.

12 Gold Thick Film Pastes

Kenichiro Takaoka

CONTENTS

Introduction..279
Thick Film Process in Forming an Electrode 279
Features of Thick Film Paste 280
Types of Gold Paste 282
Applications and Future Trends in Gold Pastes..284
References..286

INTRODUCTION

The growth of the microelectronics industry during the 1970s resulted in the use of thick film pastes based on gold being firmly established in commercial use for various electronic devices and sensors [1–4]. There are many types of thick film conductor pastes which are usually composed of a metal powder, a metal oxide including a glass frit, and an organic binder. The metal powder can be silver, gold, platinum, palladium, etc., used either individually or in combination, depending on the properties and functions required. In developing new devices, a critical issue for manufacturers is product reliability, and therefore gold, which is chemically inert, is often selected as the metal powder despite the relatively high price. In this chapter the principles of thick film technology and the use of gold paste in this application is described.

THICK FILM PROCESS IN FORMING AN ELECTRODE

Gold thick film conductors not only provide printed wiring in devices, but also resistor terminations, electrodes, and solderable pads to facilitate the packaging and assembly of circuits. However, to simplify the terminology used in this chapter, the generic term *electrode* will be used to describe the final printed and heat-treated paste.

The thick film process in forming an electrode generally consists of printing, drying and firing, as shown in Figure 12.1. First, the conductor paste for the electrode is printed onto the substrate using a screen with a certain pattern. Next the printed substrate is processed by drying, before firing to form the electrode. As mentioned above, a metal powder and metal oxide, including a glass frit, are well dispersed into an organic binder to produce a thick film conductor paste. The paste is adjusted to the appropriate viscosity according to the substrate material, targeted film thickness, printing screen and pattern. In the screen printing process, variables can include the type of squeegee, the screen mesh size and the thickness of the emulsion, which in turn will depend on the required film thickness and line width. Due consideration needs to be given to the substrate which may be alumina, glass, low temperature co-fired ceramic (LTCC), or a polymer, depending on the application. The conductor paste and the substrate should be well matched to the end application. The printed substrate is usually put in a drier at 50–150°C to vaporize the solvent and is subsequently fired in a conveyor furnace at 500–900°C to burn out the organic binder. The temperature profiles of the conveyor, i.e., ramp rate and dwelling time, will affect the final quality of the electrode.

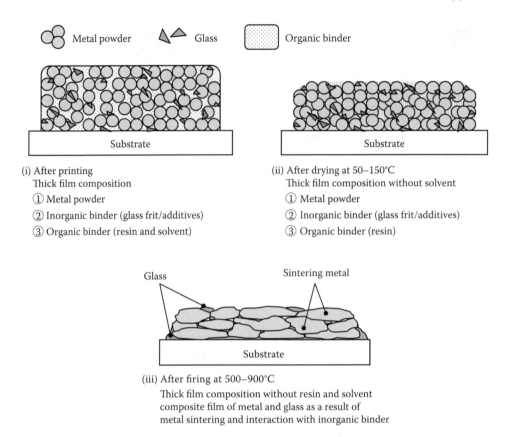

FIGURE 12.1 Schematic of the thick film process in forming an electrode.

A significant change of conductor thickness is observed in proportion to the elevated temperature. This is due to metal sintering and interactions with the inorganic binder. The thickness after firing is generally reduced to around 30% of the thickness after printing, as shown in Figure 12.1.

An example, representative of the thick film process including the sintering behavior during firing, is shown in Figures 12.2 and 12.3. In Figure 12.2, gold metal particles tend to grow slowly and the individual particles can still be distinguished below 600°C. The sintering rate is accelerated above 600°C and the individual particles cannot be distinguished at 800°C. It is also possible to monitor the advance of sintering from the viewpoint of the particle size. Smaller particles are less stable and the sintering of 0.5 μm particles occurs even at 600°C, when the individual particles cannot be distinguished (Figure 12.3). The spherical gold particles shown in Figures 12.2 and 12.3 are produced by reducing the metal ions in solution while controlling temperature, pH, reducing agent, additives, etc. Manufacturers can fine-tune the size of the gold particles by changing the gold ion concentration in the solution.

FEATURES OF THICK FILM PASTE

There are two well-established technologies—thick film and thin film—that can produce comparable metallizations for a range of applications. The thin film process consists of masking and etching followed by sputtering or vapor deposition using vacuum equipment. This equipment can be expensive to purchase, install, and maintain and also requires expert operator skill. It is considered

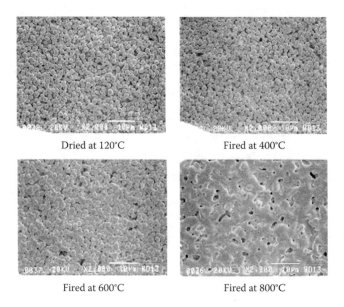

Dried at 120°C   Fired at 400°C

Fired at 600°C   Fired at 800°C

FIGURE 12.2   Sintering behavior of gold paste during firing (SEM images).

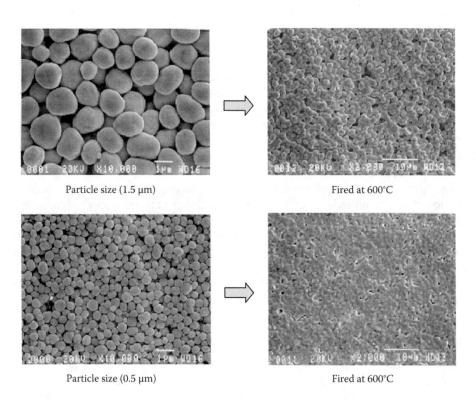

Particle size (1.5 μm)   Fired at 600°C

Particle size (0.5 μm)   Fired at 600°C

FIGURE 12.3   Sintering comparison in different particle sizes (SEM images).

TABLE 12.1
Comparison between Thin Film and Thick Film Technologies

	Thin Film	Thick Film
Film formation	Sputtering	Screen printing
	Vapor deposition	Dipping
Patterning	Photoetching	Spraying
Film thickness	<3 µm	1–100 µm
Line resolution	Submicron pitch	100 µm pitch
Film composition	Homogeneous and controllable at the atomic scale	Uncontrollable at the atomic scale
		Easy to vary composition
Film quality	Dense film	Density, smoothness, and hardness are controllable
	Extremely smooth surface	
Adhesion strength with substrate	Inferior to thick film	Good (<50 N/mm^2)
Heatproof	Inferior to thick film	Good
Others	Equipment is expensive	Equipment is inexpensive
	Manufacturing is expensive	Manufacturing is inexpensive
	Difficult to form three-dimensional pattern	Easy to form three-dimensional pattern

that, in principle, however, thin film technology can give finer and more precise line resolution than thick film technology. Table 12.1 is a comparison of the thin film and thick film processes.

A wide range of thick film conductor compositions, based on different metal powders and inorganic and organic binders, can be produced depending on the end use. Several metal compositions of conductor paste are represented in Table 12.2.

The metal choice dominates the characteristics of conductor paste. Consideration also needs to be given to conductivity, reliability against ion migration, soldering, etc., but it will also be necessary for the engineer to consider whether the material cost is acceptable for the electrode. When both high conductivity and reliability under high temperature and humid conditions are required for a certain electrode, a gold composition could be the best choice if the material cost can be accepted. If soldering is added as a further requirement, it maybe necessary to consider a gold alloy composition (such as gold and a certain ratio of platinum) that will have a higher cost than pure gold or a silver alloy composition (such as silver and a certain ratio of palladium) with a compromise on conductivity and reliability.

Although the quantity of inorganic binder in the paste is not significant in terms of final composition, it can be quite influential on metal sintering. The action of the inorganic binder in forming an electrode is as follows. Initially, the binder softens during firing, diffusing and interacting with the metal particles as well as the substrate. As a result, metal sintering is promoted and the electrode adheres to the substrate. In this way, engineers can control and fine-tune the sintering as well as the adhesion strength by selecting the appropriate binder. The constituent parts of the inorganic binder (glass frit or metal oxide) diffuse deeply into the substrate creating a so-called anchor effect.

TYPES OF GOLD PASTES

The glass frit type of thick film gold paste that has just been described is commonly used for electronic devices and sensors. The particle size of the gold powder is 3 µm or less, and particles will tend to be spherical in shape in order to achieve a dense film after firing. The finished electrode thickness after printing will usually range from 1 to 20 µm, depending on particle size, metal content of the paste, etc.

TABLE 12.2
Features of Thick Film Conductors

	Examples of the Paste	Sheet Resistivity	Features	Applications
Silver compositions	TR-3025 (fired at 850°C)	1.5–3.5 $\mu\Omega$ cm	High conductivity Ion migrates easily Sulfurated easily	General wiring Resistor terminal
Silver-palladium compositions	Ag/Pd ratio 87/13 TR-2637 (fired at 850°C)	15–19 $\mu\Omega$ cm		
	Ag/Pd ratio 79/21 TR-4846 (fired at 850°C)	18–25 $\mu\Omega$ cm	Compatible with solder	General wiring Resistor terminal Lead terminal
	Ag/Pd ratio 70/30 TR-4865 (fired at 850°C)	25–35 $\mu\Omega$ cm		
Gold compositions	TR-1531 (fired at 850°C)	3–4.5 $\mu\Omega$ cm	Excellent wire bondability Incompatible with solder Expensive Good chemical stability	High reliable wiring Wire bonding pad Lead terminal
Copper compositions	TR-8901 (fired at 900°C)	2–4 $\mu\Omega$ cm	High conductivity Low material cost Need to be fired in nitrogen	General wiring Lead terminal

There is another type of gold paste called gold resinate or liquid gold. This paste, which is composed of organometallic compounds, has long been used for decorating tableware [5]. This traditional technology has also been applied to the electronics industry, where highly reliable fine tracks are required. Metal-organic pastes are not particulate in nature but in a completely liquid state. The constituent metal powder and inorganic and organic binders are homogeneously dissolved in the organic solvent. The differences in the two types of gold materials are shown in Table 12.3. There is a noticeable difference in the thicknesses of the resulting films.

TABLE 12.3
Features of Frit-Type and Metal-organic Gold Pastes

	Frit-type Gold Paste	Metal-organic Gold Paste
Gold metal element	Gold metal powder	Organic compound of gold
Inorganic binder	Glass frit, oxidation bismuth, copper oxide, etc.	Organo-metallic compound (bismuth, silicon, lead, etc.)
Organic binder	Resin	Resin
	Solvent	Solvent
Thickness of the film	1–100 μm Thickness is easy to control	<1 μm
Dispersion state of the paste	Particulate state: Inorganic particles are dispersed in organic binder	Liquid state: Organic compounds are dissolved in organic binder

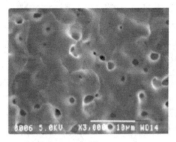
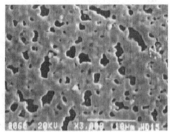
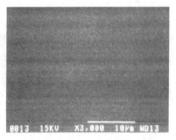

| (a) Frit-type gold paste
Gold content 85 wt%
Fired thickness 10 μm | (b) Frit-type gold paste
Gold content 40 wt%
Fired thickness 1.2 μm | (c) Metal-organic gold paste
Gold content 20 wt%
Fired thickness 0.3 μm |

FIGURE 12.4 Comparison of gold pastes fired at 850°C.

An example of the typical electrode structure obtained using each material type after firing at 850°C is shown in Figure 12.4. Three types of gold pastes are shown. One is a typical frit type of gold paste in which gold powder of 1.5 μm is used and has a gold content of 85 wt% gold. After printing and drying, this paste is fired at 850°C, as shown in Figure 12.4a. This can be compared to another specially prepared frit-type paste for which gold powder of 0.3 μm is used and is thinned with organic binder, resulting in 40% gold content. The surface of this paste, again fired at 850°C, is shown in Figure 12.4b. Finally, a metal-organic paste that is composed of organic compounds plus gold and an organic binder is shown in Figure 12.4c. The fired thicknesses of these three pastes are 10, 1.2, and 0.3 μm, respectively. It can be seen that the metal-organic paste has a smoother surface and allows a very dense electrode structure to be achieved, despite a large difference in the thickness. Although the frit-type paste is designed to be comparable to the metal-organic paste, the film density certainly deteriorates as the gold content decreases. In spite of a much lower gold content than the frit type, no voids on the surface of the metal-organic paste are observed. There seems to be no doubt that frit-type and metal-organic pastes involve separate mechanisms for forming an electrode, although a metal-organic paste is processed in the same manner as the frit type. In the case of the metal-organic paste, however, the organic compounds decompose into a form of colloidal particle during firing. These particles, which are not micron but nanometer in size, are unstable and subsequently sinter to form a smooth, thin, and dense electrode. The constituent ingredients of the inorganic binder should be optimized to successfully control the sintering and adhesion strength with the substrate. In addition, because the metal-organic paste contains a great deal of organic binder, it is necessary to pay particular attention to furnace ventilation during firing in order to prevent discoloration on the surface of the electrode.

APPLICATIONS AND FUTURE TRENDS IN GOLD PASTES

Gold pastes have been commercially used for highly reliable electrodes in electronic devices and circuits, in particular for applications such as medical, space, and military equipment. Figure 12.5 shows a gold wire bonded to such an electrode.

As already mentioned, gold's chemical inertness allows it to strongly resist ion migration, which makes it particularly suitable for the production of ultrafine lines in electronic circuits, and so it may be expected that gold will contribute in the future to the miniaturization of electronic devices. However, what is needed to make the most of the reliability of gold as a paste material is a simple and commercial method for producing fine line resolutions. Although it is known that screen printing is one of the simplest methods to make an electrode, it is also known that the method has a limit in terms of line resolution. A frit-type gold paste printed using a screen is shown in Figure 12.6. It is apparent that the process can be used to make a 50-μm line although the line definition is not of optimum quality, but that a 40-μm line is clearly not achievable.

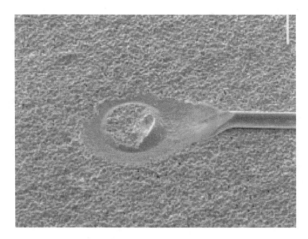

FIGURE 12.5   Gold wire bonding on an electrode of gold paste.

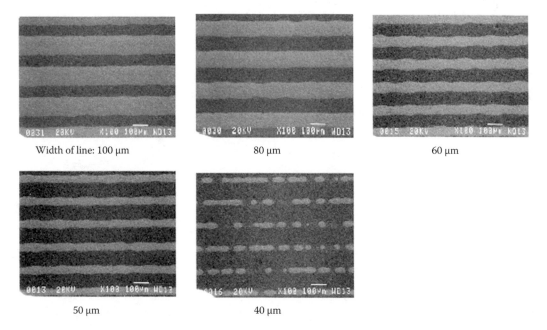

Width of line: 100 μm   80 μm   60 μm

50 μm   40 μm

FIGURE 12.6   Fine line resolution in screen printing by using a frit-type gold paste.

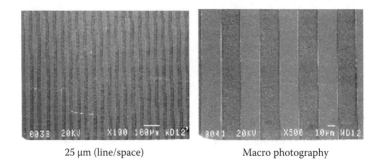

25 μm (line/space)   Macro photography

FIGURE 12.7   Fine lines of a metal-organic paste by a chemical etching technique.

A typical application for metal-organic pastes is for electrodes in thermal print heads. Figure 12.7 shows an electrode with a 25-μm linewidth. In this case, a square pattern of printed paste is converted into fine lines by means of both photolithography and chemical etching techniques (iodine and potassium iodide solution are the usual etchants). In Figure 12.7, the clear definition of the electrode can be observed. With respect to the electrode for thermal print head applications, it is important to tune the ingredients of the inorganic binder to achieve the required wire bonding capability, adhesion strength with the substrate, as well as compatibility with resistor material.

One further trend in the use of gold thick film pastes is the increasing need to use materials without negative health and environmental issues, such as lead, cadmium, etc. Recently, metal-organic gold pastes based on lead-free compositions have been developed in response to these new environmental requirements.

REFERENCES

1. Gold in thick film hybrid microelectronics, R. G. Finch, *Gold Bull.*, 1972, **5**, 26.
2. A new generation of thick-film gold conductor pastes, Anon., *Gold Bull.*, 1975, **8**, 13.
3. Gold in thick-film conductors, R. H. Caley, *Gold Bull.*, 1976, **9**, 70.
4. Gold thick film conductors, M. V. Coleman and G. E. Gurnett, *Gold Bull.*, 1977, **10**, 74.
5. Gold in decoration of glass and ceramics, G. Landgraf, in *Gold Progress in Chemistry, Biochemistry and Technology*, ed. H. Schmidbaur, 143–171. John Wiley & Sons Ltd., New York.

13 Gold Bonding Wire

Koichiro Mukoyama

CONTENTS

Introduction..287
Wire Bonding Process...287
Gold Wire for the Ball Bonding Method 289
    High Strength Type 289
    Fine Pitch Bonding Type..291
    High Bond Reliability Type 291
Bonding Wire Production Process 292
    Melting and Casting...292
    Drawing 293
    Annealing 293
    Winding..293
    Final Inspection..293
Suggested Reading...293

INTRODUCTION

Wire bonding technology is one of the most widespread and reliable interconnection techniques in microelectronics. Gold, aluminum, and copper and their alloys can all be used as wire materials. High purity gold bonding wire has been widely used in semiconductor devices since the assembly of semiconductor packages began approximately 40 years ago. Gold is used because of its ductility, low electrical resistance and resistance to oxidation. Recently, as the applications for semiconductor devices have expanded, the requirements for gold bonding wire have become more diverse and many different kinds of gold bonding wire are now produced for wiring the electrodes of an integrated circuit (IC) and the inner leads of semiconductor packages. In addition, the ongoing trend in miniaturization of electronics products in turn requires semiconductor packages to become smaller, thinner and lighter, and gold bonding wire products able to meet these challenges are continually being developed. This chapter provides an overview of the use of gold bonding wire in the electronics industry (Figure 13.1).

WIRE BONDING PROCESS

There are essentially two forms of wire bonds: wedge bonds and ball bonds. The fact that gold wire can be bonded in the shape of a wedge or a ball makes it extremely versatile. The basic process for making a gold wire ball bond is shown in Figure 13.2. In the ball formation stage, a spark or small flame is used to locally melt the end of the gold wire so as to form a spherical ball that is approximately twice the diameter of the wire. The ball is thermosonically welded to a metallized pad on the semiconductor during the first bonding stage. A loop of wire (defined as high, middle, or low) is formed as the bonding capillary moves across to the contact pad of the device package or circuit board. During the second bonding stage the wire is welded to the metallized pad of the package.

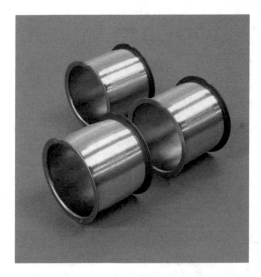

FIGURE 13.1   **(See color insert following page 212.)** Gold bonding wires wound on spools, ready for use in the electronics industry.

Finally, the wire is cut, leaving a length protruding to form the next ball, before the process continues making numerous ball bonds within a single device (Figure 13.3).

The quality requirements for gold bonding wire are very strict and a number of important issues must be controlled by manufacturers. Issues include the straightness of bonded wire, i.e., ensuring no contact between wires, avoiding cracking or delamination of the IC chip due to the bonding process, good reliability of the first and second bonding, consistent loop formation and height and

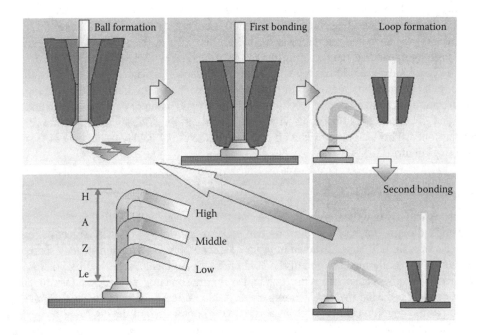

FIGURE 13.2   Schemmatic of the ball bonding process, indicating the length (Le) of the heat affected zone (HAZ).

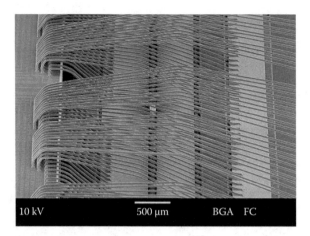

FIGURE 13.3 SEM micrograph highlighting the large number of gold bonding wire connections made within a typical integrated circuit.

long-term reliable bonding. As an example of these issues we can observe examples of poor wire straightness. There are essentially three types of deviation from satisfactory wire straightness: curving, leaning, and so-called snake-wire. These are shown in Figure 13.4.

GOLD WIRE FOR THE BALL BONDING METHOD

Bonding wire is used to connect the electronic circuit of a semiconductor chip with a leadframe or a substrate and to transmit an electric signal between them. Gold wire with a diameter of 15–100 µm is used for this purpose. Despite the high cost of gold itself, the ball bonding method using gold wire remains the major interconnection technology in semiconductor assembly processes, largely because it permits the highest manufacturing productivity.

The ball bonding process is also a very flexible technique and can be used in a wide range of applications from low pin count devices such as LEDs to high pin count devices such as microprocessors. Recently, advanced wire bonding technology with 20 µm gold wire has been adopted in mass production, and the assembly evaluation of 15 µm gold wire for the next generation of devices has been completed. Many types of gold wire are manufactured by the leading wire suppliers. Typical gold wires are shown in Table 13.1.

HIGH STRENGTH TYPE

To achieve the dual aims of a high density mounting and material cost reduction, high strength gold bonding wire with the same wire bonding performance as thinner wire is demanded. To make pure gold harder, selected dopants are added to the metal.

The selected wire diameter is generally determined by the electrical specification of the IC package. It is necessary to exercise caution in selecting wire diameter because wire strength lowers in proportion to the square of wire diameter, and wire sweep resistance lowers as a direct result. Within the industry bonding wire manufacturers are set challenges by the electronics industry to reduce wire diameter, e.g., from φ 25 µm to φ 23 µm or φ 20 µm to φ 18 µm (depending on the application), by using high strength types of gold wire. In order to achieve wire diameter reduction, leadframe, substrate, molding resin, and die bonding material also need to be carefully selected at the same time.

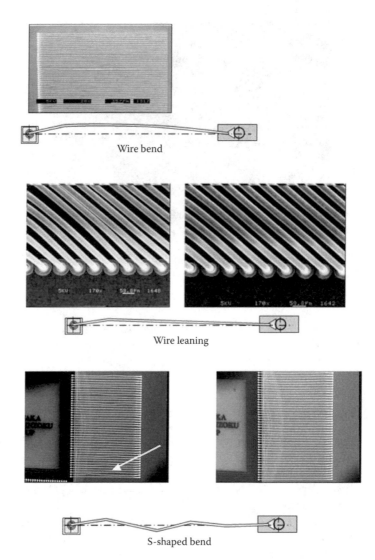

FIGURE 13.4 Examples of problems in wire straightness.

TABLE 13.1
Types of Gold Bonding Wire Used in the Electronics Industry

Type/Name	Feature
Au Bonding Wire (Purity 4N: 99.99% or More)	
Y	Soft wire, applied to IC packages molded by epoxy resin with high heat expansion coefficient
FA	General purpose wire for standard leadframe; improved second bondability; reduce second bond heel crack issues
GFC,GFD	Semihard and hard types, widely used for fine pitch applications
GMG,GMH-2	Highest mechanical strength and hardness, for wire sweep resistance in the molding process
Au Alloy Bonding Wire	
GPG,GPG-2	Au-1%Pd wire with high bond reliability
LC	Au alloy wire for material cost reduction

FIGURE 13.5   Squashed ball shape of conventional type and advanced type.

FINE PITCH BONDING TYPE

New 99.99% (4N) gold wire (GFC and GFD in Table 13.1) has been developed for fine pitch bonding applications. The distinguishing feature of this fine pitch bonding type of wire is the good circularity of the squashed ball after bond formation. This enables the selection of a larger wire diameter, with the advantage of increased wire sweep resistance. The shape of squashed balls of the fine pitch type and the conventional gold wire is shown in Figure 13.5. This type of gold wire is generally adopted in high pin count BGA package or in multitier stacked packages.

HIGH BOND RELIABILITY TYPE

In the ball bonding process, the first ball bond results in the formation of gold–aluminum (Au-Al) intermetallic compounds between the gold wire and the aluminum electrode on an integrated circuit die heated at 150–200°C. It was originally thought that thick intermetallic compound or void formation caused by mutual solid diffusion in this Au-Al region resulted in first ball bond failure due to an increase in electrical resistance. However, it has recently been shown that this is probably not the main factor causing electrical deterioration in the semiconductor package. The real cause is the corrosion of aluminum oxidized by a halogen, especially bromine, that liberates from the flame retardant used in the chip's molding resin. It is possible that even small amounts of bromine can destroy the ball bond area and so mold resins without halogen are now widely used to ensure very high bond reliability. In addition to this, gold-1% palladium alloy bonding wire is used as a high bond reliability gold wire and has already been mounted in many kinds of IC package. This 1% Pd doping in Au is effective in restraining corrosion and crack formation (Figure 13.6). This so-called 2N gold wire with 99% Au purity is widely used in any semiconductor package

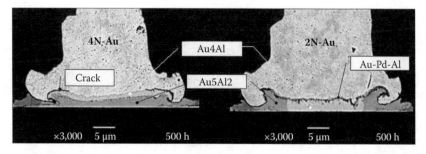

FIGURE 13.6   Cross sectional view of the first bond aged at 175°C for 500 h, showing crack formation in a 4N wire bond.

TABLE 13.2
Comparison in Wire Bond Reliability between 2N-Au, 3N-Au, and 4N-Au Stored at 175°C

Time (Hours)	0	500	1000	1500	2000	3000
4N-Au (99.99% Au)	✓	✗	✗	✗	✗	✗
3N-Au (99.9% Au)	✓	✗	✗	✗	✗	✗
2N-Au (GPG, GPG-2)	✓	✓	✓	✓	✓	✓

where joint reliability is more strictly demanded, especially in the electronic components for automobiles. In addition, 3N gold wire with 99.9% Au purity has been introduced as an additional high bond reliability type of wire by some wire suppliers. Its performance is slightly better than conventional 99.99% gold wires on long-life bond reliability, but does not match that of 2N, Au-1%Pd alloy wire (Table 13.2).

BONDING WIRE PRODUCTION PROCESS

The manufacturing of gold bonding wire is comprised of five processes: melting and casting, drawing, annealing, winding, and final inspection (Figure 13.7).

MELTING AND CASTING

As specified by the American Society for Testing and Materials (ASTM), most gold bonding wire is 99.99% pure gold. To achieve this level of purity, gold with 99.999% or even 99.9999% purity, refined from commercial 99.99% gold, is used as the raw material for bonding wire. During the

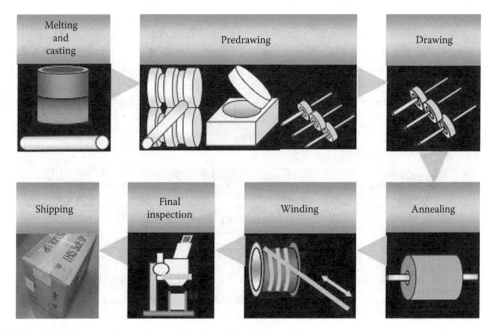

FIGURE 13.7 The manufacturing process for gold bonding wire.

melting and casting process, a small amount of additional alloying elements is added to improve hardness, tensile strength and the crystal grain size of the final products.

DRAWING

The drawing process from a thick-diameter rod to a thin-diameter wire is achieved by pulling the gold alloy through numerous diamond dies with decreasing hole diameters to achieve a final diameter specified by the wire user. This drawing technology takes advantage of high-speed drawing machines and customized chemical lubricants to achieve the required quality. In addition, plant cleanliness must be strictly controlled during all phases of the manufacturing process.

ANNEALING

Wire drawing down to small diameters accumulates strain in the metal, and therefore after drawing all wire products need to have a heat treatment to remove the strain. This is also necessary to protect the cold-worked gold wire from changes in mechanical properties due to aging at normal temperature. The standard specification of mechanical properties after annealing, provided for each diameter of gold bonding wire, is prescribed in terms of breaking load and elongation.

WINDING

After annealing, gold wire is wound onto a spool according to the end-users specifications. To allow for maximum productivity in the end user's process, a long winding length is generally required, and recently 3 km or 5 km of gold wire per spool has become standard. Wound wire quality affects wire bondability, and it is important that winding conditions (e.g., tension, width, etc.) are precisely controlled.

FINAL INSPECTION

From the initial drawing process to the final winding process, gold bonding wire is inspected by an in-line quality system. In addition, a random sampling inspection is carried out in a final inspection process. This includes visual inspection, measurement of wire diameter and assessment of mechanical properties. After this final inspection, products are packed and shipped to the end user.

SUGGESTED READING

L. S. Benner, T. Suzuki, K. Meguro, and S. Tanaka, eds., *Precious Metals Science and Technology*, 1991, IPMI, Florida, p. 570.

A. Bischoff et al., Recent developments in gold bonding wires for microelectronics, Proc. IPMI conference, Tucson, 2001, IPMI.

Y. H. Chew, C. C. Wong, C. D. Breach, F. Wulff, and S. Mhaisalkar, Effects of calcium on the mechanical properties of ultra-fine grained gold wires, *Journal of Alloys & Compounds*, 2006, **415**, 193–197.

Y. H. Chew, C. C. Wong, C. D. Breach, F. Wulff, T. T. Lin, and C. B. He, Effects of Ca on grain boundary cohesion in Au ballbonding wire, *Thin Solid Films*, 2006, **504**, 346–349.

Y. H. Chew, C. C. Wong, C. D. Breach, F. Wulff, and T. Y. Lew, Doping induced simultaneous improvement of strength & ductility in ultrafine grained gold wires, *Journal of Materials Research*, 2006, **21**(9), 2006, 2345–2353.

T. Ellis, The future of gold in electronics, *Gold Bulletin*, 2004, **37**(1/2), 66–71.

P. Goodman, Current and future uses of gold in electronics, *Gold Bulletin,* 2002, **35**(1), 21–26.

G. Harman, *Wire Bonding in Microelectronics: Materials, Processes, Reliability and Yield*, McGraw-Hill Professional, New York, 1997.

G. Humpston and D. M. Jacobson, New (A) High Strength Gold Bond Wire, *Gold Bulletin*, 1992, **25**(4), 132–145.

T. Müller, E. Milke, and E. K. Chung, New developments in wire bonding for future packages, presented at *IMAPS 2007, the 40th International Symposium on Microelectronics Celebrating 40 Years of Excellence!*, November 2007, San Jose, CA.

Ch. Simons, L. Schräpler, and G. Herklotz, Doped- and low-alloyed gold bonding wires, *Gold Bulletin*, 2000, **33**(3), 89–96, 102.

S. Tomiyama and Y. Fukui, Gold bonding wire for semiconductor applications, *Gold Bulletin*, 1982, **15**(2), 43–50.

F. Wulff and C. D. Breach, Measurement of gold ballbond intermetallic coverage, *Gold Bulletin*, 2006, **39**(4), 175–184.

14 Gold in Dentistry

Paul J. Cascone

CONTENTS

Foreword 296
Background and Terminology 296
  Dental Terminology 296
  Historical Review 296
    Dentures 296
    Wires 298
    Fillings 298
    Inlays, Crowns, and Bridges 299
    Porcelain-Fused-to-Metal Restorations 299
Requirements for Dental Materials 299
  Chemical Properties 299
    Inertness in the Oral Environment 299
    Biocompatibility 300
    Galvanic Reactions 300
    Tarnish Films 300
  Mechanical Properties 301
    Single Restorations 302
    Bridgework 302
  Thermal Properties 303
  Fabrication Properties 304
  Alloying Elements for Gold-Based Dental Alloys 304
    Gold–Platinum Alloys 305
    Gold–Palladium Alloys 305
Impact of Commodity Prices on Alloy Development and Gold Usage 305
  United States 305
  Germany 308
  Japan 308
Development of Standards, Specifications, and Regulations in the Dental Industry 308
  Early Attempts at Developing Standards for Gold Alloys 308
  Other Dental Materials 310
  Beginnings of Regulations 310
  Modern Regulations and Standards 310
  Other Countries 310
Future of Gold in Dentistry 311
  Traditional Casting Alloys 311
  Electroforming 311
  Sintering 312

3D Printing..312
Other Unique Opportunities for Gold 314
Summary...315
Acknowledgments...315
References...315

FOREWORD

The journal *Gold Bulletin* has published many excellent reviews on the use of gold in dentistry [1–4]; this chapter is written to supplement, not duplicate, the work of those authors.

Historically, the use of gold in dentistry has been reflected in the economic cycles of boom and bust. Increased usage occurs when innovative ways to exploit its attributes are found. Decreased usage occurs when the cost outweighs the benefits. The story then becomes one of material substitution. How and why gold was used in fabricating different types of dental appliances, and what material replaced gold and why is our story. In each case we shall see the introduction of a new process or technology that uses gold in yet another way. The two reasons for this are the biocompatibility and versatility of gold.

BACKGROUND AND TERMINOLOGY

DENTAL TERMINOLOGY

Humans have two sets of teeth. The first deciduous set gives way in our youth to a permanent adult set. This chapter discusses the role of gold in replacing one or more of the permanent set of teeth when they are lost due to accident or disease.

The human jaw is split into two parts: the upper jaw, or maxillary and the lower jaw, or mandible. The teeth are distributed symmetrically about a median line in the form of an arch. The space within the arch is called the palate. Our front teeth (incisors) cut or incise the food while our back teeth (molars) grind the food into small particles ready for swallowing. When we chew or masticate, there is some (but limited) movement of the jaw laterally. Bringing the top and bottom jaws together shows how the teeth meet or occlude. For the most part, each tooth has two antagonists so that if one tooth is lost the other can keep the opposing tooth from moving. The occlusal surfaces of the teeth must have the proper contours in order to function effectively. How well the upper and lower teeth interact determines the efficiency of mastication. Our food must be chewed in order to be properly digested.

The loss of a tooth not only adversely affects our well being but also our speech and appearance. It is no wonder that throughout history man has attempted a myriad of ways to restore lost teeth. The section of dentistry concerned with the restoration of oral function is called prosthodontics [5]. The field is divided into three branches: removable, fixed, and maxillofacial. The first two focus on replacement of missing teeth while maxillofacial involves the jaws and soft tissues as well. Prior to the twentieth century replacing missing teeth fell into the removable category (i.e., dentures). The fixed category required the ability to have the substitute tooth function similar to the lost tooth. This technology only developed in the early 1900s.

HISTORICAL REVIEW

Dentures

Any material in the oral environment needs to contend with a variety of conditions: hot, cold, acidic, basic, and even anaerobic. A chemically inert material, like gold, has distinct advantages.

Buccal view

FIGURE 14.1 Denture from the Roman Imperial Age. Gold wire supported the lost tooth. (Image courtesy of Archaeological Superintendence of Rome.)

The ancients recognized this and used gold wires to hold "replacement" teeth in place. Every sort of material was used as the replacement tooth, including carved ivory, the original tooth, and shells. The technique was employed in Phoenicia before 2000 BC. The Etruscans used gold wires and gold bands about 700 BC [6]. It is believed that the Romans learned these techniques from their Eutruscan neighbors, but because the Romans cremated their dead little evidence of the technical diffusion exists. Figure 14.1 shows a rare example of such a restoration from the Roman Imperial Age [7]. The root of the lost tooth had been cut off and the tooth drilled to accommodate the gold wire. The wire was wrapped around adjacent teeth for support. The same basic technique was still being used in the 1700s.

After porcelain teeth were introduced in 1780, dentures were fabricated using swaged gold foil in bands to improve the structural integrity. A denture fabricated for George Washington has a full set of porcelain teeth and gold palette for comfort (Figure 14.2). Gold springs provided alignment

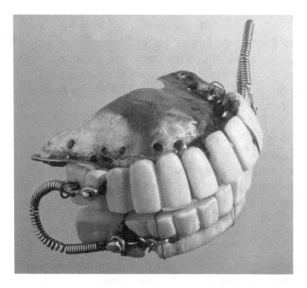

FIGURE 14.2 Dentures belonging to George Washington. (Object Courtesy of National Museum of Dentistry, Baltimore, MD. Image Courtesy of National Museum of American History, Smithsonian Institution.)

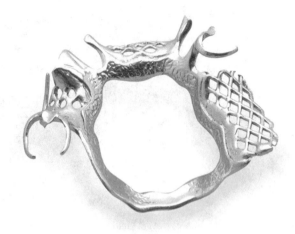

FIGURE 14.3   Modern partial denture casting. (From The Argen Corporation. With permission.)

and mechanical assist during chewing. Teeth were riveted onto the gold band. No wooden teeth for President Washington!

In 1851 vulcanite, a hard rubber, was discovered and used as a denture base for replacing swaged gold in full dentures. Porcelain teeth were set and aligned within the rubber. The rubber was less expensive and more comfortable than the swaged gold.

Cast gold alloys were used to attach the denture to the remaining teeth for partial dentures. These alloys were replaced with the introduction of the cobalt–chromium alloy Vitallium® in 1930. A methyl methacrylate resin was also introduced in the 1930s replacing the vulcanite rubber for denture bases. This resin was easier to use and softer in the mouth, and could be made pink in color for esthetics. The partial denture casting has metal webbing (Figure 14.3) to support the resin base which in turn supports the porcelain teeth.

Wires

Gold wires were used for making clasps for partial dentures and orthodontic applications (braces). In the 1950s stainless steel wire began replacing gold in orthodontic appliances because of their low cost, high strength, and easy manipulation. The short length of time the wires were in the mouth alleviated concerns over long-term corrosion. As with gold wires, the patient needed to return to the dentists for continual adjustment as the teeth moved into the desired position.

The introduction of the pseudoelastic nickel—titanium alloy (Nitinol) wires in the 1980s changed orthodontic treatments forever. The wires "remembered" their previous length, resulting in a consistent and continual force eliminating the need for constant adjustment. The tight oxide on the material also provided for the necessary corrosion resistance in the mouth. Today gold wires find limited use for fabricating clasps and special orthodontic appliances.

Fillings

Our ancestors recognized the benefit of filling cavities in teeth and were very imaginative in the materials employed. Fabric, lead, tin, and cork are some examples. Gold foil came to be the material of choice. Much effort went into developing a technique to properly fill the cavity with gold.

In 1833, an amalgam made up of silver and mercury was introduced into the marketplace. For a time there was some resistance in the dental community with some dentists swearing never to use amalgam. The benefits provided by the material, however, were substantial, especially for those who could not afford gold. After a short time most dentists supported the use of amalgam.

Cohesive gold was introduced in 1855. This material was a substantial improvement over gold foil and, for a brief time, the use of gold for fillings rose. By 1895, however, an amalgam composition, essentially the same formulation as used today, was introduced. The ease of use and low cost of amalgam resulted in market dominance well into the 1970s. Today, tooth-colored composites are the materials of choice for filling cavities although the use of amalgam is still extensive throughout the world.

Inlays, Crowns, and Bridges

Inlays are used to repair the surface of a tooth. Attempts to improve the accuracy of gold foil inlay restorations in the 1890s resulted in the adaptation of the lost-wax method of casting for the production of inlays in the early 1900s [8]. As is typical with such developments many contributors were responsible for the successful processing technique which is still in use today. The technique was rapidly adopted for the production of crowns (single tooth restorations) and bridges (multitooth restorations). The lost-wax process produced a casting whose interior surface was an exact match of the stump cut by the dentist in the patient's mouth. When the restoration was cemented in place full function for chewing was restored. This was a substantial improvement over the amount of function restored by dentures. The materials of choice for these restorations were gold alloys consisting of 70–80% gold with silver, copper, and small platinum or palladium additions.

By 1932, there were many different gold alloys on the market. The need for some standardization was evident. Ultimately, the American Dental Association Specification Number 5 for dental gold alloys (based on the review by the National Bureau of Standards) emerged.

Porcelain-Fused-to-Metal Restorations

While the lost-wax technique expanded the use of gold, there was a desire to improve the esthetics of the restorations. Although porcelain restorations had been used for years, they did not fit as well as the gold castings and were difficult to fabricate. During the 1950s, tooth-colored resins were introduced that could be attached to the facings of the gold castings, but the resins wore away and discolored.

The next significant development in porcelain occurred in 1962. Weinstein described a system of baking a porcelain onto a gold alloy (U.S. Patent 3052982). The alloy was cast as a shell (called a coping) using the lost-wax process so the fit was accurate. The alloy provided the strength and the porcelain provided the esthetics of natural teeth. Initially, the most popular alloy contained 87% of gold with some platinum and palladium additions. The alloy was yellow in color. This porcelain-fused-to-metal (PFM) restoration still dominates the market although, as we shall see, the alloy composition is no longer primarily gold.

REQUIREMENTS FOR DENTAL MATERIALS

CHEMICAL PROPERTIES

Inertness in the Oral Environment

In order to maintain its structural integrity over the life of the restoration, the dental alloy cannot react with the fluids in the oral environment. Reactions can be minimized by two mechanisms: the alloy can either be noble or become passive. If the material is noble, it does not react to the oral fluids. Gold and the platinum group elements (platinum, palladium, rhodium, ruthenium, and iridium) are intrinsically inert in the oral environment. Iridium, ruthenium, and rhodium are difficult to fabricate. The cost of platinum has restricted its widespread use as a dental material (except for one brief period of time when gold was more expensive) thus leaving gold and palladium as the primary metals for use in dentistry.

In order for the alloy to become passive the material must develop a protective surface to prevent reactions from occurring. Cobalt- and nickel-based alloys become passive by the formation of a chromium oxide layer on the alloy surface. Titanium develops a very tight oxide that does not require any additional alloying elements to develop passivity.

Biocompatibility

Despite their nobility or passivity there may be some metallic ions released from the alloy over the years of expected use. Elements that are released cannot be toxic (i.e., poisonous). This does not mean that the materials are nonallergenic. Consider that a toxic material *will* cause harm in a species while an allergen *may* cause harm in an individual. It is possible to test for toxicity while it is very difficult to predetermine an allergen. A given material may trigger an allergic reaction of varying degree in some individuals but not in others. This issue can be addressed by performing a risk analysis. The accepted protocol is given in ISO 14971. A lively discussion on this topic can be found in Lang et al. [9].

Galvanic Reactions

Any combination of dissimilar metallic alloys gives rise to the possibility of galvanic reactions. It is crucial to determine when these reactions become clinically significant. Coupling most of the various types of dental alloys does not produce a significant current density [10]. This is not the case with reactions between silver amalgam and the other metallic materials. Coupling gold alloys with amalgam results in significant current densities. Thus silver amalgams are sacrificed. The silver is deposited on the gold alloys resulting in a dark film on their surfaces.

Coupling titanium with other dental materials was studied because titanium is used for dental implants [11,12]. On the electrochemical scale, titanium is less noble than the gold and palladium alloys but is nobler than the cobalt and nickel alloys. The studies confirmed the expected sequence. Coupling titanium with gold or palladium alloys resulted in a small current flow that rapidly diminishes due to the formation of titanium oxide. The adherent oxide shuts down the electron flow preventing any deterioration. Coupling titanium with nickel or cobalt alloys, however, resulted in these alloys corroding. This is especially true for the nickel-based alloys. For this reason, the use of nonprecious alloys in conjunction with titanium implants is discouraged.

Tarnish Films

Any dark discoloration that adheres to the alloy surface is considered a tarnish film (an example is shown in Figure 14.4). The source of the film formation may be interaction with other alloys in the mouth, as discussed previously, or intrinsic to the alloy formulation. Tests on a series of gold–silver–copper alloys containing small amounts of palladium and platinum demonstrated that the gold content alone did not provide sufficient protection to prevent the film formation [13]. With the proper formulation, however, a 42% gold alloy performed as well as a traditional 68% gold alloy. The root cause of the film formation is the miscibility gap in the gold–silver–copper system. Upon cooling from the liquid state, two phases develop: one contains gold and silver, the other, gold and copper. This microsegregation sets up the galvanic cell. The only other metallurgical system that is prone to tarnish film is the silver–palladium–copper system. This system also has a miscibility gap that results in microsegregation. The silver–palladium alloys with no copper may form a thin film if

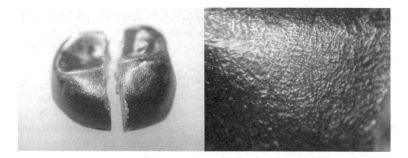

FIGURE 14.4 Example of a tarnished full gold crown and a magnified area. (From The Argen Corporation. With permission.)

the palladium content is below 20 wt%. Corrosion resistance and tarnish resistance are two separate phenomena as demonstrated by Treacy and German [14]. It is relatively easy to attain a corrosion-resistant material by increasing the alloy's nobility (i.e., gold, platinum, and palladium content) but tarnish resistance is dependent on the details of the alloy's microstructure.

MECHANICAL PROPERTIES

As the amount of replaced tooth structure increases, the strength of the prosthetic material needs to increase. Figure 14.5 show three typical restorations: an inlay (lower left), a full cast restoration of one crown (lower right), and a three unit bridge done in the PFM technique. The strength of the alloys used must increase in the same order. For the porcelain-fused-to-metal technique the challenge for the laboratory technician is to have the porcelain overlay appear as close as possible to natural teeth. The greater the porcelain thickness, the easier this is to accomplish. This means that the alloy coping should be as thin as possible. In order to withstand the mastication stress, the elastic

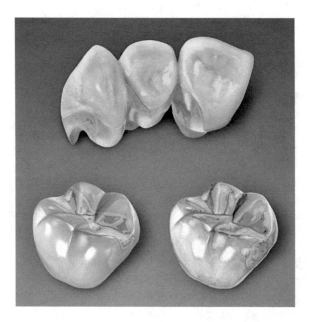

FIGURE 14.5 Examples of an inlay, full crown, and PFM. (From The Argen Corporation. With permission.)

modulus needs to increase as the coping thickness decreases. Although the importance of the elastic modulus is recognized, there is little guidance from the standards.

Single Restorations

Inlays need to fit a tapered cavity cut by the dentist in the diseased tooth. The margins are sealed by physically moving the alloy against the tooth structure, thus requiring a material with a low yield strength and high elongation. The malleability of gold makes it perfectly suited for this application. A single crown, cast entirely of gold alloy, requires a more rigid structure to resist the stresses of mastication. This is accomplished by adding copper and silver, thus increasing the yield strength. The PFM technique requires alloys with high strength levels because the majority of the restored tooth structure is porcelain. The thin copings form a support structure for the porcelain and provide for an accurate fit to the "stump" remaining in the patient's mouth. All of the standards address these application-dependent requirements. What has not been addressed, however, is the variation in biting stress going from the anterior teeth to the posterior teeth. Figure 14.6 shows that an average biting force changes from 90 to 400 Newtons [15]. What also changes is the nature of the stress. The front teeth have a single edge to shear food. The stress on the restoration is compressive and relatively low. The rear teeth grind food up into smaller particles. The cusps and valleys of the opposing teeth mate resulting in a mortar and pestle action that produces alternating tensile and compressive forces on the restoration. These forces can be quite high. Restorations for the anterior portion of the mouth then do not need to be as strong as those made for the posterior portion of the mouth. Very high gold materials such as electroformed copings (100% gold), CAPTEK™ copings (87% gold), and GOLDTECH® BIO2000 (99.7% gold) can be used for anterior restorations because of the low strength requirements. These materials are discussed in the following sections.

Bridgework

As the number of units in a restoration increases, the strength of the alloy must also increase. The two end members of the three-unit bridges are called abutments. They are supporting the middle

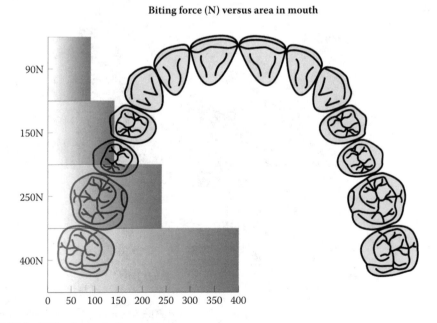

FIGURE 14.6   Biting forces in the mouth.

Thermal parameters

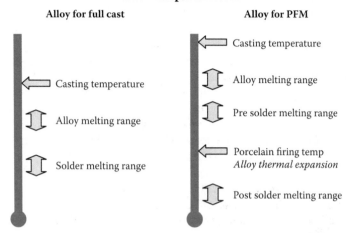

FIGURE 14.7 Temperature considerations for dental alloys.

unit called a pontic (this is the tooth being replaced). When the bridge is loaded during mastication, the units do not experience the stress at the same time, or with the same force. This produces a torque on the restoration. Also, the interproximal areas—the areas that join the pontic to the abutments—will experience alternating tensile and compressive forces during mastication as the stress is applied and released. The standards address the minimum strength requirements for alloys to tolerate the complex stress distribution experienced by bridgework.

THERMAL PROPERTIES

The melting range of the alloy must allow for its fabrication as well as any subsequent thermal processing, i.e., soldering (the term soldering in dentistry is actually torch brazing) or porcelain baking. Figure 14.7 shows the criterion for the alloys used for full cast and those used for the PFM technique.

High temperature creep becomes a concern because the porcelain firing temperature is relatively high (about 900°C). The restoration must be properly supported in the porcelain oven during the multiple porcelain firing cycles. A high solidus temperature is desired. The thermal expansion of the alloy must be compatible with one or more porcelains on the market. While thermal expansion is measured on heating the material, in practice it is the cooling phase and the contraction of the material that is operative.

The same parameters for successful glass-to-metal seals are active for baking porcelain onto dental alloys. Dental porcelain is essentially a glass with a high crystalline content. This means that the alloy must contract at the same rate as the porcelain (glass) below the transition temperature of the porcelain. The transition temperature is considered to be the point at which the porcelain becomes rigid (or sets). A significant deviation in contraction between the alloy and porcelain once the porcelain glass has set will result in the porcelain cracking. The firing of the dental porcelains is very rapid: heating to a temperature at a fast rate, holding at temperature for a minute or two, and then cooling. Because the thermal conductivity of the alloys is greater than that of the porcelains the thermal expansion of the alloys needs to be higher than that of the porcelains.

Recently porcelains have been introduced that can be pressed onto the alloy. The processing cycle requires the alloy to be held at a high temperature for a long time (20 min). For these materials, the alloy's thermal expansion must be closer to that of the porcelain than for the normal firing cycle. Thermal compatibility then is dependent not only on the thermal expansion coefficient but also on the thermal conductivity of the alloy. In addition the design of the restoration also plays an important role in determining the amount of residual stress in the final restoration.

All bridgework will contain one or more pontics. If the porcelain is completely covering the pontic then the heat within the alloy cannot escape. Within hours of being fabricated the porcelain will continue to develop stress around its circumference until the interior of the alloy reaches room temperature. Occasionally the stress may be sufficient to produce a crack in the porcelain. This time-delay type of fracture can be alleviated by leaving a collar of metal on one side of the pontic. This band will prevent the formation of "hoop" stresses in the porcelain that cause the cracks. Similar to the glass-to-metal seals, adherence of the porcelain to the alloy is via a chemical bond between the porcelain and alloy. The critical sequence is as follows:

- Upon heating, a metal oxide is formed on the alloy surface.
- The oxide is dissolved by the glass in the porcelain.
- The porcelain is saturated with the oxide at the interface with the alloy.

For the noble metals small additions of base elements produce the necessary oxide and achieve adherence.

Fabrication Properties

The processing of dental restorations is typically performed in air. Melting the alloy is done in air by using a gas-oxygen torch, induction furnace, or an electrical resistance furnace. Firing dental porcelain is also done in an oxidizing atmosphere, despite the term "vacuum" furnace. This situation has restricted the metallurgical systems used for dental restorations and is one reason for the success of gold-based materials.

Alloying Elements for Gold-Based Dental Alloys

Table 14.1 shows the effects of adding alloying elements to select gold properties required for the PFM technique. Starting with pure gold the first row shows the desired modifications: the strength and the solidus temperature need to increase, the thermal expansion should decrease, and the color should remain as close as possible to that of gold. The next series of rows show the effects of adding

TABLE 14.1
Effects of Alloying Elements on Select Gold Properties

Element	Weight %	Strength	Color	Solidus	Expansion
Desired		Increase	Same	Increase	Decrease
Pt	<10	+	-	+	-
Pt	10–20	++	--	++	--
Pd	<10	++	--	++	--
Rh	<1	+	+		
Ag	<5				+
Zn, In	1–3	++++		-	+
Fe, Mn	0.5–1.5	++++			+
Ta	0.1–1.0	+	+		
Cu, Sn	<0.5	+			
Re, Ir, Ru	<0.2	+			

TABLE 14.2
Typical Dental Gold Alloy Compositions

Alloy Composition	Proof Stress Mpa	Youngs Modulus Mpa	Solidus Temp.°C	Color
88 Au, 9.58 Pt, In, Fe, Ag	345	85,000	1055	Rich yellow
88.4 Au, 9.5 Pt, In, Mn	510	90,000	1050	Yellow
84 Au, 9.9 Pt, 3.3 Pd	600	92,000	1085	Pale yellow
65 Au, 26 Pd, In, Ga	552	120,700	1140	White
40 Au, 40 Pd, 10 Ag, In, Ga, Sn	550	125,000	1120	White

certain elements at the typical weight percentages. Gold provides the biocompatibility and nobility. The yellow color also aids the porcelain esthetics. Platinum and palladium additions provide additional nobility while decreasing the thermal expansion and increasing the solidus temperature. The base elements increase the alloy's strength, improve the casting accuracy, and provide for porcelain adherence via a metal oxide. Table 14.2 shows some typical dental alloy compositions for the PFM technique.

Gold–Platinum Alloys

The strengthening mechanism for the gold–platinum alloys is precipitation hardening [16]. The base elements additions are iron, indium, manganese, and zinc. The alloys are generally yellow in color. A great variety of alloys can be developed by judicious base element additions to a 90 gold–10 platinum alloy. The addition of small quantities of palladium improves the strength and, by raising the solidus temperature, improves the resistance to high temperature creep.

Gold–Palladium Alloys

The addition of 10 wt% palladium renders the alloy white (or silver) in color and shifts the hardening mechanism to solid-solution strengthening. Base element additions are indium, tin, gallium, and zinc. These are the most popular dental alloys because they can be easily modified to suit different porcelains and economic needs.

IMPACT OF COMMODITY PRICES ON ALLOY DEVELOPMENT AND GOLD USAGE

The dental market in the United States shows the most sensitivity to commodity prices while the markets in Germany and Japan show the effect of government-sponsored programs.

UNITED STATES

In 1959, essentially all fixed restorations were made from gold alloys; the average gold content of the alloys was between 60% and 70%. The advent of the PFM technique resulted in ever-increasing demand for gold due to the excellent esthetics of the final restoration. The years 1960 through 1968 saw annual increases of gold usage. The most popular alloy was a yellow (gold-colored) gold–platinum–palladium alloy with a gold content of 87%.

At this time, the economies of the world were tied together by the currency formalization of the Bretton Woods agreement of 1946. This system stabilized currency exchange rates, and because the United States was on the gold standard [17] it effectively fixed the price of gold at US$35 per

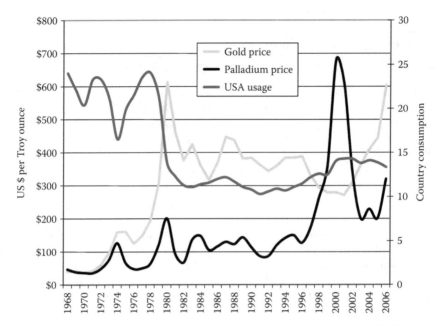

FIGURE 14.8 Gold usage for the United States from 1968 to 2006. (Data from GFMS Limited. With permission.)

Troy ounce. Official government transactions between countries were pegged at this exchange rate and only the U.S. dollar was convertible into gold. By 1968 the system no longer was viable and a two-tier system was adopted. Official transactions were carried out at the $35 rate but the price of gold was allowed to float for commercial transactions. The price of gold rose to $42 per Troy ounce. Nonprecious alloys were introduced as a substitute for gold for dental restorations.

The United States went off the gold standard in 1971, eliminating the convertibility of U.S. dollars to gold. The price of gold was allowed to float, and it rose steadily. Figure 14.8 shows the price of gold and the U.S. consumption of gold for dental restorations from 1968 to 2006. In the 1970s the increasing price of gold resulted in the introduction of white (silver colored) alloys, substituting palladium for gold, thus decreasing the usage of gold. The lowered usage, however, was offset by a growth in the marketplace as more people were being covered by dental insurance. In fact the graph belies how fast the marketplace for dental restorations was growing in the 1970s. Figure 14.9 shows the market share of the three basic alloy types from 1968 to 2001. The growth of the nonprecious segment follows the increasing price of gold. The same is true for the palladium-based alloys.

The appearance of many unproven gold substitutes caused much confusion in the marketplace [18,19]. The ADA attempted to bring some order by developing a classification system for dental alloys. Three categories were established:

- *High noble*—Alloys containing a minimum of 60% gold and platinum group elements with at least 40% being gold
- *Noble*—Alloys containing at least 25% gold and platinum group elements
- *Predominantly base*—Alloys containing less than 25% gold and platinum group elements

The system was adopted by many insurance carriers in the United States and Canada.

Referring again to Figure 14.8, we see that the decrease in the price of gold in the 1980s did not result in an increase in the demand for gold. The market is inelastic. Once a substitute is found to function there is little reason to return to gold. In 1997, however, a singular event occurred that

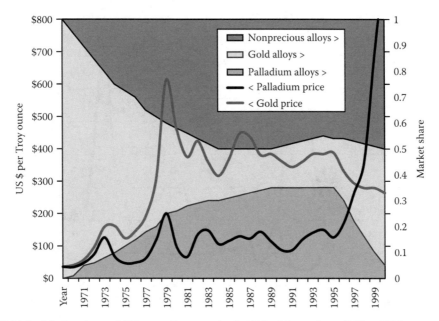

FIGURE 14.9 Market share of different alloy types in the United States from 1968 to 2000.

moved some of the market, however briefly, back to gold. A rapid rise in palladium price, resulting from a physical shortage of the material, increased the demand for gold and reduced the market share of the palladium-based alloys. The increased demand was small and short lived as the price of palladium decreased. As the price of gold rose its use declined once more. By 2002 the use was back to the early 1990s level.

The fragmentation of the alloy marketplace is summarized in Table 14.3. The year of each alloy system and the market reason for the introduction are shown. The alloys are those used with the PFM technique.

In the late 1990s, the introduction of milled zirconia copings resulted in further erosion of the high gold yellow PFM alloy market share. Some believe white zirconia produces a better esthetic result than the yellow golds. As the price of gold rose through 2008 the use of zirconia and other all-ceramic restorations increased.

TABLE 14.3
Alloy Developments for the Porcelain-Fused-to-Metal Technique

Year	Alloy System	Comments
1960	87 Au 6 Pt 6 Pd	Gold colored alloy for porcelain-fused-to-metal
1968	Nickel–Chromium	Nonprecious alloy
1970	52 Au 27 Pd 16 Ag	Silver colored alloy; platinum content eliminated, gold content reduced
1972	60 Pd 28 Ag	Gold content eliminated
1976	52 Au 38 Pd	Silver content eliminated from gold alloy to address porcelain discoloration problems
1982	80 Pd 10 Cu	Silver content eliminated from the palladium alloy to address porcelain discoloration problems
1995	86 Au 10 Pt	Palladium content eliminated to address cost of palladium
2000	65 Au 26 Pd	Gold alloy to address differing porcelain expansions
2004	40 Au 40 Pd 10 Ag	Gold alloy to minimize cost

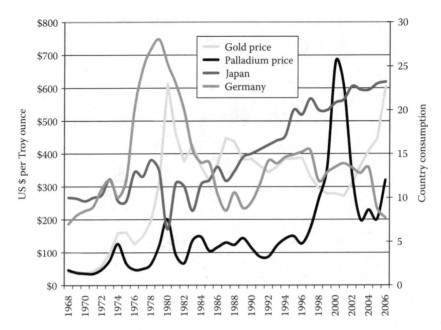

FIGURE 14.10   Gold usage for Germany and Japan from 1968 to 2006. (Data from GFMS Limited. With permission.)

GERMANY

Figure 14.10 shows the demand for gold by German companies for dental restorations with time. The German socialized medicine program of the 1970s initially covered most, if not all, of the cost of a dental restoration. Gold usage rose steadily. The rising cost of gold forced the government to modify the program and it began to require some payment by the patient. Demand for gold fell as the patient was required to pay an increasing share of the cost of the restoration. The German market increasingly used nonprecious alloys as a substitute for gold. An unusual marketing campaign in the 1980s against palladium prevented any significant market share of these materials from developing.

JAPAN

The Japanese use of gold in dental restorations seems to defy the price of gold. Figure 14.10 shows the consumption of gold to increase steadily over time with little relationship to the price of the material. Restrictions on the use of mercury contributed to more gold inlays being fabricated in place of amalgams. In addition, unlike the United States, the Japanese market did not readily adopt the use of nonprecious alloys.

DEVELOPMENT OF STANDARDS, SPECIFICATIONS, AND REGULATIONS IN THE DENTAL INDUSTRY

EARLY ATTEMPTS AT DEVELOPING STANDARDS FOR GOLD ALLOYS

The first specification of dental alloys was published by the American Dental Association (ANSI/ADA Specification No. 5) in the 1930s. This specification defined four types of alloys based on their chemistry and hardness values as shown in Table 14.4. The basis of this specification was a survey

TABLE 14.4
ADA Specification No. 5 (1961 Rev.)

Type	Au Plus Pt Group Minimum	Brinell Hardness Quenched Min	Brinell Hardness Quenched Max	Brinell Hardness Hardened Min	Tensile Strength Hardened Min	Percent Elongation Quenched Min	Percent Elongation Hardened Min	Fusion Temperature Minimum
I	83 wt%	40	75			18		927°C
II	78 wt%	70	100			12		899°C
III	78 wt%	90	140			12		899°C
IV	75 wt%	130		200	6,328 kg/cm²	10	2	871°C

Note: This specification became defunct in December of 2007.
Source: Data from ADA Specifications No. 5 (1961 Rev.)

of the alloys used in the U.S. market for different types of restorations. The dental applications are shown in Table 14.5. No performance standards were established based on scientific studies. The hardness values were used as a surrogate for the strength of the material. This rule of thumb is valid only because the metallurgical system of the different types was the same: gold-based alloys containing silver and copper with small quantities of platinum and palladium. For Types I and II the hardness limitations resulted in compositional restrictions as well. These types of restorations require the dentist to physically move the alloy in order to seal the margin upon seating in the mouth. This can only be accomplished using high gold alloys.

Once accepted by the dental community, amalgam quickly substituted for Type I alloys. The silver-based amalgams have a lower material cost and can be immediately placed by the dentist, eliminating the need for sending an impression to a dental laboratory. Natural tooth colored composites (mixtures of plastic and glass) started to replace amalgams in the 1970s for esthetic reasons. This trend continues today.

Type II alloys remain substantially unchanged. Their use, however, is declining. Increasingly, ceramic materials are used as substitutes to improve esthetics.

Type III alloys are still used today for full cast crowns. This type of alloy has the most metallurgical alternatives because of economic concerns. Initially alloys were developed with ever-lowering gold content and then silver–palladium alloys were introduced having no gold at all. The challenge has been to reduce the cost of the material without sacrificing the functional and fabrication characteristics of the alloy.

Type IV alloys declined in use rapidly as the cobalt–chromium alloys were seen to be functionally superior to the gold alloys for partial denture applications. Apart from their lower cost,

TABLE 14.5
Dental Application by Type

Type	Application
I	Single-surface inlays
II	Multisurface inlays
III	Portions of crowns, full crowns and multiple unit bridges
IV	Thin wall structures like partial dentures

restorations made from the cobalt alloys are lighter in weight, providing more comfort when worn. Also, the cobalt alloys are stronger allowing for thinner castings.

OTHER DENTAL MATERIALS

The increased use of amalgam and cobalt–chromium alloys resulted in ADA specifications for these materials as well. The demonstrated success of the lower gold alloys and the introduction of porcelain-fused to metal restorations, however, did not result in the publication of timely specifications leaving the market with no guidance for almost two decades. The ADA attempted to fill this void in the mid 1970s with the introduction of an Acceptance Program that required clinical studies to demonstrate efficacy and toxicity testing to demonstrate biocompatibility. This program met with limited success because the ADA is not a regulatory body. Manufacturers could voluntarily comply, as most did with the introduction of the previous standards, but because dentists were not obligated to use the products that the ADA certified or accepted, it was difficult for manufacturers to justify the cost of clinical evaluations, and because the products had already been in the marketplace for many years, there was little incentive or perceived need.

BEGINNINGS OF REGULATIONS

In 1976, the U.S. Congress passed the Medical and Dental Devices Act which brought oversight to the dental industry for the first time. The manufacturers were required to comply with Good Manufacturing Practices under the auspices of the Food and Drug Administration. All of the products on the market at the time were automatically grandfathered and considered safe for distribution. No product standards were defined. The International Organization for Standardization, however, was very busy with a number of committees defining new standards for all forms of dental materials.

MODERN REGULATIONS AND STANDARDS

In 1996, the European Union (EU) required all manufacturers to have a quality system compliant with ISO 9000 and demonstrate that the products met the requirements of the Medical Device Directive (for safety) in order to gain the right to carry the CE mark. All dental products sold within the EU are required to have the CE mark and to comply with the appropriate ISO standard if one existed. The EU adopted ISO 13485, a quality system for the medical and dental industries superseding ISO 9000; the FDA accepted the use of ISO 13485.

Effective 2006 the ISO working groups adopted one standard (ISO 22674:2006) to cover all forms of dental prosthetic alloys eliminating compositional requirements. This standard brings consistency in evaluating metallic materials for dental devices. Six types were established defined by the yield strength and elastic modulus of the materials: Type 0 covers the electroformed and sintered materials. These materials had no specification up until this point. Types 1 through 4 are defined by their application requiring minimum yield strength and elongation values. Type 5 covers essentially the alloys meant for partial dentures and requires a minimum yield strength and elastic modulus.

The ADA terminated the certification and acceptance programs for dental alloys in 2007.

OTHER COUNTRIES

Many countries initially adopted specifications for dental alloys similar, if not identical, to ADA Specification No. 5 and, with time, have changed to systems similar to the FDA or the European Union.

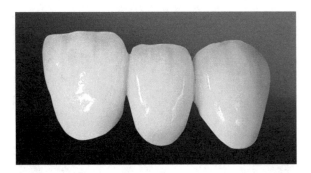

FIGURE 14.11   (See color insert following page 212.) Goldtech Bio2000 three unit bridge. (From The Argen Corporation. With permission.)

THE FUTURE OF GOLD IN DENTISTRY

TRADITIONAL CASTING ALLOYS

As discussed previously, the anterior portion of the mouth is a low stress area. This allows for the use of very high gold alloys. GOLDTECH BIO 2000® is a 99.7% gold PFM alloy that was designed to take advantage of the exceptional esthetics of pure gold and its unsurpassed bio-compatibility. The alloy is fabricated using traditional dental laboratory techniques providing a reliable and assured fit. The unique formulation (Cascone U.S. Patent 5,922,276) results in excellent adhesion and compatibility to all porcelains. The finished restoration exhibits exceptional aesthetics. The deep rich yellow color of the alloy eliminates any shadows at the gingival and graying of the porcelain shades producing superior vitality. Unlike other high gold materials GOLDTECH BIO 2000 can be cast into bridgework in one step. Figures 14.11 and 14.12 show typical results, and many industry professionals consider GOLDTECH BIO 2000 to be an excellent alternative to metal-free restorations.

ELECTROFORMING

The process for fabricating dental copings using electroforming techniques has been discussed by many authors [20,21]. The technique produces excellent results both functionally and esthetically [22]. Figure 14.13 shows a typical porcelain crown, while Figure 14.14 shows a telescopic crown. Telescopic crowns are used to connect two restorations.

FIGURE 14.12   (See color insert following page 212.) Obverse of Goldtech Bio2000 three unit bridge. (From The Argen Corporation. With permission.)

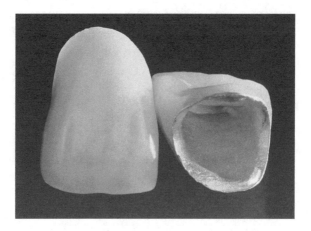

FIGURE 14.13   **(See color insert following page 212.)** Crowns made from electroplated gold. (From Wieland Dental + Technik GmbH & Co. With permission.)

SINTERING

Captek is a unique composite metal system for esthetic crowns and bridges. The Captek material and lab process is designed to combine the benefits of high purity gold and the strength of platinum and palladium in a thin (less than 0.3 mm) nonoxidizing, warm yellow coping (or bridge understructure). Capillary attraction is used to draw the high purity gold in and through a skeleton of predominately platinum/palladium particles (Figure 14.15). The resulting composite metal microstructure is also referred to as an "internally reinforced gold," strong, well fitting, and clinically proven to repel harmful plaque and bacteria. The excellent results are shown in Figure 14.16.

3D PRINTING

The digital manufacturing process popularly known as 3D printing originated at MIT in the early 1990s. Today, it is one of the most widely used solid freeform fabricating (or additive) technologies,

FIGURE 14.14   A telescopic crown fabricated from electroplated gold. (From Wieland Dental + Technik GmbH & Co. With permission.)

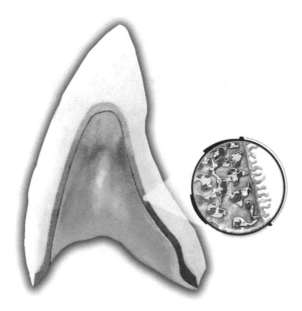

FIGURE 14.15 A schematic representation of the CAPTEK® composite structure. (From Precious Chemicals, Ltd. With permission.)

transforming the powder form of polymers, ceramics, and metals into solid and semisolid objects, as defined by an electronic solid model file. The shapes that can be formed by digital manufacturing processes are not constrained by the geometric limitations inherent in conventional shape-making processes such as extrusion, molding, or machining.

The process that produces these 3D structures begins with an electronic, three-dimensional solid model file, generally called a CAD model. The 3D printer uses an inkjet printhead to deposit controlled droplets of specially formulated binder onto thin layers of powdered material. The binder forms the desired part, which is built in a multilayer process, one "digital" slice at a time. Once the printing is complete, the shapes are removed from the unbound, loose powder and chemically or mechanically bonded or solidified.

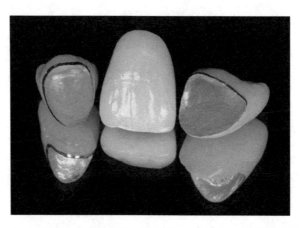

FIGURE 14.16 **(See color insert following page 212.)** Crowns fabricated from CAPTEK®. (From Precious Chemicals, Ltd. With permission.)

FIGURE 14.17 **(See color insert following page 212.)** Sintered crowns fabricated by 3D printing. (Image courtesy of imagen. With permission.)

The Ex One Company, through its ProMetal®, ProMetal RCT™, and imagen® businesses, utilizes the printing process to produce molds, tools, and other objects from metal and ceramic powders. In 2005, imagen was formed to focus on the production of components used in the dental and medical industries. Dental products, such as copings for use in crowns and bridges, are now formed through the printing of high noble gold alloys and other metal powders. This process begins through the use of 3D scanning equipment that can create a "digital impression" of a tooth structure. The image, converted into a solid model of the coping, is then printed on a 3D printer, sintered (Figure 14.17) and solidified through a thermal process (Figure 14.18), ready for porcelain.

OTHER UNIQUE OPPORTUNITIES FOR GOLD

As was stated in the beginning of this chapter, the unique properties of gold continually bring the material to the forefront for use in new technologies. The newest area of technology is the

FIGURE 14.18 **(See color insert following page 212.)** Crowns infiltrated with gold ready for porcelain. (From imagen. With permission.)

investigation of the properties of materials of extremely small size: nanoparticles. The melting point of gold nanoparticles is dependent on the particle size; the smaller the particle, the lower the melting point. Researchers [23] have utilized this property of gold nanoparticles to reduce the sensitivity of teeth. People with sensitive teeth have dentin tubules that are larger and more numerous than those who do not show sensitivity. Gold nanoparticles were brushed into the open tubules and then irradiated. The gold particles fused and blocked the tubules. This development fulfills one prediction of the potential possibilities of nanodentistry [24].

SUMMARY

The last century has seen the usage of gold in dentistry grow in the industrialized countries. Even as substitutes reduced gold's market share in one technology, another technology adopted its use. As dental care increases throughout the world gold will continue to play an important role in dental health.

ACKNOWLEDGMENTS

Vitallium® is a registered trademark of Howmedica Osteonics Corp., 325 Corporate Dr., Mahwah, NJ 07430.
GOLDTECH BIO 2000® is a registered trademark of The Argen Corporation, San Diego, CA.
Captek® is a registered trademark of Precious Chemicals, Ltd. ATIDIM Industrial Park, Tel Aviv, Israel.

REFERENCES

1. The use of gold in dentistry, J. A. Donaldson, *Gold Bull.*, 1980, **13**, 117.
2. The use of gold in dentistry – Part II, J. A. Donaldson, *Gold Bull.*, 1980, **13**, 160.
3. Oral golds, D. Brown, *Gold Bull.,* 1988, **21**, 24.
4. Gold in dentistry: Alloys, uses and performance, H. Knosp, R. J. Holliday, and C. W. Corti, *Gold Bull.*, 2003, **36**, 93.
5. *Mosby's Dental Dictionary*, Thomas J. Zwemer, 1998. St. Louis: Mosby, Inc.
6. *Dentistry: An Illustrated History*, Malvin E. Ring, 1992. New York: Harry N. Abrams, Inc.
7. A gold dental prosthesis of Roman Imperial Age, S. Minozzi, G. Fornaciari, S. Musco, and P. Catalano, *Am. J. Med.*, 2007, **120**, e1-e2.
8. The long history of lost wax casting, L. B. Hunt, *Gold Bull.*, 1980, **13**, 63.
9. *International Workshop Biocompatibility, Toxicity and Hypersensitivity to Alloy Systems Used in Dentistry*, B. R. Lang, H. F. Morris, and M. E. Razzoog, eds., 1986. Ann Arbor: The University of Michigan School of Dentistry.
10. Study of galvanic corrosion between dental alloys, M. H. Zürcher, in *Biocompatibility, Allergies and Resistance to Corrosion: A Global Scientific Approach*, 1993, 43–53, Metalor, Neuchâtel Switzerland.
11. Galvanic corrosion behavior of titanium implants coupled to dental alloys, M. Cortada, et al., *J. Mater. Sci. Mater. Med.*, 2000, **11**, 287.
12. Galvanic corrosion behavior of implant suprastructure dental alloys, N. M. Taher, Al Jabob, A. S., *Dent. Mater.*, 2003, Jan. **19**(1):54–59.
13. Tarnish and corrosion behaviour of dental gold alloys, L. W. Laub and J. W. Stanford, *Gold Bull.*, 1981, **14**, 13.
14. Chemical stability of gold dental alloys, D. J. L. Treacy and R. M. German, *Gold Bull.*, 1984, **17**, 46.
15. *Phillips' Science of Dental Materials*, ed. Anusavice, 2003. 11th Edition. St. Louis: Saunders.
16. Gold alloys for porcelain-fused-to-metal dental restorations, R. German, *Gold Bull.,* 1980, **13**, 57.
17. *The Library of Economics and Liberty,* M. D. Bordo, 2002. Gold Standard. http://econlib.org/library/Enc/GoldStandard.html.

18. Casting metals in dentistry: Past-present-future, K. Asgar, *Adv. Dent. Res.*, 1988, **2**(1), 33–43.
19. An evaluation of casting alloys used for restorative procedures, K. F. Leinfelder, *JADA* 1997, **128**, 37.
20. Electroforming technique, A. J. Raigrodski, C. Malcamp, and W. A. Rogers, *J. Dent. Technol.*, 1998, **15**, 13.
21. Gold electroforming system: GES restorations, F. Behrend, *J. Dent Technol.*, 1997, **14**, 31–37.
22. Treatment behavior and complete-mouth rehabilitation using AGC crowns: A case report, J. Dölger, C. Gadau, and R. Rathmer, *Int. J. Peridontics Restorative Dent.*, 2001, **21**, 373.
23. Filling in dentinal tubules, M-H. Liu, C-H. Chan, J-H. Ling, and C. R. C. Wang, *Nanotechnology*, 2007, **18**, 475104 (6pp).
24. Nanodentistry, R. A. Freitas, *JADA*, 2000, **131** (Nov.), 1559–1565.

15 Decorative Gold Materials

Peter T. Bishop and Patsy A. Sutton

CONTENTS

Introduction.. 317
Decorative Gold Films 319
    Overview 319
    History of Gold Film Decoration 321
    Chemistry of Decorative Gold Formulations 326
        Gold Compounds.. 326
        Gold Nanoparticles.. 340
        Burnish Gold Formulations 341
        Self-burnish Gold Formulations 343
        Microwave-resistant Gold Formulations 344
        High-firing Gold Formulations.. 344
    Influence of Minor Metals on Film Properties and Structure 345
    Application Techniques.. 348
        Brushing 348
        Machine Banding 348
        Screen Printing 348
        Pad Printing 350
        Ink-jet Printing 350
Purple of Cassius Gold Enamels 350
    Overview 350
    History.. 351
    Color Generation of Colloidal Gold.. 353
    Preparation of Purple of Cassius 354
Gold Leaf 357
    Overview 357
    History.. 358
    Gilding Techniques 361
    Gold Beating 362
Conclusions.. 363
References... 363

INTRODUCTION

Of all the precious metal elements, gold has been used the most extensively for decoration onto porcelain, pottery and glass. This is either to give metallic gold films or colloidal pink colors, with a current estimated annual usage of 10 tonnes for just decorative applications. These applications have a long history, with gold leaf first applied to ceramics in ancient China. In the 1700s, chemically reduced gold powder was used. Then in 1827, a completely new type of thin bright gold film formed from a liquid gold was developed for application to porcelain and glassware. This discovery enabled

(a)

(b)

FIGURE 15.1   (See color insert following page 212.) (a) Depicts a bowl having bright gold bands carried out by hand gilding for the handles, and machine banding where the gold is aligned with the color. (b) Depicts a combination of burnish and raised matt gold effects with multicolored enamel decoration. Application is by a combination of decal and machine banding.

the industry to develop both technically and commercially. Historically, two types of decoration were used, namely, ground coats, which are fired after application onto unglazed and unfired ware, and overcoats, which are applied to a glazed surface of ware and then fired, usually at relatively low temperatures. The latter give rise to lustrous metallic gold effects and are the dominant decorative effect today (see Figures 15.1 and 15.2). Also seen in Figures 15.1 and 15.2 is the purple-to-red colloidal gold effect, commonly known as Purple of Cassius, seen in Figure 15.1 as rose colors, and in the glass flute as a luster or stain. Purple of Cassius is routinely used for red- and purple-colored decoration on ceramic ware.

Detailed reviews for the reader on the application technology, materials used [1,2], and history [3] are available. The scope of this chapter is to introduce the materials and application technologies associated with decorative gold materials, highlighting their origin, history, and chemistry. It is aimed at the reader with a scientific background who is curious about the development of this technology, and who may wish to pursue this area further.

FIGURE 15.2   (See color insert following page 212.) An ornately decorated flute glass containing bright-banded gold and raised matte gold on the thick filmed area.

DECORATIVE GOLD FILMS

OVERVIEW

Two general classes of gold compositions exist for the thermal application of gold to porcelain, ceramic, and glassware; these are termed liquid bright golds and burnish golds. The key gold-containing building block for liquid bright gold formulations is traditionally a gold-organo-sulfur compound in terpene-type solvents. This can be between 4 and 25 wt% gold concentration, and is a brown to black viscous liquid often having a characteristic odor of natural terpene solvents such as pine oil. From such solutions, thin gold films can be obtained between 0.1 and 1 µm in thickness. The basic synthesis, invented by H. G. Kühn (1827) in Germany [4], involves reacting elemental sulfur with oils (such as lavender) to produce polysulfide solutions, referred to as balsam sulfide. The sulfided oil is mixed with chloroauric acid, and this produces the so-called terpene gold sulfide. This is thought to be a polymeric gold-mercaptide species, also referred to as gold resinate or gold sulforesinate. As will be seen later, well-defined gold(I) mercaptides can be used as alterative gold sources; very recently gold nanoparticles stabilized with N and S donors have been employed.

Because the early gold precursors were soluble in terpenes and compatible with natural resinous materials (such as asphalt and colophony), the viscosity could be well controlled. This allowed for application onto porcelain and glass using methods like brushing (gilding), spraying, screen-printing, or spinning. After the applied gold film has gone through a mild drying process to remove solvent, it can be fired. A gradual decomposition takes place up to 300°C before metallization occurs between 450 and 750°C. In the early days, the adhesive strength of the gold film was poor and, above 500°C, agglomeration of the gold film to give an off-bright appearance often resulted. A solution to both problems was to introduce "minor metal" additions. The addition of resinates of bismuth, chromium, cobalt, and vanadium and many other elements (often in complicated combinations) gave enhanced adhesion. Addition of rhodium resinate prevented the agglomeration of gold at the higher fire temperatures by producing Rh_2O_3. This allowed for bright reflective gold films at temperatures above 500°C. Even today, these formulations still exist commercially. Other precious metals can be added to these formulations to influence the fired film color. Silver is commonly used to lighten the gold color to obtain a more yellow or lemon effect. Addition of palladium or platinum turns the golden film into a gray metallic luster. This color tone is very effective in enhancing or complementing dark enamel co-decorations, such as deep cobalt blue pigments.

The fired, as-purchased gold film is at least 22 karat and has a thickness typically between 50 and 125 nm, as thicker films can result in film blemishes giving way to off-bright effects. The bright commercial formulations are often extremely complicated resulting in anywhere between 10 and 100 different raw materials! In part, this is caused by the more traditional gold formulations containing many natural products, such as terpenes or gums in small quantities. Simplified gold formulations are now commercially available, and these have synthetic resins and pure solvent systems rather than naturally occurring materials.

The other class of gold decoration is commonly referred to as burnish gold. The gold is in the form of a powder traditionally made by fragmenting or grinding gold foil, or more recently by chemical precipitation (Figure 15.3). The gold powder can be fired onto porcelain or pottery with the addition of bismuth fluxes and other minor metals that influence color and durability. The fired films are matte in appearance and must be polished with a mild abrasion instrument to produce a lustrous effect, hence the term burnish gold. The skill is not to remove gold but to flatten the textured gold surface into smooth domains. This gives the lustrous metallic gold look that is neither mirror reflective nor dull (Figure 15.2 best illustrates this effect). The combination of flat reflective sections and textured segments gives rise to multiple reflective surfaces, and these can be observed under an optical microscope. An extension to the burnish gold appearance is the so-called self-burnish or semimatte gold finish. This is achieved without the need for mechanical treatment and is observed straight after the firing process. Ideally

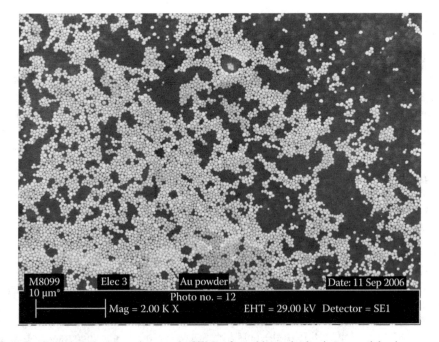

FIGURE 15.3 Scanning electron micrograph (SEM) of a gold powder having a particle-size range of 0.2–0.5 μm. This is a precipitated gold powder made in the Johnson Matthey laboratories and is used as a building block in burnish and semimatte gold decoration. Typical concentration of powder in commercial inks varies between 10 to 22 wt%.

a homogeneous matte appearance is required and is achieved by adding mica-type pigments to the bright gold phase, or indeed a variety of thermally stable flat or thin pigments or compounds. The main driving force for this alternative gold decorative appearance is the desire to remove labor and equipment costs associated with physical burnishing, and also the desire to produce homogeneous matte films, which can be a problem when mechanical treatment is used. Typical film thickness ranges for burnish golds are between 0.1 and 2 µm; for self-burnish the range is 0.1–0.5 µm. The former gives rise to lustrous golds that are often used on expensive top-of-the-range ware or collector plates.

As discussed, the major use of liquid golds is for the decoration of ceramic and glassware. As already demonstrated, decorations can range from simple bands on the edges of ceramic ware to lavish designs on glassware (as seen in Figures 15.1 and 15.2). Gold is commonly used in conjunction with ceramic colors (sometimes up to six or seven colors) to produce multicolored effects. It can be applied over a ceramic color or on top of raised white enamels to give a frosted finish. Glass and ceramic packaging uses gold decoration, for example, in the cosmetics and liquor industry with decoration of bottles. The flat glass industry is also an area in which gold is used successfully on panels, clock faces, furniture glass, and giftware. Gold is used on architectural features such as ceramic tiles, glazed brick, enamelled steel, and structural glass.

The major manufacturers of these wares maintain large decoration warehouses as gold film decoration is still labor intensive. A trend over the last 20 years has seen manufacturers move away from European locations to the Far East where labor expenses are significantly lower. This has resulted in traditional tableware manufacturing centers, such as the United Kingdom's Stoke-on-Trent, becoming less dominant forces than they once were 30 years ago.

Liquid golds also have functional uses [1,5]. The property of gold as an efficient reflector of *infra-red* radiation is seen for the thin golds produced from liquid golds. The effectiveness stems from gold's ability to reflect heat in the wavelength range of 0.2–15 µm and its very low emissivity value of 0.05. Liquid golds have been successfully applied to aluminum, magnesium, and titanium for heat-reflective coatings. Thin, lightweight gold films have been used to reduce heat transmission on aircraft engine shrouds, drag parachute containers, tail cone assemblies, blast shields, ducts, and tubing, and for protection of heat-sensitive parts in space equipment from solar radiation.

The economics of gold decoration suffers from some misconceptions as it is thought to be expensive and to add excessive cost to the decorated article, even if the value of the ware is considerably enhanced in most instances. Only when the mathematics is looked at is the true cost realized. The fired film is essentially 22 karat gold with an average thickness of 125 nm. For 100 g of liquid gold containing 10 g of gold metal, a surface area of 5 m^2 can be covered. The price of liquid gold formulations depends on metal price, which during 2008 saw record highs. Nevertheless, the gold does go a long way and will be a minor fraction of the overall price structure of a single article or ware.

HISTORY OF GOLD FILM DECORATION

The gold decoration of porcelain and bone china, such as edging on plates or placing intricate patterns or pictures onto vases and ornaments, has been a common and accepted decorative feature of the ceramic industry. It has been nearly 300 years since the introduction of chemical methods for the preparation of gold powder and its successful application onto porcelain. This provided a superior replacement for the older decorative method of gold leaf applied over a layer of linseed oil and litharge. These older methods originated from China, and texts that describe these ancient applications are available. As far back as 180 years ago, early liquid gold preparations suitable for gilding were introduced. A great deal of secrecy over recipes was common, giving the early porcelain makers' advantages over their competitors. A summary of the history of gold film decoration up to present-day developments follows.

FIGURE 15.4   Johannes Friedrich Böttger (1685–1719). The old secrets of Chinese porcelain were redis-
covered by Böttger. He produced the first unglazed porcelain in 1709 and in the following year the Meissen
factory was established. He prepared some of the first gold powders for use in decoration but died when only
34 and was not able to develop his work.

One of the earliest recorded pioneers of gold decoration was Johannes Friedrich Böttger (1685–
1719) (Figure 15.4). Böttger was a young apothecary (pharmacist) and alchemist who fled from the
court of King Friedrick I of Prussia and was then commissioned in Dresden by Augustus the Strong,
Elector of Saxony, who was a keen collector of china. His new master instructed Böttger to attempt
the production of porcelain in around 1707 and this is where the history begins. After several failed
attempts, Böttger finally built a kiln in 1709 that would reach high enough temperatures to fire his
ware. Once he started to produce porcelain (the Meissen factory was established the following year),
his attentions turned to decoration, and he experimented with the ancient Chinese technology of
using gold leaf. Unfortunately, these articles were not durable enough as they were not fired and virtually no examples have survived to the present day. However, Böttger's early experience gave him
the ideas that would result in early versions of liquid golds that could be gilded onto porcelain.

Born in Schless in Saxony, Böttger was apprenticed at the age of 12 years to an apothecary
named Zorn from Berlin. While in this household, he met with Johann Kunckel, an alchemist
and author of several texts on advanced chemistry during the seventeenth century. Kunckel had
published in 1679 a textbook *Ars Virtaria Experimentalis oder vollkommene Glasmacher-Kuns*t.
The text included a section on the gilding and enamelling of glass. A variety of processes were
described including simply sticking on gold leaf and also heating gold leaf-decorated porcelain over
hot coals to make it more durable. Böttger was aware of this book, and it is assumed that he used
this for some of the initial experiments in Saxony in his early porcelain decorative attempts. Some 7
years after his initial experiments, Böttger came into possession of another work by Kunckel called
Laboratoria Chymica. This book motivated Böttger to find a solution to his gilding problem, for in
that book was a section on gold powder preparation detailing how the addition of ferrous sulphate
to gold in aqua regia could produce "a quite beautiful and very fine precipitate." This form of gold
powder enabled Böttger and his successors at Meissen to produce simple liquid gold preparations in
a reproducible manner for gilding onto porcelain. Unfortunately, Böttger died not long after this at

FIGURE 15.5   Picture of Johannes Gregorius Höroldt (1696–1775). In 1720, Höroldt went to the Meissen factory where his talents allowed him to produce a complete palette of enamel colors and to devise a reliable method of gilding that was used for many years.

the age of 34 before his key discovery could be developed. It is said he died from his exertions and hardships during his studies; he was often under significant duress.

The influential figure of Johannes Gregorius Höroldt (1696–1775), (Figure 15.5) developed this work initially by expanding on Böttger's work. It was not until 1723 that a reliable, durable gilding process became reality in the Meissen factory, some years after Böttger's death. Höroldt had arrived from Vienna and worked in the Meissen factory. Expanding on Böttger's work, he set about producing a range of colors, including gold, that eventually became a dominant feature in European porcelain decoration. The Meissen hierarchy instructed Höroldt to document his procedures, and the small parchment notebook used for this task is still within the Meissen factory, dated December 24, 1731. The section on gilding, Chapter 12, consisted of ten pages and is titled, "How Gold Is to Be Dissolved and Precipitated." It describes how the gold would be fired first and then burnished with flint. The next step was to apply the colors and then refire, which caused further dulling of the already burnished gold. However, reburnishing the gold was then carried out to yield "even more beautiful" gold decoration, according to Höroldt's notes. This technique formed the basis for gilding at Meissen, and more elaborate patterns were designed and used until around 1740 when the Seven Years' War ended the domination of Meissen.

A very different approach was taken in the porcelain factories in France, where the precipitation technology was absent. The earliest example of recorded French gold decoration was in 1695 in a book by Dr. Martin Lister, later physician to Queen Anne. On his visit to Paris, he is quoted as reporting "I saw the pottery of St Clou with which I was marvellously well pleased…. They had arrived at burning on gold in neat Chequer Works." Essentially this was applying gold leaf with a hot iron as used by bookbinders. The biggest technological advancements in gold gilding came from the Vincennes porcelain works, founded in 1738. Francois Gravant, a founding figure in the Vincennes factory, secured a recipe for gilding. This described the use of ground/powderized gold leaf in starch and egg white, which was washed and dispersed into a mixture of fine sand and litharge before being gilded and fired onto porcelain. This was later superseded by a new composition purchased by the Benedictine monk, Brother Hypolite, around 1748 for the sum of three thousand livres. This formulation later became known as the "Monk's secret recipe."

However, the Vincennes factory soon suffered severe financial difficulties, resulting in King Louis investing heavily to save the factory. In 1751, a new appointment was made in the form of Jean Hellot (1685–1766), a chemist, and the factory relocated to Sèvres in 1756. Hard-paste porcelain was first made at Sèvres in 1772, and its chemistry is attributed to the famous chemist Pierre-Joseph Macquer (1718–1784) who had joined Hellot as his assistant in 1757. Because of the introduction of hard porcelain glaze, a new problem was to find a suitable decorative gold formulation to sit on the topside (airside). The new director of Sèvres, Melchior-Francois Parent, appointed in 1772, devised a method for producing fine gold powder. This involved forming an amalgam of gold-mercury by dissolving gold into excess mercury then carefully back evaporating free mercury to leave the amalgam product. The material was then dissolved into nitric acid to remove most of the mercury as a nitrate salt, leaving behind fine gold powder. Not unsurprisingly, and even in this era, there were concerns about the health and safety of workers using such a procedure. Even so, the fine gold powder resulting from the process gave a very pleasing gold film when applied and fired. However, despite this, factories and workers still favored the less hazardous method of fine-grinding gold leaf.

A further modification was then made, which involved precipitating a gold(III) solution with mercury nitrate. A finer gold powder was produced with less mercury present; however, the traces that remained contributed to the superior adhesion of the gold to the hard porcelain. It is widely thought (but not proven conclusively) that this modified gold powder preparation was first described around 1781 by Jean-Jaques Bachelier, who was a protégé of Madame de Pompador, the artistic director at Sèvres for over 40 years, and this new method was adopted in the mid 1780s at Sèvres. The gold powder was typically washed with turpentine, and both bismuth oxide and borax were added to further aid adhesion to the porcelain.

The Sèvres factory was the leader in the use of hard-paste porcelain in France, and others soon followed, several of them under the patronage of nobility keen to associate themselves with porcelain manufacture. One of the original factories was Limoges, which initially followed the early Sèvres procedure of mixing in gold leaf and honey and applying with borax and water gum. They then used ferrous sulfate-precipitated gold and, during the mid-nineteenth century, moved onto the mercury nitrate method for gold powder production.

Meanwhile, the United Kingdom was also discovering gilding technologies. Josiah Wedgwood founded the famous factories of his name in Staffordshire in 1759 and was soon producing earthenware, both for practical and decorative use. He was given his first assignment by royalty, namely Queen Charlotte, the wife of King Louis III, for a complete set of "tea things," which had a mixture of gold background and raised green flowers. After several months of in-house experimentation looking at techniques to "burn gold" onto porcelain, Wedgwood came up with the production of fine gold powders. His *Commonplace Book* describes one of the recipes as "Gilding on Porcelain-10 parts of Gold powder and 1 of ceruse-with gum water-Burnt in then polish with agate rubbing always one way."

The porcelain factory with the longest history in England is the Worcester Royal Porcelain Company, founded in 1751. They followed similar trends and used gold leaf and honey in their initial workings but combined it with a pre-made flux, which was ground into the gold to produce decoration that was thick and lustrous. Interaction with the Paris factories influenced their gilding, and they moved to the gold powder preparation methods of ferrous sulfate or mercury nitrate as reducing agents to produce the fine gold powders discussed above. The mercury-containing gold powder, combined with turpentine and fluxes to aid adhesion when fired at temperatures 700 to 900°C, was the most common form of decoration in the Staffordshire factories until the late-nineteenth century.

The development of the first true "liquid gold" formulations was largely driven by the need to produce cheaper decorative gold products. At the start of the nineteenth century, times were

FIGURE 15.6 Heinrich Gottlob Kühn (d. 1870). The above is a commemorative plaque made at Meissen depicting Kühn as a young man. Trained as a metallurgist he came to the factory in 1814 and became director in 1849 until his death in 1870. He developed a successful liquid gold preparation using soluble gold materials to give bright film. They were not very durable but formed the basis for modern bright gold decoration.

economically very tough and the Meissen factory had only just reopened after a period of closure. Hence, economy was a high priority, and a new and cheaper method for gold gilding needed to be found. Heinrich Gottlob Kühn (Figure 15.6) was assigned this task, and he finally succeeded in 1827 in devising a new gold formulation that, unlike other gilding media, was entirely soluble, that is, not gold powder but a "liquid" form of gold. This was produced by taking a solution of gold chloride in "balsam of sulfur," an oily substance, which in itself was prepared by taking sulfur and reacting it with turpentine.

In fact, this liquid gold form had been known for sometime and is described by Neri in *L'Arte Vetraria* of 1612 and by many other writers of handbooks and encyclopedias in later years. Indeed, the Staffordshire potters such as Spode and Wedgwood describe using such materials to give thin gold lusters around 1812. A full and comprehensive preparation appeared in 1824 published by the widow of Thomas Lakin, herself a partner in Lakin and Poole in Staffordshire. In this process, labeled "102," a recipe is described that takes gold and dissolves it into aqua regia and then reacts this with balsam of sulfur. This produced a pink or purple coloration, which is in fact colloidal gold. It can be assumed that Kühn of Meissen was familiar with this work and modified the recipe by effectively concentrating up the gold from colloidal (micromolar) to percent concentrations. This had the effect of depositing enough gold to give a coherent fired bright gold film and appears to be the birth of modern day liquid gold compositions. The gold was named *Glanzgold* in German and, as forementioned, "liquid bright gold" in English.

The now-secret liquid bright gold formulations produced by Kühn (Meissen) leaked out first to Paris, where a French patent was secured by two brothers (Dutertre, French patent No. 5336 dated 1850). The Dutertre workshop was said to be using 4 kg of gold per month and employed about 500 workers. This seems to be the earliest evidence of mass-produced gilded bright gold decoration onto porcelain. Other workers in Europe started to adapt these early formulations. For example, a professor of mathematics at Passau, named Bergeat, investigated the Kühn recipes and improved them to the extent that modified compositions could be cofired with colored enamels and had exceptionally good resistance to high fire temperatures. Within Bergeat's product, small amounts of bismuth, chromium, and cobalt were present to aid adhesion to the ware.

Meanwhile in England, George Matthey (cofounder of Johnson Matthey) provided the Minton factory with a number of samples of liquid gold preparations in around 1878, albeit with apparently very little success. He later found out that a key precious metal was missing from his liquid gold formulations. After Matthey, the Minton business turned to the Roessler brothers, Hector and Heinrich, with their company "die Deutshe Gold und Silber Scheide-Anstalt" established in Frankfurt in 1873 (later to become Degussa). During their work, they were offered a recipe devised by Dr. G. C. Wittstein, a private tutor in chemistry in Munich. Although Heinrich Roessler had very little success with this recipe, his curiosity was aroused and he decided to analyze other recipes, such as the Passau gold of Bergeat and the Nürnberg gold from Leuchs (now well-known products). In Bergeat's product, small amounts of bismuth, chromium, and cobalt were detected, designed to assist in the adhesion of the gold film to ceramic, but nothing similar could be found in the superior gold from Leuchs. By chance, on a visit to Johnson Matthey in London, Heinrich mentioned his difficulties in competing with the Nürnberg firm, only to be told that Johnson Matthey were themselves puzzled by the purchase of small amounts of rhodium chloride by Leuchs. This gave the Roessler brothers the information they needed. Thus they, too, added rhodium to their products and recipes, and by 1879 were producing reliable liquid golds. These minor metal recipes still form the basis of modern liquid golds in the twenty-first century, and it is interesting to note it was a European collaboration that produced these modern-day formulations.

This was the first entry into the gilding of porcelain and pottery by an established refiner and fabricator of gold. They were now equipped with a comprehensive knowledge of the chemistry of gold and in 1885 arrangements were made with both Bergeat and Leuchs for the Roessler brothers to take over their products and processes. The Roessler brothers (Degussa) used all the recipes and, more importantly, solved the problems of formulation stability and so they became the main supplier of liquid golds in 1885, operating from plants in Frankfurt and New York.

Commercial production of liquid golds had by now been started by other refiners of gold in Germany, and refiners in Holland, England, and France soon embarked on similar activities. The bright golds expanded into burnish gold formulations by adding percentage amounts of gold powder into the formulations. This resulted in a much thicker deposit and a matte finish that could be burnished. The basic composition of liquid golds was now well and truly established. A significant amount of research has been carried out over the last 100 years and has resulted in an extensive range of products and formulations varying in gold content, color, and viscosity for a wide range of hand and machine-type applications. The use of synthetic resins, control of gold particle size, and utilization of environmentally friendly solvent systems has further evolved the product range. The last 10 years has seen advances in new gold materials by utilizing preformed nanoparticles, and the use of new application techniques, such as inkjet printing, all of which are discussed in subsequent sections.

CHEMISTRY OF DECORATIVE GOLD FORMULATIONS

Gold Compounds

Gold Sulforesinates

As described in the previous section, the major breakthrough in the quality of bright gold decoration was made by the development of "liquid gold" formulations by Heinrich Gottlob Kühn, with his discovery that "sulfur balsam" could be reacted with gold(III) chloride solution. This resulted in soluble gold sulforesinates that metallized on heating to form bright, reflective gold films. Due to the strict secrecy of the "liquid gold" production processes, little synthetic information was published. A detailed description of the sulforesinate process was first reported by Chemnitius [6] in

1927, outlining the reaction between turpentine oil, Venice turpentine, and sulfur at 165°C to form a sulfur balsam, which was then reacted with potassium tetrachloraurate to form a gold sulforesinate. This was diluted with natural oils such as lavender, rosemary, and pine, and natural resins such as asphalt and colophony were added to adjust the viscosity of the formulation. He also reported the importance of the minor amounts of rhodium, bismuth, and chromium metal oxides to film quality and adhesion. Despite many further modifications [7–13] over the last 80 years with respect to the terpenes, solvents, resins, and minor metal additions, the basic formulation described by Chemnitius is still used commercially in some present liquid gold formulations. This is evidenced by recent patent activity in this area with Schulz et al. [12] reporting the use of specific resin combinations of polyamides, sulfurized dammar, modified rosin, alkylphenol, and other resins within printing formulations to produce high-quality bright gold films with improved dishwasher resistance. Lukas and Kühn [13] report advantages of using indium as minor metal addition over the traditional rhodium-containing formulations.

Despite their commercial success, a detailed understanding of the gold sulforesinate formulations has been hindered by the complexity of the formulations, the use of natural products, and the unknown chemical nature and structure of the gold species formed on reaction between gold chloride and sulfur balsams. Many investigations into the gold sulfur chemistry have been carried out and have been summarized by Henning et al. [14]. Results were inconclusive, but it is thought that reactions between gold(III) and thioether, sulfide, thioketone, and mercaptan groups may all possibly be involved [15].

The traditional gold sulforesinate liquid gold technology has many commercial disadvantages, such as poor yield, limited shelf-life stability (gold powder precipitation), and significant issues with product reproducibility because of the use of mainly natural raw materials and poorly characterized gold precursors. In light of this, alternative gold sources were sought by the industry.

Gold Mercaptide Systems

Overview The first patent relating to the formation of gold(I) mercaptides for use in liquid golds was published by Ballard in 1947 [16]. This disclosed the reaction of auric chloride solution with isolated cyclic terpene mercaptans (most specifically pinene mercaptans) to form gold mercaptides. These could be used instead of gold sulforesinates to produce light-colored, stable liquid gold formulations. A closely related gold mercaptide prepared from thioborneol [17] was reported in 1935, but no reference was made to its use for decorative applications. More recently, the synthesis of gold bornylmercaptide and its use in compositions for integrated microcircuit production has been reported [18].

During the 1960s, a series of patents was filed by Fitch and co-workers advocating the use of primary [19], secondary [20], tertiary [21] and aryl [22] gold(I) mercaptides as the gold source within organic liquid gold formulations. This work demonstrated the huge potential of gold mercaptides within decorative applications because of the almost limitless ability to vary the R group in the "AuSR" compounds, enabling control over important properties such as solubility in a wide range of solvents. A more recent patent discloses the use of a new class of odorless gold(I) mercaptocarboxylic acid esters that show significant advantages with respect to drying time of formulations compared to traditional products [23].

The growing drive in the 1990s within the formulation industry toward the use of environmentally acceptable resins [such as polyacrylates, cellulose, and poly(vinyl alcohol)], and solvents (such as propylene glycols, glycol ethers, and preferably water), combined with the gradual restrictions being imposed on the use of volatile organic compounds (VOCs) in the workplace, was widely believed to pose a threat to the traditional liquid gold industry. This led to the synthesis of a range of gold(I) mercaptides conferring water solubility and the development of aqueous based formulations containing these compounds [24–27]. Several gold(I) mercaptides containing carboxylic acid and amino-acid mercaptide ligands were previously known because they are used to treat rheumatoid arthritis, but these had not been evaluated for use in decorative formulations.

Synthetic Routes   The high nobility of gold has meant the element itself is not a particularly useful starting material for the exploration of its chemistry. Instead, tetrachloroauric acid, $HAuCl_4$, formed by the dissolution of gold in *aqua-regia*, is most commonly used as a precursor to gold(I) compounds.

Gold(I) mercaptides can be prepared by adding three molar equivalents of the desired mercaptan (HSR) to an aqueous solution of $[AuCl_4]^-$, which causes reduction of gold(III) to gold(I) and the formation of the desired gold(I) mercaptide (Equation 15.1).

$$n\,[AuCl_4]^- + 3n\,RSH \rightarrow \{Au(SR)\}_n + n\,RSSR + 3n\,HCl + n\,Cl^- \qquad (15.1)$$

The above method was adopted by Ballard [16] for the synthesis of the cyclic terpene gold(I) mercaptides and also by Fitch [19–22] during his studies. However, this common early route is not recommended when using expensive, or difficult to prepare mercaptan ligands, because of the use of the mercaptan as a sacrificial reductant with two molar equivalents required to effect the gold reduction. This formation of the disulfide as a by-product also means that careful separation of the desired product is required.

An improvement to this synthetic route involves the use of a wide variety of different reagents to reduce gold(III) to gold(I), typically in aqueous solution to form an *in situ* gold(I) intermediate. Fitch disclosed the use of alkyl sulfides (typically methyl) for this purpose [19–22] with the thioether acting as a two electron reducing agent for the conversion of gold(III) to gold(I) as represented in Equation 15.2.

$$[AuCl_4]^- + 2\,SR_2 + H_2O \rightarrow [ClAuSR_2] + R_2S=O + 2\,HCl + Cl^- \qquad (15.2)$$

This intermediate can then be converted to the desired gold(I) mercaptide by the addition of only one equivalent of mercaptan ligand (Equation 15.3).

$$n\,[ClAuSR_2] + n\,HSR \rightarrow \{Au(SR)\}_n + n\,SR_2 + n\,HCl \qquad (15.3)$$

Some claimed that the methyl sulfide and methylsulfoxide by-products are sufficiently soluble in water that they can be readily removed from the gold(I) mercaptide by water washing, enabling high-purity products to be isolated in very high yield. The kinetics of the reduction by alkyl sulfides has been investigated and the reaction is shown to occur by a two-stage process dependent on the α-donor ability of the sulfide ligand and the concentration of the water present [28].

Further improvements to the synthetic route have since been reported. These include the use of methionine as a reducing agent [23,29] and also the use of water-soluble hydroxyl or carboxy-substituted thioethers [30]. The water-soluble sulfide, thiodiglycol [31,32] and the closely related ethyl 2-hydroxyethyl sulfide [24] have also been found to be particularly suitable because of the high solubility and stability of the gold(I) intermediate in aqueous solution, and the ease of removal of their reaction by-products.

The gold cyano complexes $M[Au(CN)_2]$, where M = alkali metal [33,34], and bis(halide)gold(I) ions [35] have also been used as precursors for gold(I) mercaptides, but these routes have not been utilized by the decorative industry.

Physical Properties   The gold(I) mercaptides $\{Au(SR)\}_n$ are generally isolated in high yield as white or yellow amorphous powders. They are polymeric in nature with the favored linear S-Au-S coordination of gold(I) achieved by donation of a lone pair of electrons from sulfur to form sulfur-bridged polymeric chain (Figure 15.7) or ring species.

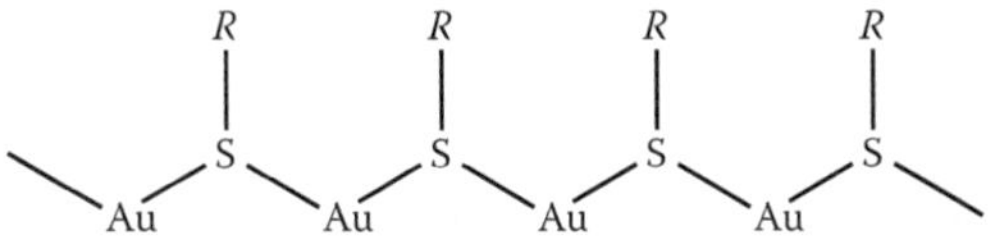

FIGURE 15.7 Linear coordination of gold(I) mercaptides.

Chains are also observed for other gold(I) complexes including gold(I) halides $\{AuX\}_n$ [36], and the $[RS]^-$ anion has been regarded as a pseudohalogen by some workers [37].

As discussed in the overview, the nature of the mercaptan group has been widely varied and is known to strongly influence the solubility properties of the prepared gold compounds. Solubility trends can be clearly established from the patent literature reported by Fitch with respect to the nature of the mercaptide ligand and the solubility of $\{Au(SR)\}_n$, where R is a simple alkyl group. These reveal a general increase in solubility with increasing substitution on the α-carbon adjacent to the sulfur. Primary gold(I) mercaptides exhibit extremely poor solubility in all solvents [19], secondary mercaptides show variable solubility depending on the nature of substituent R [20] and tertiary gold(I) mercaptides generally exhibit high organic solubility (Figure 15.8) [21].

Hence, the primary gold(I) mercaptide $\{Au(S\text{-}n\text{-}C_{12}H_{25})\}_n$ shows no solubility in any organic solvents, but the tertiary isomer $\{Au(S\text{-}t\text{-}C_{12}H_{25})\}_n$ is highly soluble in nonpolar solvents such as xylenes, essential oils, and terpenes. The insolubility of the primary gold(I) mercaptides is believed to be a consequence of the polymeric structure of $\{Au(SR)\}_n$ compounds, and it is assumed that the primary gold mercaptides form long S-Au-S chains. In contrast, it is thought that the steric bulk on the α-carbon within tertiary gold mercaptides prevents chain formation and favors the formation of discrete cyclic units. Molecular weight determinations for $\{Au(S\text{-}t\text{-}C_{12}H_{25})\}_n$ in benzene gave a value of 2211 g/mol, correlating well with a cyclic molecule of six mercaptide units. Similar correlations between the degree of substitution at the α-carbon and the association of polymer units in solution was made by Åkerstrom for silver(I) mercaptide compounds [38]. Despite the significantly improved solubility of the tertiary mercaptides over their primary and secondary analogs, only tertiary mercaptides with 9–18 carbon atoms are sufficiently soluble enough in organic solvents to be useful in the preparation of bright gold films [21].

The solubility of aryl gold(I) mercaptides has also been shown to vary significantly depending on the nature of the ring substitution. The p-tert-butylphenyl gold(I) mercaptide possesses extremely high solubility in solvents such as chloroform and xylene, exceeding 30% by weight of compound, and is particularly useful as the gold source within bright gold formulations [22,39]. Another example showing high solubility is o-methyl-p-tert-butylphenyl gold(I) mercaptide. In contrast, the solubility of p-methyl substituted analogs is poor, suggesting a strong influence of the para-substituted group on solubility. For the series $\{Au(SC_6H_4R\text{-}p)\}_n$ where R = H, methyl, ethyl, i-propyl, n-butyl, t-butyl, the solubility in nonpolar solvents was shown to increase dramatically with increasing steric bulk at the para position [32]. Solution NMR studies showed three sets of alkyl and aromatic resonances in the ratio 2:2:1. Further NMR, chromatographic, and molecular weight studies by Parish et al. [32] indicated a single cyclic species with a degree of association

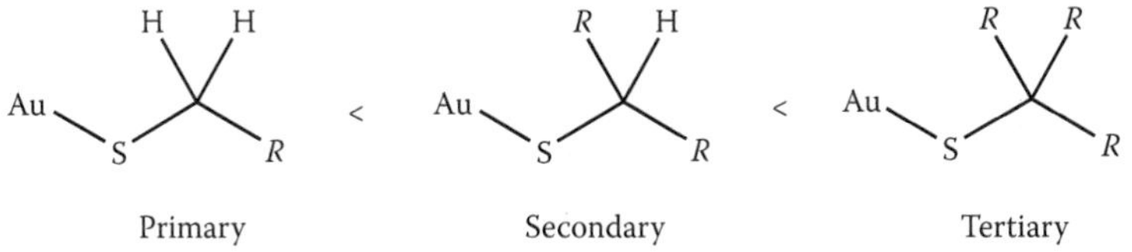

FIGURE 15.8 Trends in solubility for gold(I) mercaptides.

of five AuSR units, and a cyclic pentameric structure in solution consistent with the 2:2:1 intensity pattern observed by NMR. Wiseman et al. reported similar NMR spectra in their studies on $\{Au\{SC_6H_4\text{-}p\text{-}CMe_3)\}_n$ [40].

Although many of the tertiary and aryl gold(I) mercaptide compounds disclosed by Fitch are highly soluble, their solubility is limited to nonpolar organic solvents and natural oils and they show little solubility in more polar solvents such as terpineol, which are commonly used in printing pastes. This is also the case for the cyclic terpene mercaptides. A new class of gold–halogen–benzyl mercaptide compounds of the general formula

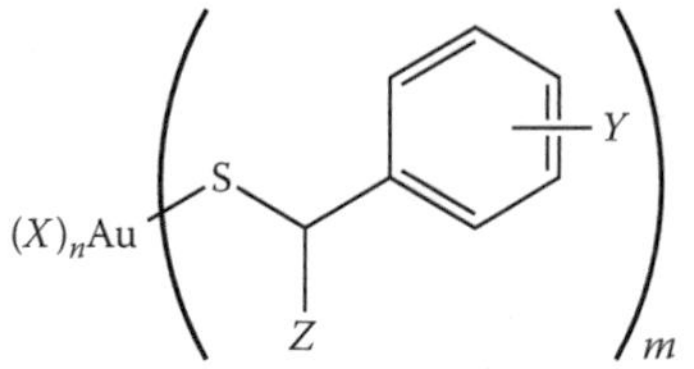

where $n = 0.1 - 0.3$, $m = 0.7 - 0.99$, and total $m + n = 0.8 - 1.2$, and X is a halogen and Z and Y are hydrogen or hydrocarbon groups were disclosed by Maeda et al. [41] and are reported to exhibit excellent solubility in both terpineol and aromatic solvents, as well as exceptional solution stability. The exact nature of these halogen-mercaptide compounds is not known, but they are considered to be coordination compounds with several gold mercaptide groups positioned around the halogen group via the gold atom.

Some of the most recent developments in liquid gold precursors involve the synthesis of water-soluble gold(I) mercaptides as reported by Bishop [24]. The mercaptide ligands in these compounds contains a water-solubilizing functional group, having the general formula

$$Au \longrightarrow S \longrightarrow R \longrightarrow X$$

where the mercaptide can be aliphatic, aromatic, or heterocyclic, and X = a nitro group, $-COOH$, $-SO_2OH$, $-OH$, $-C(O)NH_2$, $-NH_2$ or $-OP(O)(OH)_2$ and R represents a divalent organic group, with gold(I)-N-(2-mercaptopropionyl)glycine (Figure 15.9) shown to be particularly promising.

The water/polar solvent solubility is achieved for these compounds by the generation of an ammonium or quaternary ammonium salt, for example, triethyl-ammonium, via the carboxylate group within the mercaptide ligand (Equation 15.4).

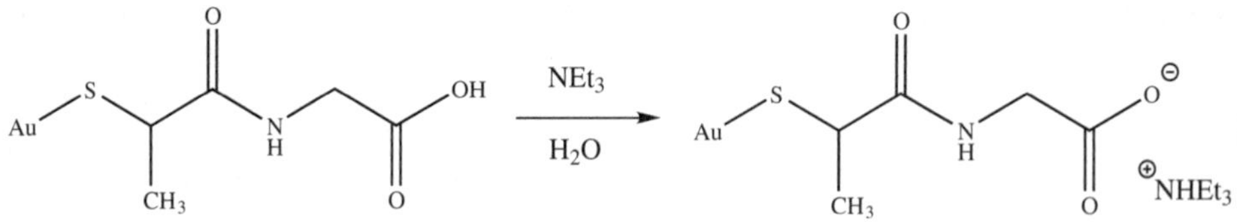

(15.4)

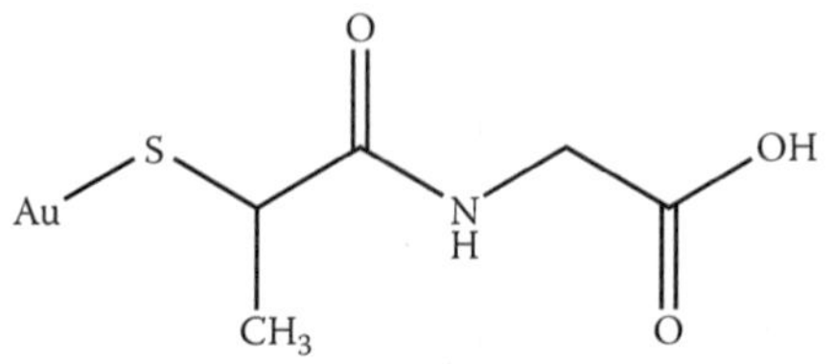

FIGURE 15.9   Structure of gold(I)-N-(2-mercaptopropionyl)glycine.

The use of water-soluble gold(I) dimercaptide compounds has also been reported [26]. These have the general formula

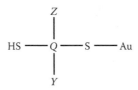

where Q = a tetravalent organic radical containing 2–10 carbon atoms, Y = hydrophilic group from the series –COOH, –COO⁻(cation) or esters of the carboxylate containing ethylene oxide backbones, and $Z = Y$ or H, or –OR, –SR, –SO$_3$R, –NR$_2$ –NR$_3$X. It is reported that, on reaction with one equivalent of gold, the mercaptide ligand retains a free mercaptan group. A specific example of these compounds is monogold(I) dimercaptosuccinic acid, which shows excellent water solubility on salt formation.

Structural Studies   Despite the large number of gold(I) mercaptides known, little is known about the structural aspects of the compounds utilized within liquid gold formulations. Simple nonsterically hindered gold(I) mercaptides are polymeric, amorphous, and exhibit poor solubility, which has precluded crystal growth from solution for single crystal structure analysis. Hence, only five X-ray crystallographic structures are reported in the open literature [40,42–44], and these examples all involve the use of sterically hindered mercaptide ligands.

The highly sterically hindered compounds [Au{SC(SiMe$_3$)$_3$}]$_4$ [42] and [Au{SSi(O-t-C$_4$H$_9$)$_3$}]$_4$ [43] both contain planar eight-membered Au$_4$S$_4$ squares in the solid-state with the gold(I) in a two-coordinate linear geometry linked by bridging mercaptide ligands. The steric demands of the mercaptide ligand prevent close approach of, and further aggregation of, the tetrameric units, as also observed for the silver analogs [45].

Crystals of the highly substituted aryl gold(I) mercaptide [Au(SC$_6$H$_2$-2,4,6-i-propyl$_3$)]$_6$ were successfully grown by Schroter from THF [44]. The solid-state structure consists of six formula units forming a centrosymmetrical 12-membered Au$_6$S$_6$ ring in the chair conformation. Gold–gold contacts are observed for all three compounds and, although they are shorter than van der Waals contacts [46], they are not regarded as significant.

Crystals of the highly soluble gold(I) aryl mercaptide, {Au(SC$_6$H$_4$-p-CMe$_3$)}, originally disclosed by Fitch in the 1960s, have been grown from ethoxybenzene nearly 30 years later [40]. Structural analysis revealed an Au$_{10}$S$_{10}$ core which is illustrated in Figure 15.10. What is most intriguing about this structure is the dominance of aurophilic gold-gold interactions that leads to the formation of a gold-catenane structure. This involves two interpenetrating pentagons defined by five bridging mercaptides and five gold(I) ions, with a sixth gold(I) ion at the center of each pentagon. The pentagon rings show gold(I) to exhibit linear geometries, ligated by bridging mercaptide anions. The structure is stabilized by nine short gold–gold contacts (average 3.05 Å) involving the Au(I) and Au(6) atoms (Figure 15.11), and also by longer gold–gold contacts (3.59 Å). The dominance of gold–gold interactions has been observed for other gold(I) complexes containing phosphine ligands, but has not been previously observed for homoleptic gold(I) mercaptides.

Structural analysis of the ortho-tert-butylphenyl substituted isomer, {Au(SC$_6$H$_4$-o-CMe$_3$)} revealed an Au$_{12}$SR$_{12}$ core, demonstrating the influence of substituent ring position on the solid-state structure (Figure 15.12). The central core can again be described as a [2]catenane, with the two interpenetrating hexagonal rings each consisting of six bridging mercaptide ligands and six gold(I) ions exhibiting linear coordination. Although gold–gold interactions are longer than those seen for the p-tert-butylphenyl structure, many gold–gold contacts less than the van der

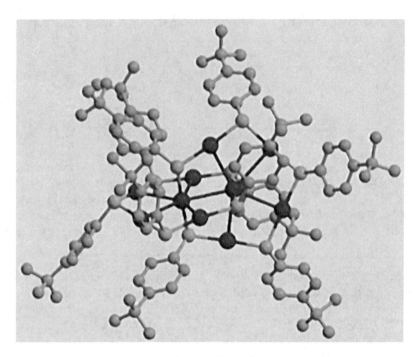

FIGURE 15.10   Solid-state structure of [Au(SC$_6$H$_4$-p-CMe$_3$)]$_{10}$.

Waals radius of gold are observed. The steric bulk in the ortho position appears to be responsible for the higher nuclearity compound, with the steric constraints relieved by the larger
ring sizes.

These structural studies on gold(I) mercaptides, although limited in numbers, suggest that a wide
range of oligomeric/polymeric arrays are possible for gold(I) mercaptides. The steric nature of the
mercaptide ligand has a significant influence on solid state structure, as well as on the presence of
gold–gold aurophilic contacts within some compounds which further influences structural arrangements and stability.

Thermal Decomposition Studies   Gold metallization temperatures were reported by Fitch for
selected secondary [20], tertiary [21] and aryl [22] gold(I) mercaptide compounds, and compared
to the metallization temperature obtained for gold sulforesinate (Table 15.1). Under the testing

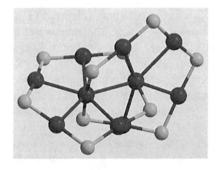

FIGURE 15.11   Interpenetrating pentagonal core of [Au(SC$_6$H$_4$-p-CMe$_3$)]$_{10}$.

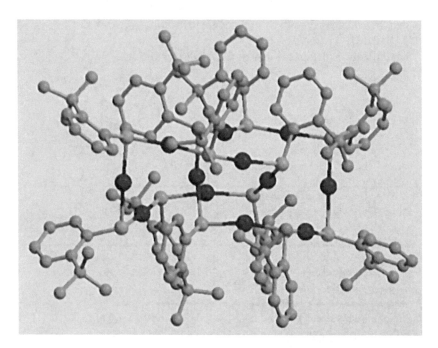

FIGURE 15.12 **(See color insert following page 212.)** Interpenetrating hexagonal core of [Au(SC$_6$H$_4$-o-CMe$_3$)]$_{12}$.

conditions used, the gold(I) mercaptides all formed conductive films at significantly lower temperatures (100–250°C) than the traditional sulforesinate materials (270°C). This ability of gold(I) mercaptides to form conductive films at relatively low temperatures was described as "an outstanding advantage" for these gold film precursors, enabling decoration of plastics, wood, paper, textiles and leather. Note that the metallization temperatures quoted refer to the temperature at which the heated films became conductive and do not necessarily imply that bright, reflective films were formed at this stage.

Limited thermal data are available for other gold(I) mercaptide compounds. Gold(I) bornyl mercaptide decomposes between 134 and 240°C [14], gold(I) di-mercaptosuccinic acid (Au-dmsa) starts to decompose at 194°C [26], and gold(I)-N-(2-mercaptopropionyl)glycine is reported to melt at 230°C, with decomposition [24].

On studying the data available, there does not appear to be any obvious trend between mercaptide ligand and metallization/decomposition temperature. This could be because of the possible variations in polymeric structure of these compounds; interestingly, the gold(I)-p-tert-butylphenyl mercaptide compound, shown by crystallography to exhibit strong gold–gold contacts within the solid state, has one of the highest metallization temperatures recorded by Fitch.

Another influential parameter for consideration is the tendency for aliphatic mercaptides to undergo metal assisted C-S bond cleavage resulting in the formation of metal sulfide products [47]. The thermal decomposition of a series of water-soluble gold(I) mercaptides, including {Au(SCH$_2$CO$_2$H)}$_n$, {Au(SCH$_2$CH$_2$CO$_2$H)}$_n$, and {Au(SCMe$_2$CO$_2$H)}$_n$ under flowing air, was investigated using simultaneous thermal analysis (STA) and thermal gravimetric–evolved gas analysis (TG-EGA) techniques [48]. Data obtained from these thermal analytical studies indicated that all three gold(I) mercaptide compounds decompose via a two-stage weight loss process with the initial major weight loss (26–35 wt%, endothermic) complete by 250°C, followed by a small weight loss (3–5 wt%, exothermic) up to 500°C. Final residues are consistent with the formation of gold metal.

TABLE 15.1
Metallization Temperatures of Gold(I) Mercaptides

Compound	Metallization Temperature,°C
Gold sulforesinate	270
Gold(I) o-methylphenyl mercaptide	215
Gold(I) m-methylphenyl mercaptide	130
Gold(I) p-methylphenyl mercaptide	135
Gold(I) mixed methylphenyl mercaptide	100
Gold(I) phenyl mercaptide	115
Gold(I) p-tert-butylphenyl mercaptide	230
Gold(I) o-methyl-p-tertbutylphenyl mercaptide	105
Gold(I) 2-napthyl mercaptide	250
Gold(I) pinene mercaptide	205
Gold(I) isopropyl mercaptide	165
Gold(I) sec-butyl mercaptide	150
Gold(I)-1-methylheptyl mercaptide	165
Gold(I)-1-methyldodecyl mercaptide	170
Gold(I)-1-methyloctadecyl mercaptide	180
Gold(I) tert-dodecyl mercaptide	160
Gold(I) tert-amyl mercaptide	190
Gold(I) tert-octyl mercaptide	170
Gold(I) tert-butyl mercaptide	195

Note: Metallization temperature refers to the temperature required to obtain conductive films by heating a solution of gold mercaptide in solvent (10 wt% gold) on a glass slide for 1 h in a temperature-controlled oven, with testing repeated at 5°C intervals.

For $\{Au(SCH_2CO_2H)\}_n$, there is a close correlation between the weight losses observed by STA and TG-EGA and those predicted theoretically for a two-stage process involving a gold(I) sulfide intermediate species. EGA showed evolution of carbon dioxide and carbon disulfide during the initial weight loss and evolution of sulfur dioxide during the small exothermic secondary weight loss. Weight loss and EGA data for $\{Au(SCH_2CH_2CO_2H)\}_n$ and $\{Au(SCMe_2CO_2H)\}_n$ compounds again clearly indicated thermal decomposition via a two-stage process. For $\{Au(SCH_2CH_2CO_2H)\}_n$ the evolution of propenoic acid is observed during the first weight loss, again indicating a gold–sulfur intermediate, followed by evolution of sulfur dioxide at higher temperatures. A similar decomposition profile was observed for $\{Au(SCMe_2CO_2H)\}_n$ with the evolution of 2-methyl-2-propenoic acid during the initial weight-loss step.

Controlled thermolysis of a series of anionic gold(I) mercaptide compounds, $[R(CH_3)_3N][Au(SR)_2]$ where $R = C_{14}H_{29}$, $C_{12}H_{25}$ and $R = C_{12}H_{25}$, C_6H_4-p-C_8H_{17}, C_6H_4-p-CH_3 has been studied by Nakamoto et al. [49]. Solid-state thermolysis of $[C_{14}H_{29}(CH_3)_3N][Au(SC_{12}H_{25})_2]$ at 180°C for 5 h under nitrogen, resulted in the formation of gold nanoparticles (average size 26 nm). Further mechanistic studies showed that reductive elimination of the mercaptide ligand occurred during thermolysis forming disulfide and gold metal nanoparticles. The nanoparticles were stabilized by the alkyl groups arising from the quaternary ammonium cation of the precursor compound.

Although differing significantly in their decomposition conditions and in gold nuclearity, the studies mentioned show that different pathways to metallic gold exist for gold(I) mercaptide complexes.

Formulations   As outlined in the introduction, liquid gold formulations traditionally comprise four major components: gold source (gold compound, gold powder or gold flake), minor metal additions to improve film quality, solvents and polymers/resins. The combined resin/polymer and solvent system employed enables control over ink rheology so that formulations suitable for application to ware by a variety of methods (typically brushing and screen printing) can be prepared. Firing temperatures typically range from 500 to 950°C depending on the substrate. This process removes the organic residues to leave lustrous bright or burnish gold films.

The first gold(I) mercaptides reported for use within liquid gold formulations utilized cyclic terpene mercaptans as ligands [16]. The high solubility of these compounds in traditional organic solvents, (for example, essential oils), and their compatibility with minor metal fluxes and sulfided rosin resins enabled the preparation of pale colored, highly stable formulations that produced bright, reflective films on firing. Further improvements to this system were made by Morgan and Wagner [11], whose introduction of synthetic epoxy polymers into the formulations enabled the films to be "cured" at 75–200°C after application to ware in order to produce hard films that were not damaged by handling prior to firing.

The trends in solubility observed by Fitch during his systematic studies into the preparation of primary, secondary, tertiary, and aryl gold(I) mercaptide compounds significantly influences the use of these compounds within formulations. The insolubility of the primary gold(I) mercaptides in all organic solvents precluded their use as precursors for bright films, but they were found to be suitable as metalizing precursors for burnished gold films, providing an alternative gold source to the fine gold powders traditionally used within these compositions. The primary gold(I) mercaptides could also be used in combination with gold sulforesinates to produce semi-matte gold films on firing. The influence of mercaptide ligand on the fired film properties of secondary gold(I) mercaptides is also clearly shown in Fitch's patent, with compounds exhibiting low solubility, such as gold(I) isopropyl mercaptide, only suitable for use in burnish gold formulations. In contrast, the soluble gold(I) alpha-methyl benzyl mercaptide can be used within bright, matte and burnish gold formulations. Similar trends were observed for tertiary gold(I) mercaptides, with only those compounds containing nine or more carbon atoms soluble enough to be used as precursors for bright film formation. Gold(I) tert-dodecyl mercaptide was found to be a particularly suitable precursor both for producing bright films both at low temperatures, and for use within liquid gold compositions. For the gold(I) aryl mercaptides, only highly soluble compounds such as gold(I)-*p*-tert-butylphenyl mercaptide and gold(I)-*o*-methyl-*p*-tert-butylphenyl mercaptide were suitable for bright film production.

One disadvantage of the gold(I) mercaptide compounds was reported to be their undesirable odor on application and particularly during firing to produce gold films. To overcome this, gold(I) mercaptocarboxylic acid esters of the general formula

where *X* is a $C_1 - C_3$ alkylene group and *Z* is a group from the series tricyclo$(5,2,1,0^{2,6})$decane-8- or −9-yl or tricyclo$(5,2,1,0^{2,6})$decyl-3- or −4-methyl; for example,

have been developed which have more acceptable odors [23].

These compounds were mixed with sulfurized colophonium resin to produce formulations that were found to be particularly suitable for screen printing applications because of the unexpected reduction in drying times observed for these gold ester inks. Reductions in drying times of over 90% were observed compared to analogous formulations containing the previously known gold mercaptide ester, gold(I) mercaptoacetic acid-(2-ethylhexyl)-ester.

It is interesting to note that all of the formulations containing soluble gold(I) mercaptides still utilized the natural raw materials such as rosin, asphalt, and sulfurized balsam resins found within the traditional gold sulforesinate systems. This meant that the compositions still had some disadvantages in terms of product consistency, especially with the formation of black spot defects on firing which cannot be envisaged beforehand, and with undesirable red coloration on the reverse of transparent substrates such as glass. Improved gold decorating compositions were achieved by the use of synthetic acrylate polymers of the specific formula:

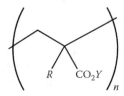

where R = H or a C_1–C_4 alkyl group and Y = a C_5–C_{40} hydrocarbyl group, which can be monocyclic, bicyclic, or tricyclic, and which may carry further ring substituents. Y is most suitably a terpenyl group derived from a bicyclic monoterpene, examples of which are bornyl, isobornyl, thijyl, fenchyl, pinocamphyl, and isopinocamphyl groups [39]. The replacement of traditional natural resins with poly(isobornyl methacrylate) gave bright gold films from both gold(I) p-tert-butylphenyl mercaptide and gold sulforesinate systems which were free of defects and with good reverse color. The choice of synthetic polymer was shown to be crucial as widely available acrylate polymers such as poly(butyl methacrylate) also give homogenous formulations, but result in poor quality, off-bright films on firing.

There is very little published information on the thermal decomposition process for "liquid gold" films. Deram et al. recently published their work on the thermal decomposition and thin-film formation of a commercial thermoplastic liquid gold ink [50]. The exact composition of the formulation studied was not determined, but it was believed to consist of a gold mercaptide containing aryl or ester functional groups in combination with resins and microcrystalline paraffin wax.

Thermal TGA analyses, coupled with IR spectroscopy and mass spectrometry studies, shows an initial low temperature melting of the waxes within the formulation (55–90°C), and this was also confirmed by optical microscope studies. On heating between 100 and 300°C, a weight loss of ~70 wt% occurs because of volatilization of the low molecular-weight waxes followed by decomposition of more complex resins and the mercaptide compounds. The transition from organic-based formulation to metallic coating occurs in a narrow temperature range from 300 to 350°C, corresponding with a change in color from matte brown to shiny metallic yellow, and with the emergence of XRD peaks for crystalline gold. Further heating above 350°C is required for the complete decomposition of the organic medium and elimination of residual carbon, and a second smaller weight loss of 15–20 wt% is observed together with the emission of water and carbon dioxide caused by the combustion of the gold mercaptide.

A strong exotherm is observed by DSC (Differential Scanning Calorimetry) at 450°C, and this is correlated with a morphological reorganization of the film from a porous inhomogeneous layer to a smooth, virtually nonporous film consisting of crystalline gold particles, with an average particle size of 150 nm.

Some of the most recent developments in liquid gold formulations have involved the synthesis of water soluble gold(I) mercaptide compounds, and the development of aqueous formulations containing them [24–27]. These formulations contain water as a major solvent, in combination with co-solvents such as glycols and glycol ethers, and water soluble polymers are used to control the rheological properties of the inks. Small amounts of water-soluble minor metal compounds such as rhodium acetate and bismuth citrate must also be added to aid film formation and adhesion.

The gold compounds used within these formulations achieve water solubility via the presence of a carboxylic acid functional group within the mercaptide ligand which is reacted with an organic base (commonly triethylamine) to form the soluble ammonium salt. Gold(I)-N-(2-mercaptopropionyl) glycine is reported to be particularly suitable for bright film formation [24]. The polymer and co-solvents within inks containing this, and closely related gold compounds, must be carefully formulated to ensure that homogeneous systems are obtained, both as formulations and as dried films after application to the ware. On firing, the polymer and the gold compound must decompose simultaneously to ensure that bright films are obtained. Particularly suitable polymers for these formulations are poly(acrylic acid), poly(vinyl alcohol), poly(vinyl pyrrolidone), and hydroxyalkyl cellulose. It has also been reported that careful selection of water compatible surface active agents, together with very low levels of co-solvent can enable even formulations of very high water content to be successfully applied to ware and retain homogeneity on drying and firing, resulting in bright film formation [25].

Similar aqueous formulations to those described above have also been reported that utilize water soluble gold(I) dimercaptides such as gold(I) dimercaptosuccinic acid to form bright films on firing [26]. Workers have shown that the presence of the extra thiol group in these compounds is essential for bright film formation, with the corresponding gold(I) mercaptosuccinic acid compound producing dull films in comparison. It is believed that the extra mercaptide group stabilizes the gold compound during the drying and firing process.

Note that although the aqueous-based formulations are well suited to decoration of ceramic ware by brushing and direct screen printing applications (onto flat ware such as tiles or glass), the decoration of ware by decal transfers is more complex for these systems because they use water to release the transfer from the decal paper. It is reported that cross-linking aqueous polymer systems can be utilized to produce water-resistant films suitable for water-slide decal use, but no specific examples of any such formulations are recorded [26]. Alternatively, the aqueous gold formulations can be printed onto heat release transfer paper and used to decorate ware using an automated application process.

Gold-Silver Mercaptides

As mentioned in the introduction, it is well known that the color of the bright gold films prepared from gold sulforesinate and gold(I) mercaptide formulations can be altered by the inclusion of silver into the formulation. The color can be tuned from a red-yellow gold ("red gold") through to yellow and greenish yellow ("lemon gold") by simply increasing the level of silver within the formulations. To achieve the desirable green-yellow hue of the "lemon gold" films, the silver content of the fired alloy film must be high, often 20–30% silver relative to the total precious metal content.

Traditionally, formulations contained silver sulforesinates as the source of silver. These were typically prepared via the reaction of silver nitrate or carbonate with a sulfur balsam such as Venice turpentine to form deep brown materials, as observed for the analogous gold systems [51]. Although these silver sulforesinate systems gave adequate results in terms of decorative effect, problems with shelf-life stability, reproducibility, and the limited solubility range of these materials led to the search for alternative silver precursors.

Many silver(I) mercaptide compounds, $\{Ag(SR)\}_n$, are known, and Åkerstrom [38] reported a systematic study of primary, secondary, and tertiary silver(I) mercaptides highlighting the relationship between substituent R, polymeric nature, and the solubility of the silver(I) mercaptides. As observed for the analogous gold compounds, the solubility of many silver(I) mercaptide compounds is very limited. However, some sterically hindered compounds, such as silver tert-alkylmercaptides have been shown to be soluble enough to be used within liquid gold formulations in combination with the equivalent gold compounds [52].

Fitch expanded on his work on gold(I) mercaptides in combination with silver(I) mercaptides to include the synthesis of novel gold-silver mercaptide precursor compounds [53]. These were prepared by simply reacting the gold(I) mercaptide compounds with an equimolar amount of silver(I) mercaptide or silver(I) carboxylate at elevated temperatures in a suitable solvent (for example, toluene) (Equation 15.5).

$$Au\text{-}SR + Ag\text{-}SR \rightarrow AuAgSRSR \qquad\qquad (15.5)$$

Typically, temperatures of 50–120°C are required to ensure that the reaction goes to completion within a reasonable time frame, with higher temperatures needing to be avoided to prevent the thermal decomposition of the mixed gold–silver compounds. Although not generally isolated from solution, the gold-silver mixed mercaptide compounds can be obtained as well-characterized solids by evaporation of the solvent or precipitation from solution, indicating that they are not simply a mixture of the two compounds.

Of particular note for these compounds is the ability to incorporate previously insoluble gold and silver mercaptides into formulations suitable for bright film formation. For example, the reaction between the soluble gold(I) alpha-methylbenzyl mercaptide and insoluble silver(I) n-octadecyl mercaptide at 50°C in toluene leads to the formation of a highly soluble gold-silver compound that can be used to prepare bright, lemon gold films on firing. This observation provides further evidence for the formation of novel bimetallic compounds, and the improved solubility of reaction products has dramatically increased the number of gold and silver mercaptide precursors available to the liquid gold formulators, with approximately 145 new precursor compounds reported.

One important advantage of these equimolar gold–silver compounds is that they enable the incorporation of silver at higher levels than achieved via the traditional use of silver compounds as minor metal additions to the formulation. Bright lemon gold films with a lower gold content, and, therefore, lower cost can thus be prepared. A further advantage is the observation that the mixed compounds generally metallize at lower temperatures than the individual gold and silver(I) mercaptide compounds. This enabled the formation of very desirable metallic, lemon yellow films onto nonrefractory substances such as plastics and textiles, offering significant advantages over the traditional sulforesinate systems.

Sulfur-free Gold Compounds

Nonsulfur containing gold compounds have found limited use as precursors for decorative bright gold films. The main driving force for removal of sulfur from such systems was to improve the odor of the formulations and to remove the toxic and corrosive fumes of hydrogen sulfide and sulfur oxides that are released during the firing of gold sulforesinate and gold(I) mercaptide based liquid gold formulations.

The preparation of 1,2,4 triazolyl gold compounds, in which the 3-position can be hydrogen or hydrocarbon substituted (1-11 carbons), was reported by Chambers [54]. The compounds are synthesized via the reaction of dimethyl sulfide gold(I) chloride complexes [also used as intermediates for gold(I) mercaptide] with the desired 1,2,4-triazolyl in the presence of a tertiary amine. Although these compounds are reported to be suitable precursors for decorative bright films no formulation details were reported, although it was noted that the decomposition of 3-nonyl-1,2,4-triazolyl gold(I) occurs between 230 and 330°C when heated at 10°C/min.

Liang [55] proposed the use of gold compounds isolated from the reactions between ammonium salts of carboxylic acids and gold amine complexes (prepared by the reaction of gold chloride with an amine). A specific example is the reaction of gold chloride solution with 1,2-diaminopropane which results in the precipitation of a white solid that is collected and redissolved in water. This gold intermediate is then added to a solution of decanoic acid and 1,2-diaminopropane. Stirring and standing results in the formation of an orange oil that separates from the aqueous phase. This non-sulfur gold resinate is isolated as a xylene-soluble viscous red-orange oil with a gold assay of only 15–17 wt%. As for the triazolyl gold compounds, no specific examples of formulations, or bright films, from this class of gold precursors are reported.

Gold carboxylates of the general formula, $[Au_x(RCO_2)_y]_z$, where x, y, and z are integers, such that $x = y \cdot z/3$, have also been reported to be suitable precursors for gold film formation [56]. They are prepared by reacting an alkali or alkaline earth organic carboxylate salt with a gold salt in an organic liquid at a temperature necessary to form the desired organic gold carboxylate. The products are removed by filtration and then purified by redissolving in a second organic solvent. In a specific example, gold trichloride, sodium 2-ethyl hexanoate, and 2-ethyl hexanoic acid were heated to 50°C and the resulting slurry filtered and rinsed with pentane to leave a yellow solution. This was warmed to remove the pentane leaving the product as an oil. The gold carboxylate compounds typically decompose at temperatures of 250°C and below making them advantageous for application to plastic substrates. Again, no formulation or film quality details are actually reported for these compounds, and it is believed that none of these or the other nonsulfur compounds discussed earlier have been successfully used in commercial liquid gold formulations.

More recently, Harada and Okamoto have reported the synthesis of gold acetylide compounds [57]. The compounds are of the formula, $Au(-C \equiv C-R)$, where R indicates a hydrocarbon group containing 1–8 carbon atoms and may or may not contain an oxygen atom. The compounds are prepared by the reduction of tetrachloroauric acid with sodium sulfite in the presence of potassium chloride. The desired acetylene derivative is then added, in combination with sodium acetate as an exchange promoting substance, to precipitate the gold acetylide compound as a brown sludge-like substance. A particular example using 3,5-dimethyl-1-hexyn-3-ol as acetylide ligand is prepared according to the equation:

$$HAuCl_4 + Na_2SO_3 + HC \equiv C(OH)(CH_3)CCH_2CH(CH_3)_2 + 4\,CH_3COONa + H_2O \rightarrow$$

$$Au\text{-}C \equiv C(OH)(CH_3)CCH_2CH(CH_3)_2 + 4\,NaCl + 4\,CH_3COOH + Na_2SO_4 \qquad (15.6)$$

The isolated products have been characterized by thermal analysis, atomic absorption analysis, and elemental analysis. The decomposition temperature of these compounds is approximately 400°C enabling them to be used for the decoration of glass and ceramic substrates. A further patent by the same workers provides detailed examples of the use of these gold acetylide compounds within formulations suitable for both brushing and printing applications that form metallic lustrous golds on firing [58]. In contrast to previous non-sulfur precursors, these gold acetylide compounds and formulations have been successfully commercialized by Daiken Chemical Company Ltd, and the sulfur-free nature of their products is strongly promoted within their product literature.

Very recently, novel amine-metal hydroxide complexes have been reported to produce high quality gold films on thermal decomposition [59]. These nitrogen-containing metal hydroxide complexes are formed by a process which involves partitioning an aqueous solution of chloroauric acid in the presence of base, with an organic solvent system containing an amine. In the preparation of a gold-hexylamine complex, the pH of the chloroauric acid solution was adjusted to 10.5 with sodium hydroxide. The aqueous solution was then added to a solution of n-hexylamine in toluene. Vigorous stirring of the reaction mixture resulted in an orange-red organic layer, which was separated and then reduced by rotary evaporation to obtain a glassy brown-red solid. The products were characterized by elemental, thermal, IR, and NMR analysis. Analysis showed evidence for the presence of coordinated amine, and NMR analysis showed both free and coordinated amine, and also the presence of imines (formed when the excess amine used in the reaction partially reduces the gold in the reaction mixture resulting in oxidation of some of the amine functional groups). The gold amine complexes have been found to exhibit very low metallization temperatures making them very suitable for deposition on to plastic substrates. Although no formulations are reported, AFM images of gold films on PET substrates prepared from the gold hexylamine complexes by spin coating and k-bar coating, and subsequent heat-treatment at 130°C show very smooth, continuous films of very low surface roughness.

Gold Nanoparticles

Colloidal "Purple of Cassius" gold stains have been known for centuries, and are still used today to produce a range of enamel colors for the decoration of ceramics. The preparative route to traditional gold colloid "sols" via the reduction of gold chloride by tri-sodium citrate and heat has been utilized for over three centuries to produce citrate-stabilized colloidal suspensions containing particles of 15–20 nm in size [60]. The sols can only be prepared at very low gold concentrations, and show limited stability and solubility. However, in the 1990s, a synthetic route to the formation of mercaptan-stabilized gold nanoparticles of increased stability was reported by Brust et al. [61]. The gold particles formed are typically of size 1–10 nm and contain a central metallic core stabilized by mercaptan groups. This protecting shell prevents agglomeration by means of steric and electrostatic barriers and imparts a greater stability to these nanoparticles than to those previously generated.

A recent patent discusses the use of gold nanoparticles stabilized with mercaptan ligands within liquid gold formulations [62]. Preparation involves phase transfer of gold chloride, $AuCl_4^-$ into an organic solvent, such as toluene, using a phase transfer agent, for example, Aliquat 336 ($[CH_3(CH_2)_7]_3NCH_3^+Cl^-$). Mercaptan ligand, for example, HSC_6H_4-p-CMe_3, is added to the organic layer, and the resulting solution is then added to an aqueous solution of excess sodium borohydride with vigorous stirring and cooling. The addition results in the instant formation of a dark brown organic layer, which is separated and washed with water. Addition of methanol affords precipitation of the gold nanoparticle product as a black powder, which can be purified by recrystallization from toluene/methanol. The gold assay is typically ~78 wt% gold, and the majority of the nanoparticles were shown to be between 1 and 2.5 nm as determined by TEM analysis. The metallic nanoparticles are further characterized by a plasmon resonance absorption which, for gold, occurs in the visible region. This surface plasmon resonance phenomenon (see Chapter 2) is a collective excitation of free electrons at the interface between a metallic core and the insulating shell of ligands, and is often very weak for small nanoparticles, as observed by weak surface plasmon resonances at 500 nm for this product. This method is a modification of the Brust method, as it was found that the addition of the borohydride solution into the organic layer (as reported by Brust) results in excessive foaming and increases in reaction temperature, which are very difficult to control during larger-scale synthesis. Surprisingly, simply reversing this addition step overcomes the problems of the prior art. The described synthetic route can be adapted to prepare gold nanoparticles stabilized by a wide range of different mercaptan ligands, and this new group of materials were found to be precursors to bright film formation that possess a number of advantages over more traditional gold(I) mercaptide compounds.

As discussed previously, the use of many of the known gold(I) mercaptide compounds was precluded by their poor solubility in common organic solvents. The gold nanoparticles show a significant increase in solubility over their molecular analogs. This means that the formulator has a significantly increased choice of the mercaptan ligand that can be used, and also a significant increase in the choice of solvents available for dissolution of the nanoparticles, ranging from the traditional nonpolar organic solvents through to more polar, environmentally friendly solvents such as propylene glycol ether solvents, which are increasingly being used in the formulation industry.

The use of decals or transfers as an application vehicle for bright gold decorative inks was, until recently, restricted because of the interaction of the covercoat materials used [poly(butylmethacrylate) and aromatic solvents] with the gold ink paste. Dissolution of the gold materials into the covercoat layer occurred, often resulting in off-bright films and a purple haze at the edge of the gold decoration. Advances in resin and gold gelation formulation technology in the 1980s reduced the amount of interaction observed on firing, and bright gold decals became a standard method of gold decoration worldwide. An advantage of the gold nanoparticles is the reduction of such interactions as compared to their molecular gold(I) mercaptide analogs. This is believed to be caused by the lower mobility of the nanoparticle precursors during the early stages of heat-treatment (< 150°C), and also the tendency for the nanoparticles to self-assemble and form "loose" gold films prior to the thermal decomposition of the stabilizing ligand. A further advantage with respect to the removal of "covercoat interaction" is their compatibility with certain synthetic polymers such as Scripset 540 and 550, which are poly(styrene-co-maleic acid) partial butyl esters, as these acid based resins also provide a further barrier to attack of the gold layer by the covercoat solvents. Simple screen printing formulations prepared from gold nanoparticles stabilized with HSC_6H_4-p-CMe_3 in combination with Scripset 540 show no covercoat interaction whereas compositions containing the gold(I) analog result in gold films with a removable surface scum and purpling round the edge of the print.

One further area of particular interest is the influence of the stabilizing ligand on the thermal decomposition and metallization temperatures of the gold nanoparticle materials. Gold nanoparticles stabilized with HSC_6H_4-p-CMe_3 decompose to give bright gold films at about 220°C, whereas those stabilized with ethyl-2-mercaptopropionate decompose to give bright metallic films at 100°C on polyester sheets. This demonstrates their huge potential, not just for the decoration of plastics and paper, but also for use as conducting tracks in organic electronic devices.

The ability of ligand-stabilized gold nanoparticles to metallize at very low temperatures has been further explored. The synthesis of gold nanoparticles prepared by the reduction of gold chloride solution with tertiary amine in the presence of protective polymers has been reported [63]. The protective polymers employed are comb-shaped block co-polymers containing multiple amino groups on the main chain, and side chains with different chemical structures. Depending on the polymer used, aqueous or toluene compatible gold nanoparticles can be prepared. These polymer-stabilized systems have been shown to be suitable for decoration of glass, plastic, ceramic, and photo paper resulting in metallic gold films on heating to low temperatures (250°C) or simply drying at room temperature for higher gold assay pastes.

The synthesis of amine-stabilized gold nanoparticles, particularly suited to low temperature applications, has been recently reported. These can be prepared either by the reduction of amine-gold hydroxide complexes by sodium borohydride or photochemical reactor [59], or by a simple thermal treatment of gold(I) acetylide compounds in the presence of amines and organic solvent [64]. A particular example is the preparation of oleylamine-stabilized gold nanoparticles by heating gold phenylacetylide and oleylamine in Shellsol solvent to 120°C to form a red colloidal solution. On cooling, excess methanol was added to precipitate the stabilized nanoparticles. Recrystallisation from Shellsol and methanol produced a dry, brown powder in near quantitative yield. TEM analysis confirms the preparation of small (<10 nm) monodisperse gold nanoparticles (Figure 15.13).

These new amine-stabilized gold nanoparticles can be solubilized in organic solvents to produce formulations capable of forming a decorative, shiny, gold film on paper at room temperature [65]. The printing has been achieved using a Dimatix, highly configurable piezo inkjet printer, and the decorative effect can be achieved in one printing pass. The printed gold surface exhibits strong specular reflection on smooth paper, and an improved decorative effect viewable at a wider angle can be achieved by printing onto special textured paper (Figure 15.14). These materials offer significant potential to open up new worldwide markets in the decoration of printable "real gold" greeting cards and religious texts, together with the potentially significant market for low-temperature printable electronic components.

Burnish Gold Formulations

As discussed in the introduction, burnish gold is one of the oldest known methods of decorating ceramic ware with gold. Typically, a gold powder is mixed with a suitable medium and applied to the desired substrate and fired to remove the organic material leaving the gold powder adhered to the substrate. The "as-fired" gold film surface is very rough, and visually looks matte light brown in color. The gold layer is then carefully polished or "burnished" using sand or other polishing agents to produce a lustrous semibright gold film.

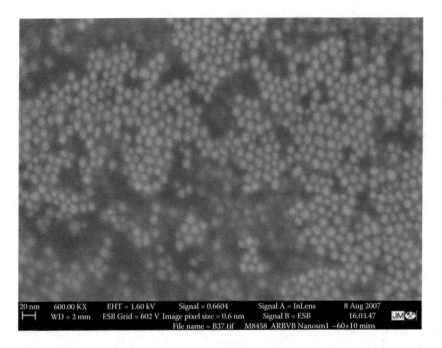

20 nm	600.00 KX	EHT = 1.60 kV	Signal = 0.6604	Signal A = InLens	8 Aug 2007	JM
	WD = 2 mm	ESB Grid = 602 V	Image pixel size = 0.6 nm	Signal B = ESB	16.03.47	
			File name = B37.tif	M8458 ARBVB Nanosm1 −60+10 mins		

FIGURE 15.13   Transmission electron micrograph (TEM) of oleylamine-stabilized gold nanoparticles.

The gold powders used in burnish golds are prepared by the chemical reduction of tetrachloroauric acid by suitable reducing agents (such as iron sulfate, sulfur dioxide, sodium sulfite, and hydroxylamine hydrochloride), often in the presence of a protective colloid such as gum acacia or other surfactant materials. The choice of reducing agent and protective colloid can significantly influence the powder particle size and morphology. Powders formed on reduction with iron(II) sulfate are often coarse in nature, whereas the reduction of gold chloride by sodium or potassium sulfite can be controlled to produce powders of a size range between 0.8 and 5.8 μm [66, 67]. The particle size and morphology of the gold powder can significantly influence the appearance of the fired burnish gold films.

FIGURE 15.14   **(See color insert following page 212.)** Ink-jet printed gold oleylamine nanoparticle films printing onto paper at room temperature.

Burnish gold formulations contain gold powder dispersed in an organic medium (resins and solvents) and the heavy gold powder must be dispersed fully within the ink either by ball milling or triple-roll milling immediately prior to use to ensure reproducible fired film properties. Although burnish golds usually contain gold powder, formulations can also be prepared using other insoluble gold precursors such as gold sulfides and selected gold(I) mercaptide compounds. Traditionally, burnish gold formulations contained gold powder in combination with mercury compounds such as mercury(II) oxide or chloride [68]. On firing, a gold/mercury alloy forms giving rise to a dense, homogenous, smooth surface, and a high quality fired appearance. However, volatilization of highly toxic mercury from the film occurs during the firing process, which is obviously very undesirable from both health and environmental perspectives, and mercury-containing formulations have largely been replaced by mercury-free formulations. Ballard reports the replacement of mercury oxide with elemental sulfur [69]. A more recent example involves the use of gold sulfide as gold source [70]. This use of a low-density gold precursor avoids the problems of gold powder dispersion and the polymethacrylate-based inks print well at low gold contents. The fired decoration can be polished to a high-quality finish.

As for the bright liquid gold products, small amounts of other metals and metal oxides are often also incorporated into the film to improve film adhesion and influence film color [71]. Silver is added in the form of carbonate, phosphate, oxide, mercaptide, or simply silver powder, to alloy with the gold during firing and produce a greenish yellow film color. Bismuth and vanadium compounds or oxides are typically added to the formulation to aid adhesion to ware on firing. It was also common for burnish gold formulations to contain low melting lead borosilicate glass frits to achieve good film adhesion at low firing temperatures. However, today the use of lead-containing glasses is undesirable because of lead release issues, and special lead-free flux systems are now available [72].

Water-based burnish gold formulations are also available to the decorator. Lotze describes a burnish gold formulation containing organic-soluble gold sulforesinate or gold(I) mercaptide precursors in combination with gold powder or sulfide, together with heterocyclic ketone solvents, particularly 1-methyl-2-pyrrolidone [73]. The formulations are water-thinnable (up to 20 wt%) and may optionally contain water dilutable resins. Landgraf also reported aqueous-based burnish gold formulations [74]. These contain gold powder or an insoluble gold compound in combination with polyvinylpyrrolidone and an aqueous acrylic resin dispersion, together with water and alcohol/glycol-based solvents.

Self-burnish Gold Formulations

High-quality burnished gold films are expensive because of the high gold content of the ink and the labor-intensive polishing process after firing to reveal a lustrous semibright gold finish. Therefore, much work has been undertaken by the industry to develop formulations that form a lustrous, semimatte gold film directly on firing, hence the term "self-burnish." This effect is not achieved by simply mixing together bright and burnish gold products together, but involves addition of flat platelet-shaped inorganic materials to bright gold inks.

Homogenous, self-burnish films can be obtained by the addition of low levels (0.1–3 wt%) of nonmetallic luster pigments, such as mica and bismuth oxychloride (widely used by the cosmetics industry) into bright liquid gold formulations [75]. High quality satin-matt films can also be prepared by the addition of metallic flake particles, especially gold flakes, as matting agents within the formulations. The particle size of such inorganic matting agents can vary considerably from 1 to 50 μm, and this obviously affects the surface finish of the fired gold film. As for burnish gold formulations, the matting agent must be effectively dispersed throughout the formulation prior to use to ensure reproducible results on firing.

Homogenous self-burnish gold films can also be prepared by the addition of organic materials as matting agents [76]. Micronized polyethylene waxes have been reported to be particularly effective. Bright gold formulations containing 0.5–2.0 wt% of finely dispersed wax of particle size 3–10 μm form semimatte gold films with good adhesion, even at low firing temperatures (< 500°C), making them particularly suitable for glass decoration.

Microwave-resistant Gold Formulations

Ceramics and glassware decorated with traditional bright and burnish decorative gold films are not compatible with microwave usage due to arcing (sparking) of the continuous metallic film upon microwave irradiation. Various approaches have been adopted to overcome this problem. One such approach does not involve modification of the gold formulation, but involves printing the decorative ink as a non-continuous layer [77]. Dots of a maximum diameter of 5 mm are used to print the desired pattern ensuring that the distance between dots is at least 0.2 mm. This approach ensures that the gold dots do not absorb enough energy from the microwave radiation to lead to electric arcing, and if the sum of the areas of the individual dots is at least 60% of the total surface area of the decoration, then the gold area still appears as a continuous gold film to the human eye.

Another approach involves reducing the conductivity of the decorative gold films so that electric arcing does not occur on microwave heating. It has been determined that if the gold film has a specific area resistance of more than 1 MΩ/sq, it remains unaffected by microwave radiation, because of its low conductivity. The conductivity of gold film can be decreased by the addition of increased amounts of certain minor metal additions to the liquid gold formulation, most notably bismuth and silicon compounds which form their respective nonconductive oxides during firing [78]. Organosilicon compounds containing three hydrolysable functional groups on the silicon atom, which decompose to form silicon dioxide on thermal decomposition were found to be particularly useful for enabling microwave-resistant decoration when used at levels of 0.1–2 equivalents of silicon per equivalent of gold [79]. The use of other nonconductive metal oxide forming compounds in addition to silicon and bismuth have been shown to be beneficial for forming nonconductive gold films on firing. In particular, the use of organic indium compounds [80,81], and barium resinate [82] have been patented specifically for their use in forming microwave-resistant gold films.

All the microwave-resistant liquid gold formulations require the addition of high levels of the non-conductive metal oxide compounds to the inks. This is generally advantageous to certain film properties, such as abrasion resistance, because of the high levels of glassy metal oxide flux within the films, but is detrimental to the fired film appearance. Thus, microwave-resistant gold films are notably darker and redder than traditional liquid gold films, making them less desirable for decorative purposes. Goebel and Landgraf report the use of a special combination of flux elements which contain nickel compounds that, when combined with aluminum and chromium compounds, fire to give a more natural color and high gloss [83]. More recently the addition of zirconium and/or aluminum, in combination with bismuth and silicon to gold compositions, has been reported to afford microwave-resistant gold films that maintain a high-quality, golden color on firing [84].

High-firing Gold Formulations

Recent trends in tableware decoration have led to the firing of ceramic colors onto porcelain using very fast but high temperature firing cycles (typically 1180–1250°C) to produce color decoration that shows the high dishwasher durability required for today's market. Traditional liquid gold decorations fire well at lower firing temperatures (700–900°C), but at temperatures above 900°C, the gold can tend to agglomerate into larger particles, particularly if the melting point of gold (1063°C) is exceeded. This results in the formation of a dull, brown layer with poor adhesion to the ceramic substrate. Therefore, both burnish and bright gold compositions capable of firing at these elevated temperatures have been developed.

The addition of Purple of Cassius gold, tin oxide, tin compounds, iron oxide, or aluminum oxide to burnish gold compositions was shown to increase the firing temperature up to 1400°C, and also increased the mechanical and chemical durability of the resultant gold film [85]. Best results were obtained on firing in combination with an intermediate layer consisting of one or a mixture of high melting-point compounds such as zirconium oxide or silicate, tin oxide, alumina, or silica. Burnish gold films were also found to be stabilized at firing temperatures of 1100–1400°C by the combination of the gold powder or gold compound source with base elements such as silicon, titanium, zinc, zirconium, and also additional noble metals of high melting point, such as platinum or rhodium [86]. Analyses of fired films suggested that alloys are formed between the gold and other metals on firing, and that the base metals are only partially oxidized during the firing process, but in a particularly finely divided form. This fine dispersion allows for firm bonding of the oxides to both the metal and to the ceramic substrate and prevents the molten metal coalescing during the firing cycle.

Bright liquid gold compositions suitable for high temperature firing up to 1350°C have been achieved by the use of special fluxing minor metal additions to the gold formulation, preferably at levels of 0.5–3 mol of fluxing agent per mole of precious metal. A wide range of suitable fluxing agents and levels are listed, but formulations containing niobium and tantalum compounds appear to be particularly suited to this application [87].

Bright gold films stable up to temperatures of 1250°C can be achieved by the addition of a combination of two or more of the following glass forming oxides of boron, lead and silicon to the formulation in the form of resinates at a level of 1–28 wt% glass oxides to gold. These films were also found to be of exceptional abrasion and dishwasher resistance [88].

INFLUENCE OF MINOR METALS ON FILM PROPERTIES AND STRUCTURE

As discussed in the introduction, it was determined very early on during liquid gold development that, although the choice of gold compound, resin and solvent was essential to obtaining a high-quality bright gold film on firing, the addition of small amounts of other elements was required to improve the optical, adhesion and durability properties of the films [89]. Such elements are often termed fluxes or "minor metals" and, although most elements of the periodic table have been reported to be useful within liquid gold formulations, the most common minor metals encountered within commercial systems are bismuth, chromium, palladium, platinum, rhodium, silicon, silver and vanadium. These metals are usually incorporated into the formulations as compounds (carboxylates, mercaptides), or as salts of natural or sulfurized resins (resonates). These compounds decompose during the firing process to leave their oxide residues finely dispersed throughout the gold film that forms simultaneously during the heat treatment. The metal-oxide type and level within the gold film has an important role in film formation influencing film morphology, color, chemical durability, abrasion resistance and firing stability. The role of some of the minor metal elements has been studied in detail and is relatively well understood, whereas the role of lesser-used elements remains undetermined.

The role of silver, palladium and platinum within liquid gold formulations is essentially to alter the color hue of the fired film. Silver alloys with the gold during the firing process and alters the color from a red-yellow gold through to yellow and greenish yellow ("lemon gold") on increasing the level of silver within the formulations to high levels (20–30 % metal). As discussed in the section on gold–silver mercaptides, either traditional silver carboxylates/resinates or specifically designed mixed gold–silver mercaptide compounds are used as silver source within the formulations.

The inclusion of palladium or platinum within liquid gold formulations again alters the color of the bright gold film through alloy formation. Films that typically contain 20–40 wt% of palladium or platinum within the alloy produce reflective "white gold" films which can be yellowish white, silvery white, or metallic gray depending on the palladium or platinum content. Many formulations contain platinum or palladium in the form of sulfurized resonates traditionally utilized by the liquid gold industry. However, as for the analogous gold systems, many of the sulforesinates showed poor solubility and poor product reproducibility, and mixed halide mercaptide complexes, $RS\text{-}Pt(Cl)SR_2$ [90], palladous salt *bis*-thioether complexes [91], and bis-chelate palladium derivatives [92] have been utilized within "white gold" formulations.

Rhodium is included at low levels in virtually all commercial liquid gold formulations as it is essential for maintaining a continuous gold film at the firing temperatures used in the decorating industry. It is believed that pyrolysis of films leads to the formation of rhodium oxide at the grain boundaries of the gold particles and prevents further agglomeration of the gold particles at increased firing temperatures. Electron microscopy studies on films formed from liquid gold formulations prepared with and without a rhodium minor metal addition showed a dramatic influence on the grain size of the gold particles within the film [93]. Rhodium-containing films were shown to consist of a complete covering of very fine gold particles when fired at 400°C. On continued firing through to 700°C the gold crystals were shown to have increased in size, reaching 3–4 μm by 850°C, and on heating to 1000–1100°C a dull, discontinuous gold film with very large grains was observed. The absence of rhodium within the formulation significantly affected the grain size of the gold particles within the film and lowered the temperature at which dull discontinuous films were obtained from 1000–700°C making them unusable at typical firing temperatures.

Similar results were observed by Milgram in his studies on the thermal decomposition of gold pinenemercaptide formulations [94], with rhodium-free films decomposing between 185–240°C to give small colloidal particles which increased with size on continued heating to reach ~6 μm at 600°C to give dull, discontinuous films. Inclusion of rhodium within the formulation led to the formation of rhodium oxide at 300–325°C on the grain boundaries of the gold particles, preventing the diffusion of one gold particle to the next, thus inhibiting grain growth. Above 900°C, the rhodium oxide starts to decompose and the gold particles increase in size as if no rhodium was present. This work also demonstrated the influence of the firing atmosphere on the rhodium addition, showing that an oxidizing atmosphere is essential to prevent reduction of rhodium oxide to rhodium metal on pyrolysis.

Many have attempted to replace rhodium with a wide range of other cheaper minor metal additions such as lithium, cadmium, tin, lead, uranium, and antimony [95,96], but high temperature stability was not achieved. Promising results were obtained by the replacement of rhodium with thorium [97]. However, due to its radioactive nature this was not considered to be a suitable commercial replacement for rhodium. Further rhodium-free formulations containing mixtures of iridium, chromium, tungsten, and bismuth [98] have been reported, as has the replacement of rhodium by indium within traditional sulforesinate formulations [13].

The roles of other minor metal additions such as bismuth, chromium, and vanadium are somewhat less clearly understood. The addition of bismuth is strongly associated with an improvement in film adherence and durability, specifically abrasion resistance, chromium is widely believed to improve adhesion and the optical appearance of the gold films, and vanadium is widely used within liquid gold formulations designed for low temperature firing onto glass substrates. Several theories have been put forward as to how gold films adhere to the ceramic surface during firing, including the formation of gold-oxygen bonds [99], the formation of Au-Si bonds [100], and the formation of Au-Al compounds [101]. Diffusion between the gold layer and the ceramic substrate has also been postulated and Katz et al. reported the diffusion of glass components into a coated gold film at 370°C [102]. There is very little published information regarding the composition and characterization of the films formed on the thermal decomposition of liquid gold formulations. In order to try and determine how these adhere to the ceramic substrate during firing, and how the addition minor metal additions influence this adhesion process, two recent studies have used a wide range of characterization methods to investigate the nature of liquid gold films fired onto ceramic substrates.

Darque-Ceretti et al. used the combined characterization techniques of scanning electron microscopy (SEM), x-ray diffraction (XRD), energy dispersive X-ray spectrometry (EDS), particle induced X-ray emission (PIXE), and Rutherford back-scattering spectrometry (RBS) to analyze the structure and chemistry of liquid gold films fired onto a coffee cup and saucer [103]. The gold film on the saucer edge was shown to be contaminated by carbon and hydrogen due to incomplete decomposition of the organometallic liquid gold. Segregation of the glaze-whitening elements, zinc and zirconium, was shown to have occurred at the glaze-gold interface, and diffusion of a small

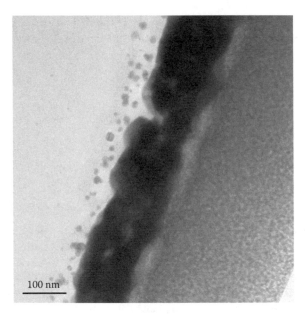

FIGURE 15.15 TEM (FIB cross-section) image showing gold nanoparticles dissolved into the glass substrate, resulting in a red coloration of the gold film when viewed through the glass side (from the reverse).

quantity of gold metal into the substrate was observed. A red mark on the glaze near the gilded area was clearly shown by high-resolution SEM to consist of gold nanoparticles between 25 and 40 nm. This red marking is a common problem in gold decoration, and a red reverse color is commonly observed when liquid gold formulations are fired onto glass. This red reverse color has been shown by TEM (Figure 15.15) to also be caused by the presence of colloidal gold particles that have diffused into the glass just below the glass-gold interface [104]. The gilded layers on the cup were quite different in nature, with porous layers, no carbon or hydrogen contamination, and deep penetration into the substrate, leading to an important mechanism of adhesion. The differences between the cup and saucer gold films are most likely to be caused by the reported differences in substrate composition of the two articles.

Popescu et al. carried out comprehensive SEM, EDS, and XRD analysis of five liquid gold compositions (three commercial, two known compositions) fired onto glazed white faience tiles for 20 minutes at 750°C [105]. Although all the fired films were bright and continuous by eye, SEM analysis revealed differing crystalline granular microstructures, with crystallite size dependent upon the minor metal elements and levels present. They observed that films containing Bi, Pd, Pt, Rh and Cu in greater concentrations had very fine microstructures with a grain size of 0.05–0.1 μm, because of a high number of gold nucleation sites created by the high concentration of metal oxide precursor elements. Films with lower levels of these metal oxides had larger gold grain sizes caused by lower numbers of nucleation sites resulting in rapid crystallite growth. They also concluded that, for the liquid gold compositions of this study, the presence of rhodium was not sufficient to obtain fine grains of gold and that a synergistic effect of the oxide precursor materials was very important.

The data obtained from film composition studies also highlighted the important role of fluxes in the transport phenomena that takes place during film formation. The minor metal flux elements migrate towards the film-substrate interface. Mutual diffusion was seen to occur with elements from the substrate (Al, Si, Ca, Mg) migrating from the substrate toward the surface of the gold film. The formation of a specific interfacial layer between the gold film and the substrate was observed. This was shown to be composed of mainly zirconium, silicon, and gold, and XRD studies showed this interface layer to consist of crystalline gold and crystalline zircon, $ZrSiO_4$, in the form of a solid-solution within this region. The good adherence of these films was explained by the formation of the Au-ZrSiO$_4$ solid-solution interfacial layer and by the mutual diffusion processes of the minor metals and the glaze elements.

APPLICATION TECHNIQUES

Liquid gold formulations are used to produce a decorative effect on a wide range of substrates such as porcelain, china, earthenware, tiles and glass. A wide variety of application techniques are available for use by the decorator. These range from traditional methods such as hand brushing, which has been used for hundreds of years and relies on the skill of the applicator, through to modern, technology-driven application methods such as ink-jet printing.

Brushing

Brushing or gilding was the original method of application for the first liquid gold formulations developed by Kühn in the 1820s, and remains a very common application method to the present day. Brushing gold formulations are typically low viscosity inks (40–150 mPa at 20°C), which are applied by brush to the desired ware. This application technique is widely used for the decoration of the rims of plates and cups with a narrow band of gold. The hand decorator places the object to be decorated on a turntable and rotates the turntable whilst applying the liquid gold formulation to the rim of the substrate with a fine brush. Brushing gold formulations are also applied onto ware using broader brushes for the decoration of larger areas with gold, especially onto handles and ceramic ornaments. This application technique is labor-intensive and requires a high level of skill by the hand decorator to ensure that uniform surface thickness is achieved, and that a consistent fired decorative effect is achieved for every piece of ware decorated.

Machine Banding

Three different types of machine banding methods are commonly used for the decoration of ceramic ware with gold films and are briefly described here.

Machine brush applicator systems are very similar to the hand brushing technology and utilize the same low viscosity liquid gold formulations.

Steel wheel or tungsten carbide wheel applicator systems are used for the decoration of plates with very thin or "verge" lines. The wheel rotates in a low viscosity liquid gold formulation (similar to brushing golds) and the plate to be decorated is adjusted and rotated in contact with the rotating wheel, which transfers the gold ink to the plate during the time that the wheel and plate are in physical contact.

For edge-line decoration of ware, the use of neoprene wheel systems is often preferred. In contrast to the other machine banding methods, a highly viscous and "tacky" ink rheology is required. This viscous gold paste is transferred by a system of steel wheels onto the neoprene wheel which then transfers the gold paste onto the rotating article to be decorated. The gold formulations must be designed using slow evaporation rate solvents to avoid changes in ink viscosity and behavior, ensuring that the paste applies consistently during long application runs.

Screen Printing

The screen printing process has been utilized for the mass production of liquid gold films since the 1950s. It is one of the most versatile of all printing processes, and can be used to print on a wide variety of substrates, including paper, plastics, glass, metals, and ceramic tiles.

The three essential elements of screen printing are: the screen which is the image carrier; the squeegee; and ink. The screen is simply a porous mesh which is stretched tightly over a metal frame. The mesh is typically nylon, polyester, or stainless steel. An emulsion is used to cover the non-printing area of the screen (blocking the screen holes) and thus defines the image to be printed. The emulsion process works because of the use of a light-sensitive material that hardens when exposed to ultraviolet light, and becomes impermeable to the screen printing ink. The squeegee is a flexible blade which is used to force the ink through the image design on the screen. At one time, squeegees were typically made of rubber, but today most are made from polyurethane which is less prone to wear and damage, and can produce up to 25,000 prints without degradation. Many further factors such as size and form, angle of use, pressure and speed of the squeegee blade can influence the properties of the print obtained.

The screen printing process simply involves placing the screen printing ink onto the screen above the stencil image (which is positioned over the article requiring decoration). The ink is then drawn across the screen using the squeegee which applies pressure thereby forcing the ink through the open mesh areas of the screen and forming the desired image on the substrate below. The thickness of the print obtained is essentially controlled by the diameter of the screen threads and the thread count of the mesh. The screen mesh refers to the number of threads per inch of fabric, with the more numerous the threads the finer the screen and the thinner the deposit onto the substrate below. For bright gold films, screens containing 350–450 mesh per inch are typically used, whereas burnish gold films need thicker applications and typically use screens containing 270–350 mesh per inch.

For optimum printing properties, the screen printing formulations should be designed to be thixotropic in nature so that the inks have significant structure at rest, but low viscosity and good fluidity under shear (when forced through the screen with the squeegee) and little flow once printed on the substrate.

Direct Printing onto Glass

Within the glass decoration industry "hot melt inks" are applied directly to glass substrates through screens that are heated to temperatures of around 70°C. Special thermoplastic gold pastes have been specifically designed using waxes and thermoplastic resins so that they remain fluid on the heated screen, but form a hard, dry film on making contact with the cold glass substrate [106]. This enables the decorated glass to be handled immediately after printing and makes it possible to print a complete wraparound design on non-flat substrates such as bottles.

Decals

Decals or transfers are substrates (typically paper) printed with a pattern which is then available to be applied or transferred to decorate another surface. Decals offer a distinct advantage over the direct decoration of ware in the ability to print the design onto flat paper and then apply the transfer onto curved or irregular surfaces that cannot be decorated by direct screen printing techniques.

Liquid gold decals are prepared by screen printing the ink through the screen onto special paper. Once dry, the printed image is overprinted with a layer of polyacrylate polymer dissolved in aromatic solvents known as a covercoat. This protects and holds together the delicate gold ink films, enabling intricate designs to be applied to ware. The application process is dependent on the type of decal paper used. Most gold decals are water-slide decals in which the gold ink is printed onto a base paper coated with a gum that dissolves in water. After a short soaking period (30 s), the decal image can be slid off the paper and applied onto the substrate. A squeegee or lint-free cloth is then used to ensure that no excess air or water is trapped under the decal before the firing process is carried out. Ideally, applied decals should be dried prior to firing. Water-slide decals must be applied by hand to glass and ceramic ware, making this a labor-intensive process.

Heat-release transfers offer an alternative process in which the gold ink is applied to a wax coated paper. The application of these decals to ware is an automated process where machinery melts the wax and transfers the decals to the ware to be decorated. Although this method is much faster, and requires no drying time, it requires significant investment in machinery and is suited to much higher-volume industrial applications, and most of the gold decal business remains as water-slide decals.

The use of decals for bright gold decoration was hindered for a long time by the fact that dissolution of the gold layer and subsequent reactions occurred between the gold ink and the components of the protective covercoat layer resulting in the formation of matt, dark, or cracked gold films on firing. The first successful approach to overcoming the problems of covercoat interaction involved the printing of an extra protective layer between the gold ink and covercoat, and this was followed by the introduction of bright gold printing formulations prepared from special systems of resins and gold compounds which were highly flexible and did not require a protective layer. This brought about a significant breakthrough for gold decal decoration and enabled it to become the standard method of decoration.

Pad Printing

This printing process uses steel or polymer plates onto which the design to be printed has been etched. During the automated process, the gold ink is drawn by a blade across the engraved plate, and the excess ink then removed to leave gold ink within the etched areas. A silicone "pad" is then pressed onto the engraved plate to pick up the gold ink and this is then transferred to the object to be decorated. Specific properties of the gold ink such as viscosity, tackiness, and drying time must be carefully controlled to achieve high quality printed images.

Ink-jet Printing

Direct-write ink-jet printing is fast becoming the prime technique for the deposition and patterning of functional and decorative materials in the liquid phase onto a substrate. Ink-jet printing has many advantages over the traditional printing methods such as screen printing, offset lithography, engraving, flexography, and gravure. Since only the ink is in contact with the substrate, it is possible to print an image on many different types of substrates, such as plastics, paper, or textiles, and even onto substrates with irregular shapes or surfaces. Designs are generated quickly and can be changed over in a matter of seconds, since the whole system is driven primarily by computer-based software. The digital nature of the image design process means that short production runs or unique printing are now far more cost-effective as no expensive screens need to be constructed. The new image file simply has to be loaded onto the computer controlling the printer. This also means there can be a much faster change-over between different production batches. Furthermore, as ink-jet printing is computer-controlled, both printing resolution and drop-placement accuracy are very well defined.

There are two basic types of ink-jet printers: bubblejet and piezoelectric. Bubblejet printers rely on the ink being heated at the printhead. This creates a bubble that expands and forces ink out of the nozzles. In piezoelectric printing, which is how most large-scale industrial ink-jet printers work, piezoelectric crystals located behind printhead nozzles expand and contract forcing ink out of the printhead. This is carried out in a controllable way to produce a desired image or pattern.

Gold ink-jet inks have been formulated for both bubblejet and piezoelectric printing. Molecular and nanoparticle gold compounds are synthesized and then dissolved in suitable solvents. The selection of solvent is essential to good printing. Polymeric additives can then be added to the ink formulation if required to ensure the correct surface tension and viscosity for printing. Once the gold ink formulation has been successfully printed onto the glass substrate, the ink droplets must not spread or contract over the surface as this decreases the resolution of the printed image and adversely affects the quality of the decorative print on firing.

PURPLE OF CASSIUS GOLD ENAMELS

Overview

In general, the coloring of glazeware fired at high temperatures (above 600°C) has required the use of thermally stable mixed metal oxide combinations to maintain color stability. These oxide pigments are mixed with glass frits (which in their own right can contain up to 20 different oxides) to impart durability and gloss on the final color. The collective term for a composition containing a glass frit and pigment is an enamel. Enamels are normally applied using similar techniques to

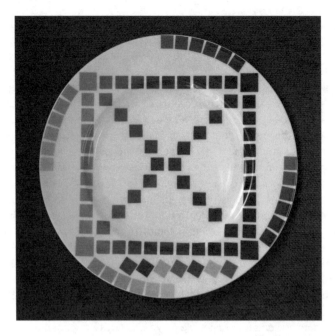

FIGURE 15.16 **(See color insert following page 212.)** Display plate depicting the differing shades of Purple of Cassius available to a decorator. The light shades on the left have low concentrations of gold, 0.5–1 wt%, whereas the darker shades have 2–2.5 wt% gold present. The color hue can be varied by addition of silver during the preparation. The silver will alloy with gold and cause a red shift in the colored appearance.

decorative golds, such as screen printing, machine banding, or gilding. Of all the enamel compositions available, the best compositions for delivering colors ranging from pink-red to purple rely on the combination of colloidal gold and stannic hydroxide, referred to as Purple of Cassius (Figure 15.16). This section highlights the preparation and history of Purple of Cassius and the color effects that can be achieved.

History

It is widely believed when dealing with the chemistry of gold with glass and ceramics, and in particular Purple of Cassius, that the inventor of the color effect was Andreas Cassius, often quoted as "Purple of Cassius was discovered by Andreas Cassius of Leyden in 1685 and is described in his work *De Auro*." More specifically Andreas Cassius is credited with discovering the purple color by preparation of colloidal gold and stannous hydroxide that bears his name and is still the most effective means of producing enamel colors ranging from pink to maroon [107].

However, there is strong evidence that the purple colors were discovered sometime before the publication of *De Auro* (1685). In fact a detailed method for preparing the purple compound of gold was published in 1659, and by around 1679 was in use as a colorant in the glass factory at Potsdam. This initial invention is attributed to Johann Rudolf Glauber (1604–1670), born in Karlstardt am Main and the son of a barber-surgeon. By all accounts, he was a skilful practical chemist and a prolific author on chemistry. In 1648, he left his native Bavaria to settle in Amsterdam, the thirty-year war had just ended and the aftermath of the war inspired him to write his ideas on the uses of chemicals and minerals and metals in four parts titled *Des Teutschlandts Wohlfahrt* (Prosperity of Germany). Glauber was familiar with several means of precipitating gold from aqua regia solution and is the first to document the precipitation with stannic hydroxide. He states in part IV of his document in 1659 the following: "Take 1 loth of fine gold powder, dissolve it with 3 to 4 loth of strong rectified spiritu salis; to the solution add 12 to 15 loth of pure water and put in a piece of tin weighing 2 loth. Put the vessel on a warm sand bath and let it stay warm for 1 or 2 hours but do not boil it, and the gold will precipitate from the solution in the form of a brilliant purple-color powder."

There is no evidence that Glauber used the above to color glass. However, there is evidence that some 25 years before Cassius discussed purples of gold with tin for coloring ceramics that the basic material was already documented. The use as decoration can be traced back to a chemist discussed already in the liquid gold section, namely Johann Kunckel who was in charge of the glass factory in Potsdam around 1678. Kunckel's successful use of the purple precipitate to produce ruby glass, often referred to as Kunckel's glass, was not revealed until the posthumous publication of his work, *Laboratorium Chymcium,* in 1716 [108], 13 years after his death. A key passage in the text is as follows: "It began in this way. There was a doctor of medicine called Cassius who discovered how to precipitate Solis cum Jove (gold with tin); I believe that Glauber may have given him the idea, or so it seems to me. Dr Cassius tried to introduce it into glass, but when he tried to form glass from it or when it was taken from the fire, it was clear as any other crystal and he was unable to form a stable red.... As soon as I had heard of this, I immediately took it up, but only I know what trouble I had to find the composition, to get it right and to obtain a durable red."

If we turn our attention to Cassius, in fact there were two, a father and son both named Dr. Andreas Cassius. There is some confusion as to who actually documented the purple coloration initially. However, it is clear that the younger (son) wrote *De Auro*. Andreas Cassius, Jr., was born in Hamburg in 1645, became an M.D. in 1668 and practiced as a physician in Lubeck, where he died at the end of the century. He wrote several medical articles and in 1685 published *De Auro*, an alchemical work and it makes several references to the "salt" and "fixed sulfur" of gold. The title page of *De Auro* is often quoted as the first reference to Purple of Cassius and is translated as "Thoughts Concerning that Last and Most Perfect Work of Nature and Chief of Metals, Gold, its Wonderful Properties, Generation, Affections, Effects and Fitness for the Operations of Art." Various texts from *De Auro* illustrate Cassius's thoughts and experiments involving Purple of Cassius, as it is known today and will always be.

By 1719, Purple of Cassius was routinely being produced and used in the Meissen works. At first sight, the formulas are simple and the first documented recipe for production came from the Meissen color technologist Samuel Stöltzel. The recipe is summarized as follows: "Take 3 loth of best aqua regia and dissolve in it half a gold ducat, and the liquid will turn yellow. Take 4 loth of English tine and dissolve in as much aqua regia as is necessary. It will likewise produce a fine ticture. When both solutions are ready, add a few drops of each simultaneously to a quart glass of pure water ... if it is bigger it is even better ... and continue until it has all gone in, when a red tinge will appear and the purple will fall to the bottom."

The emerging Sèvres factory, under the patronage of Louis XV and Madame de Pompadour, appointed Jean Hellot in 1751, who expanded on the purple enamel and produced the so-called Rose Pompadour (the name used by collectors), this being a particularly pleasing rose color. The preparation was somewhat temperamental. However, it did not stop it being produced in the United Kingdom in the Worcester and Chelsea factories around this time. To overcome issues with the reproducibility of the preparation, the reaction conditions and their effect on color was studied extensively. More recently, the Sèvres factory produced in 1985 a particularly clean purple color (Figure 15.17), and together with a new frit system, Sevres No. 601 was produced.

The identity of the color center was hotly disputed over a span of about 150 years from the initial work on Purple of Cassius. Some schools of thought were that gold existed as gold oxide, whereas others thought finely divided gold. Of note was the work carried out by Alphonse Bouisson, who was the assistant to Alexandre Brongniart, the director of the Sèvres factory. Bouisson embarked on a long research campaign on Purple of Cassius and dismissed it as gold oxide but reported it as finely divided metallic gold, publishing his findings in 1830. In 1857, Michael Faraday delivered

FIGURE 15.17 (See color insert following page 212.) Sèvres No. 601 developed at the *Manufacture de Sèvres*.

his Balerian lecture to the Royal Society, "Experimental Relationships of Gold and Other Metals to Light," and in passing stated, "I believe the Purple of Cassius to be essentially finely divided gold associated with more or less of oxide of tin" [109]. It was not until the early 1900s that the true nature of Purple of Cassius was experimentally verified. Richard Zsigmondy won a Nobel Prize in chemistry for his work associated with this. He was a Viennese chemist who spent several years studying gold colors and joined the Achott Glassworks in Jena in 1897. Together with an optical physicist at Jena, Heinrich Siedentopf, he developed the slit microscope for the detailed study of colloids, an instrument capable of resolving gold particles down to nanometer dimensions. Zsigmondy successfully prepared Purple of Cassius from a deep red colloidal gold solution and showed by microscopy that the purple was colloidal gold in nanometer dimensions deposited onto colloidal stannic acid [110]. The Nobel Prize was awarded in 1925 for his work [111].

COLOR GENERATION OF COLLOIDAL GOLD

The intense red to purple color seen for the Purple of Cassius pigment is achieved by precipitating a colloidal gold solution onto a solid oxide substrate, typically tin oxide. To maximize color intensity, the gold particles need to be unagglomerated and less than 40 nm in size. The color effect is caused by a narrow adsorption band at 520 nm, referred to as the surface plasmon resonance band. If particles agglomerate together and increase above 40 nm, a blue shift occurs and light scattering starts to dominate, resulting in muddy color hues. In 1902, Gustav Mie [112], using classical electromagnetic theory, calculated from bulk properties of metallic gold the absorbance of colloid gold particles as a function of the particle size. His theoretical work also predicted that the wavelength of the plasmon band depends on the shape, surface composition, and dielectric environment of the gold particles.

These concepts are used in modern-day production and research and development to predict and control the color effects associated with Purple of Cassius. For example, most gold colloid solutions seen in the literature are as dispersions in water with fairly consistent dielectric constants. When

FIGURE 15.18 **(See color insert following page 212.)** Gold colloids (5–35 nm in size) deposited onto three metal oxides with differing dielectric constants; red (on silica) dielectric constant = 4.5; purple (on alumina) dielectric constant = 9–11.5; blue (on titania), dielectric constant = 110. The color shifts from red to blue with increasing dielectric constant.

considering glass enamels, the glass media in which the colloid is dispersed may consist of many different components, all with vastly different dielectric constants. This can lead to massive differences in hue and color strength for the final gold enamels. A practical demonstration illustrating the effect of dielectric constant of the surrounding metal oxide on color shift is demonstrated in Figure 15.18. Gold nanoparticles of approximately the same size (before and after heat treatment) have been deposited onto three oxides having differing dielectric values. It can clearly be seen that the oxides with low dielectric constants such as silica have red shifts whereas high dielectric materials such as titania shift the color to blue shades which is in accordance with predictions made by Mie. Another strategy for controlling the color is to alloy silver with gold. According to Mie, additions of silver will shift the plasmon band to a shorter wavelength and hence make the particles appear redder. Theory predicts that a single plasmon band with alloys having up to 25% silver will exist, but above this level two peaks appear, one caused by silver and the other gold. In practice, an excess of silver is used to redden off the gold.

PREPARATION OF PURPLE OF CASSIUS

Preparation of Purple of Cassius has evolved over the years, and it is certainly true that variation in preparation conditions can change the hue and strength of color. Two key stages are involved in its preparation, according to the scheme in Figure 15.19. The first stage involves the formation of a gold sol and then its subsequent stabilization. The initial reaction is a redox reaction reducing gold chloride to gold metal via stannous chloride to stannic form, shown below in Equation 15.7.

$$2Au^{3+} + 3\,Sn^{2+} \rightarrow 2Au^{o} + 3\,Sn^{4+}$$ (15.7)

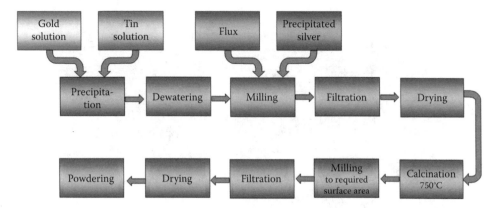

FIGURE 15.19 Production of gold enamel based on Purple of Cassius.

Of note is the use of three mole equivalents of tin(II) for the three-electron reduction of gold chloride to the metallic state. A key parameter is the stabilization and controlled particle growth of the resultant gold sol. Electron optical examination via TEM indicates the optimal particle size of the gold colloid to be 10–15 nm, as particles larger than this will result in weaker colors and tend toward the blue region. Key parameters that require careful control are reaction temperature, concentration, stirring efficiency, and trace elements present. Current-day technology uses pure reagents such as stannic and stannous chlorides. Historically, a mixture has always been used, freshly prepared by dissolution of tin metal in aqua regia (nitric and hydrochloric acid mixture) to produce Sn(IV) species. This is then reduced with tin metal to produce a Sn(II)/Sn(IV) mixture. The oxide ratio of this mixture is important for the efficient reduction of the gold.

The second stage of this process is effectively the stabilization of the gold colloid. If it is not deposited onto an inert support, the colloid will keep growing during the reduction phase, resulting in agglomeration of gold and loss of color. The initial stabilization is achieved by deposition of the freshly reduced gold onto stannic acid, produced via the hydrolysis of stannic chloride during the reduction stage. Hence the ratio of stannous Sn(II) and stannic Sn(IV) chlorides are important for reaction completion. Typical batch sizes would involve 0.5 kg of gold metal within a reaction volume on 1000 L vats. The resulting flocculated gold/tin sol is next de-watered and then wet milled with the desired glass fit composition. This milling stage is crucial to color intensity as poor milling will result in low-color strength-fired enamels caused by poorly dispersed frit/gold constituents. Also if the gold sol is dried at this stage agglomeration of gold particles can occur, resulting in loss of color. Milling is carried out in conventional ball or bead mills. The key is always is to obtain a high dispersion of the gold sol around the glass frit, typically a 1 to 2% gold composition would be used. The resulting material can then be dried without loss of color. Indeed color enhancement can be achieved by additional calcinations and wet milling. Significant research has gone into achieving reproducibility of color and more recently into reproducing high tint strength. Recent studies have centred around the reduction stage and use of well-defined tin materials rather than the mixtures discussed earlier.

As mentioned earlier, Sèvres has long been at the forefront of Purple of Cassius technology. In 1985, a particularly bright red purple was produced in the Sèvres laboratories and, as a consequence, a new glass frit system was developed resulting in the Sèvres no. 601 red gold purple (Figure 15.17). Studies were initiated on the best synthetic conditions to produce the red hues and characterization of the resulting colloidal gold [113]. Key findings show the use of a pure tin precursor in the form of anhydrous stannous chloride together with an understanding of the influence of reaction times

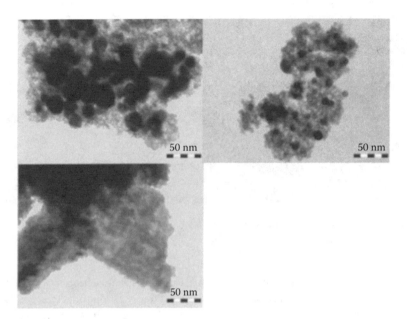

FIGURE 15.20 TEM of gold colloids prepared via the Sèvres method with addition times of 30 s at pH = 1 (top left), pH = 6.4 (top right), and pH = 8 (bottom left).

and pH control, allowing for more reproducible clean red purple tints. In addition, better gold assay control was achieved resulting in more consistent end color intensity.

The Sèvres studies highlighted the variability of the colloidal nature of the gold under differing reaction time and pH conditions. For example under short reaction times of 30 s and at pH 1, gold particles of 12 nm were obtained (Figure 15.20) with a particle distribution in the range 4–52 nm (Figure 15.21). The resulting fired enamels were too dark and blue. On moving to pH 8, gold oxide was formed which when processed and fired gave gold metal and the color was a light violet color. However, when the precipitation reaction was buffered to pH 6.4 with sodium hydrogen carbonate, a bright red precipitate was obtained. The TEM indicates the particle size distribution to be much narrower between 3 and 20 nm. So, although similar in average particle size to the precipitate formed under acidic conditions, the narrow size distribution of the pH 6.4 precipitate allows for redder-fired enamels, whereas at pH 1 the muddier color is caused by the presence of larger particles causing light-scattering according to Mie theory predictions.

An advantage of inorganic pigments (and in particular colloid gold) is their inherent stability. For example, some of the original Faraday colloidal gold suspensions still exist today. The high color stability has led to studies using colloid gold in organic type coating. The basis of the technology is the production of highly stable gold nanoparticles having diameters in the range of 10–30 nm that do not agglomerate [63]. Stabilizing organic groups have been used to prevent agglomeration, in particular a cone shaped polymer having multiple amino groups on the main and side chains (see Figure 15.22 for a pictorial representation). The particle-size control during their production is achieved by carefully adding tertiary amines as mild reductants. Gold paints and hence red/purple-colored films were fabricated from these gold building blocks. Gold pastes suitable for application onto glass were produced by mixing the gold solutions with a sol-gel binder system. The resulting films retain their intense red color when heat-treated at 280°C for 125 h (Figure 15.23). These gold paints can be modified to produce flip-flop type colors by mixing with aluminum flake, or indeed conductive films after baking at 250°C. It is envisaged that these paints will be used as new colorants to compliment the high temperature Purple of Cassius

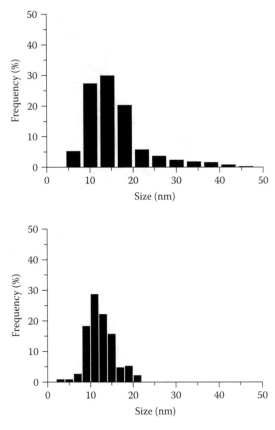

FIGURE 15.21 Particle size distribution histograms from TEM imaging for samples prepared with addition times of 30 s at pH = 1 (top) and pH = 6.4 (bottom).

materials for low fire applications, and the flip-flop pigments used as novel car paints or for decorative effects.

GOLD LEAF

OVERVIEW

The high malleability of gold has led to its application as decorative gold leaf. Gold leaf is produced by hand- or machine-beating and requires up to seven production stages to produce the thin gold. The history and origins of gold leaf are rich and fascinating, and are best read from previous review papers and on the primary references contained within [114–123]. This section will overview some of its history and gives a flavor of traditional production techniques to make and apply these materials. Special tools are required to handle the application of leaf onto articles and highly traditional fabrication procedures for beating gold and making into transfers ready for use are established (Figure 15.24).

Using appropriate techniques, foils of gold with a thickness of less than 100 nm can be prepared. A possible reason for gold's malleability has been proposed by Nutting and Nuttall [124] (see Chapter 2). Typical thickness of gold leaf purchased on today's markets ranges from 0.05 to 0.8 μm and when held up to the light it appears translucent. An Internet search reveals a multitude of companies selling these products and the required specialist tools for their application onto articles. The skill of the worker

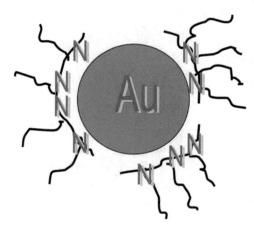

FIGURE 15.22 Comb-shaped copolymer stabilized gold nanoparticles. Particle sizes for gold can be obtained between 10 and 30 nm. The gold nanoparticles are produced in 100% yield.

is key to the final finish of the leaf. Twenty-four karat gold leaf is used for gilding statues, churches, public buildings, tombstones, weather vanes, clocks, heraldic shields, or where other metals would tarnish. Lower karat golds (23–23.5) are used to gild leather such as book binding [116], furniture, and picture frames.

HISTORY

The origins of gilding, that is, the application of a layer of gold onto another less rare surface, goes back at least 5000 years to the beginning of the third millennium BC. This is evidenced by the existence of some silver nails, in the care of the British Museum, that have been gilded by

FIGURE 15.23 **(See color insert following page 212.)** A colored glass film prepared from the comb-shaped copolymer stabilized gold nanoparticles and combined with an oxide sol gel to produce a durable low temperature "type" Purple of Cassius.

FIGURE 15.24 Basic tools of the gold beating trade are displayed in this picture, together with a few books of finished leaf. The large double faced hammer (made of cast iron) is the only tool used by the beater. The cross shaped weights below the hammer, hold the cutches, shoders and molds together between the various manufacturing steps. After the first stage of beating, boxwood pincers, two sizes of which are seen in the picture, exclusively manipulate the gold leaf. Foil is placed on the calf leather cushions (square objects where gold leaf pad is sat) to be cut with steel skewing knives, whilst the thinner gold leaf is cut or quartered with a wagon on the cushion, two shown above. The two objects on the left of the picture are in fact hares' feet, which were traditionally used to apply brime (fine powder) onto the goldbeaters, skins. Parchment is essential in the whole process and is used to encase the cutches, shoders, and molds during the beating process.

wrapping gold foil around the silver nail heads. The nails were obtained from the site of the Tell Brak in Northern Syria [125]. This appears to be the earliest form of gilding, and did not depend on physical or chemical bonding with the substrate surface but simply on the mechanical effect of wrapping the foil over the nail. Gilding of this type was never going to be very durable; however, there were ways of improving it, by either applying an adhesive to the surface where the gold is to sit, or by overlapping the edges of the foil and then joining them by burnishing. Gold foil gilding was applied in both of these ways to give durable coatings and these techniques persisted at least until the middle of the first millennium BC. The Treasure of the Oxus (Figure 15.25), found in Central Asia in the middle of the nineteenth century, is an example that displays these gold foil decorative techniques.

Apart from the above improvements in gold foil gilding, the application techniques with foil started to develop in two ways. First, the foil started to become thinner, between 0.1 and 0.8 μm, to the extent where it should correctly be called gold leaf. The second was that the gold foil could be attached to grooves that were etched into the surface to be decorated. The use of gold leaf was not possible until the introduction of a refining method to produce pure malleable gold, which is estimated to be around 2000 BC [123]. Fabrication techniques such as gold beating to produce gold leaf go back as far as the new Kingdom in Egypt (see Figure 15.26). The British Museum has a funerary papyrus of Neforronpet [126] of the fourteenth century BC, who is described as "chief of makers of thin gold." Hammering out gold or electrum, a naturally occurring gold-silver alloy, is depicted on Egyptian tomb paintings as far back as 2500 BC, (Figure 15.27); however, it is not clear if this is gold leaf or the thicker gold foil. The picture is an illustration from the tombs at Saqqara and Thebes and show goldbeaters working with gold founders and goldsmiths.

FIGURE 15.25   (See color insert following page 212.) The head of a silver statuette of a youth with a gilded headdress from the Treasure of the Oxus (fifth century BC). It is not clear from the picture, but the seam around the edge of the head-dress where the two gold sheets come together has been burnished (shinier area).

FIGURE 15.26   (See color insert following page 212.) Funerary papyrus of Neforronpet (fourteenth century BC), who is depicted with his wife Hunro. The papyrus was originally decorated with gold leaf on the head-bands, collars, armlets, and ankles of the two figures, but most is unfortunately now lost.

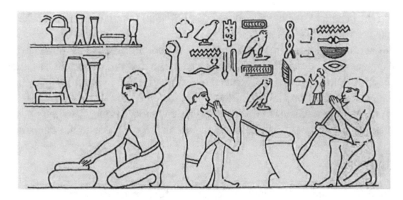

FIGURE 15.27 This illustration is from the ancient Egyptian tomb at Saqqara (2500 BC). This shows the melting of gold with the assistance of blowpipes (right) and the beating of gold with a rounded stone held in the hand.

What is clear is that gold leaf was insufficiently strong to support its own mass and so new methods of attaching it needed to be found. One method was to use an adhesive material to glue the leaf to the article and this process is still used today (described in more detail later) to decorate wood and stone and for the decoration of leather book bindings. Unfortunately, the naturally occurring animal or vegetable glues, which were initially used, have long disappeared due to biological decomposition and in many cases the gilding is lost or only traces are left. Another form of gilding developed was the method of burnishing gold onto pure silver and copper. To enhance the adhesion the gold film was heated. This method is limited to applications onto pure silver and copper, however, examples of it are seen from Roman times in the form of object by Lins and Oddy around the first millennium BC and AD [127].

GILDING TECHNIQUES

In architectural gilding, two main types occur, namely, oil and water. Oil gilding is used for general decoration and can be applied to most interior and exterior building surfaces for instance porous materials such as stone. The surface is pretreated with finely ground gesso (gypsum) or a similar finely ground material. The new surface is then rubbed down to yield a very fine texture. Next gold "size," a glue traditionally based on linseed oil, is applied to the surface to be gilded. As the gold leaf is extremely thin it is manipulated onto the surface with tools specially developed for the purpose, such as a gilders knife, tip, and cushion. After gilding, the surface can be brushed with a slewing mop to remove loose fragments of gold and optionally the surface can be buffed with fine cotton wool.

Water gilding is a more complicated process requiring more time and preparation; however, it is considered to yield an improved glossier finish compared to oil gilding. Water gilding is mainly used for picture frames, furniture, religious articles, and sculpture. The process involves applying six to twelve coats of gesso to the substrate to produce a very fine finish followed by four to five coats of a refined clay material termed bole, which is available in several colors. This surface is once more polished to a finely textured finish. As the clay material is nanoparticulate in nature, it yields a surface with minimal texture ideal for gold leafing. The surface is next coated with "size" and allowed to dry before being wetted with water and the gold leaf is then applied immediately. When dry any loose fragments are cleaned off, as for oil-based gilding. An additional coat of lacquer is added (termed Ormola) together with a red resin like material, historically referred to as dragon's blood. Commonly, bright and matte finishes are used within the same article; this is achieved by not over lacquering the bright areas but instead double gilding and then using a burnisher in the form of a rounded agate to produce the finish.

Because of the translucent nature of the gold leaf, the color of the background surface will influence the final appearance. For example, on picture frames a dull red color will be noticed where the gold is worn and the colored bole underneath shows through. The applied surface coating will also affect the finish. Oil applications are often left unvarnished but on areas exposed to wear and tear it can be treated with a light varnish. Gilding surfaces can have lacquers applied or even pigmented materials applied as an overcoat to distress the surface or impart a particular decorative effect.

GOLD BEATING

True beaten gold leaf remains the material of choice for prestige gilding in which durability is called for. The first recipes for gold leaf production via beating appeared in the twelfth century as evidenced by the emergence of a comprehensive text appearing in *De Diversis Artibus*, part of the *Encyclopaedia of Christian Arts by Theophilus* [128]. Essentially, the recipe describes the preparation of gold leaf and is still recognizable by goldbeaters today. Another valued text is by Dr. William Lewis in *Commercium Philosophico-Technicum* of 1763, whose description of the production in France highlights the effect damp and cold weather has on the gold fabrication process, and this is realized in modern times with humidity-controlled production.

The traditional materials and terminology used for the gold beating process require some definition. Often animal materials are used in contact with the gold during the beating process. For example, in the early stages of beating, where a four-meter gold ribbon is prepared, "vellon" is used to interface with the gold. This material is a treated calf's skin and is today substituted with carbon paper. Goldbeater's skin is also referred to, and this is a thin transparent membrane made from the large intestine of an ox, which has high tensile strength, ideal for taking a force such as hammering. It is now substituted with plastics such as PET. Montgolfier paper is also discussed and was produced in France. The original material made by the Montgolfier brothers was a composite of silk and paper that was used as the key material in early hot air ballons; however, substitutes are readily used. The exact use for this paper is not clearly defined but again it was used in contact with gold in the middle stages of beating.

Terms such as *cutch* and *shoder* are commonly used within gold beating. A shoder is a package of goldbeater's skins used in the second beating process; effectively, they are holders for the gold bound within inter-leafs, whereas the cutch is used in the first beating process. Hand beating is still used today. However, automated machines are now used routinely for today's production in the initial annealing stages and hammering to process the micronized gold leaf.

The traditional production begins by melting and alloying together gold, silver, and copper, depending on the color desired of the leaf, using about 1800 g of gold. An ingot is produced with dimensions in the region of 280 by 40 by 13 mm, which is forged to a required dimension so that it can be then reduced in thickness by placing through steel rollers with repeated annealing cycles (about 700°C) or softening to yield ribbon of about 25 μm thick and 4 m long. From the resultant ribbon, about 300 g of gold is cut into squares and about 200 are inserted into a cutch made up of sheets of vellum encased with bands of parchment. After 30 min of beating, the gold squares extend to the edge of the cutch and are about 100 mm in dimension; these are then quartered and placed into a second cutch. The gold is now supported by sheets of Montgofier paper and bound with parchment bands. Another 30 min beating results in thinner squares, which are quartered again. These are placed into a shoder made of goldbeater's skins. The quartering process requires the leaf, now less than 6 μm thick, to be cut using a skewing knife requiring skill and patience. All manipulations are carried out with wood pincers as steel would cause sticking. These thin slices of gold are now placed into a mold, between beaters' skins, in fact about 1200 of them. A machine is now used to cut the gold leaf as it is too delicate to cut by hand. The mold and the skins require some attention before the final reduction in thickness can be achieved. The skins are cleaned with a gypsum powder known as brime, and then placed in a hot press to dry using a process termed "flying," in which the skins are flicked in a fanning motion to remove any residual moisture. The next stage involves about 4 to 5 h beating depending on the gold purity. The leaf is then subjected to a number a cutting procedures and finally 25 pieces of gold leaf end up in a book containing tissues to keep them separate. One book of this would contain 0.325 g of gold. Typical thickness would be 0.1 μm if reduced by machine or if reduced in thickness by skilled beating then down to 0.05 μm can be achieved. As can be imagined, handling of the delicate materials requires great skill by the craftsperson.

CONCLUSIONS

This chapter was written to enable the reader to obtain a broad overview of the materials and processes used in the decorative gold industry. Some observations can be made from the technology discussed, and some thoughts from the authors follow.

Decoration with gold is still a traditional technology, with its origins dating back 300 years for fired gold decoration, and thousands of years for decoration with gold leaf. Notably, despite the development of new classes of gold materials (namely, water-soluble gold mercaptides and gold nanoparticles) over the last 25 years, these have not made a large impact on the gold decorating products commercially available. This is because gold sulforesinate technology is still extensively used within liquid gold products, despite the availability of new, well-characterized, gold materials. In part, this is because of the gelled nature of the sulforesinate systems, and the influence that this has on ink rheology and application properties. When substituting with the new gold materials, formulations exhibit newtonian type rheology as no sulfur-based gelation takes place, and it is difficult to match the rheology of the sulforesinate systems by the incorporation of synthetic polymers within the formulations.

The traditional nature of the liquid gold industry has also meant that decorators are not keen to switch to a new product that "feels" different on application (especially for gilding) despite the potential advantages of using well-characterized, stable, synthetic formulations. This problem was particularly encountered with the attempted introduction of water-based liquid gold formulations, and was further compounded by the lack of legislation to push these novel, environmentally acceptable products into the market.

However, it is hoped that recently developed gold materials (in particular gold nanoparticles) will allow for an expanded future product range capable of branching out into new markets such as the electronics industry for conductive applications, and the plastic/paper decorative sector, where low processing temperatures are required.

Key cost factors for gold decorating products are metal price and labor costs (as this industry has remained labor intensive with gilding and decal transfer application still carried out by hand). Recent economic factors have influenced the geographical location of the decorative gold manufacturing, with the relocation of factories from Europe to the Far East in order to benefit from lower labor costs.

REFERENCES

1. Liquid golds: The products and their applications, A. N. Papazian, *Gold Bull.*, 1982, **15(3)**, 81.
2. Gold in Decoration of Glass and Ceramics. G. Landgraf, in *Gold, Progress in Chemistry, Biochemistry and Technology*, ed. H. Schmidbauer, John Wiley & Sons, New York, 1999, 143–173.
3. Gold in the pottery industry. The history and technology of gilding processes, L. B. Hunt, *Gold Bull.*, 1979, **12**, 116 and references therein.
4. Königlich Sächsische Porzellanmanufaktur, K. Berling, Meissen, 1910, 128.
5. Gold films for the control and utilization of radiant energy, R. Langely, *Gold Bull.*, 1979, **8(2)**, 34.
6. The preparation of liquid bright and burnished gold, F. Chemnitius, *Sprechsaal*, 1927, **60**, 182.
7. Action of sulfur and a few sulfur compounds on terpenes, P. Budnikow and A. Schilow, *Ber. Dt. Keram. Ges.*, 1922, **55**, 3982.

8. The action of sulfur on the terpenes and the utilization of this reaction for the preparation of a liquid solution of gold, Budnikow, *Compt. Rend.*, 1933, **196**, 1898.

9. Producing gold compound for organic gold paste – by dissolving chloroaurate crystals in alcohol solvent, adding sulfur balsam, decanting, adding halogenated organic solvent, and then filtering, H. Onishi, Japanese patent JP 07232918, 1995.

10. Improvements relating to gold resinate solutions, E. Dürrwachter and E. Bosch, GB patent 1048145, 1964.

11. Decorating compositions, J. Morgan and C. Wagner, U.S. patent 2,842,457, 1958.

12. Noble metal preparation and its use to produce noble metal decorations by direct and indirect printing, A. Schulz, E. Zimmerbeutel, S. Keyn, and S. J. Hackett, U.S. patent 6,355,714 B1, 2002.

13. Glanzedelmetallpraparat, A. Lukas and H. C. Kühn, European patent 1,295,855 A1, 2003

14. Gold and compounds in ceramics, O. Henning, H. Heinke, and J. Steinwachs, *Wissenschaftliche Zeitschrift der Hochschule für Architektur und Bauwesen Weimar*, 1967, **14**, 385.

15. Reaction of organic sulfur compounds with gold chloride, A. Nakatsuji and H. Yamaguchi, *Kogyo Kagaku Zasshi*, 1957, **60**, 694

16. Gold compounds and ceramic decorating compositions containing same, K. H. Ballard, U.S. patent 2,490,399, 1949.

17. Sulfur compounds of terpenes. VIII. Actions of gold chloride on thioborneol and thiocamphor, A. Nakasuchi, *J. Soc. Chem. Ind., Jap.*, 1935, **38**, Suppl. 617 B.

18. Composition for making leads in integrated microcircuits and method of making same using said composition, V. B. Baltrushaitis and K. V. Sadauskas, U.S. patent 4,221,826, 1980.

19. Gold decorating compositions containing gold primary mercaptides, H. M. Fitch and N. J. Millburn, U.S. patent 2,994,614, 1961.

20. Gold secondary mercaptides, H. M. Fitch and N. J. Summit, U.S. patent 3,163,655, 1964.

21. Gold tertiary mercaptides and method for the preparation thereof, H. M. Fitch and N. J. Millburn, U.S. patent 2,984,575, 1961.

22. Gold aryl mercaptides and decorating compositions containing same, H. M. Fitch and N. J. Summit, U.S. patent 3,245,809, 1966.

23. Gold(I) mercaptocarboxylic acid esters, method of their preparation and use, M. Lotze and H. Mehner, U.S. patent 5,235,079, 1993.

24. Precious metal composition, P. T. Bishop, European patent 0,514,073, 1994.

25. Aqueous precious metal preparations and their use for manufacturing precious metal decorations, A. Schulz and M. Höfler, U.S. patent 5,545,452, 1996.

26. Mononoble metal dithiolates, preparations containing them and their use, A. Schulz and M. Höfler, U.S. patent 5,705,664, 1998.

27. Organosulfur gold compounds, a process for their production and their use, A. Schulz and M. Höfler, U.S. patent 5,721,303, 1998.

28. Reduction of gold(III) to gold(I) by dialkyl sulfides. Evidence for an atom-transfer redox process, G. Annibale, L. Canovese, L. Cattalini, and G. Natile, *J. C. S. Dalton Trans.*, 1980, **7**, 1017.

29. Process for the preparation of gold(I) mercaptides, M. Lotze and M. Bauer, German patent 4040447, 1992.

30. Verfahren zur herstellung von goldmercaptiden, G. Paret, German patent 1216296, 1966

31. A general synthesis for gold(I) complexes, A. K. Al-Sa'ady et al., *Inorganic Synthesis*, 1985, **23**, 191.

32. Mossbauer and nuclear magnetic resonance spectroscopic studies on "Myocrisin," "Solganol," "Auranofin" and related gold(I) thiolates, A. K. Al-Sa'ady, K. Moss, C. A. McAuliffe and R.V. Parish, *J. C. S Dalton Trans.*, 1984, **8**, 1609.

33. Sulphauro compounds and processes for their production, N. R. Trenner and F. A. Bacher, U.S. patent 2,370,593, 1945.

34. Gold complexes of L-cysteine and D-penicillamine, D. H. Brown, G. C. McKinley, and W E. Smith, *J. C. S Dalton Trans.*, 1978, 199.

35. Auromercaptobenzenes and process of making same, A. Feldt and P. Fritzche, U.S. patent 1,207,284, 1916.

36. X-ray structural investigations of gold, P. G. Jones, *Gold Bull.*, 1981, **14**, 102.

37. The structural chemistry of metal thiolate complexes, I. G. Dance, *Polyhedron*, 1986, **5**, 1037.

38. Silver(I) alkanethiolates, A. Akerstrom, *Arkiv. Kemi.*, 1965, **24**, 505.

39. Liquid gold compositions, P. Marsh, European patent 1,244,753, 2001.

40. Homoleptic gold thiolate catenanes, M. R. Wiseman, P. A. Marsh, P. T. Bishop, B. J. Brisdon, and M. F. Mahon, *J. Am. Chem. Soc.*, 2000, **122**, 12598.

41. Organic gold compounds and method of preparing the same, K. Maeda, H. Takamatsu, G. Sakata, and T. Mita, GB patent 2,238,544, 1991.

42. Synthesis and characterization of gold(I) thiolates, selenolates, and tellurolates: X-ray crystal structures of $Au_4[TeC(SiMe_3)_3]_4$, $Au_4[SC(SiMe_3)_3]_4$, and $Ph_3PAu[TeC(SiMe_3)_3]$, P. J. Bonasia, D. E. Grindleberger, and J. Arnold, *Inorg. Chem.*, 1993, **32**, 5126.

43. Contributions to the chemistry of silicon-sulfur compounds. 65. Synthesis, crystal and molecular structure of cyclo-tetrakis[tri-tert-butoxysilanethiolatogold(I)], $[(t-C_4H_9O)_3SiSAu]_4$, the first examples of a Au_4S_4 ring system, W. Wojnowski, B. Becher, J. Sassmannshausen, E. M. Peters, K. Peters, and H. G. von Schnering, *Z. Anorg-Allg. Chem.*, 1994, **620**, 1417.

44. Thiolatokomplexe des einwertigen golds. Synthese und struktur von $[(2,4,6-iPr_3C_6H_2S)Au]_6$ und (NH_4) $[(2,4,6-iPr_3C_6H_2S)_2Au]$, I. Schroter and J. Strahle, *Chem. Ber.*, 1991, **124**, 2161.

45. Chemistry of silicon sulfur compounds. XXXIV. Tetrameric silver(I)-tri-tert-butoxysilanethiolate, W. Wojnowski, M, Wojnowski, K. Peters, E. M. Peters, and H. G. von Schnering, *Z. Anorg-Allg. Chem.*, 1985, **530**, 79.

46. Van der Waals volumes and radii, A. Bondi, *J. Phys. Chem.*, 1964, **68**, 441.

47. Dimolybdenum complexes with sulfide and thiolate ligands as precursors to mixed-metal clusters: crystal structure of $[Mo_2Ru_2(\mu_3-S)_2(\mu-SPr^i)_2(CO)_4(\eta-C_5H_5)_2]$, H. Adams, N. A. Bailey, S. R. Gay, L. J. Gill, T. Hamilton, and M. J. Morris, *J. Chem. Soc. Dalton Trans.*, 1996, **12**, 2403.

48. Metal complexes as precursors for film deposition processes, P. A. Marsh, Thesis, Open University, UK, 1998.

49. Thermolysis of gold(I) thiolate complexes producing novel gold nanoparticles passivated by alkyl groups, M. Nakamoto, M. Yamamoto, and M. Fukusumi, *J. Chem. Soc., Chem. Commun.*, 2002, 1622.

50. Study of "liquid gold" coatings: Thermal decomposition and formation of metallic thin films, V. Deram, S. Turrell, E. Darque-Ceretti, and M. Aucouturier, *Thin Solid Films.*, 2006, **515**, 254.

51. Verfahren zur herstellung einer silberverbindung fuer den einsatz in edelmetallpraparaten zu dekorationszwecken, W. Keller and D. Wiemann, German patent DD269525, 1989.

52. Decorating composition and method of decorating therewith, H. M. Fitch, German patent 1969, 1286867.

53. Gold-silver coordination compounds and decorating compositions containing same, R. C. Langley and H. M. Fitch, GB patent, 1009539, 1963.

54. Triazolyl golds, W. Chambers, U.S. patent 3,803,158, 1974.

55. Precious metal nitrogenous organo reaction products, A. Liang, U.S. patent 4,201,719, 1980.

56. Gold carboxylates and process for preparing the same, T. E. Nappier, WO 92/10459, 1992.

57. Metal acetylide compound and process for preparing the same, A. Harada and Y. Okamoto, European patent EP 0,969,006 A1, 2000.

58. Metal composition containing metal acetylide, blank having metallic coating formed therewith, and method for forming the metallic coating, A. Harada and Y. Okamoto, European patent EP 0,976,848 A1, 2000.

59. Nitrogen containing metal hydroxide complexes, P. T. Bishop and V. Buche, WO 2006/131766 A2, 2006.

60. Coagulation of colloidal gold, B. V. Enustun and J. Turkevich, *J. Am. Chem. Soc.*, 1693, **85**, 3317.

61. Synthesis of thiol-derivatised gold nanoparticles in a two-phase liquid-liquid system, M. Brust, M. Walker, D. Bethell, J. Schiffrin, and R. Whyman, *J. Chem. Soc. Chem. Commun.*, 1994, 801.

62. Gold nanoparticles, P. T. Bishop, P. A. Marsh, B. J. S. Thiebaut, and A. M. Wagland, U.S. patent 2003/0,118,729 A1, 2003.

63. Coating materials containing gold nanoparticles, A. Iwakoshi, T. Nanke, and T. Kobayashi, *Gold Bull.*, 2005, **38(3)**, 107.

64. Process for producing metal nanoparticles and process for producing acetylides process, P. T. Bishop, A. Boardman, and V. Buche, WO 2007/110665 A2, 2007.

65. Nanotech Gold News, Special Edition Boston 2008, World Gold Council, 2008, 5.

66. Gold powder, V. Daiga, U.S. patent 3,768,994, 1973.

67. Gold metallising compositions, O. Short, U.S. patent 3,717,481, 1973.

68. Silk screen printing paste for decoration of heat-resisting base, F. E. Kerridge and C. S. Couper, Canadian patent 546066, 1957.

69. Burnish gold, K. H. Ballard, U.S. patent 2,383,704, 1945.

70. Stovable non-toxic screen-printing paste for production of gold-coloured transfer designs, J. Strauss, German patent 2111729, 1972.

71. Decorative burnish gold composition, W. Wild and G. Landgraf, U.S. patent 4,594,107, 1986.

72. Improvements in or relating to liquid burnish gold for the preparation of ceramic transfers or for direct application to a heat resisting base, S. Chandra, GB patent 721,906, 1955.

73. Gold polish preparation, M. Lotze and D. Frembs, U.S. patent 4,780,502, 1988.

74. Burnish gold decorating composition, G. Landgraf and W. Gobel, GB patent 2216536, 1989.

75. Non-burnished precious metal composition, D. Cuevas, F. R. Russo, and F. E. Schindler, U.S. Patent 4,418,099, 1983.

76. Gold preparation for decoration of glass and ceramics – comprise adding e.g., gold mercaptide derivative to hot solution of polyethylene wax in non polar solvent, G. Landgraf, German patent DE4122131, 1992.

77. Microwave stable tableware, W. Wild, J. Strauss, and G. Landgraf, U.S. Patent 4,713,512, 1987.

78. Mikrowellenbestandige Edelmetall-Dekoration, J. Strauss, W. Gobel, G. Landgraf and W. Wild, European patent 0,296,312, 1988.

79. Verwendung von glanzedelmetallpraparaten fur mikrowellenbestandige dekore auf geschirrteilen, M. Lotze and H. Mehner, European patent 0,313,890 B1, 1992.

80. Paste gold for finish painting, T. Giichi and T. Kazuyoshi, JP 62138379, 1978.

81. Gold colour glaze for finish painting, T. Giichi and T. Kazuyoshi, JP 62105990, 1978.

82. On-glaze decorating liquid gold, S. Masato and Y. Ryuta, *Jpn. Kokai Tokkyo Koho.*, JP 6048779, 1994.

83. Glanzedelmetallpraparat, W. Goebel and G. Landgraf, European patent 0,440,877 B1.

84. Paste gold and golden ornaments, K. Ito and K. Sugita, U.S. patent 6,428,880 B1, 2002.

85. Gold preparation for producing a gold or gold alloy decoration upon a siliceous base, J. Strauss and W. Wild, GB patent 1,403,482, 1975.

86. Coating ceramic articles using compositions containing gold, B. Apelt, A. Siebert, D. Sattler, and F. Anthofer, GB patent 1,524,701, 1975.

87. Method and composition for decorating glass-ceramics, P. J. Murphy, U.S. patent 266,912, 1966.

88. Method of depositing an adherent gold film on the surfaces of a suitable substrate, G. B. Fefferman, U.S. Patent 3,653,946, 1972.

89. Brilliant gold, F. Chemnitus, *J. Pract. Chem.*, 1927, **117**, 245.

90. Halogenoplatinous mercaptide-alkyl sulphide complexes, H. M. Fitch, U.S. patent 3,022,177, 1962.

91. Palladium decorating compositions, H. M. Fitch, U.S. patent 3,216,834, 1965.

92. Bis-chelate derivatives of palladium, S. Trofimenko, U.S. patent 3,876,675, 1975.

93. A. Boettcher, *Ber. D. Keram. Ges.*, 1955, **32**, 175.

94. Properties of gold films formed from organometallic solutions, A. A. Milgram, *J. Electrochem. Soc.*, 1971, **118**, 287.

95. The preparation of liquid gold for glass and china, S. G. Tumanov, *Steklo i Keram.*, 1961, **6**, 26.

96. Factors affecting the quality of liquid gold resinate preparations, P. A. Levin, *Steklo i Keram.*, 1968, **(25)3**, 36.

97. Bright yellow golden water without rhodium and its making method, S. Huanyang, L. Xianru, and H. Zehui, CN 1,069,995, 1993.

98. Noble metal containing overglaze colour composition for ceramics containing also thorium resinate, N. Ochiai, GB patent 2,248,635, 1991.

99. Fundamentals of glass to metal bonding: VIII, Nature of wetting and adherence, J. A. Pask and R. M. Fulrath, *J. Am. Ceram. Soc.*, 1962, **45**, 592.

100. Au and Al interface reactions with SiO_2, R. S. Bauer, R. Z. Bachrach, and L. J. Brillson, *Appl. Phys. Lett.*, 1980, **37(11)**, 1006.

101. Improved adhesion of gold coatings on ceramic substrates by thermal treatment, T. P. Nguyen, J. Ip, P. Le Rendu, and A. Lahmar, *Surf. Coatings Technol.*, 2001, **141**, 108.

102. *In-situ* stress measurements of gold films on glass substrates during thermal cycling, A. Katz, S. Nakahara and M. Geva, *J. Appl. Phys.*, 1991, **70(12)**, 7343.

103. An investigation of gold/ceramic and gold/glass interfaces, E. Darque-Ceretti, D. Helary, and M. Aucouturier, *Gold Bull.*, 2002, **35(4)**, 118.

104. Internal report, Johnson Matthey, P. Bishop, C. Williams, and A. Boardman, Johnson Matthey Technology Centre, UK, 2004.

105. The characteristics of gold films deposited on ceramic substrate, V. Popescu, I. Vida-Simiti, and N. Jumate, *Gold Bull.*, 2005, **38(4)**, 163.

106. Thermoplastic precious metal decorating compositions, R. C. Langley and D. B. Kellam, U.S. patent 3,092,504, 1963.

107. The true story of Purple of Cassius, L. B. Hunt, *Gold Bull.*, 1976, **9(3)**, 134 and references therein.

108. Collegium Physico-Chymicum Experimentale, oder Laboratorium Chemical, J. Kunckel, Hamburg, 1716, 650.

109. Experimental relationships of gold and other metals to light, M. Faraday, *Philos. Trans. R. Soc.*, 1857, **17**, 147.

110. Purple of Cassius, R. Zsigmondy, Justus Liebigs *Annalen der Chemie*, 1898, **301**, 361; see also Colloids and the Ultramicroscope, Translation, Alexander, New York, 1909, 65–67.
111. *Nobel Prize Winners in Chemisty*, E. Farber, New York, 1953, 95.
112. Contribution to the optics of turbid media, particularly of colloidal solutions, G. Mie, *Ann. Phys. Lpz.*, 1908, **25**, 377.
113. A new procedure for the production of red gold purples at the 'Manufacture Nationale de Ceramiques de Sevres', O. Dargaud, L. Stievano, and X. Faurel, *Gold Bull*, 2007, **40(4)**, 283.
114. Gilding through the ages, A. Oddy, *Gold Bull.*, 1981, **14(2)**, 50.
115. The ancient craft of gold beating, E. D. Nicholson, *Gold Bull.*, 1979, **12(4)**, 161.
116. Gold in book binding, G. Bologna, *Gold Bull.*, 1982, **15(1)**, 25.
117. Transferable gold coatings, B. J. Stitch, *Gold Bull.*, 1979, **12(2)**, 110.
118. The ages of gold, T. Green, Publ. GFMS Ltd., London 2007, ISDN 978-0-9555411-1-7.480.
119. Gold technology in ancient Egypt, T. James, *Gold Bull.*, 1972, **5(2)**, 38.
120. Gold in mosaic art and technique, G. Bustacchini, *Gold Bull.*, 1973, **6(2)**, 52.
121. The gilding of Lorenzo Ghibetis's 'Doors of Paradise', E. Mello, *Gold Bull.*, 1986, **19(4)**, 123.
122. On the origin of a gilding method of the ancient Baghdad silversmiths, G. Eggert, *Gold Bull.*, 1995, **28(1)**, 13.
123. Ancient Egyptian gold refining, J. H. F. Notton, *Gold Bull.*, 1994, **7(2)**, 50.
124. The malleability of gold. An explanation of its unique mode of deformation, J. Nutting and J. L. Nuttall, *Gold Bull.*, 1977, **10**, 2.
125. British Museum, Department of Western Asiatic Antiquities. Part of Inventory No 127430.
126. British Museum, Department of Egyptian Antiquities, Papyrus No. 9940.
127. P. A. Lins, W. A. Oddy, *J. Archaeol. Soc.*, 1975, **2**, 365.
128. 'On Divers Arts: The Treatise of Theophilis', The Chigaco University Press, Chicago, 1963,39; and Theophilus, *De Diversis Artibus*, Book I, Chapter 23, translated in English by J. G. Hawthorne and C. S. Smith.

16 Nanotechnological Applications of Gold

Jonathan A. Edgar and Michael B. Cortie

CONTENTS

Introduction: Why Gold? 369
Nanoscale Particles and Films 370
    Gold Nanospheres 370
    Gold Nanorods 371
    Nanoshells and Semishells 373
    Other Particle Shapes 374
    Mesoporous Gold Sponges 374
    Thin Films 374
Industrial Applications 376
    Pigments and Colorants 376
    Metallic Glazes and Sinter Inks 376
    Spectrally Selective Coatings 376
    High-Density Data Storage 378
    Optoelectronic Devices and Switching 379
    Catalysis 379
Sensor and Analytical Applications 380
    Colorimetric Sensors 380
    Refractometric Sensors 380
    Electrochemical and Other Sensors 382
    Surface-Enhanced Raman Spectroscopy 382
    Use of Gold in Medical Diagnostics 384
        General 384
        Two-Photon Luminescènce 384
        Magnetic Resonance Imaging 385
        Photoacoustic Microscopy 385
        Optical Coherence Tomography 385
Therapeutic Possibilities for Nanoscale Gold 386
Conclusions 388
References 388

INTRODUCTION: WHY GOLD?

The scientific and technological properties of gold nanoparticles and nanoscale coatings have been the subject of much research in the last decade. There are several stand-alone reviews of these topics in the literature (see, for example, Daniel and Astruc [1] and Glomm [2]), while a whole issue of *Chemical Society Reviews* was recently targeted at these topics [3], and there is at least one other recent text devoted to gold. So why another review here? The reason is that the sources mentioned generally focus on some preferred aspect of the field, usually from a scientific perspective. In contrast, we will attempt here to provide an overarching synopsis of those *technological* attributes and applications of gold nanostructures that we consider interesting or significant. We will not amplify any particular application or its associated science in great detail and instead will provide references to sources in which the interested reader may find more information. Unlike most of the prior reviews, we include the patent literature within the scope of our review. Our list of existing and possible applications for gold at the nanoscale is not exhaustive, but we have selected those applications that seem especially promising.

Why is it gold that is so popular for use in nanotechnology rather than, for example, copper or silver or aluminum or iron? The answer is that gold in the elemental form possesses a suite of properties that make it uniquely suitable for use at nanoscale dimensions. The first and possibly foremost reason is that gold is one of the very few metallic elements that can be prepared in a stable metallic form with nanoscale features or dimensions under ambient oxygen pressure. Other metals, with the possible exception of platinum, will oxidize unless specially protected. Because surface oxides may grow to several micrometers thickness before becoming self-passivating, their presence will generally destroy any nanoscaled shape or feature. This is even true of some other metals considered to be relatively noble; nanoscale copper sponge, for example, is pyrophoric unless passivated with an oxide layer, while naked silver nanoparticles survive in air for only several hours. A second reason for the popularity of gold lies in its chemical properties. It is sufficiently noble to not oxidize under atmospheric conditions but can nevertheless be selectively bonded to sulfur-terminated organic molecules. This permits the exploitation of a nanoscale gold surface as a platform on which to assemble all manner of molecular structures. In particular, the concept of self-assembled monolayers on gold has received much attention [4]. A third reason is that gold nanoparticles and nanostructures are generally readily synthesized. Procedures for producing near-monodisperse gold nanospheres were introduced in the second half of the twentieth century [5–7] while synthetic methods for the high-yield synthesis of other shapes, such as gold nanorods [8], have progressed significantly over the last 15 years. There are also host of lithographic techniques available for creating nanoscale gold structures [9].

A fourth reason for the interest in gold is its interesting optical properties, in particular, the surface plasmon resonance that gold nanoparticles exhibit with light. This causes the optical extinction cross-section of the nanoparticle to considerably exceed its geometric cross-section. Of course, gold is not unique in this respect and the effect is observed in the other coinage metals (Ag, Cu), the alkali metals (Li, Na, K, Cs), and in aluminum [10]. It can be appreciated, however, that production and retention of nanoparticles of the other elements mentioned is challenging due to their reactivity, and gold will be favored because of this. Despite equivalent nobility to gold, platinum does not exhibit a strong optical response and hence is limited in its application. In addition, thin continuous coatings of gold have the highest known reflectivity in the infrared, a feature that leads to application of nanoscale gold coatings in applications as diverse as architectural windows and the space program.

This chapter will first describe the important physical forms of nanoscale gold, after which their many proposed or actual technological applications will be described. Some emphasis will be placed on the properties and applications of nanorods and nanoshells, as exploration of these topics is currently at the forefront of scientific endeavour. Applications for conventional nanospheres will be noted, too, but with less detail as this form of nanoscale gold is already quite well known.

NANOSCALE PARTICLES AND FILMS

GOLD NANOSPHERES

The production of gold nanospheres, whether deliberate or accidental, has a long history. The plasmon resonance peak of gold nanospheres (see Chapter 2 for a discussion of the phenomenon of plasmon resonance) has been used to color glass since Roman times [11], and gold nanospheres were certainly present in some of the ancient medical potions used in China [12], India [13] or Europe [1]. Famous nineteenth century scientists such as Faraday and Zsigmondy were intrigued by the strong colors of colloidal gold, and the famous optical theory of Mie was inspired by their colors. Turkevich developed the very convenient "citrate" synthesis method for gold nanospheres in 1951 [5], which was further improved by Frens in 1973 [6]. These methods produced an aqueous sol of 15–30 nm diameter gold spheres. In 1994 Brust et al. provided a method by which even smaller spheres, 4–10 nm in diameter, could be produced and kept in an organic phase [7]. Among the smallest gold nanoparticles that can be made by chemical means are the ligand-capped clusters, such as $Au_{55}(PPh_3)_{12}Cl_6$ and its derivatives, first pioneered by Schmid in 1981 [14]. These are about a nanometer in size, and so small that they are characterized by how many gold atoms they contain rather than their diameter. Even/smaller gold clusters can be made by physical means, such as by laser ablation [15].

There have been a multiplicity of additional recipes published for all sizes (and shapes) of gold nanoparticles. However, not all proposals are equally useful. For example, there seems to be no merit to preparing gold nanoparticles in vegetable or herbal broths. This is because a gold nanoparticle ensemble is completely characterized by the size, shape, and crystal structure of its gold nanoparticles, the extent of the dispersion in these parameters, and the nature of the surface-capping ligands. Therefore, it does not matter whether the reductant used to prepare it was "artificial" or not. Use of chemically pure reductants and surfactants would always be preferred as it has the advantage of offering reproducibility (in principle) and the opportunity for a rational analysis of the mechanisms occurring.

Gold Nanorods

Gold nanorods (Figure 16.1) are an especially suitable platform for the exploitation of all of gold's useful properties. In particular, they exhibit a significantly greater optical extinction than nanospheres, with efficiencies of up to an order of magnitude greater [16]. Therefore, a desired optical density can be achieved with a smaller volume of nanorods than nanospheres. Even more useful, however, is the fact that nanorods are optically characterized by dual extinction peaks (see Figure 16.2), where the peak at ~515 nm is fixed and attributed to the transverse mode (width) and the dominant peak is tunable (>515–1500 nm) and attributed to the longitudinal mode. The longitudinal mode is dependent on the ratio of the length and width, not just length alone. Nanorods /are frequently qualified in terms of their aspect ratio (length/width) for this reason.

Due to their geometry, nanorods exhibit polarization-dependent optical properties in which excitation of a selected mode is possible. This effect may be observed macroscopically by embedding nanorods in a poly(vinyl alcohol) (PVA) film: the rods become aligned by stretching the film [17,18]. Incident light with electric field parallel to the longitudinal axis will excite the longitudinal mode whereas incident light with the electric field at right angles to the length will excite the transverse mode. The optical response will exhibit features of both modes for randomly oriented nanorods in solution, or for unpolarized light propagating at right angles to the long axis of the rod.

Numerous methods have been applied for the synthesis of nanorods, starting historically with electrochemical deposition within nanoporous alumina membranes [19], followed by photochemical reduction in the presence of a cationic surfactant [20,21] (cetyltrimethylammonium chloride, $C_{16}TAC$), electrochemical reduction in the presence of a cationic surfactant [22] (cetyltrimethylammonium bromide, $C_{16}TAB$), a bioreduction method [23], and finally the most widely used synthesis: seed-mediated wet-chemical reduction in the presence of $C_{16}TAB$ [24] (sometimes also with the cosurfactant benzyldimethylammonium chloride, BDAC [8]). The most noteworthy advancements of the nanorod syntheses have been the surfactant-mediated control of shape, use of $C_{16}TAB$ in place of $C_{16}TAC$, and the recognition of the useful effect of Ag^+ ions. Currently almost all rods are made by batch processes, but there have also been reports of successful rod growth in a continuous reactor [25].

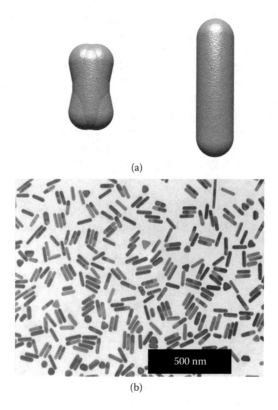

(a)

(b)

FIGURE 16.1 Gold nanorods. (a) Dog bone–shaped rod produced using addition of Ag ions (left); classic right-cylindrical nanorods produced in absence of Ag additions (right). (b) Transmission microscopy image showing an ensemble of right-cylindrical nanorods. (Modified from Tunable infrared absorption by metal nanoparticles: The case for gold rods and shells, N. Harris, M. J. Ford, P. Mulvaney, and M. B. Cortie, *Gold Bull.,* 2008, **41**, 5.)

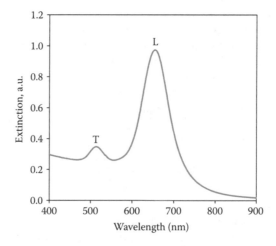

FIGURE 16.2 Typical optical extinction spectrum of dog bone–shaped nanorods, measured for an aqueous suspension (authors, unpublished work).

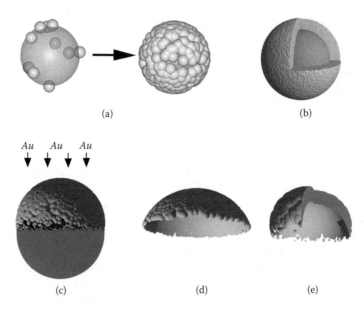

FIGURE 16.3 Nanoshells and semishells. (a) Nanoshells can be made by aggregating gold nanospheres on the surface of a suitable dielectric core. The aggregation can then be "developed" to produce a continuous shell, shown in cutaway view in (b). (c) Semishells can be readily produced by the physical vapor deposition of gold onto a polymer microparticle. (d) Nanocaps are merely one of the possible forms resulting. (e) For geometric reasons the thickness of the shell in these shapes is tapered because more gold is deposited at the top of the templateing sphere than on its sides.

NANOSHELLS AND SEMISHELLS

Another nanoparticle shape that has attracted much interest is the nanoshell, or "core-shell" particle, and its derivatives, the family of semishells, which include hemispheres as well as so-called nano-cups and nanocaps [26] (Figure 16.3). This family of shapes has interesting and complex plasmon resonances, a point which initially attracted only sporadic attention [27–29] before the nanoshells in particular were picked up by a group at Rice University (Houston, Texas) in the late 1990s [30,31]. There is a range of sizes for which both nanorods and nanoshells have very suitable structures in terms of maximizing extinction efficiencies [16,32]; however, as the dimensions of each respective particle fall below their ideal range, peak broadening occurs due to surface scattering of electrons [33,34]. Another attenuating effect on the peaks in the extinction spectrum occurs for dimensions above the ideal range, caused in this instance by increased scattering of light or radiation damping [33,35].

Nanoshell structures are based on two main architectures: core shell and hollow. Core-shell particles consist of a spherical dielectric core, which is commonly SiO_2 [31,36], a polymer [37] or Au_2S [38], and an external metallic coating or shell. The metallic coating of the core-shell particles based on SiO_2 or polymers may be grown chemically by functionalizing the surface of the core so that small Au nanoparticle seeds can attach to it. Additional gold ions and reductant are then added, and the existing nanoparticles act as nucleation sites on which a thicker and continuous coating of Au is grown. A practical point is that SiO_2 and polymer core particles are generally limited to minimum diameters of ~100 nm. This means that the nanoshells that they produce will exhibit considerable scattering of light. Originally the Au_2S core/Au shell was thought to form by a diffusive reduction mechanism in which the outer layers of the Au_2S were reduced to Au [38]. This was followed by an adjusted two stage growth mechanism, supported by Mie theory calculations [30].

Hollow gold nanoparticles can be formed by galvanic replacement of a sacrificial spherical tem-plate, usually silver [39,40] or cobalt [41]. Gold salt, typically $HAuCl_4$ solution, is added gradually to a suspension of the template particles at an elevated temperature (with respect to room temperature). Gold metal is then epitaxially deposited on the surface of the particle while the electrochemically active core dissolves. The final particle geometry is a hollow gold analogue of the template particle, so the nanoshell product will express the same shape and symmetry as the template particle. In the case of silver templates, the shell will initially be an alloy of gold and silver. Excess addition of gold salt will result in pin-hole formation in the shell and ultimately fragmentation [40].

Ideal nanoshells with diameters of less than 50 nm will exhibit a single tunable extinction peak. An increase in the size of the nanoshell size beyond 50 nm leads to the introduction of multipole resonances; in addition, scattering effects begin to dominate the extinction spectrum. The peak in optical extinction for nanoshells may be tuned from ~520 nm into the near-infrared (NIR).

Spherically symmetric nanoshells do not exhibit a polarization-dependent optical response. As with nanorods, anisotropy introduces polarization dependence to nanoshell particles. Anisotropic nano-shells, termed semishells [42,43] or nanocaps [44,45], may be synthesized by shielding part of the core (nanocups) or depositing metal at an oblique angle onto a monolayer of polymer spheres (nanocaps).

Due to the nature of the syntheses for nanoshells there has been some doubt as to the source of the observed optical effects. For example, Au_2S/Au syntheses form mixed populations of nano-shells and nanospheres, as is evident from representative extinction spectra. This has caused some debate whether the observed red-shifted peak was due to nanoshells or aggregates of nanospheres [34,46–48]. However, evidence leans more toward nanoshells with two important examples. The first is sourced from characterization of the growth kinetics of the reaction supported by Mie theory as presented by Averitt et al. [30]. Within this study the development of two extinction peaks is monitored; one peak is characteristic of spherical gold nanoparticles, the other is attributed (by the authors) to the development of a nanoshell of gold. As a nanoshell thickens the extinction peak blue shifts and narrows for low size distributions. Mie theory was used to support this growth mechanism by calculating the combined extinction spectrum of spherical gold nanoparticles and a gradually thickening nanoshell. This model fits the measured growth kinetics closely. Additional evidence for the nanoshell geometry was provided by Raschke et al. who measured single-particle spectra by dark field analysis [34]. Their data provide good evidence for nanoshell geometry because the spectra displayed a narrow linewidth and proved to be polarization independent.

OTHER PARTICLE SHAPES

Although we have emphasized nanospheres, nanorods, and nanoshells, it is worth noting that other particle shapes such as triangles [49,50], hexagons, boxes [51], "cages" [52], semishells [42,45], nanorings [53–55], nanoribbons [56], nanocaps [57,58], nanocups [42,59], rattles [60], and disks [61] can also be made. However, except for the triangles, the yield is currently rather low. In general, selectivity is a problem in the synthesis of the more exotically shaped nanoparticles, with mixed populations being common (Figure 16.4).

MESOPOROUS GOLD SPONGES

De-alloying of Au alloys or intermetallic compounds can be used to produce mesoporous gold sponges [62] of gold, also known as nanoporous gold [63–65], or Raney gold [66]. This material is fully percolated but with a very high surface area (Figure 16.5). Applications that have been investigated include optical coatings [67], catalysts [68–70], substrates for surface enhanced Raman spectroscopy [71,72], biosensor electrodes [73], and high-efficiency supercapacitors [74].

THIN FILMS

The oxidation resistance of Au makes it especially suitable as a coating material. Films of Au can be deposited by electrochemical plating or by one of the several kinds of physical vapor deposition.

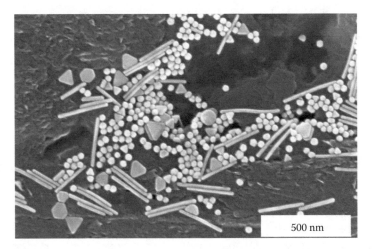

FIGURE 16.4 Scanning electron microscope image of a mixed sample of gold spheres, long aspect ratio nanorods, nanotriangles, and hexagonal plates produced by reduction of HAuCl$_4$ using ascorbic acid and CTAB. (J. Edgar, unpublished work.)

Nanoscale films of gold that are thinner than about 50 nm are partially transparent to light over the visible wavelengths but highly reflective in the infrared wavelengths. This has led to the use of such coatings on glass for spectral selectivity, about which more will be discussed later. The surface of gold films can also support a propagating surface plasmon resonance (SPR) (see Chapter 2). Another attribute of thin gold films is that they can be annealed to show a strong (111) crystal texture with respect to the plane of the substrate. The comparatively large (111) facets on the surface of such films may be used as a platform on which to build a self-assembled monolayer (SAM) or other organic surface structures [4]. Attachment of organic or other molecules to the surface of a gold film, or even to the top of an existing SAM on the surface of a gold film, will modulate the surface plasmon resonance referred to earlier, and is used as the basis of SPR spectroscopy, a sensitive analytical

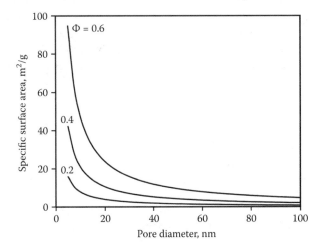

FIGURE 16.5 Specific surface area of mesoporous gold as a function of volume fraction of pores and diameter of pores. This simplified calculation neglects issues of interconnectivity and pore shape. Note also that sponges with values greater than ~0.5 are very fragile and sinter readily. The best experimental value of specific surface area for this system is 20 m^2/g. (Reproduced from Electrochemical capacitance of mesoporous gold, M. B. Cortie, A. I. Maaroof, and G. B. Smith, *Gold Bull.*, 2005, **38**, 15. With permission.)

technique for immunochemical reactions. The field is now relatively mature [75,76] and will not be addressed in any depth here.

INDUSTRIAL APPLICATIONS

PIGMENTS AND COLORANTS

The use of gold nanoparticles to color glass is now very well known [1,11]. The color is due to the plasmon resonance peak at about 520 nm, which absorbs the green portion of white light, and to interband transitions at 300–400 nm, which absorb the blue. What is left is red light, and this is what we see. If the nanoparticles are big enough to also scatter some of the light (as they are for the famous Roman Lycurgus cup) [11], the color of the glass in reflection will be greenish, corresponding to photons scattered off the particles by radiation damping of the plasmon resonance. The gold is introduced during glassmaking in the form of a salt such as $AuCl_3$. After the glass is solidified, there is initially no color, and the precipitation of the Au nanoparticles must generally be induced by annealing [77]. The dye and glaze known as Purple of Cassius consists of gold nanoparticles precipitated onto tin oxide particles. It has been used since 1659 to produce red to purple vitreous enamel glazes for use on high value pottery and porcelain [78].

Gold nanoparticles have also occasionally been used to dye textiles, a practice that dates back to at least 1794 [1] but which is occasionally still reported today with regard to silk [79], wool, and cotton [80]. Reddish to purple colors are obtained. A more novel approach has been to incorporate gold nanoparticles into a paint system which then provides an interesting range of optical effects [55]. None of these technologies, however, are in the commercial mainstream at this time.

METALLIC GLAZES AND SINTER INKS

Gold nanoparticles may also be used as a precursor in mixtures designed to produce thin, continuous coatings of metallic gold. Originally, interest seems to have been directed toward the producing of decorative glazes that could be applied by painting or printing and then fired to produce a golden sheen. More recently, however, there has been keen interest in sinter inks for electronic applications. These are ink systems comprised of nanoparticulate metallic gold (or platinum) in a liquid carrier that can be applied by some conventional technology, such as ink-jet printing. They are then baked at a suitably low temperature to produce a continuous, electrically conductive metallic track or device [81]. Alternatively, the conductive track can be obtained by rastering the ink layer with a laser of the appropriate wavelength and power [82]. This is a rapidly evolving technology and the temperatures required to sinter such deposits have been brought down to less than 140°C [81], which allows for their use on polymer substrates.

SPECTRALLY SELECTIVE COATINGS

Spectrally selective coatings, as the description suggests, are intended to block or transmit a selected range of the electromagnetic spectrum. For example, coatings on the windows of buildings should ideally transmit the entire visible range but block all remaining wavelengths of the solar spectrum. Figure 16.6 shows the standard AM 1.5 solar spectrum and the photo-optic response of the human eye [83]. However, regular window glass blocks the majority of the ultraviolet anyway so, predominantly, the role of any coating on it is to attenuate the near-infrared (NIR) and infrared (IR) wavelengths. There are other applications for spectrally selective coatings, such as the dichroic filters used in optics [84], but of course architectural windows represents by far the largest potential market.

Currently, commercial exploitation of precious metals as spectrally selective coatings manifests in the form of continuous thin films. These films are used for their high reflectance of NIR and IR

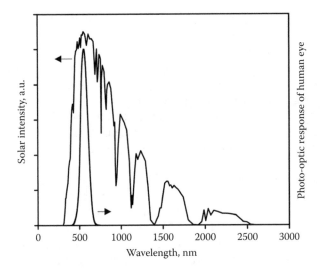

FIGURE 16.6 Intensity of the standard "AM 1.5" solar spectrum at various wavelengths, with the photo-optic response of the human eye superimposed. It is desirable that window coatings maximize transmission within the photo-optic range but minimize it elsewhere in the spectrum.

wavelengths, but in the case of Au films they also reflect some of the upper visible, giving them a reddish tinge in reflection and a green–blue color in transmission. However, some issues with these films include the high cost of the vacuum systems required to produce them, while their high reflectance can cause intolerable quantities of glare in surrounding environments. The coatings are commonly characterized in terms of the ratio of transmitted visible light (weighted for ophthalmic response) to total solar radiation transmitted, a ratio designated as T_{vis}/T_{sol}. Under standardized conditions the maximum possible value for T_{vis}/T_{sol} is 2.08 [85] whereby the entire visible spectrum is transmitted and all outlying wavelengths reflected or absorbed. The ratio possible for thin films of gold varies between 1.0 and 2.0 [86] but brightness in transmission is very low at the upper range of the ratio.

Coatings of hemispherical gold nanoparticles deposited in high density or aggregated films have also been demonstrated as spectrally selective coatings [85,87], but in this case the selectivity is achieved by selective absorption rather than reflection. Whereas isolated gold nanoparticles exhibit a distinct plasmonic absorption peak at about 520 nm, which is nearly in the middle of the visible range of the spectrum, high densities of hemispherical nanoparticles exhibit a broadened and red-shifted absorption peak that extends from the upper visible into the NIR. This increases the ratio of T_{vis}/T_{sol}. So far, however, these films are not yet commercially competitive [85].

Spectrally selective coatings consisting of gold nanorods or nanoshells could increase the optical efficiency of nanoparticulate gold films due to their enhanced optical response and the more precise control possible over peak absorption maxima. It has been determined that the ideal nanorod-based spectrally selective coating would consist of a mixture of nanorods with varying aspect ratios of 3–10 [88]. An analogous optimization is possible with nanoshells [89]. A spectrally selective coating consisting of nanorods with aspect ratios of 3–10 would give a maximum T_{vis}/T_{sol} rating of 1.44 at a gold loading of 0.26 g/m². With gold prices currently approaching ~US$1000/troy ounce this corresponds to a cost of about US$8/m² for the coating without taking glass and production costs into account. This still presents a competitive option as commercially available coatings of similar ratings retail at a much greater price, commonly $50–$100/m².

It is known that particle plasmons begin to couple strongly for identical adjacent nanoparticles separated by less than their diameter. Coupling effects depend on the size, geometry, and orientation of adjacent particles [90]. Calculations show that for axial [84,90,91] (end-to-end) and lateral [84,90] (side-to-side) orientations of nanorods of equal aspect ratio a red shift and blue shift of the particle plasmon will be observed, respectively. In the latter case the intensity of the longitudinal peak is also attenuated [84]. The coupling effects of both arrangements can be closely approximated by the net structure approaching that of a single nanorod of double the original aspect ratio in the axial case or half the original aspect ratio in the lateral case.

Coupling effects in nanoshells have also been determined, where strong red shifts are observed for nanoshells in close contact [92]. Nanoshell coupling effects in which individual nanoparticles exhibit distinct peak positions due to varied aspect ratios appear to have not yet been explored.

Despite these effects, the assumption that interparticle interactions are absent or minimal is not unreasonable for a film of nanorods expressing distinct aspect ratios. This is because the plasmon resonances of nanorods of different aspect ratios (and therefore different frequencies) do not interact much [84]. Calculations of lateral arrangements of two nanorods of different aspect ratio, 2 and 4 (but equal volume), maintain distinct longitudinal peaks even at zero spacing, i.e., touching. In the case of zero spacing between nanorods the intensity of the longitudinal resonance of the aspect ratio 2 nanorod is slightly enhanced whereas the intensity of the aspect ratio 4 nanorod is attenuated. In both instances, however, the position of the peak is red shifted noticeably.

Due to the dependence of the optical properties of nanorods on the polarization of incident light, their orientation also needs to be considered for optimal spectral selectivity. Nanorods lying flat on the surface of the glass will give minimum absorbance at NIR or IR wavelengths (depending on longitudinal peak position) for angles of incidence approaching 0°. This means at times of maximum solar intensity (e.g., at midday) heat transfer due to NIR-IR transmittance will not be attenuated as much as at other times of the day (for a vertically oriented window). However, if the nanorods are oriented perpendicular to the plane of the window then the absorbance at times of high solar angle will be maximized.

In contrast to nanorods, nanoshells do not exhibit polarization dependence due to their spherical symmetry. This could be a pro or a con, depending on the situation.

Laminates or polymer coatings containing dispersed rods or nanoshells could provide a cheap and effective solution for application of spectrally selective coatings of nanoparticles to existing structures. A novel implementation of this is to impart NIR/IR extinction properties of nanoshells to protective eyewear or contact lenses [93].

In addition to spectral shielding applications, mixtures of nanorods and/or nanoshells [89] may be prepared that would absorb the vast majority of the solar spectrum. Such a coating would have utility in solar harvesting for photothermal applications.

HIGH-DENSITY DATA STORAGE

The ever-increasing requirement for storage of digital data has stimulated the introduction of a broad range of innovative technologies. Spectral encoding, that is, the exploitation of optical rather than magnetic properties, is one of the modalities of data storage of interest. In this case, data are written and read by exploiting the optical properties of the data bit. Laser irradiation is used to induce a spectral change in the target region of the recording medium. Depending on the nature of the recording medium, the change may be folding of a protein or molecule, a phase change, or fragmentation or melting of a nanoparticle. In all cases the optical response of the effected region is altered.

Photoresponsive organic and polymeric molecules have been reported as options for spectrally encoded multilayer data storage devices. For multilayer devices a two-photon induced spectral change gives high three-dimensional spatial resolution because only data bits within the focal volume experience an effect [94]. Gold nanorods are an attractive option for optical data storage, too [95]. This is because their optical properties are polarization dependent and tuneable, they are thermally and chemically stable, and they do not bleach or age. Very high data densities are in principle possible due to the nanoscale dimensions of nanorods (typically the diameters are 10–20 nm, and lengths are in the range 30–100 nm). However, with conventional optics a dark field spectroscope is required to detect data bits of single nanoparticles, and this is only applicable to a two-dimensional system. While this already represents an exceedingly high data density, current technology additionally provides the option of detecting single nanoparticle information from multilayer devices using confocal dark-field spectroscopy [96,97]. Similar to the two-photon excitation technique, confocal microscopy images only focused light and so can access different depths of a vertical data storage column.

Data storage devices based on gold nanorods are necessarily limited to the write-once/read-many paradigm as the structural change responsible for the spectral modification is irreversible. However, these ideas can be extended to patterning and encryption applications [98]. Polymer films doped with nanorods could be patterned with characteristic or coded features to prevent counterfeiting of documents or currency.

A nonoptical application of gold nanorods appears as a method for self-assembling and setting magnetic nanoparticles for high-density data storage [99]. Self-assembly at a polar/nonpolar liquid interface is suggested where orientation is assured by coating gold–nickel composite nanorods with hydrophobic thiols and an hydrophilic charged polymer species on the gold and nickel ends, respectively. Dispersion of nanorods at the interface is assumed to be due to electrostatic repulsion of like charges. Nanorods are set by polymerization of an organic monomer solution, for example, methyl methacrylate, where the nickel ends remain free to allow access by a read/write head.

Yet another, quite different type of data storage has been suggested, one which exploits the capacitance of tiny nanoparticles surrounded by an insulating matrix. Flow of charge onto or off the nanoparticles causes the current-voltage characteristics of the material to display a hysteresis during cycling. This is what is essentially required for an electronic memory. Further details of this as yet rather experimental technology may be found in the literature [100].

OPTOELECTRONIC DEVICES AND SWITCHING

There has been interest for some years in developing dielectric matrices containing gold (or other) nanoparticles for their nonlinear optical properties. Switching or control of optic fiber devices at infrared or terahertz frequencies is the most attractive market for these technologies. A review of the subject is available [101], while a paper by Cattarin et al. [102] is an example of recent work in this area. However, in broad summary it seems that the figure-of-merit for third-order optical effects in these systems is not yet sufficiently high to be commercially attractive. Gold nanospheres and nanorods have also been investigated for their potential use in liquid crystal displays [103,104].

CATALYSIS

The fact that gold nanoparticles could, under some circumstances, be very effective catalysts was first noted in 1980s in two quite independent publications by Hutchings [105] and Haruta et al. [106,107]. There has been a growing activity in this field since then, driven by both scientific curiosity and commercial interest [108]. Gold nanoparticle catalysts are interesting because they offer both the prospect of better yield or selectivity in some reactions; and they appear to be effective at temperatures lower than many other catalysts. Gold can, for example, catalyze the oxidation of carbon monoxide at 217 K (–56°C) [109]—a remarkable attribute. More information on the field may be found in Chapter 6, and in the literature [108,110–113]. Here we will merely add that it is not only the nanoparticles that can be efficacious catalysts, but that gold mesoporous sponges have also been shown to be active [68–70].

SENSOR AND ANALYTICAL APPLICATIONS

There is an enormous market available for sensors or devices that can measure or simply detect some highly diluted chemical analyte. Detection of hormones, DNA, or toxins at the level of a few molecules at a time has significant value. This is reflected in the great number of papers and patents in this area. Here we will survey the more popular schemes. Spherical metallic nanoparticles, generally as a proof of concept, have demonstrated remarkable utility as substrates for these applications [114–116]; however, gold nanorods and nanoshells could be even better [117–119] due to their more tunable optical properties.

COLORIMETRIC SENSORS

There are a number nanoparticle sensor schemes that rely on a variation in optical response to indicate detection of an analyte. The simplest (and oldest) one relies on aggregation of a nanoparticle sol when a particular analyte is present. The sol is pink prior to aggregation and turns blue or black after the analyte is detected. This is a colorimetric sensor and only qualitative in nature, but it has been quite widely investigated [120]. The idea is based on a variation of the immunogold reactions established in 1971 by Faulk and Taylor [121] in terms of which gold nanoparticle is functionalized with a molecule that binds to a selected target. Simplified colorimetric assays reliant on aggregation of nanoparticles present a nonquantitative but rapid, indicative sampling method for confirmation of the presence of an analyte [122–124]. There is a home pregnancy test kit based on this principle [125], and the scheme can also be used to detect low levels of any particular type of DNA [122,126] or, indeed, any other analyte, such as lead [127] or cocaine [128], for which a suitable targeting chemistry can be established.

The agglomeration process can also be detected by nonoptical methods [129]. Further information on the underlying mechanism of the aggregation process may be found in the literature (e.g., Kim et al. [130]). A related idea is the use of gold nanopowders in forensic science, in particular for their ability to selectively bind to fingerprints of various types (e.g., Choi et al. [131]).

REFRACTOMETRIC SENSORS

A more sophisticated type of system uses an array of gold nanoparticles attached to a substrate. Each nanoparticle could in principle be conjugated to detect a different analyte. This increases the efficiency of any potential assay by allowing multiple simultaneous analyses from one sample. Arrays such as these would be particularly useful in providing indicative results in emergency rooms or for field analysis in forensics or environmental sampling. However, in order to achieve high sensitivity, a simple detection method capable of probing the optical response of a single nanoparticle is required. Dark-field optical microscopy is emerging as a possible candidate for these requirements as it allows the spectrum of a single particle to be measured using conventional optical techniques [33,119,132–135]. Dark-field microscopy utilizes scattered light from samples to form an image. It cannot resolve the nanoparticles themselves but analysis of the optical signal can provide information about the scattering component of their extinction properties. An alternative method to obtain scattering spectra of single particles is total internal reflection microscopy [136].

The optical properties of gold nanorod sensors can be exploited in at least three ways in such a sensor array: polarization dependence, where the orientation of a tethered nanorod is probed [133]; local refractive index change resulting from surface environment properties changing with binding events of an analyte [117,124,137–139]; and aggregation induced by cross-linking of nanorods due to analyte/detecting molecule binding [123]. Gold nanoshells have also demonstrated three possibilities for sensing architectures. These are SERS detection of adsorbed analytes [140–142]; scattering spectra dependence on local refractive index [118,119]; and aggregation by cross-linking nanoshells resulting in extinction peak broadening and attenuation [143].

The polarization dependence of the optical properties of gold nanorods affords the possibility of orientation based sensors [133,144]. (In a reverse sense, the nanorods could be tethered to a substrate by a conformation switching molecule or polymer to produce color change coatings [84,104].) Change in chemical environment (e.g., pH) or the tether species (e.g., oxidation or specific binding) would trigger a conformational response in the tether. This switch of conformation directly affects the orientation of the bound nanorod and the optical properties in polarized incident light. Dark-field microscopy can be used to detect the orientation change of single gold nanorods by probing the polarized optical response of the sensing substrate and analyzing the signal of individual particles.

As noted, plasmonic response not only depends on polarization and shape but also on the local refractive index of the surrounding medium [145]. A change in local refractive index can be induced by varying the bulk medium in which the nanorod or nanoshell is dispersed, or by an increase or displacement of adsorbates at the interface of the nanoparticle and bulk medium. Variations in the local refractive index have a pronounced effect on the position of the plasmon resonance peaks of gold nanorods and nanoshells. This effect can be exploited to construct trace chemical or biosensors, where binding events in proximity to the nanoparticles trigger a shift in local refractive index and thus optical response.

As an example, a nanorod sensor based on refractive index change has been suggested for the biotin–streptavidin couple [117,137]. In the case of detection of streptavidin, which has a very high affinity for the thiolated biotin molecule, an absorbance peak shift of 366 nm per refractive index unit (RIU) was observed [117]. This corresponded to a limit of detection of 0.42 nM (25 ng/mL). This is superior to the results reported for a similar scheme based on gold nanospheres in which a shift of only 76.4 nm/RIU was achieved, corresponding to a limit of detection of 16 nM (~950 ng/mL) [146]. Yu and Irudayaraj report much higher sensitivities and, importantly, that local refractive index sensitivity of gold nanorods increases with aspect ratio [124]. Nanorods with aspect ratios of 2.8–7 gave sensitivities from 445 to 1100 nm/RIU. It has been claimed that nano-bipyramids are even better than parallel-sided nanorods for this type of sensor [147].

Sensitivity to local refractive index of hollow gold nanoshells has also been reported by Sun and Xia. A maximum response of 408.8 nm/RIU was obtained by immersing nanoshells attached to a glass substrate in varying solvents and measuring extinction spectra [118]. However, the source of this value is uncertain as a calculated peak shift of 306.6 nm/RIU was determined for nanoshell dimensions approximated as representative. This discrepancy has been suggested as being due to polydispersity of the sample [118] supported by broad peak widths of ~325 nm (906 meV) as displayed by the spectrum. SEM and TEM images provided typical particle dimensions of 50 nm total diameter and 4.5 nm shell thickness, dimensions that were utilized in the calculated value.

Dark-field single-particle spectroscopy of gold nanoshells with an Au_2S dielectric core has provided somewhat different results than those for hollow gold nanoshells [119]. A measured value of 123.6 nm/RIU is reported in contrast to a calculated value of 162 nm/RIU. No specific particle dimensions were provided; however, as indicated by the narrow peak width of the scattering spectra, ~60 nm (180 meV), a low gold volume is assumed where ~50% of the damping associated with such a peak width is attributed to surface scattering of electrons when compared with Mie theory. In any case, an increased response to local refractive index variations is evident for nanoshell geometries with respect to spherical gold nanoparticles. Dependence of the plasmon peak position on the chain length of an alkanethiolate adsorbate on a nanoshell has been reported with a shift of ~2.94 nm per methylene unit [118]. From this value the limit of detection for hexadecanethiol was estimated to be ~27 nM.

Assigning a peak shift to the binding event of an analyte could be subject to interference by subtle changes in bulk medium, particularly at trace levels. This problem could be circumvented by the inclusion of unconjugated nanoparticles of distinct extinction peak to that of the conjugated particles. The presence of these particles would provide a reference peak to which the variations due to the medium could be assigned and removed from the detecting peak of the conjugated nanoparticles, leaving the shift due to adsorbed/bound species. In addition to providing a reference peak, a mixture of particles with distinct aspect ratios would allow multiplexing for the detection of multiple species simultaneously [124,148]. Alternatively the single particle spectra could be monitored pre- and postaddition of the sample using dark-field analysis where a "large" area can be monitored with detection achieved through digital analysis of the optical signal from the particles.

Surface plasmon resonance spectroscopy (SPRS) is yet another instance of a refractometric sensor, one that has already become highly developed in a commercial sense, with several instruments available. Excitation of a propagating surface plasmon polariton on the surface of a continuous gold film can be achieved by the use of special techniques that couple incident light onto the gold. Because the properties of the propagating wave depend acutely on the local dielectric constants immediately above the gold film, this technique can be used for some kinds of chemical analysis. Typically, a gold film of only a few tens of nanometers in thickness is used, and it is pre-coated with suitable antibody molecules designed to bind to the desired analyte. Binding causes a detectable and quantitative shift in the optical properties of the film, in particular to details of the reflected spectrum. The field is now relatively mature and several commercially available instruments exist. Further details are available in the literature [76–75].

For extension of nanorod sensors into portable applications miniaturization of the detection method is required. Portability of a sensor is particularly important for fields such as forensics and environmental studies.

ELECTROCHEMICAL AND OTHER SENSORS

As mentioned, gold films are a very suitable platform on which to assemble organic layers of various types. This has ensured their use as electrode substrates for a large variety of electrochemical sensing technologies. This is a large field, which we will not attempt to survey except to note that the transduction of such a sensor may be based on resistive, current (faradaic), or capacitive principles. The very high surface area of nanoparticle aggregates or mesoporous sponges is useful in terms of amplifying the double layer capacitance of electrochemical sensors (e.g., Mortari et al. [73]). Sensors using nanoparticulate aggregates have been proposed for mechanical strain based on variation of electrical resistance [149] or optical properties [150], and for chemical environment based on the phenomenon of chemiresistance [151]. Inversion of the phenomena can give mechanical actuation from a change in electrochemical or chemical environment [152].

SURFACE-ENHANCED RAMAN SPECTROSCOPY

Analytical techniques that exploit nonlinear optical properties have the capacity to gain from the intense near-field enhancement observed in the vicinity of gold nanostructures. This implies that previous detection limits may be surpassed up to the point of single molecule detection [153].

Raman scattering spectroscopy is one such technique to exploit near-field enhancements; detection of a *single* DNA base molecule has been demonstrated using clusters of colloidal *silver* nanoparticles [154]. Clusters of nanoparticles exhibit strong plasmon coupling, with near-field enhancements generally located at the points of contact between particles. Raman spectroscopy probes the inelastic scattering of photons by vibrational modes of chemical bonds (e.g., C–C). Different vibrational modes will give rise to characteristic shifts in the incident radiation. Signal intensity depends on the number of bond species present and Raman scattering events. Unfortunately only a small fraction of incident photons are Raman scattered (~1 in 10^8 photons) with the majority undergoing elastic or Rayleigh scattering. Surface-enhanced Raman scattering (SERS) was first observed in 1974 on a silver electrode roughened by cycled redox reactions [155]. Originally the enhancement of the Raman signal was assigned to an increase in the surface area and thus an increase in adsorbed analyte; however, subsequent work has shown that the Raman effect is amplified by the near-field enhancement of the electric field at the metal surface.

Following a suggestion by Moskovits, Creighton et al. demonstrated SERS from adsorbates on silver and gold sols [156]. Within this work, supporting evidence for the near-field enhancement mechanism was presented as the maximum enhancement coincided with the plasmon resonance of the metallic nanosphere. It should be noted that, following addition of the analyte to the sol, the absorbance spectrum developed a shifted peak most likely due to coagulation. This would have resulted in coupling of the particle plasmons potentially giving slightly inaccurate but elevated enhancements.

An additional order of magnitude enhancement (with respect to spherical nanoparticles) was predicted for substrates comprised of prolate spheroids of silver and gold [157]. Following recent progress in the syntheses of nanorods, these predictions have been demonstrated experimentally [158]. Analogous to the problems of agglomeration experienced in the case of nanospheres, replacing the shape-directing/capping surfactant on nanorods can be difficult. However, SERS has been recorded for the surfactant itself [158] and some foreign analytes [159,160] in solution giving enhancements of approximately five orders of magnitude for plasmon resonant excitations.

SERS on nanoshell substrates is relatively new but large enhancements are also observed [161]. Plasmon resonance dependence of the SERS signal is also evident with nanoshells [161,162]. A single Au nanoshell, with a rough surface, has a calculated SERS enhancement of $\sim 4 \times 10^4$ whereas a Au nanoshell dimer gives a polarization-dependent enhancement of $\sim 5 \times 10^7$ [163]. Arrays of Au or Ag semishells seem to offer considerable potential as a reproducible SERS substrate [164].

Off-resonance Raman enhancement has also been demonstrated for isolated nanorods [158], but no such effect seems to have been observed for isolated nanospheres or nanoshells. SERS from the capping agent of the nanorods was probed resulting in an enhancement of 10^4 to 10^5. This presented an additional one to two orders of magnitude enhancement over the Wang and Kerker model at the same excitation wavelength. Discrepancy with the model in the off-resonance condition is attributed to the increase in the polarizability of the molecule adsorbed to less stable facets of gold nanorods [158], i.e., (110) or (100). Another consideration for discrepancies between experimental results and early theoretical models is the exclusion of the effects of radiation damping which has a negative effect on the SERS enhancement [165].

Field calculations of isolated Au nanoshells and nanoshell dimers illustrate the regions of greatest enhancements (the "hot spots"). For a dimer configuration the strongest field enhancement is in the region of nearest separation of the nanoshells [163,166]. It is also apparent that for dimer pairs of equal aspect ratio but varied volume the enhancement is slightly higher again [166]. Calculations of surface roughened nanoshells indicate that they should cause greater enhancements than smooth nanoshells [163].

Films of aggregated nanospheres and nanorods have also been compared [167]. This provides a convenient medium for exposure to the analyte and (potentially) reuse when coupled with self-cleaning substrates such as TiO_2 [168]. SERS of analytes on aggregated nanorods are expected to give extremely large enhancements.

In general, Ag substrates give a better enhancement than those of Au. However, a test substrate of Au would be more reliable after protracted storage due to Au's resistance to oxidation. Overall, there is considerable commercial interest in SERS requiring that factory-made substrates be stable for extended times. Shifting of the wavelength of the plasmon resonance to the near-infrared by use of Au is also possibly useful because it allows the SERS measurement to be taken, in principle, through the "tissue window" [169] in live patients.

As a final example of a SERS-based sensor, we note the "all-optical nanoscale pH meter" utilizing a pH-sensitive adsorbate on SiO_2/Au (core/shell) nanoshells [140]. In this case, the value of the pH is determined by analyzing the SERS signal of 1,4-mercaptobenzoic acid (para-MBA) which is in a protonated, partially protonated or deprotonated state at low, intermediate or high pH, respectively (in the range pH 5.80–7.60). An analogous functionality could be anticipated with regard to para-MBA conjugated nanorods.

USE OF GOLD IN MEDICAL DIAGNOSTICS

General

Gold nanoparticles have been the basis of diverse cell or protein-specific staining techniques for a few decades, a capability that was evidently pioneered by Faulk and Taylor in 1971 [121]. Being a heavy element, Au nanoparticles in tissue sections are particularly visible in an electron microscope [170]. This fact is used in combination with the ability of gold nanoparticles to bind to selective antibodies to produce highly selective stains for biological tissue. Actually, because the gold nano-particles also scatter visible light, such selective binding can often also be detected in an ordinary optical microscope, possibly using the dark-field techniques described earlier. Many of the relevant principles and procedures are summarized in an authoritative text on the subject [171]. There has been continuing development of the technology, with the enhanced scattering properties of large gold nanoshells [172] and nanorods [173] having been recently exploited for the early diagnosis of cancer.

A different mode of detection is exploited in the aggregation type of bioassay, which has already been discussed earlier.

Biomedical imaging techniques, in particular, multiphoton luminescence microscopy [174,175], optical coherence tomography [176], and photoacoustic imaging [177–179], have utilized gold nanorods as contrast-enhancing agents. SERS has been demonstrated as a complimentary feature to diagnostic applications of conjugated nanorods whereby analysis of the Raman signal allows for distinction between benign and cancer cell lines [160]. Nanoshells have also demonstrated contrast enhancement for photoacoustic imaging [180], two-photon luminescence microscopy [181], and contrast enhancement for optical coherence tomography [182]. These enhancements of existing medical diagnostic techniques have implications for improved procedures which could lead to improvement in treatment due to early detection. Each will be discussed in turn in the following sections.

Two-Photon Luminescènce

Two-photon induced luminescence (TPL) is a multiphoton process whereby absorption of two photons causes an electron to transit from ground to excited state by way of an intermediate virtual state [183]. Relaxation results in the recombination of the electron-hole pair and emission of a photon. Probability of two-photon absorption is very low, so high intensity irradiation is required for a measurable effect. Although this seems like a disadvantage it signifies that TPL will be produced only within a localized volume around the focal point [183]. This means that by scanning through x, y, and z a detailed three-dimensional image may be formed with boundaries on resolution being primarily limited to the spot size of the laser. Diagnostically, imaged data presented in two-dimensional slices is highly useful as it provides cross-sectional information of irregularities or areas of interest. Rendering the cross-sectional information into three-dimensional reconstructions provides additional information, which can guide surgical procedures.

Plasmon-enhanced TPL of roughened metal substrates was characterized in 1986 by Boyd et al. [184]. Gold nanorods [175] and nanoshells [181] exhibit a plasmon-enhanced TPL signal where the intense near-field, present under resonant conditions, enhances the relatively weak luminescence. A single gold nanorod produces a TPL signal ~60 times brighter than the two-photon fluorescence from a single rhodamine 6G molecule, a common contrast agent (2320 GM and 40 GM for gold nanorod and rhodamine 6G, respectively) [175]. (Two-photon emission intensities are quantified in terms of an action cross-section given in Göppert-Mayer (GM) units [183].) Of course, semiconducting CdSe-ZnS quantum dots (QDs) have demonstrated TPL intensities of 2000–47000 GM [185] which is far greater than for gold nanorods. However, such QDs are highly toxic and although *in vivo* experiments have been performed [185], the protective layer coating the QDs could break down prior to excretion resulting in heavy metal (Cd) poisoning. Gold nanoparticles are generally considered as bioinert [14,186].

In vivo demonstrations of TPL with gold nanorods have been published utilizing live mouse models. In this case unmodified nanorods were injected intravenously upon which single nanorod TPL signals were observed within the vasculature [175]. Targeted TPL microscopy is also possible where conjugated nanorods are injected for specific location of, for example, tumor cells [174]. Once located the laser excitation can be focused on the tumor cells to induce localized plasmonic heating and hence cell death (see the section on medical therapeutics to follow). Targeted TPL with gold nanoshells has also been shown *in vitro* trials [181].

Magnetic Resonance Imaging

Due to their geometry, incorporation of super-paramagnetic iron oxide nanoparticles (SPIO) into the core of metallic nanoshells allows for contrast enhancement for magnetic resonance imaging (MRI) and subsequent photothermal treatment [187]. Super-paramagnetic properties can also be imposed on gold nanorods by coating with iron oxide nanoparticles [188]. MRI is a broadly applied diagnostic tool for discerning abnormal tissue deposits or for studying body mechanics at a fine scale. SPIO nanoparticles have been shown to provide strong contrast in MRI for numerous systems (e.g., bowel, liver, lymph nodes, etc.) [189]. Super-paramagnetism refers to the absence of a net magnetic moment without an externally applied magnetic field. This property of the nanoparticles prevents them from self-aggregating due to magnetic attraction and protective coatings allow for separation after termination of the external magnetic field. Depending on their size SPIO nanoparticles can be used as negative or positive contrast agents.

Gold nanoshells with an inner SPIO nanoparticle core have demonstrated high contrast efficiency for MRI [187,190]. In addition, the metallic shell can be conjugated with antibodies raised against a particular cellular trait (e.g., tumor cells), and upon NIR irradiation photothermal treatment may be administered and subsequently tracked using MRI [187,190].

Photoacoustic Microscopy

Photoacoustic microscopy is a sensitive diagnostic tool for detecting abnormalities in soft tissues such as skin and vasculature. This technique is noninvasive and, unlike X-ray imaging, nonionizing so there is no undue damage caused by the imaging process. A photoacoustic signal is induced by irradiating the target region with pulses from a laser. Body tissue within this region will absorb the incident radiation and undergo rapid thermal expansion. This rapid expansion induces an acoustic signal that is detected by an ultrasonic transducer.

Formation of the image is analogous to conventional ultrasound. Two-dimensional [191] and three-dimensional [192] images can be formed using photoacoustic microscopy, with the three-dimensional images being reconstructed from two-dimensional slices of incremental depth, a process known as tomography. As mentioned previously, to increase the penetration of the laser, wavelengths in the red-NIR range (650–900 nm) must be used. However, as the strength of the photoacoustic signal is proportional to the absorption coefficient and the absorption by unmodified soft tissue is minimal at these wavelengths, formation of acoustic images is difficult. Gold nanorods [177–179,193] or nanoshells [180] tuned to express maximum absorption within the soft tissue window can be used to increase the absorption dramatically and thus increase the acoustic signal through plasmonic heating effects. Specific cell line targeting can be achieved by conjugating the surface of the nanoparticles with antibodies. This process can be used for diagnostics or observation of retaliatory systemic processes, such as nanorod enhanced imaging of early inflammatory response [178]. Also systemic imaging can be performed such as improved contrast in brain images as provided by intravenous injections of nanoshells [180]. This could result in early detection, and thus prevention, of life-threatening conditions.

Optical Coherence Tomography

Optical coherence tomography (OCT) was first modified for biological systems by Huang et al. in 1991 [194]. OCT provides high-resolution, *in vivo* images of biological systems rapidly and noninvasively. The principal of operation is analogous to ultrasonic imaging but uses NIR light in place of sound waves. A two-dimensional cross-section is formed by analyzing differential scattering signals using low-coherence interferometry. Time-of-flight information is determined by use of a Michelson interferometer where the low-coherence light source is split into two paths. One path reflects off a reference mirror, the other is incident on the tissue sample. By scanning the reference mirror position the amplitudes and delays from the sample reflections are quantified. OCT has proven to be a useful diagnostic tool and has made significant advancements for ophthalmic diagnoses [195], for conditions such as glaucoma and macular degeneration.

Nanoshells tuned to exhibit high extinction in the soft tissue and OCT illumination window have been shown to provide high contrast enhancement for healthy tissue and tumorous tissue [182]. These nanoshells were synthesized to exhibit both high absorption and scattering cross-sections. High scattering properties are essential for providing contrast in the OCT image.

THERAPEUTIC POSSIBILITIES FOR NANOSCALE GOLD

For medical applications gold nanoparticles are synthesized to express peak absorbance within the soft tissue window, ~650–900 nm [169]. Within this range hemoglobin and water, primary absorbers of visible and IR wavelengths, respectively, exhibit their lowest absorption coefficient allowing penetration of light up to 10 cm [169]. Implementation of gold nanoparticles in *in vivo* medical applications relies on characterization of their effect on biological systems. In general, studies utilizing gold nanoparticles have revealed that any cytotoxicity that they might have is due to species, usually surfactants or capping ligands, adsorbed onto their surface. There is broad (but not complete) agreement in the field that the gold itself is inert. For example, in the case of gold nanorods, the shape directing surfactant (C_{16}TAB) is a known cell-lysing agent [196]. Therefore, pretreatment of the nanorods to remove excess C_{16}TAB is required [197]. In contrast, C_{16}TAB-coated nanorods show low cytotoxicity [198–200], but by using various layer-by-layer polymer coating architectures the toxicity can be further reduced and the surface properties engineered to produce a tunable cell uptake [198]. Nanoshells are also reported to show low toxicity for a range of cell lines [201,202].

The rational therapeutic use of gold nanoparticles [203] is based on one of three modalities. The particles can be used (1) to deliver either a localized heating effect, (2) as a platform onto which to attach cytotoxic or other organic molecules, or (3) as radioactive sources. Combined therapeutic and diagnostic applications of gold nanorods [123,175,178,199,204–207] and gold nanoshells [180,182,187,208—213] have been published [182,214–217]. Particles with such dual functionality are sometimes described as "theranostic" [218]. The complex range of possibilities for the medical use of gold nanoparticles is summarized in Figure 16.7, which provides a suggested taxonomy of applications.

The application of nanorods and nanoshells in the photothermal mode can be both primary and secondary. Both utilize the localized heating of nanoparticles induced under plasmon resonant conditions [219] but they differ in terms of the effect sought from this heating. Primary nanoparticle treatment involves two main stages: First, nanoparticles are conjugated with antibodies for specific cell line targeting; upon injection the conjugated particles will bind to cells expressing targeted characteristics. Induction of cell death, stage two, is achieved by localized plasmonic heating at nanoparticle binding sites, that is, tumor site (target cells), using laser irradiation. This localized heating causes thermal stress resulting in cell death.

Primary nanoshell applications mainly utilize photothermal tumor ablation in demonstrating the efficacy of nanoshells as therapeutic agents [220] and were pioneered by the group of N. Halas at Rice University in Houston, Texas [172,209,211]. Targeting excessive or inappropriate neovasculature, which is commonly a function of tumor development, is a suggested demonstration of hyperthermic nanoshell treatment [221]. Examples of nanoshell photothermal therapy tend to utilize laser intensities of ~4×10^4 W.m^{-2} [182,209,211], which are comparatively high. Postsurgical treatment has also seen improvement from plasmonic heating properties of nanoshells. Near-infrared laser

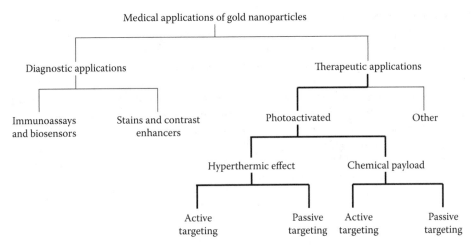

FIGURE 16.7 Classification of the biomedical uses of gold nanoparticles. (Reproduced from Gold nanosphere-antibody conjugates for hyperthermal therapeutic applications, D. Pissuwan, C. H. Cortie, S. M. Valenzuela, and M. B. Cortie, *Gold Bull.*, 2007, **40**, 121. With permission.)

tissue welding exploits a "bio-solder" to improve healing of surgical incisions [208]. The bio-solder consists of a moderate concentration of nanoshells in a bovine serum albumen (BSA) solution. Near-infrared wavelengths are used to minimize damage to surrounding areas and increase penetration which ultimately results in a stronger, faster healing weld.

In the case of gold nanorods, laser intensities as low as 5×10^2 W·m^{-2} have been used to induce cell death where dependence on total laser fluence rather than flux is emphasized (a minimum fluence of 30 J·cm^{-2} seems required [206]). At such low power less than 1% of healthy cells were killed whereas 81% of the target cells (to which the gold nanorod had been conjugated) were terminated. Several applications of primary nanorod therapeutics have been demonstrated ranging from targeted destruction of immune cells or macrophages [206] to destruction of tumor cells [214], bacteria [222,223], and even of protozoan parasites [199]. Although the destruction of macrophage cells may seem counterintuitive in some cases it may prove beneficial [224] assuming antibodies can be raised to differentiate between healthy and infected macrophages.

Secondary nanoparticle therapeutic applications also utilize plasmonic heating properties. Nanorods [225] or nanoshells [226,227] are embedded in a thermoresponsive polymer matrix doped with a therapeutic agent. The thermoresponsive polymer, commonly NiPAAm (N-isopropylacrylamide), could be injected as a colloidal solution of micrometer-scaled beads or implanted as a pellet [213, 228]. Alternatively the core of the nanoshell could constitute the thermoresponsive polymer/therapeutic agent for diffusive release upon plasmonic heating [229]. Sequestering the composite within the nanoshell leaves the surface for adsorption of antibodies etc. for targeting particular regions or species in the body to maximize therapeutic efficacy. Once inserted, NIR laser irradiation excites the plasmon of the nanoparticles whereby the surrounding medium is heated. This causes a conformational change of the polymer molecules, leading to a decrease in volume of the composite. This decrease is accompanied by release of the therapeutic agent into the surrounding environment, e.g., the bloodstream. By controlling the laser intensity or exposure of the polymer composite the rate of drug release could be controlled to achieve prolonged, constant administration to maximize therapeutic efficacy.

Another somewhat unrelated mode of therapy is to use the gold nanoparticles simply as platforms with which to transport a drug [230] or genetic material [231] into a cell. In this case there is no plasmonic heating required and the property of the gold that is being exploited is its surface chemistry.

A major issue with the application of nanoparticles to medical applications is the relatively unknown effect that the particles themselves have over long periods. The broadest view is that they are excreted as waste in an unspectacular manner and are generally nontoxic [197]. Dynamic light scattering has been demonstrated as a quantitative method for monitoring the concentration of nanoshells remaining in the bloodstream prior to administration and exhaustion of their therapeutic role [232]. This method could also be applied to monitoring concentrations of nanorods, although interpretation of the results requires additional consideration due to the rod geometry [233]. The uptake of gold nanoparticles into live cells seems to depend on their size, shape, and the nature of any capping ligands present [186].

Finally, we note in passing that it has been claimed that gold nanoparticles can serve as an active adjuvant (a substance that increases the efficacy of a vaccine) [234].

CONCLUSIONS

Gold nanoparticles and nanostructures have physical and chemical properties that make them uniquely suited for exploitation in diverse forms of nanotechnology. Applications of gold nanoparticles have been greatly strengthened by the advancement in synthetic procedures for control of their geometry. In particular, nanorods and nanoshells provide a simple yet effective means to exploit the features of nanoscale gold. The variations in physical dimensions and morphologies provide seemingly limitless options for exploitation of these effects. In general, it appears that synthesis of nanorods provides the simplest route to a reliable, monodisperse product. Unfortunately, current synthetic procedures are limited to producing rods of aspect ratios of less than about five in practical quantities. Nevertheless, this is long enough to shift the longitudinal plasmon peak of the rods into the near-infrared, an attribute which is desired in many applications.

The applications of the nanoparticles and structures that we have surveyed here have extended from various kinds of sensors and analytical devices, through spectrally selective coatings and sinter inks to medical diagnostics and therapeutics. Nanoscale gold must surely be one of the most versatile and valuable materials systems available!

REFERENCES

1. Gold nanoparticles: assembly, supramolecular chemistry, quantum-size-related properties, and applications toward biology, catalysis, and nanotechnology, M.-C. Daniel and D. Astruc, *Chem. Rev.,* 2004, **104**, 293.
2. Functionalized gold nanoparticles for applications in bionanotechnology, W. R. Glomm, *J. Dispersion Sci. Technol.*, 2005, **26**, 389.
3. Gold—an introductory perspective, G. J. Hutchings, M. Brust, and H. Schmidbaur, *Chem. Soc. Rev.*, 2008, **37**, 1746.
4. Self-assembled monolayers of thiolates on metals as a form of nanotechnology, J. C. Love, L. A. Estroff, J. K. Kriebel, R. G. Nuzzo, and G. M. Whitesides, *Chem. Rev.*, 2005, **105**, 1103.
5. A study of the nucleation and growth processes in the synthesis of colloidal gold, J. Turkevich, P. C. Stevenson, and J. Hillier, *Discuss. Faraday Soc*, 1951, 55.
6. Controlled nucleation for the regulation of the particles size in monodisperse gold suspensions, G. Frens, *Nat. Phys. Sci.*, 1973, **241**, 20.
7. Synthesis of thiol-derivatized gold nanoparticles in a 2-phase liquid-liquid system, M. Brust, M. Walker, D. Bethell, D. J. Schiffrin, and R. Whyman, *Chem. Commun.*, 1994, 801.
8. Preparation and growth mechanism of gold nanorods (NRs) using seed-mediated growth method, B. Nikoobakht and M. A. El-Sayed, *Chem. Mater.*, 2003, **15**, 1957.
9. Preparation of nanoscale gold structures by nanolithography, N. Stokes, A. M. McDonagh, and M. B. Cortie, *Gold Bull.*, 2007, **40**, 310.
10. Plasmon absorption in nanospheres: A comparison of sodium, potassium, aluminium, silver and gold, M. G. Blaber, M. D. Arnold, N. Harris, M. J. Ford, and M. B. Cortie, *Phys. B,Condensed Matter*, 2007, **394**, 184.

11. The Lycurgus Cup – a Roman nanotechnology, I. Freestone, N. Meeks, M. Sax, and C. Higgitt, *Gold Bull.*, 2007, **40**, 270.

12. China's ancient gold drugs, Z. Huaizhi and N. Yuantao, *Gold Bull.*, 2001, **34**, 24.

13. Nanogold pharmaceutics. (i) The use of colloidal gold to treat experimentally-induced arthritis in rat models; (ii) Characterization of the gold in Swarna bhasma, a microparticulate used in traditional Indian medicine, C. L. Brown, G. Bushell, M. W. Whitehouse, D. Agrawal, S. Tupe, K. Paknikar, and E. R. Tiekink, *Gold Bull.*, 2007, **40**, 245.

14. The relevance of shape and size of Au_{55} clusters, G. Schmid, *Chem. Soc. Rev.*, 2008, **37**, 1909.

15. Structures of neutral Au_7, Au_{19}, and Au_{20} clusters in the gas phase, P. Gruene, D. M. Rayner, B. Redlich, A. F. G. v. d. Meer, J. T. Lyon, G. Meijer, and A. Fielicke, *Science*, 2008, **321**, 674.

16. Calculated absorption and scattering properties of gold nanoparticles of different size, shape, and composition: applications in biological imaging and biomedicine, P. K. Jain, K. S. Lee, I. H. El-Sayed, and M. A. El-Sayed, *J. Phys. Chem. B.*, 2006, **110**, 7238.

17. Optical properties of aligned rod-shaped gold particles dispersed in poly(vinyl alcohol) films, B. M. I. Van der Zande, L. Pages, R. A. M. Hikmet, and A. Van Blaarderen, *J. Phys. Chem. B*, 1999, **103**, 5761.

18. Optical control and patterning of gold-nanorod poly(vinyl alchohol) nanocomposite films, J. Pérez-Juste, B. Rodriguez-Gonzalez, P. Mulvaney, and L. M. Liz-Marzan, *Adv. Func. Mater.*, 2005, **15**, 1065.

19. Transparent metal microstructures, M. J. Tierney and C. R. Martin, *J. Phys. Chem.*, 1989, **93**, 2878.

20. Preparation of rodlike gold particles by UV irradiation using cationic micelles as a template, K. Esumi, K. Matsuhisa, and K. Torigoe, *Langmuir*, 1995, **11**, 3285.

21. Method for manufacturing metal nanorods and use thereof, Y. Y. Niidome, S. K. Nishioka, H. Kawasaki, H. Hirata, Y. Takata, J. Satoh, D. Mizoguchi, M. Ishihara, M. Nagai, and M. Murouchi, 2006, US 2006/0196309.

22. Gold nanorods: electrochemical synthesis and optical properties, Y.-Y. Yu, S.-S. Chang, C.-L. Lee, and C. R. C. Wang, *J. Phys. Chem. B*, 1997, **101**, 6661.

23. Multiple twinned gold nanorods grown by bio-reduction techniques, G. Canizal, J. A. Ascencio, Gardea-Torresday, J., Yacaman, and M. Jose, *J. Nanoparticle Res.*, 2001, **3(5/6)**, 475.

24. Seed-mediated growth approach for shape-controlled synthesis of spheroidal and rod-like gold nanoparticles using a surfactant template, N. R. Jana, L. A. Gearheart, and C. J. Murphy, *Adv. Mater.*, 2001, **13**, 1389.

25. Microfluidic continuous flow synthesis of rod-shaped gold and silver nanocrystals, J. Boleininger, A. Kurz, V. Reuss, and C. Sönnichsen, *Phys. Chem. Chem. Phys.*, 2006, **8**, 3824.

26. Investigation of the optical properties of hollow aluminum "nano-caps", J. Liu, B. Cankurtaran, G. McCredie, M. Ford, L. Wieczorek, and M. Cortie, *Nanotechnology*, 2005, **16**, 3023.

27. Scattering of electromagnetic waves from two concentric spheres, A. L. Aden and M. Kerker, *J. App. Phys.*, 1951, **22**, 1242.

28. Elastic scattering, absorption, and surface-enhanced Raman scattering by concentric spheres comprised of a metallic and a dielectric region, M. Kerker and C. G. Blatchford, *Phys. Rev. B*, 1982, **26**, 4052.

29. Composite structures for the enhancement of nonlinear-optical susceptibility, A. E. Neeves and M. H. Birnboim, *J. Opt. Soc. Am. B*, 1989, **6**, 787.

30. Plasmon resonance shifts of Au-coated Au_2S nanoshells: insight into multicomponent nanoparticle growth, R. D. Averitt, D. Sarkar, and N. J. Halas, *Phys. Rev. Lett.*, 1997, **78**, 4217.

31. Nanoengineering of optical resonances, S. J. Oldenburg, R. D. Averitt, S. L. Westcott, and N. J. Halas, *Chem. Phys. Lett.*, 1998, **288**, 243.

32. Plasmon absorption in gold nanoparticles: shells versus rods, N. Harris, M. B. Cortie, M. J. Ford, and P. Mulvaney, *Gold Bull.*, 2008, **40**, 5.

33. Contributions from radiation damping and surface scattering to the linewidth of the longitudinal plasmon band of gold nanorods: a single particle study, C. Novo, D. Gomez, J. Perez-Juste, Z. Zhang, H. Petrova, M. Reismann, P. Mulvaney, and G. V. Hartland, *Phys. Chem. Chem. Phys.*, 2006, **8**, 3540.

34. Gold nanoshells improve single nanoparticle molecular sensors, G. Raschke, S. Brogl, A. S. Susha, A. L. Rogach, T. A. Klar, J. Feldmann, B. Fieres, N. Petkov, T. Bein, A. Nichtl and K. Kürzinger, *Nano Lett.*, 2004, **4**, 1853.

35. Evaluation of the limits of resonance tunability in metallic nanoshells with spectral averaging method, S. Schelm and G. B. Smith, *J. Opt. Soc. Am. A*, 2005, **22**, 1288.

36. Optically-absorbing nanoparticles for enhanced tissue repair, J. L. West, R. Drezek, S. Sershen, and N. J. Halas, 2004, United States Patent 6,685,730.

37. Gold nanoshells on polystyrene cores for control of surface plasmon resonance, W. Shi, Y. Sahoo, M. T. Swihart, and P. N. Prasad, *Langmuir*, 2005, **21**, 1610.

38. Controlled synthesis and quantum-size effect in gold-coated nanoparticles, H. S. Zhou, I. Honma, H. Komiyama, and J. W. Haus, *Phys. Rev. B*, 1994, **50**, 12052.

39. Template-engaged replacement reaction: A one-step approach to the large-scale synthesis of metal nanostructures with hollow interiors, Y. Sun, B. T. Mayers, and Y. Xia, *Nano Lett.*, 2002, **2**, 481.

40. Alloying and dealloying processes involved in the preparation of metal nanoshells through a galvanic replacement reaction, Y. Sun and Y. Xia, *Nano Lett.*, 2003, **3**, 1569.

41. Synthesis, characterization, and tunable optical properties of hollow gold nanospheres, A. M. Schwartzberg, T. Y. Olson, C. E. Talley, and J. Z. Zhang, *J. Phys. Chem. B.*, 2006, **110**, 19935.

42. Reduced symmetry metallodielectric nanoparticles: chemical synthesis and plasmonic properties, C. Charnay, A. Lee, S. Man, C. E. Moran, C. Radloff, R. K. Bradley, and N. J. Halas, *J. Phys. Chem. B*, 2003, **107**, 7327.

43. Partial coverage metal nanoshells and method of making same, N. J. Halas and R. K. Bradley, 2003, US Patent 6,660,381.

44. Anisotropic optical properties of semi-transparent coatings of gold nano-caps, J. Liu, B. Cankurtaran, L. Wieczorek, M. J. Ford, and M. B. Cortie, *Adv. Func. Mater.*, 2006, **16**, 1457.

45. Fabrication of hollow metal nanocaps and their red-shifted optical absorption spectra, J. Liu, A. I. Maaroof, L. Wieczorek, and M. B. Cortie, *Adv. Mater.*, 2005, **17**, 1276.

46. Near infrared optical absorption of gold nanoparticle aggregates, T. J. Norman, J. C. D. Grant, D. Magana, J. Z. Zhang, J. Liu, D. Cao, F. Bridges, and A. V. Buuren, *J. Phys. Chem. B*, 2002, **106**, 7005.

47. Reply to "Comment on 'Gold Nanoshells Improve Single Nanoparticle Molecular Sensors'," G. Raschke, S. Brogl, A. S. Susha, A. L. Rogach, T. A. Klar, J. Feldmann, B. Fieres, N. Petkov, T. Bein, A. Nichtl, and K. Kurzinger, *Nano Lett.*, 2005, **5**, 811.

48. Comment on "Gold nanoshells improve single nanoparticle molecular sensors," J. Z. Zhang, A. M. Schwartzberg, T. Norman, C. D. Grant, J. Liu, F. Bridges, and T. v. Buuren, *Nano Lett.*, 2005, **5**, 809.

49. Electric-field-assisted growth of highly uniform and oriented gold nanotriangles on conducting glass substrates, P. R. Sajanlal and T. Pradeep, *Adv. Mater.*, 2008, **20**, 980.

50. Biological synthesis of triangular gold nanoprisms, S. S. Shankar, A. Rai, B. Ankamwar, A. Singh, A. Ahmad, and M. Sastry, *Nature Mater.*, 2004, **3**, 482.

51. Shape-controlled synthesis of gold and silver nanoparticles, Y. Sun and Y. Xia, *Science*, 2002, **298**, 2176.

52. Gold nanocages: bioconjugation and their potential use as optical imaging contrast agents, J. Chen, F. Saeki, B. J. Wiley, H. Cang, M. J. Cobb, Z.-Y. Li, L. Au, H. Zhang, M. B. Kimmey, X. Li, and Y. Xia, *Nano Lett.*, 2005, **5**, 473.

53. Preparation of mesoscopic gold rings using particle imprinted templates, F. Yan and W. A. Goedel, *Nano Lett.*, 2004, **4**, 1193.

54. Synthesis of hollow gold nanoparticles and rings using silver templates, J. A. Edgar, H. Zareie, A. Dowd, and M. Cortie. *International Conference on Nanoscience and Nanotechnology (ICONN2008)*, Melbourne, Australia, 25th-29th Feb. 2008, ARC Nanotechnology Network, Canberra.

55. Coating materials containing gold nanoparticles, A. Iwakoshi, T. Nanke, and T. Kobayashi, *Gold Bull.*, 2005, **38**, 107.

56. Aqueous-phase room-temperature synthesis of gold nanoribbons: soft template effect of a gemini surfactant, M. S. Bakshi, F. Possmayer and N. O. Petersen, *J. Phys. Chem. C*, 2008, **112**, 8259.

57. Fabrication and wetting properties of metallic half-shells with submicron diameters, J. C. Love, B. D. Gates, D. B. Wolfe, K. E. Paul, and G. M. Whitesides, *Nano Lett.*, 2002, **2**, 891.

58. Surface-adsorbed polystyrene spheres as a template for nano-sized metal particle formation. Optical properties of nano-sized Au particle, H. Takei, *J. Vac. Sci. Technol. B*, 1999, **17**, 1906.

59. Applications of surface textures produced with natural lithography, H. W. Deckman and J. H. Dunsmuir, *J. Vac. Sci. Technol. B*, 1983, **1**, 1109.

60. Synthesis of nanorattles composed of gold nanoparticles encapsulated in mesoporous carbon and polymer shells, M. Kim, K. Sohn, H. Bin Na, and T. Hyeon, *Nano Lett.*, 2002, **2**, 1383.

61. Optical properties of short range ordered arrays of nanometer gold disks prepared by colloidal lithography, P. Hanarp, M. Kall and D. S. Sutherland, *J. Phys. Chem. B*, 2003, **107**, 5768.

62. Mesoporous gold sponge, M. B. Cortie, A. I. Maaroof, N. Stokes, and A. Mortari, *Australian J. Chem.*, 2007, **60**, 524.

63. Nanoporous gold leaf: "Ancient technology"/advanced material, Y. Ding, Y.-J. Kim, and J. Erlebacher, *Adv. Mater.*, 2004, **16**, 1897.

64. Evolution of nanoporosity in dealloying, J. Erlebacher, M. J. Aziz, A. Karma, N. Dimitrov, and K. Sieradzki, *Nature*, 2001, **410**, 450.

65. Volume change during the formation of nanoporous gold by dealloying, S. Parida, D. Kramer, C. A. Volkert, H. Rösner, J. Erlebacher, and J. Weissmüller, *Phys. Rev. Lett.*, 2006, **97**, 035504.
66. The characterization of porous electrodes by impedance measurements, J. P. Candy, P. Fouilloux, M. Keddam, and H. Takenouti, *Electrochim. Acta*, 1981, **26**, 1029.
67. Optical properties of mesoporous gold films, A. I. Maaroof, M. B. Cortie, and G. B. Smith, *J. Opt. A: Pure Appl. Opt.*, 2005, **7**, 303.
68. Gold catalysts and methods for their preparation, L. Glaner, E. van der Lingen, and M. B. Cortie, 2003, Australian Patent 2003/215039.
69. M. B. Cortie, E. v. d. Lingen and G. Pattrick, in *Proceedings of the Asia Pacific Nanotechnology Forum 2003* (ed. Schulte, J.) 79, World Scientific, Singapore, Cairns, Australia, 2003.
70. Low temperature CO oxidation over unsupported nanoporous gold, C. Xu, J. Su, X. Xu, P. Liu, H. Zhao, F. Tian, and Y. Ding, *J. Am. Chem. Soc.*, 2006, **129**, 42.
71. Surface-enhanced Raman scattering on nanoporous Au, S. O. Kucheyev, J. R. Hayes, J. Biener, T. Huser, C. E. Talley, and A. V. Hamza, *Appl. Phys. Lett.*, 2006, **89**, 053102.
72. Surface enhanced Raman scattering of nanoporous gold: smaller pore sizes stronger enhancements, L. H. Qian, X. Q. Yan, T. Fujita, A. Inoue, and M. W. Chen, *Appl. Phys. Lett.*, 2007, **90**, 153120.
73. Mesoporous gold electrodes for measurement of electrolytic double layer capacitance, A. Mortari, A. Maaroof, D. Martin, and M. B. Cortie, *Sensors and Actuators B*, 2007, **123**, 262.
74. Electrochemical capacitance of mesoporous gold, M. B. Cortie, A. I. Maaroof, and G. B. Smith, *Gold Bull.*, 2005, **38**, 15.
75. Chemical gold nanosensors based on localized surface plasmon resonance, G. Barbillon, J.-L. Bijeon, J. Plain, M. L. d. l. Chapelle, P.-M. Adam, and P. Royer, *Gold Bull.*, 2007, **40**, 240.
76. Quantitative interpretation of the response of surface plasmon resonance sensors to adsorbed films, L. S. Jung, C. T. Campbell, T. M. Chinowsky, M. N. Mar, and S. S. Yee, *Langmuir*, 1998, **14**, 5636.
77. Before striking gold in gold-ruby glass, F. E. Wagner, S. Haslbeck, L. Stievano, S. Calogero, Q. A. Pankhurst, and P. Martinek, *Nature*, 2000, **407**, 691.
78. Gold based enamel colors, J. Carbert, *Gold Bull.*, 1980, **13**, 144.
79. Dyeing of silk cloth with colloidal gold, Y. Nakao and K. Kaeriyama, *J. Appl. Polymer Sci.*, 1988, **36**, 269.
80. Sorption and binding of nanocrystalline gold by Merino wool fibres - An XPS study, M. Richardson and J. Johnston, *J. Colloid Interface Sci.*, 2007, **310**, 425.
81. Studies of gold nanoparticles as precursors to printed conductive features for thin-film transistors, Y. Wu, Y. Li, P. Liu, S. Gardner, and B. S. Ong, *Chem. Mater.*, 2006, **18**, 4627.
82. Conductor microstructures by laser curing of printed gold nanoparticle ink, J. Chung, S. Ko, N. R. Bieri, C. P. Grigoropoulos, and D. Poulikakosa, *Appl. Phys. Lett.*, 2004, **84**, 801.
83. Standard Practice for Calculation of Photometric Transmittance and Reflectance of Materials to Solar Radiation, E971-88, 2003, American Society for Testing and Materials (ASTM).
84. Effect of composition and packing configuration on the dichroic optical properties of coinage metal nanorods, M. B. Cortie, X. Xu, and M. J. Ford, *Phys. Chem. Chem. Phys.*, 2006, **8**, 3520.
85. Radiative heat transfer across glass coated with gold nano-particles, H. Chowdhury, X. Xu, P. Huynh, and M. B. Cortie, *ASME J. Solar Energy Eng.*, 2005, **127**, 70.
86. "Purple glory": The science and technology of $AuAl_2$ coatings, S. Supansomboon, A. Maaroof, and M. B. Cortie, *Gold Bull.*, 2008, **41**, 296.
87. In situ precipitation of gold nanoparticles onto glass for potential architectural applications, X. Xu, M. Stevens, and M. B. Cortie, *Chem. Mater.*, 2004, **16**, 2259.
88. Spectrally-selective gold nanorod coatings for window glass, X. Xu, T. Gibbons, and M. B. Cortie, *Gold Bull.*, 2006, **39**, 156.
89. Optimized plasmonic nanoparticle distributions for solar spectrum harvesting, J. R. Cole and N. J. Halas, *Appl. Phys. Lett.*, 2006, **89**, 153120.
90. The effect of mutual orientation on the spectra of metal nanoparticle rod-rod and rod-sphere pairs, M. Gluodenis and C. A. Foss, *J. Phys. Chem. B*, 2002, **106**, 9484.
91. Synthesis and linear optical properties of nanoscopic gold particle pair structures, M. L. Sandrock and C. A. Foss, *J. Phys. Chem. B*, 1999, **103**, 11398.
92. C. E. Talley, J. B. Jackson, C. Oubre, N. K. Grady, C. W. Hollars, S. M. Lane, T. R. Huser, P. Nordlander, and N. J. Halas, *Nano Lett.*, 2005, **5**, 1569.
93. Plasmon resonant based eye protection, J. D. Payne and J. B. Jackson, 2006, U.S. Patent 2006/0275596.
94. Y. Kawata and M. Nakano, *IEEE Trans. Magnetics*, 2005, **41**, 997.
95. Spectral encoding on gold nanorods doped in a silica sol-gel matrix and its application to high-density optical data storage, J. W. M. Chon, C. Bullen, P. Zijlstra, and M. Gu, *Adv. Func. Mater.*, 2007, **17**, 875.

96. Confocal scanning dark-field polarisation microscopy, S. Kimura and T. Wilson, *Appl. Optics*, 1994, **33**, 1274.

97. Confocal three dimensional tracking of a single nanoparticle with concurrent spectroscopic readouts, H. Cang, C. M. Wong, C. Shan, A. H. Rizvi, and H. Yang, *Appl. Phys. Lett.*, 2006, **88**, 223901.

98. Patterning and encryption using gold nanoparticles, J. Pérez-Juste, P. Mulvaney, and L. M. Liz-Marzán, *Int. J. Nanotechnol.*, 2007, **4**, 215.

99. Near field optical storage mask layer, disk, and fabrication method, X. Shi and L. Hesselink, 2005, U.S. Patent 2005/0221228.

100. Organic materials and thin-film structures for cross-point memory cells based on trapping in metallic nanoparticles, L. D. Bozano, B. W. Kean, M. Beinhoff, K. R. Carter, P. M. Rice, and J. C. Scott, *Adv. Funct. Mater.*, 2005, **15**, 1933.

101. The third order nonlinear optical properties of gold nanoparticles in glasses, Part I, D. Compton, L. Cornish, and E. v. d. Lingen, *Gold Bull.*, 2003, **36**, 10.

102. Preparation and characterization of gold nanostructures of controlled dimension by electrochemical techniques, S. Cattarin, D. Kramer, A. Lui, and M. M. Musiani, *J. Phys. Chem. C*, 2007, **111**, 12643.

103. In situ synthesis and assembly of gold nanoparticles embedded in glass-forming liquid crystals, V. A. Mallia, P. K. Vemula, G. John, A. Kumar, and P. M. Ajayan, *Angew. Chem. Int. Ed.*, 2007, **46**, 3269.

104. Devices having a variable optical property and processes of making such devices, M. B. Cortie, A. McDonagh, and P. Mulvaney, 2005, Australian Patent Application AU2005/201737.

105. Vapor-phase hydrochlorination of acetylene - correlation of catalytic activity of supported metal chloride catalysts, G. J. Hutchings, *J. Catal.*, 1985, **96**, 292.

106. Novel gold catalysts for the oxidation of carbon-monoxide at a temperature far below 0°C, M. Haruta, T. Kobayashi, H. Sano, and N. Yamada, *Chem. Lett.*, 1987, **4**, 405.

107. Method for manufacture of catalyst composite having gold or mixture of gold with catalytic metal oxide deposited on carrier, M. Haruta, H.Sano, and T. Kobayasi, 1987, U.S. Patent 4,698,324.

108. Using gold nanoparticles for catalysis, D. T. Thompson, *Nano Today*, 2007, **2**, 40.

109. Gold-catalysed oxidation of carbon monoxide, G. C. Bond and D. T. Thompson, *Gold Bull.*, 2000, **33**, 41.

110. Catalytically active gold: from nanoparticles to ultrathin films, M. Chen and D. W. Goodman, *Acc. Chem. Res*, 2006, **39**, 739.

111. Gold as a novel catalyst in the 21st century: preparation, working mechanism and applications, M. Haruta, *Gold Bull.*, 2004, **37**, 27.

112. Gold catalysis, A. S. K. Hashmi and G. J. Hutchings, *Angew. Chem. Int. Ed.*, 2006, **45**, 7896.

113. Catalytic activity of Au nanoparticles, B. Hvolbæk, T. V. W. Janssens, B. S. Clausen, H. Falsig, C. H. Christensen, and J. K. Nørskov, *Nano Today*, 2007, **2**, 14.

114. Integrated nanoparticle-biomolecule hybrid systems: synthesis, properties, and applications, E. Katz, and I. Willner, *Angew. Chem. Int. Ed.*, 2004, **43**, 6042.

115. Nanoparticle arrays on surfaces for electronic, optical and sensor applications, A. N. Shipway, E. Katz, and I. Willner, *ChemPhysChem*, 2000, **1**, 18.

116. The use of gold nanoparticles in diagnostics and detection R. Wilson, *Chem. Soc. Rev.*, 2008, **37**, 2028.

117. Sensing capability of the localised surface plasmon resonance of gold nanorods, C.-D. Chen, S.-F. Cheng, L.-K. Chau, and C. R. C. Wang, *Biosens. Bioelectron.*, 2007, **22**, 926.

118. Increased sensitivity of surface plasmon resonance of gold nanoshells compared to that of gold solid colloids in response to environmental changes, Y. Sun and Y. Xia, *Anal. Chem.*, 2002, **74**, 5297.

119. Gold nanoshells improve single nanoparticle molecular sensors, G. Raschke, S. Brogl, A. S. Susha, A. L. Rogach, T. A. Klar, J. Feldmann, B. Fieres, N. Petkov, T. Bein, A. Nichtl, and K. Kürzinger, *Nano Lett.*, 2004, **4**, 1853.

120. Exploitation of localized surface plasmon resonance, E. Hutter and J. H. Fendler, *Adv. Mater.*, 2004, **16**, 1685.

121. An immunocolloid method for the electron microscope, W. P. Faulk and G. Taylor, *Immunochemistry*, 1971, **8**, 1081.

122. Selective colorimetric detection of polynucleotides based on the distance-dependent optical properties of gold nanoparticles, R. Elghanian, J. J. Storhoff, R. C. Mucic, R. L. Letsinger, and C. A. Mirkin, *Science*, 1997, **277**, 1078.

123. Fluorescence properties of gold nanorods and their application for DNA biosensing, C.-Z. Li, K. B. Male, S. Hrapovic, and J. H. T. Luong, *Chem. Commun.*, 2005, 3924.

124. Multiplex biosensor using gold nanorods, C. Yu and J. Irudayaraj, *Anal. Chem.*, 2007, **79**, 572.

125. Nanomaterials in analytical chemistry, C. R. Martin and D. T. Mitchell, *Anal. Chem. News Features*, 1998, May 1, 322A.

126. Ultra-sensitive detection of individual gold nanoparticles: spectroscopy and applications to biology, L. Cognet and B. Lounis, *Gold Bull.*, 2008, **41**, 139.

127. A simple strategy for prompt visual sensing by gold nanoparticles: general applications of interparticle hydrogen bonds, S.-Y. Lin, S.-H. Wu, and C.-H. Chen, *Angew. Chem. Int. Ed.*, 2006, **45**, 4948.

128. Visual cocaine detection with gold nanoparticles and rationally engineered aptamer structures, J. Zhang, L. Wang, D. Pan, S. Song, F. Y. C. Boey, H. Zhang, and C. Fan, *Small*, 2008, **4**, 1196.

129. A piezoelectric immunoagglutination assay for *Toxoplasma gondii* antibodies using gold nanoparticles, H. Wang, C. Lei, J. Li, Z. Wu, G. Shen, and R. Yu, *Biosens. Bioelec.*, 2004, **19**, 701.

130. Kinetics of gold nanoparticle aggregation: experiments and modeling, T. Kim, C.-H. Lee, S.-W. Joo, and K. Lee, *J. Colloid Interface Sci.*, 2008, **318**, 238.

131. Preparation and evaluation of metal nanopowders for the detection of fingermarks on nonporous surfaces, M. J. Choi, A. M. McDonagh, P. J. Maynard, R. Wuhrer, C. Lennard, and C. Roux, *J. Forensic Identification*, 2006, **56**, 756.

132. Single-target molecule detection with nonbleaching multicolor optical immunolabels, S. Schultz, D. R. Smith, J. J. Mock, and D. A. Schultz, *Proc. Natl. Acad. Sci.*, 2000, **97**, 996.

133. Gold nanorods as novel nonbleaching plasmon-based orientation sensors for polarised single-particle microscopy, C. Sönnichsen and A. P. Alivisatos, *Nano Lett.*, 2005, **5**, 301.

134. Drastic reduction of plasmon damping in gold nanorods, C. Sönnichsen, T. Franzl, G. von-Plessen, J. Feldmann, O. Wilson, and P. Mulvaney, *Phys. Rev. Lett.*, 2002, **88**, 077402.

135. Scattering spectra of single gold nanoshells, C. L. Nehl, N. K. Grady, G. P. Goodrich, F. Tam, N. J. Halas, and J. H. Hafner, *Nano Lett.*, 2004, **4**, 2355.

136. Spectroscopy of single metallic nanoparticles using total internal reflection microscopy, C. Sönnichsen, S. Geier, N. E. Hecker, G. von-Plessen, J. Feldmann, H. Ditlbacher, B. Lamprecht, J. R. Krenn, F. R. Aussenegg, V. Z.-H. Chan, J. P. Spatz, and M. Möller, *Appl. Phys. Lett.*, 2000, **77**, 2949.

137. Plasmonic detection of a model analyte in serum by a gold nanorod sensor, S. M. Marinakos, S. Chen, and A. Chilkoti, *Anal. Chem.*, 2007, **79**, 5278.

138. Surface plasmon resonance sensing system and method thereof, L. K. Chau, W. T. Hsu, and S. F. Cheng, 2007, U.S. Patent 2007/0109545.

139. A label-free immunoassay based upon localised surface plasmon resonance of gold nanorods, K. M. Mayer, S. Lee, H. Liao, B. C. Rostro, A. Fuentes, P. T. Scully, C. L. Nehl, and J. H. Hafner, *ACS Nano*, 2008, **2**, 687.

140. All-optical nanoscale pH meter, S. W. Bishnoi, C. J. Rozell, C. S. Levin, M. K. Gheith, B. R. Johnson, D. H. Johnson, and N. J. Halas, *Nano Lett.*, 2006, **6**, 1687.

141. Nanoparticle-based all-optical sensors, N. J. Halas, S. Lal, P. Nordlander, J. B. Jackson, and C. E. Moran, 2004, U.S. Patent 6,778,316.

142. Metal nanoshells for biosensing applications, J. L. West, N. J. Halas, S. J. Oldenburg, and R. D. Averitt, 2004, U.S. Patent 6,699,724.

143. A whole blood immunoassay using gold nanoshells, L. R. Hirsch, J. B. Jackson, A. Lee, N. J. Halas, and J. L. West, *Anal. Chem.*, 2003, **75**, 2377.

144. Polarization-enhanced detector with gold nanorods for detecting nanoscale rotational motion and method thereof, W. D. Frasch, and L. Chapsky, 2006, U.S. Patent 2006/0110738.

145. U. Kreibig and M. Vollmer, *Optical Properties of Metal Clusters,* Springer, Berlin, 1995.

146. A colorimetric gold nanoparticle sensor to interrogate biomolecular interactions in real time on a surface, N. Nath and A. Chilkoti, *Anal. Chem.*, 2002, **74**, 504.

147. Shape- and size-dependent refractive index sensitivity of gold nanoparticles, H. Chen, X. Kou, Z. Yang, W. Ni, and J. Wang, *Langmuir*, 2008, **24**, 5233.

148. Quantitative evaluation of sensitivity and selectivity of multiplex nanoSPR biosensor assays, C. Yu and J. Irudayaraj, *Biophys. J.*, 2007, **93**, 3684.

149. Nanoparticle films as sensitive strain gauges, J. Herrmann, K. H. Muller, T. Reda, G. R. Baxter, B. Raguse, G. de Groot, R. Chai, J. Roberts and L. Wieczorek, *Appl. Phys. Lett.*, 2007, **91**.

150. Optical strain detectors based on gold/elastomer nanoparticulated films, M. A. Correa-Duarte, V. Salgueiriño-Maceira, A. Rinaldi, K. Sieradzki, M. Giersig, and L. M. Liz-Marzán, *Gold Bull.*, 2007, **40**, 6.

151. Gold nanoparticle chemiresistor sensors: Direct sensing of organics in aqueous electrolyte solution, B. Raguse, E. Chow, C. S. Barton, and L. Wieczorek, *Anal. Chem.*, 2007, **79**, 7333.

152. Nanoparticle actuators, B. Raguse, K. H. Muller, and L. Wieczorek, *Adv. Mater.*, 2003, **15**, 922.

153. Surface enhanced Raman spectroscopy, C. L. Haynes, A. D. McFarland, and R. P. V. Duyne, *Anal. Chem.*, 2005, 338A.

154. Detection and identification of a single DNA base molecule using surface-enhanced Raman scattering (SERS), K. Kneipp, H. Kneipp, V. B. Kartha, R. Manoharan, G. Deinum, I. Itzkan, R. R. Dasari, and M. S. Feld, *Phys. Rev. E*, 1998, **57**, R6281.

155. Raman spectra of pyridine adsorbed at a silver electrode, M. Fleischmann, P. J. Hendra, and A. J. McQuillan, *Chem. Phys. Lett.*, 1974, **26**, 163.

156. Plasma resonance enhancement of Raman scattering by pyridine adsorbed on silver or gold sol particles of size comparable to the excitation wavelength, J. A. Creighton, C. G. Blatchford, and M. G. Albrecht, *J. Chem. Soc. Faraday Trans.*, 1979, **75**, 790.

157. Enhanced Raman scattering by molecules adsorbed at the surface of colloidal spheroids, D. S. Wang and M. Kerker, *Physical Review B: Condensed Matter and Materials Physics*, 1981, **24**, 1777.

158. Surface-enhanced Raman scattering of molecules adsorbed on gold nanorods: off-surface plasmon resonance condition, B. Nikoobakht, J. Wang, and M. A. El-Sayed, *Chem. Phys. Lett.*, 2002, **366**, 17.

159. Aspect ratio dependence on surface enhanced Raman scattering using silver and gold nanorod substrates, C. J. Orendorff, L. A. Gearheart, N. R. Jana, and C. J. Murphy, *Phys. Chem. Chem. Phys.*, 2006, **8**, 165.

160. Peptide-conjugated gold nanorods for nuclear targeting, A. K. Oyelere, P. C. Chen, X. Huang, I. H. El-Sayed, and M. A. El-Sayed, *Bioconjugate Chem.*, 2007, **18**, 1490.

161. Surface enhanced Raman scattering in the near infrared using metal nanoshell substrates, S. J. Oldenburg, S. L. Westcott, R. D. Averitt, and N. J. Halas, *J. Chem. Phys*, 1999, **111**, 4729.

162. Surface-enhanced Raman scattering on nanoshells with tunable surface plasmon resonance, R. A. Alvarez-Puebla, D. J. Ross, G. A. Nazri, and R. F. Aroca, *Langmuir*, 2005, **21**, 10504.

163. Nanoparticle-based surface-enhanced Raman spectroscopy, C. E. Talley, T. Huser, C. W. Hollars, L. Jusinski, T. Laurence, and S. Lane, *NATO Science Series, Series I: Life and Behavioural Sciences (Advances in Biophotonics)*, 2005, **369**, 182.

164. Surface enhanced Raman spectroscopy: new materials, concepts, characterization tools, and applications, J. A. Dieringer, A. D. McFarland, N. C. Shah, D. A. Stuart, A. V. Whitney, C. R. Yonzon, M. A. Young, X. Zhang, and R. P. V. Duyne, *Faraday Discuss.*, 2006, **132**, 9.

165. Radiation damping in surface-enhanced Raman scattering, A. Wokaun, J. P. Gordon, and P. F. Liao, *Phys. Rev. Lett.*, 1982, **48**, 957.

166. Finite-difference time-domain studies of the optical properties of nanoshell dimers, C. Oubre and P. Nordlander, *J. Phys. Chem. B*, 2005, **109**, 10042.

167. Surface-enhanced Raman scattering studies on aggregated gold nanorods, B. Nikoobakht and M. A. El-Sayed, *J. Phys. Chem. A*, 2003, **107**, 3372.

168. An approach to self-cleaning SERS sensors by arraying Au nanorods on TiO_2 layer, S. Li, M. Suzuki, K. Nakajima, K. Kimura, T. Fukuoka, and Y. Mori, *Proc. SPIE*, 2007, **6647**, 66470J.

169. A clearer vision for *in vivo* imaging, R. Weissleder, *Nat. Biotechnol.*, 2001, **19**, 316.

170. The recognition of metallic gold in tissue sections, H. E. Harding, *J. Clin. Path.*, 1953, **6**, 149.

171. M. A. Hayat, *Colloidal Gold: Principles, Methods, and Applications,* Academic Press, San Diego, CA, 1989.

172. Nanoshell-enabled photonics-based imaging and therapy of cancer, C. Loo, A. Lin, L. Hirsch, M. H. Lee, J. Barton, N. Halas, J. West, and R. Drezek, *Technol Cancer Res. T*, 2004, **3**, 33.

173. Surface plasmon resonance scattering and absorption of anti-EGFR antibody conjugated gold nanoparticles in cancer diagnostics: Applications in oral cancer, I. H. El-Sayed, X. Huang, and M. A. El-Sayed, *Nano Lett.*, 2005, **5**, 829.

174. Two-photon luminescence imaging of cancer cells using molecularly targeted gold nanorods, N. J. Durr, T. Larson, D. K. Smith, B. A. Korgel, K. Sokolov, and A. Ben-Yakar, *Nano Lett.*, 2007, **7**, 941.

175. *In vitro* and *in vivo* two-photon luminescence imaging of single gold nanorods, H. Wang, T. B. Huff, D. A. Zweifel, W. He, P. S. Low, A. Wei, and J.-X. Cheng, *Proc. Natl. Acad. Sci. U.S.A.*, 2005, **102**, 15752.

176. Multi-functional plasmon-resonant contrast agents for optical coherence tomography, S. A. Boppart and A. Wei, 2005, U.S. Patent 2005/0171433.

177. Targeted gold nanorod contrast agent for prostate cancer detection by photoacoustic imaging, A. Agarwal, S. W. Huang, M. O'Donnell, K. C. Day, M. Day, N. Kotov, and S. Ashkenazi, *J. App. Phys.*, 2007, **102**, 064701/1.

178. Photoacoustic imaging of early inflammatory response using gold nanorods, K. Kim, S.-W. Huang, S. Ashkenazi, M. O'Donnell, A. Agarwal, N. A. Kotov, M. F. Denny, and M. J. Kaplan, *Appl. Phys. Lett.*, 2007, **90**, 223901.

179. Photoacoustic imaging of multiple targets using gold nanorods, P. C. Li, C. W. Wei, C. K. Liao, C. D. Chen, K. C. Pao, C. R. C. Wang, Y. N. Wu, and D. B. Shieh, *IEEE T. Ultrason. Ferr.*, 2007, **54**, 1642.

180. Photoacoustic tomography of a nanoshell contrast agent in the in vivo rat brain, Y. Wang, X. Xie, X. Wang, G. Ku, K. L. Gill, D. P. O'Neal, G. Stoica, and L. V. Wang, *Nano Lett.*, 2004, **4**, 1689.

181. Two-photon-induced photoluminescence imaging of tumours using near-infrared excited gold nanoshells, J. Park, A. Estrada, K. Sharp, K. Sang, J. A. Schwartz, D. K. Smith, C. Coleman, J. D. Payne, B. A. Korgel, A. K. Dunn, and J. W. Tunnell, *Opt. Express*, 2008, **16**, 1590.

182. Near-infrared resonant nanoshells for combined optical imaging and photothermal cancer therapy, A. Gobin, M.-H. Lee, N. J. Halas, W. D. James, R. A. Drezek, and J. L. West, *Nano Lett.*, 2007, **7**, 1929.

183. Antecedents of two-photon excitation laser scanning microscopy, B. R. Masters and P. T. C. So, *Microsc. Res. Tech.*, 2004, **63**, 3.

184. Photoinduced luminescence from the noble metals and its enhancement on roughened surfaces, G. T. Boyd, Z. H. Yu, and Y. R. Shen, *Phys. Rev. B: Condens. Matter*, 1986, **33**, 7923.

185. Water-soluble quantum dots for multiphoton fluorescence imaging in vivo, D. R. Larson, W. R. Zipfel, R. M. Williams, S. W. Clark, M. P. Bruchez, F. W. Wise, and W. W. Webb, *Science*, 2003, **300**, 1434.

186. Determining the size and shape dependence of gold nanoparticle uptake into mammalian cells, B. D. Chithrani, A. A. Ghazani, and W. C. W. Chan, *Nano Lett.*, 2006, **6**, 662.

187. Bifunctional gold nanoshells with a superparamagnetic iron oxide-silica core suitable for both MR imaging and photothermal therapy, X. Ji, R. Shao, A. M. Elliott, R. J. Stafford, E. Esparza-Coss, J. A. Bankson, G. Liang, Z.-P. Luo, K. Park, J. T. Markert, and C. Li, *J. Phys. Chem. C*, 2007, **111**, 6245.

188. Iron oxide coated gold nanorods: synthesis, characterisation, and magnetic manipulation, A. Gole, J. W. Stone, W. R. Gemmill, H. C. zurLoye, and C. J. Murphy, *Langmuir*, 2008, **24**, 6232.

189. Superparamagnetic iron oxide contrast agents: physicochemical characteristics and applications in MR imaging, Y.-X. J. Wang, S. M. Hussain, and G. P. Krestin, *Eur. Radiol.*, 2001, **11**, 2319.

190. Use of core-shell gold nanoparticle which contains magnetic nanoparticles for MRI T2 contrast agent, cancer diagnostic therapy, T. Hyeon, J. Y. Kim, M. H. Cho, S. K. Kim, and J. Lee, 2008, patent WO 2008/048074.

191. Dual wavelength laser diode excitation source for 2D photoacoustic imaging, T. J. Allen and P. C. Beard, *Proc. SPIE*, 2007, **6437**, 64371U/1.

192. Photoacoustic imaging of the microvasculature with a high-frequency ultrasound array transducer, R. J. Zemp, R. Bitton, M.-L. Li, K. K. Shung, G. Stoica, and L. V. Wang, *J. Biomed. Opt.*, 2007, **12**, 010501.

193. High contrast optoacoustical imaging using nanoparticles, A. A. Oraevsky and P. M. Henrichs, 2005, U.S. Patent 2005/0175540.

194. Optical coherence tomography, D. Huang, E. A. Swanson, C. P. Lin, J. S. Schuman, W. G. Stinson, W. Chang, M. R. Hee, T. Flotte, K. Gregory, C. A. Puliafito, and J. G. Fujimoto, *Science*, 1991, **254**, 1178.

195. Ultrahigh-resolution ophthalmic optical coherence tomography, W. Drexler, U. Morgner, R. K. Ghanta, F. X. Kärtner, J. S. Schuman, and J. G. Fujimoto, *Nature Med.*, 2001, **7**, 502.

196. Effect of cationic liposome composition on in vitro cytotoxicity and protective effect on carried DNA, R. Cortesi, E. Esposito, E. Menegatti, R. Gambari, and C. Nastruzzi, *Int. J. Pharm.*, 1996, **139**, 69.

197. Gold nanoparticles are taken up by human cells but do not cause acute cytotoxicity, E. E. Connor, J. Mwamuka, A. Gole, C. J. Murphy, and M. D. Wyatt, *Small*, 2005, **1**, 325.

198. Assessing the effect of surface chemistry on gold nanorod uptake, toxicity, and gene expression in mammalian cells, T. S. Hauck, A. A. Ghazani, and W. C. W. Chan, *Small*, 2008, **4**, 153.

199. A golden bullet? Selective targeting of *Toxoplasma gondii* tachyzoites using antibody-functionalised gold nanoparticles, D. Pissuwan, S. Valenzuela, C. M. Miller, and M. B. Cortie, *Nano Lett.*, 2007, **7**, 3808.

200. Targeted destruction of murine macrophage cells with bioconjugated gold nanorods, D. Pissuwan, S. M. Valenzuela, M. C. Killingsworth, X. D. Xu, and M. B. Cortie, *J. Nanoparticle Res.*, 2007, **9**, 1109.

201. Immunonanoshells for targeted photothermal ablation in medulloblastoma and glioma: an in vitro evaluation using human cells lines, R. J. Bernardi, A. R. Lowery, P. A. Thompson, S. M. Blaney, and J. L. West, *J. Neurooncol.*, 2008, **86**, 165.

202. Efficacy of laser-activated gold nanoshells in ablating prostate cancer cells in vitro, J. M. Stern, J. Stanfield, Y. Lotan, S. Park, J. T. Hsieh, and J. A. Cadeddu, *J. Endourol.*, 2007, **21**, 939.

203. Therapeutic possibilities of plasmonically heated gold nanoparticles, D. Pissuwan, S. Valenzuela, and M. B. Cortie. *Trends Biotechnol.*, 2006, **24**, 62.

204. PEG-modified gold nanorods with a stealth character for in vivo applications, T. Niidome, M. Yamagata, Y. Okamoto, Y. Akiyama, H. Takahashi, T. Kawano, Y. Katayama, and Y. Niidome, *J. Controlled Release*, 2006, **114**, 343.

205. Surface modification of cetyltrimethylammonium bromide-capped gold nanorods to make molecular probes, C. Yu, L. Varghese, and J. Irudayaraj, *Langmuir*, 2007, **23**, 9114.

206. Targeted destruction of murine macrophage cells with bioconjugated gold nanorods, D. Pissuwan, S. M. Valenzuela, M. C. Killingsworth, and X. Xu, *J. Nanopart. Res.*, 2007, **9**, 1109.

207. Metal nanostructures and pharmaceutical compositions, F. Frederix and B. V. d. Broek, 2007, U.S. Patent 2007/0116773.

208. Near infrared laser-tissue welding using nanoshells as an exogenous absorber, A. M. Gobin, D. P. O"Neal, D. M. Watkins, N. J. Halas, R. A. Drezek, and J. L. West, *Lasers Surg. Med.*, 2005, **37**, 123.

209. Nanoshell-mediated near-infrared thermal therapy of tumors under magnetic resonance guidance, L. R. Hirsch, R. J. Stafford, J. A. Bankson, S. R. Sershen, B. Rivera, R. E. Price, J. D. Hazle, N. J. Halas, and J. L. West, *Proc. Natl. Acad. Sci.*, 2003, **100**, 13549.

210. Immunotargeted nanoshells for integrated cancer imaging and therapy, C. Loo, A. Lowery, N. Halas, J. West, and R. Drezek, *Nano Lett.*, 2005, **5**, 709.

211. Photo-thermal tumor ablation in mice using near infrared-absorbing nanoparticles, D. P. O'Neal, L. R. Hirsch, N. J. Halas, J. D. Payne, and J. L. West, *Cancer Lett.*, 2004, **209**, 171.

212. Metal nanoshells as a contrast agent in near-infrared diffuse optical tomography, C. Wu, X. Liang, and H. Jiang, *Opt. Comm.*, 2005, **253**, 214.

213. Optically-active nanoparticles for use in therapeutic and diagnostic methods, J. L. West, N. J. Halas, and L. R. Hirsch, 2003, U.S. Patent 6, 530, 944.

214. Cancer cell imaging and photothermal therapy in the near-infrared region by using gold nanorods, X. Huang, I. H. El-Sayed, W. Qian, and M. A. El-Sayed, *J. Am. Chem. Soc.*, 2006, **128**, 2115.

215. Diagnostic and therapeutic applications of metal nanoshells, C. Loo, A. Lin, L. Hirsch, M.-H. Lee, J. Barton, N. Halas, J. West, and R. Drezek, *Nanofabrication Towards Biomed. Appl.*, 2005, 327.

216. System and method for interacting with a cell or tissue in a body, L. van Pieterson, B. H. W. Hendriks, and G. W. Lucassen, 2007.

217. Gold nanoparticles designed for combining dual modality imaging and radiotherapy, C. Alric, R. Serduc, C. Mandon, J. Taleb, G. L. Duc, A. L. Meur-Herland, C. Billotey, P. Perriat, S. Roux, and O. Tillement, *Gold Bull.*, 2008, **41**, 90.

218. A macrophage-targeted theranostic nanoparticle for biomedical applications, J. R. McCarthy, F. A. Jaffer, and R. Weissleder, *Small*, 2006, **2**, 983.

219. Shape and size dependence of radiative, non-radiative and photothermal properties of gold nanocrystals, S. Link and M. A. El-Sayed, *Inter. Rev. Phys. Chem.*, 2000, **19**, 409.

220. Screening for cell-targeting ligands attached to metal nanoshells for use in target-cell killing, K. Angelides, 2004, patent WO 2004/020973.

221. Treatment of disease states characterized by excessive or inappropriate angiogenesis, J. L. West and J. D. Payne, International Patent Number WO/047633, 2003.

222. Targeted photothermal lysis of the pathogenic bacteria, *Pseudomonas aeruginosa*, with gold nanorods, R. S. Norman, J. W. Stone, A. Gole, C. J. Murphy, and T. L. Sabo-Attwood, *Nano Lett.*, 2008, **8**, 302.

223. Photothermal nanotherapeutics and nanodiagnostics for selective killing of bacteria targeted with gold nanoparticles., V. P. Zharov, K. E. Mercer, E. N. Galitovskaya, and M. S. Smeltzer, *Biophys. J.*, 2006, **90**, 619.

224. Macrophage-parasite interactions in *Leishmania* infections, J. Mauël, *J. Leukocyte Biol.*, 1990, **47**, 187.

225. Stable incorporation of gold nanorods into N-isopropylacrylamide hydrogels and their rapid shrinkage induced by near-infrared laser irradiation, A. Shiotani, T. Mori, T. Niidome, Y. Niidome, and Y. Katayama, *Langmuir*, 2007, **23**, 4012.

226. Temperature-sensitive hydrogels with SiO_2-Au nanoshells for controlled drug delivery, M. Bikram, A. Gobin, R. E. Whitmire, and J. L. West, *J. Control. Release*, 2007, **123**, 219.

227. Temperature-sensitive polymer-nanoshell composites for photothermally modulated drug delivery, S. R. Sershen, S. L. Westcott, N. J. Halas, and J. L. West, *J. Biomed. Mater. Res.* , 2000, **51**, 293.

228. Temperature-sensitive polymer/nanoshell composites for photothermally modulated drug delivery, J. L. West, S. R. Sershen, N. L. Halas, S. J. Oldenburg, and R. D. Averitt, 2001, U.S. Patent 6,428,811.

229. Nanoshells on polymers, F. N. Ludwig, S. D. Pacetti, S. F. A. Hossainy, and D. Davalian, 2007, United States Patent 2007/0298257.

230. Colloidal gold nanoparticles: A novel nanoparticle platform for developing multifunctional tumor-targeted drug delivery vectors, G. F. Paciotti, D. G. I. Kingston, and L. Tamarkin, *Drug Dev. Res.*, 2006, **67**, 47.

231. DNA-gold nanorod conjugates for remote control of localized gene expression by near infrared irradiation, C.-C. Chen, Y.-P. Lin, C.-W. Wang, H.-C. Tzeng, C.-H. Wu, Y.-C. Chen, C.-P. Chen, L.-C. Chen, and Y.-C. Wu, *J. Am. Chem. Soc.*, 2006, **128**, 3709.

232. Quantitative estimation of gold nanoshell concentrations in whole blood using dynamic light scattering, H. Xie, K. L. Gill-Sharp, and D. P. O'Neal, *Nanomedicine*, 2007, **3**, 89.
233. Dynamic light scattering of short Au rods with low aspect ratios, J. Rodriguez-Fernández, J. Pérez-Juste, L. M. Liz-Marzán, and P. R. Lang, *J. Phys. Chem. C*, 2007, **111**, 5020.
234. Immunogenic properties of colloidal gold, L. A. Dykman, M. V. Sumaroka, S. A. Staroverov, I. S. Zaitseva, and V. A. Bogatyrev, *Biol. Bull.*, 2004, **31**, 75.

17 Miscellaneous Uses of Gold

Richard Holliday and Christopher Corti

CONTENTS

Introduction 399
Electrical Uses 400
    Gold–Alloy Thermocouples 400
    Gold Shunt Layers on Superconducting Wires and Tapes 400
    Gold–Palladium Alloys in Spark Plug Electrodes 400
Uses of Gold as a Coating 401
    Optical Reflector Coatings on DVDs 401
    Gold-Coated Glass in the Building Industry 401
    Detection of Mercury Vapor 401
Use of Gold in Photography 401
Engineering Uses 402
    Gold Bursting Discs 402
    Gold–Alloy Thermal Fuses 402
    Thermal Protection Layers 402
    Gold–Alloy Spinnerets 402
    Engineering Use of Gold in Fuel Cells 403
    Gold in Lubrication 403
Medical Uses 403
    Implants Using Gold 405
    Gold Needles for Acupuncture Therapy 405
References 405

INTRODUCTION

In compiling a book describing the technical uses of gold, there are a number applications that clearly warrant individual chapters by merit of either the amount of gold used in that application, for example gold bonding wire (Chapter 13), or the complexity of the science underlying the application, such as catalysis (Chapter 6). However, there are also a large number of uses of gold that either do not consume large quantities of the metal or where the science and technology is much better established. A book titled *Gold: Science and Applications* would not be complete without a brief description of such uses; this following chapter provides a short summary, including some historical applications of gold where the metal no longer finds favor, due either to technology trends or the availability of cheaper materials. History tells us that changes in metal prices or further developments in technology can make seemingly obsolete applications relevant once again.

ELECTRICAL USES

The use of gold in the electronics industry is widespread and for many years has provided the largest industrial use of gold (see Chapter 1). In contrast, gold's use in electrical applications is relatively minor. In most cases this is probably a simple matter of economics with many electrical applications requiring the use of quite large quantities of metal for which gold would simply be too expensive. Of course, we should also not forget that gold's electrical conductivity is inferior to both copper and silver, metals widely used in electrical applications. Nevertheless, gold and its alloys have found use for some selected applications, examples of which follow.

GOLD–ALLOY THERMOCOUPLES

A thermocouple circuit is formed when two dissimilar metals are joined at both ends and there is a difference in temperature between the two ends. This difference in temperature creates a small current, called the Seebeck effect after Thomas Seebeck who discovered this phenomenon in 1821. Although they are not the most accurate means of measuring temperature, thermocouples have sufficient accuracy for most applications and find widespread use. Not as widely used in thermocouples as the platinum group metals, gold has found use for both low temperature and higher temperature measurements. The difficulties in measuring very low temperatures approaching absolute zero are significant. An alloy based on small additions of iron to gold displays a strong and reproducible thermoelectric effect and, as a result, has found use for cryogenic temperature measurement near absolute zero. Manufacturers such as Johnson Matthey Noble Metals have traditionally supplied thermocouple wires based on 0.03 atomic percent iron/gold v chromel, displaying almost no drift. Gold–cobalt alloys have also been used in the past, but such alloys have exhibited larger drift due to precipitation of a cobalt-rich phase under some operating regimes [1,2]. At higher temperatures, use of gold–palladium alloys (typically 60Au-40Pd) have been described, for example, in aircraft engines where good service lives up to 1000°C are required [2].

GOLD SHUNT LAYERS ON SUPERCONDUCTING WIRES AND TAPES

Superconductivity is the phenomenon whereby some materials exhibit zero resistance to the passage of an electric current when cooled to low temperatures. Various superconducting materials are used in a range of applications including MRI imaging, catheter steering, transport (magnetic levitation trains), and fault current limiters. In the fault current limiters application, the most recent generation of materials that are able to operate at higher temperatures, such as YBCO ($YBa_2Cu_3O_7$), require the use of a metallic shunt top layer to give a degree of thermal and electrical protection should a fault in the superconducting tape develop. This metallic layer is typically 50–100 nm thick and must be resistant to oxidation at the high temperatures that can arise during both processing steps and in service during current overload situation. It must also serve as a barrier to prevent ingress of moisture into the YBCO superconductor. Sputtered silver, gold, and silver–gold alloy layers have so far shown the best combination of properties for this application including electrical and thermal conductivities, contact resistance, specific heat capacities, oxidation potential, lattice constant, and coefficient of linear expansion [3].

The superconducting properties of gold intermetallics have been reviewed by Khan and Raub [4], although no intermetallic compound of gold has yet been found to exhibit sufficient superconductivity to make it industrially interesting.

GOLD–PALLADIUM ALLOYS IN SPARK PLUG ELECTRODES

The first generation of long-life spark plugs with corrosion resistant platinum or gold-palladium (typically 60Au-40Pd) electrodes were marketed in the mid-1980s. Up to this point, electrode wear usually dictated when a set of plugs had to be replaced. Using a fine wire (1.0 mm) gold–palladium center electrode, manufacturers claimed that much longer electrode life as well as improved ignitability, more power, and enhanced antifouling capabilities were achievable. In recent years it appears platinum and, more recently, iridium electrodes have found greater favor with manufacturers and consumers, although plugs containing gold–palladium electrodes can still be sourced from some outlets.

USES OF GOLD AS A COATING

OPTICAL REFLECTOR COATINGS ON DVDs

During the late 1990s a very significant use of gold (consuming at least a few metric tons of gold per year) was for optical reflector coatings on CDs and DVDs. A thin, light reflective layer is required in this application and gold is applied, typically as a 50–100 nm layer, using sputtering techniques. When this was a mainstream application in the late 1990s, the market size amounted to several billions of discs per year, each containing a few cents worth of gold. Nowadays, through further research and product development, silicon, aluminum, and other materials have largely displaced the use of gold for reasons of cost, in all but the highest quality applications such as archiving of critical information. In such cases the use of gold as the reflector layer eliminates the risk of corrosion and oxidation during long-term storage. In this case, manufacturers are reported to have performed extensive media longevity studies, with results indicating durability for many decades.

GOLD-COATED GLASS IN THE BUILDING INDUSTRY

Uncoated glass is almost completely transparent to radiation spanning the ultraviolet to infrared range. However, thin gold coatings on glass will reflect infrared radiation and transmit a proportion of visible light, an effect that is dependent on the thickness of the gold film. This property has provided scope to use gold in the coating of glazing for the control of solar radiation. This technology was widely used in the 1970s although it appears to have fallen out of favor with architects since that time. An excellent review of this application is provided by Groth and Reichelt [5]. Interest in the use of gold in the glazing industry has been revived by the potential to use nanoparticulate gold coatings in recent years (see Chapter 16).

DETECTION OF MERCURY VAPOR

Due to the health effects of mercury exposure, permitted levels of the metal in the atmosphere are regulated in many countries, and it is necessary to have portable analytical equipment capable of measuring mercury contamination with a high degree of accuracy. A good example of a situation where such measurements are required concerns levels of mercury around coal-fired power stations, a major source of mercury pollution in the United States. Gold coatings are very efficient at collecting vapor phase mercury in an airstream by amalgamation onto gold, and this principle is widely used in field and laboratory detectors. The thin gold film undergoes a significant increase in resistance upon the absorption of mercury vapor which can be easily and reproducibly measured. A personal lapel safety badge that works on the same principle was developed and marketed by the 3M Corporation during the 1970s [6].

USE OF GOLD IN PHOTOGRAPHY

There are two separate uses of gold in photographic applications. First, gold can play a minor role in the improvement of traditional silver-based photographic films. According to Gysling [7], gold compounds have played a critical role in imaging technology since it was discovered that such compounds can sensitize silver halide emulsions to give significant increases in their photographic speed. As a result gold compounds, typically gold chlorides, were first used as chemical sensitizers in commercial silver halide materials in the early 1940s. In the early 1990s, work was reported on the first successful introduction of stable gold (I) compounds as silver halide emulsion chemical sensitizers [7].

The second application concerns the use of gold for specialist archival and artistic photography through the so-called crysotype process. This printing of photographs in pure gold, rather than silver, was first achieved in 1842 by Sir John Herschel, but his innovative chrysotype process was soon consigned to obscurity, owing to its expense and uncertain chemistry [8]. In the 1980s some modern coordination chemistry of gold was applied to overcome the inherent problems, enabling an economic, controllable gold-printing process, which offers unique benefits for specialized artistic and archival photographic purposes. The color of the gold image depends on the dimensions of the gold nanoparticle used in the process, which are controlled by the parameters of the photochemical steps [9].

ENGINEERING USES

GOLD BURSTING DISCS

A bursting or rupture disk is a pressure relief device that protects a vessel or system from excess pressures. They have been commonly used in aerospace, aviation, defense, nuclear, and oilfield applications often as a backup device for a conventional safety valve. In this instance, if the pressure increases and the fitted safety valve fails to operate, the rupture disk will burst as required. The discs are usually made from thin metal foil, and gold has been used in some instances because of its ductility and resistance to corrosion. Gold discs fitted to liquid ammonia tanks, for example, have shown good durability in this application compared to other metals. The use of gold in this application was reviewed in the 1970s when this industrial application for gold was more common [10].

GOLD–ALLOY THERMAL FUSES

Heat treatment furnaces will have a maximum temperature of operation beyond which permanent damage may be done to the furnace itself. To prevent such an occurrence, furnaces have been fitted with thermal fuses that are designed to melt above a certain temperature, thereby cutting the incoming power and preventing further temperature rises. These alloy fuses need to be highly resistant to oxidation and corrosion under the operating conditions employed and, most importantly, have a narrow melting range. Gold–palladium alloys meet these requirements, and wires based on a number of different compositions have been employed to manufacture fuses operating between 1100°C and 1500°C [11].

THERMAL PROTECTION LAYERS

To reduce the weight of vehicles, the use of carbon fiber body parts is increasing in the automotive industry. However, to prevent any degradation of mechanical properties, this material should not be exposed to excessive temperatures generated by the engine or exhaust system. In some applications there is a need to protect components through the use of a thin protective shield. This has created an interesting niche application for gold either in the form of an electroplated coating directly on the carbon fiber materials or as a thin foil layer around the component. The most famous example of this application is the use of gold foil to line the engine bay of the McLaren F1 supercar.

GOLD–ALLOY SPINNERETS

Spinnerets are multipored devices through which a plastic polymer melt is extruded to form fibers. Many thousands of holes, typically 40–120 μm in diameter, are pierced into a number of different metals to manufacture spinnerets, which are used in the production of a range of manmade fibers. Gold–platinum alloys have been used in this application, where their resistance to corrosion and nonwetting characteristics under the arduous operating conditions is found to be beneficial [12]. The historical development of the alloys used in this application is described in more detail in *Gold Usage* by Rapson and Groenewald [2].

ENGINEERING USE OF GOLD IN FUEL CELLS

As well as being used as an electrocatalyst in alkaline fuel cells (see Chapter 6), gold has many applications in fuel cells due to its excellent corrosion properties, conductivity, and low contact resistance, particularly under oxidizing conditions. Gold coatings are commonly used on bipolar fuel cell separators (for example, in the Space Shuttle Orbiter fuel cells), while graphite separators for low temperature polymer electrolyte membrane cells are often gold coated to reduce contact resistance. Molten carbonate fuel cells use electrolytes that are eutectic mixtures of lithium and potassium carbonates at temperatures of around 650°C, which makes severe demands on construction materials. Gold is one of the materials best able to resist corrosion by this electrolyte, hence its use for gas and electrolyte seals, as well as improved electrical contacts. In high-temperature (600–1000°C) solid oxide fuel cells it is not necessary to use precious metal catalysts. The anodes generally consist of porous nickel alloys and the cathodes of lithiated nickel, while intercell separator plates may be made from stainless steels coated with nickel. However, gold-based inks are essential to make the air-side connections between cells that must accommodate dimensional changes due to differential thermal expansion.

GOLD IN LUBRICATION

Control of friction is a key engineering requirement in many applications. A range of additives is used in commercial lubricating oils to improve the properties of these oils, in terms of corrosion resistance and other properties. The use of organic gold compounds as possible lubricating oil additives has been explored [2] although it is not clear whether or not such materials have found commercial application. The lubricating properties of gold nanoparticles have been exploited in a niche commercial additive for petrol or diesel engines that is claimed to offer fuel savings and power improvements, as well as reduced engine noise. Traditionally graphite can be used as an ingredient of additives, along with polytetrafluoroethylene (i.e., Teflon). NanoGold Oil™ is based on a liquid additive comprised of synthetic oil and gold nanoparticles that are believed to coat a layer of gold on the engine's cylinder and piston surface, as well as on lower parts of engines such as shaft and cam shaft, thereby filling and sealing micropores and pits.

The more common application for gold as a lubricant relates to its use as a thin film, particularly in space and aerospace applications. As described by Antler and Spalvins [13], most metallic-film lubricants, e.g., Pb, Sn, and Ag, have low melting points and are not resistant to oxidation. In addition, nonmetallic lubricants may suffer from radiation damage, chemical breakdown, or evaporation under the challenging conditions of space flight. However, pure gold is an exception to this general rule, with its good oxidation resistance and higher melting point. With its low shear strength, cold welding is not the serious problem with gold that it is with other metals in such situations, and it has therefore been applied as a solid film lubricant in many space situations involving moving parts, both alone as a surface coating on the bearing surfaces and in combination with low volatility synthetic lubricants.

MEDICAL USES

The modern medicinal use of gold, termed *chrysotherapy*, is covered in Chapter 10 of this book. However, while these applications mainly deal with the use of gold compounds, there are a number of uses of the metal itself in bulk form that merit mentioning. These applications relate to the excellent biocompatibility of the metal in the human body.

IMPLANTS USING GOLD

Although not a mainstream material used in medical implants, gold is used in a number of niche surgical applications. One relatively recent use concerns a new treatment for prostate cancer that uses grains of gold, approximately the size of a grain of rice. The surgical procedure involves inserting three gold grains into the prostate using ultrasound. The position of the gold grains can be detected using X-rays (gold has high opacity to X-rays) allowing the doctors to accurately target the prostate position within 1 or 2 mm [14].

Gold-plating has a history of use in the coating of coronary stents. Inserted inside large arteries and veins, these implants act like scaffolding, propping up the blood vessels and keeping them open to allow adequate blood flow. Boston Scientific produced the Niroyal™ stent as one of the first gold-plated stents in 2001, largely in response to the need for stents that could be more accurately placed. The radiopacity of gold means that gold-plated stents offer the best visibility under an X-ray enabling them to be positioned where the surgeon wants them. The biological inertness of gold is important in this application and, according to at least one published study by Tanigawa et al. [15], gold-plated stents were found to produce the fewest macroscopic changes in surrounding intravascular tissue. In recent years, commercial availability of gold-plated stents has reduced as competing materials have gained favor.

Gold of high purity (typically 99.99%) is used as an implant material in the upper eyelid for the treatment of facial nerve paralysis (Figure 17.1a shows a typical implant). The aim of this treatment is to allow the patient's upper eyelids to close where paralysis of the eyelid muscles is preventing this motion. Following the implantation of the gold device (typically a few grams in weight), the closing is produced by the gravitational pull on the implant.

Gold is also widely used in implants for drainage and aeration of the middle ear (the tympanic cavity). In treating the condition commonly known as glue ear, it is sometimes necessary to implant a medical device that has a high infection resistance and low susceptibility to incrustation. It has been established that gold fulfills these requirements, showing a high degree of resistance to bacterial colonization in the ear and a range of gold prostheses are commercially available in high purity gold for this application. Figure 17.1b shows a range of these implants.

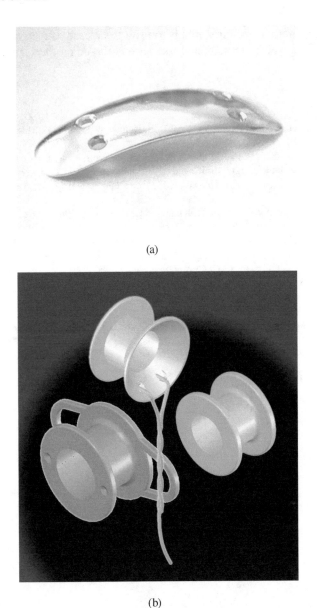

(a)

(b)

FIGURE 17.1   (a) Gold eyelid implant for the treatment of nerve paralysis. (b) Gold middle ear implants. (Courtesy of Heinz Kurz GmbH Medizintechnik.)

Gold Needles for Acupuncture Therapy

Acupuncture is a procedure based on healing through stimulation of anatomical locations on or in the skin. It is derived from Chinese medicine over 2500 years old and it is believed that the earliest acupuncture needles were sharp pieces of bone or flint and later needles were made from iron, copper, bronze and precious metals. While the needles employed by acupuncturists today are commonly stainless steel, it is not unusual for contemporary acupuncturists to employ gold acupuncture needles to treat certain ailments. Pure gold is too soft to be used as a needle material, so various alloys have been used, which are marketed like jewelry, based on the karat or carat system, e.g., K18 gold needles.

REFERENCES

1. *Precious Metals Science and Technology*, ed. L. S. Benner, T. Suzuki, K. Meguro, and S. Tanaka, 1991, IPMI, Allentown, PA, p. 530.
2. *Gold Usage*, W. S. Rapson and T. Groenewald, 1978, Academic Press, London.
3. High-power-density fault-current limiting devices using $YBa_2Cu_3O_7$ superconducting films and high resistivity alloy shunt layers, H. Yamasaki, M. Furuse and Y. Nakagawa, *Appl. Phys. Lett.*, 2004, **85**, 4427.
4. The superconductivity of gold alloys, H. Khan and C. Raub, *Gold Bull.*, 1975, **8**, 114.
5. Gold coated glass in the building industry, R. Groth and W. Reichelt, *Gold Bull.*, 1974, **7**, 62.
6. Monitoring of personal exposure of mercury vapour, Anon., *Gold Bull.*, 1978, **11**, 133.
7. Applications of Gold Chemistry in Imaging Technology, H. Gysling, Paper presented at the GOLD 2003: New Industrial Applications for Gold Conference, Vancouver, 2003.
8 Chrysotype: photography in nanoparticle gold, M. Ware, *Gold Bull.*, 2006, **39**, 124.
9. *Gold in Photography: The History and Art of Chrysotype*, M. Ware, 2006, Ffotoffilm Publishing, Brighton, UK.
10. Gold bursting discs for the protection of chemical plant, J. E. Philpott, *Gold Bull.*, 1974, **7**, 97.
11. Thermal fuses for the protection of electric furnaces, Anon., *Gold Bull.*, 1975, **8**, 112.
12. Dispersion strengthened gold-platinum, A. E. Heywood and R. A. Benedek, *Platinum Metals Rev.*, 1982, **26**, 98.
13. Lubrication with thin gold films, M. Antler and T. Spalvins, *Gold Bull.*, 1988, **21**, 59.
14. Gold in the News, Anon., *Gold Bull.*, 1995, **38**, 30.
15. Reaction of the aortic wall to six metallic stent materials, N. Tanigawa, S. Sawada, and M. Kobayah, *Acada. Radio.*, 1995, **2**, 379.

Index

Note: The letter *f* following page numbers in the index refers to figures and the letter *t* refers to tables.

A

aboveground stocks, 5–7. *See also* supply
   central banks, 5–6
   scrap, 6–7
acid Au(III) cyanide baths, 253
acid electrolytes, 251–253
acupuncture therapy, 405–406
adipic acid, 110
Ag-Au-Cu ternary system, liquidus surface of, 179*f*
age-hardening, 128
AIDS treatment, 227
air-cleaning devices, 104
alkaline electrolytes, 247–250. *See also* electrolytes
   gold alloy electrolytes, 248–250
   pure gold electrolytes, 247–248
alkenes, enantioselective hydrogenation of, 103
alkynyl complexes, 76–79
Allochrysine®, 218
alloying, 129–131
   conventional, 130–131
   microalloying, 131
alloys, 16–17
α gold oxide film, 54–55
amalgam, 299
amethyst gold, 148–149
Anglo American Research Laboratories, 105
anion exchange, 92
annealing, 126–127, 293
antiarthritic drugs, 46*f*
antimalarial gold complexes, 46*f*
antitumor gold complexes, 46*f*, 223–226
aqua regia, 31, 328
asthma, treatment of, 227
atomic radius, 18*t*
Au(0) complexes, 33–35
Au(0)/Au(I) mediator system, 57
Au(-I) complexes, 33
Au(I) complexes. *See* gold(I) complexes
[Au(1,4,7-trithiacyclononane)$_2$]$^{+2/2+/3+}$, 32*f*
Au(II) complexes, 39–40
Au-20Sn solder, 168–171
Au^{II}(SbF$_6$)$_2$, 39
[Au^{II}Xe$_4$](Sb$_2$F$_{11}$)2, 39
[Au$_2$(dcpm)$_2$]$^{2+}$, 70
Au$_2$K, 150
[Au$_{25}$(SCHCHPh)$_{18}$]$^+$, structure of, 45*f*
Au(III) complexes. *See* gold(III) complexes
Au(III) cyanide baths, 253
Au$_3$(MeN=COMe)$_3$, solvoluminescence/oxidation of, 38*f*,
   79–80
[Au$_3$(μ-Ph$_2$PAnPPh$_2$)$_3$](ClO$_4$)$_3$, 71
Au(V) complexes, 42, 43*f*
Au^{V}F$_5$, 42*f*

Au$_5$K, 150
Au$_5$Rb, 150
AuAl$_2$, 148–149
Au/Al$_2$O$_3$, 111
Au-Cu-Ag-based alloys, 136–139
Au-Cu-Al phase diagram, 149*f*
AuGa$_2$, 150
AuIn$_2$, 150
[Au(MeCS$_2$)]$_4$, 74
Au-Ni-based alloys, 139–140
Au-Pd catalysts, 94
Au-PGM alloy catalysts, 94
Au-Pt-based alloys, 139–140
Au-Pt-Pd catalysts, 105
auranofin
   administration and dosage of, 220
   structure of, 46*f*, 218–219
   in vivo transformations of, 221
Aureotan®, 218
aurocyanide, 221–222
Auromyose®, 218
aurophilicity, 15–16, 69
AuroPureH$_2$, 114
aurothioglucose, 218
aurothiomalate, 218
AuSn$_4$, 164–165
Au/TiO$_2$, 111, 114
autocatalysts, 104–106

B

Bachelier, Jean-Jacques, 324
ball bonds, 287–289
balsam sulfide, 319
bar, 10
β gold oxide film, 55
β oxide reduction, 53
binary phase diagrams, 134–135*f*
binuclear oxidative additions, 41
biomass, conversion to platform chemicals, 107–108
9,10-bis(diphenylphosphino)anthracene, 70
bis(thiosulfate)gold(I), 46*f*
blanking, 199
bonding wire. *See* wire bonding
Bond-Thompson mechanism, 94*f*
Böttger, Johannes Friedrich, 322
Bouisson, Alphonse, 352
brazing alloys, gold-based, 178–188
   gold-copper, 185–186
   gold-nickel, 186–188
   gold-palladium, 188
   industrial applications, 184–188
   karat gold, 179–184

Bretton Woods agreement, 5–6, 305–306
bridges, dental, 299
bridgework, 302–303
Brother Hypolite (monk), 324
brushing, 348
Brust method, 340
bulk metallic glass (BMGs), 152–153
bullion bank, 4
burnish golds, 319, 320–321, 341–343
bursting discs, 402

C

cable-making technology, 206
calcination, 94
caprolactam, 110
CAPTEK™, 302, 313*f*
carbon monoxide
    oxidation, 59–61, 95–97
    sensors, 115
Cassius, Andreas, 351–352
casting, 292–293
catalysis, 379
catalysts. *See* gold catalysts
catalytic wet air oxidation (CWAO), 97
cation exchange, 92
CCR5 antagonist, 227
CDs, optical reflector coatings, 401
Central Bank Gold Agreement, 6
central banks, 1, 5–6
centrifugal casting, 195, 197*t*
cetyltrimethylammonium bromide (C_{16}TAB), 32, 371,
    386–388
cetyltrimethylammonium chloride (C_{16}TAC), 371
Chagas disease, 226
chain-making, 199–200
chalcogenido complexes, 71–76
chemical processing, 107–113
    conversion of biomass to platform chemicals, 107–108
    hydrogen peroxide, 113
    methyl glycolate, 110–111
    nylon precursors, 110–111
    petroleum refining, 113
    propene oxidation, 111–113
    selective hydrogenation, 113
    selective oxidation of sugars, 111
    vinyl acetate monomer, 108–110
chemical vapor deposition (CVD), 93
China, gold production in, 4
chloride-free preparations, 93
chloroethene, 99
chrysotherapy, 218–219. *See also* gold drugs
cisplatin, 223
clusters, 43, 44*f*
coatings, spectrally selective, 376–378
Co-Hg gold layers, 268–270
coining, 199
coins, 9
cold forging, 199
cold forming
    blanking, 199
    coining, 199
    cold forging, 199
    stamping, 198–199

cold-working, 126–127
colloidal metallic gold, color generation of, 353–354
colloids, 92
color of bulk gold, 18–21
color variation, 144–150
    black, 147–150
    blue, 147–150
    purple, 147–150
    shift from yellow to reddish, greenish, and whitish,
        144–145
    white gold alloys, 145–147
colorants, 376
colorimetric sensors, 380
commodity prices, 305–308
complete oxidation, 95–97
composite electrodeposition, 257–258
computer modeling, 210–211
computer-aided design (CAD), 205–206
contact materials, 264
coprecipitation (CP), 92
core-shell nanoparticles, 373
coronary stents, 404
croton alcohol, 97
crown ether, 33
crowns, 312*f*
    dental, 299
crystal structure, 16–17
cyanide electrolyte, 241–243
cyanide heap leaching, 3
cyanide-plating baths, 247–253
    acid electrolytes, 251–253
    alkaline electrolytes, 247–250
    neutral electrolytes, 250–251
    trivalent gold cyanide electrolytes, 253
cyclohexane, 110
cyclohexanol, 110
cyclohexanone, 110
[(cyclohexyl isocyanide)$_2$Au^I](PF_6), 37*f*
cyclometalated gold(III) alkynyl complexes, structures
    if, 81–82

D

data storage devices, 378–379
decals, 349–350
decanuclear gold(I) complex, structure of, 72*f*
decorative finishes, 260–264
decorative gold materials, 317–363
    application techniques, 348–350
        brushing, 348
        ink-jet printing, 350
        machine banding, 348
        pad printing, 350
        screen printing, 348–350
    chemistry of, 326–345
        burnish golds, 341–343
        gold mercaptide systems, 327–337
        gold nanoparticles, 340–341
        gold sulforesinates, 326
        gold-silver mercaptides, 337–338
        high-firing gold formulations, 344–345
        microwave-resistant gold, 344
        self-burnish gold, 343–344
        sulfur-free gold compounds, 338–340

electroplating of, 260–264
  gold films, 319–326
  gold leaf, 357–363
  ground coats, 318
  history of, 321–326
  influence of minor metals, 345–348
  overcoats, 318
  Purple of Cassius gold enamels, 350–357
deficits, 2
deformation, 127
deformation processed metal-metal composites
    (DMMCs), 155
Degussa, 326
demand, 2
  electronics, 8
  jewelry, 7–8
density functional theory (DFT), 81
dental alloys, 9
  alloying elements, 304–305
  classification of, 306
  color of, 145–147
  and commodity prices, 305–308
  gold-palladium alloys, 305
  gold-platinum alloys, 305
  regulations, 310
  standards, 309–310
  white gold, 147
dental materials, 299–304
  biocompatibility, 300
  bridgework, 302–303
  chemical properties, 299–301
  fabrication properties, 304
  galvanic reactions, 300
  inertness in oral environment, 299–300
  mechanical properties, 301–303
  regulations, 310
  single restorations, 302
  standards, 309–310
  tarnish films, 300–301
  thermal properties, 303–304
dental wires, 298
dentistry, gold in, 296–315
  3D printing, 312–314
  alloying elements, 304–305
  bridges, 299
  crowns, 299
  dentures, 296–298
  electroforming, 311
  fillings, 298–299
  future prospects, 311–315
  historical background, 296–299
  impact of commodity prices, 305–308
  inlays, 299
  nanoparticles, 314–315
  porcelain-fused-to-metal restorations,
    299
  regulations, 310
  requirements for materials, 299–304
    chemical properties, 299–301
    fabrication properties, 304
    mechanical properties, 301–303
    thermal properties, 303–304
  sintering, 312
  standards, 309–310

traditional casting alloys, 311
  wires, 298
dentures, 296–298
deposition precipitation (DP), 92
dicyanides, 79
dicyanoaurate(I), 221–222
dielectric constant, 22
diffusion soldering, 177–178
diimine ligands, 81
dimethylamine borane (DMAB), 57
dimethylgold acetylacetonate, 93
dinuclear Au(I) alkynyl complex, structure of, 79f
diphosphines, 225
direct metal laser sintering (DMLS), 205, 208, 210, 211f
disorder-order transformation hardening, 128, 131–133
dispersion characteristics, 21–23
dispersion hardening, 128
dissolution of gold, 31–32
dithiocarbamates, 225
dodecanuclear gold(I) complex, structure of, 72f
dollar, and gold price, 11–12
double impregnation method (DIM), 92
drawing, 293
Drude permittivity, 19
Drude plasma frequency, 20–21
DVDs, optical reflector coatings, 401
dye, 376

E

electrical connections, 264
electrical resistance, 18
electrical resistivity, 18t
electrical uses of gold, 400–401
electrocatalysis of nanoparticles, 57–62
  carbon monoxide oxidation, 59–61
  formic acid oxidation, 61–62
  methanol oxidation, 61–62
  oxygen reduction, 58–59
electrochemical sensors, 382
electrodeposition, 234–238
  by electrochemical methods, 234–238
  without use of external current source, 234–236
electrodes, 279–286
electroforming, 200–201, 258–259, 311
electrolytes, 241–245
  acid, 251–253
  alkaline, 247–250
  cyanide, 241–243
  gold alloy, 248–250
  neutral, 250–251
  pure gold, 247–248
  sulfite, 243–244
  sulfite-thiosulfate, 244–245
  thiosulfate, 244–245
  trivalent gold cyanide, 253
electron paramagnetic resonance (EPR) spectroscopy, 35
electronegativity, 18t
electronic structure, 13–16
electronics, 2, 8
electroplating, gold. See gold electroplating
EMS electrochemistry, 52–53
energy dispersive X-ray spectrometry (EDS), 346–347
energy of incident light, 148f

engineering uses of gold, 402–404
    bursting discs, 402
    in fuel cells, 403
    in lubrication, 403–404
    spinnerets, 403
    thermal fuses, 402
    thermal protection layers, 403
Environmental Protection Agency (EPA), 106
equal channel angular extrusion (ECAE), 17
equilibrated metal surface (EMS), 51–52
ethene oxidation, 107
European Union, regulations of dental alloys in, 310
Ex One Company, 314
exchange-traded funds (ETFs), 10
exobidentate N^C ligands, 79–80
exobidentate N^N ligands, 79–80
explosive joining, 162
explosive welding, 162
eyelid implant, 404, 405f

F

$[F_3AsAu_I](SbF_6)$, 35
FA type bonding wire, 290t
Faraday, Michael, 352–353
federal test procedure (FTP), 105
feedstock, 107–108
fillings, 298–299
fine pitch bonding wire, 291
finishing, 203–204
first ionization potential, 15
flash gold alloy deposition, 249
flask, 194
Flowlogic™, 198
flue-gas desulfurization (FGD), 106
fluidity, 150–152
Forestier, Jacques, 217
formic acid oxidation, 61–62
Fosforcrisolo®, 218
Fourier transformed alternating current (FT-AC)
    voltammetry, 55–56
frit-type gold paste, 282–284
fuel cell cathodes, 58
fuel cells, 113–114, 403
furnaces, 402
futures market, 10

G

gemstones, 198
Germany, dental alloy market in, 308
GFC bonding wire, 290t
GFD bonding wire, 290t
gilding, 348, 361–362
Glanzgold, 325
glass, gold-coated, 401
Glauber, Johann Rudolf, 351–352
gluconic acid, 111
glucose, 57
glyceric acid, 99
glycerol, 99
GMG bonding wire, 290t

GMH-2 bonding wire, 290t
gold, 13–26
    alloying properties, 16–17
    as coating, 401–402
    crystal structure, 16–17
    dissolution of, 31–32
    electrical uses of, 400–401
    electronic structure, 13–16
    engineering uses of, 402–404
    medical uses of, 404–406
    optical properties, 18–26
    in photography, 402
    physical properties, 17–18
    prices, 11–12
    surface catalysis on, 56–57
    surface electrocatalysis on, 56–57
    surface electrochemistry of, 52–56
gold alloy electrolytes, 248–250
gold bar, 10
gold beating, 362–363
gold beating tools, 359f
gold black, 26
gold catalysts, 90–116
    commercial applications of, 103–115
        chemical processing, 107–113
        fuel cells, 113–114
        pollution control, 104–107
        sensors, 114–115
    future prospects, 115–116
    history of development, 90–91
    pharmacokinetics of, 219–220
    preparation methods, 91–95
        Au-PGM alloy catalysts, 94
        chemical vapor deposition, 93
        chloride-free preparations, 93
        colloids, 92
        coprecipitation, 92
        deposition precipitation, 92
        incipient wetness impregnation, 92
        ion exchange, 92
        liquid-phase reductive deposition, 93
        physical vapor deposition, 93
        posttreatment, 94–95
        supercritical CO_2 antisolvent technique, 93
        supported gold, 91–92
        unsupported gold, 91
    reactions, 95–103
        complete oxidation, 95–97
        liquid-phase homogenous catalysis, 101–103
        reactions of environmental importance,
            97–98
        selective hydrogenation, 100–101
        selective oxidation, 98–99
        sugars, 98–99
        vinyl chloride, 99
        water-gas shift, 99–100
gold complexes, 33–42
    antitumor properties of, 223–226
    clusters, 43
    thermodynamics of electroplating, 238–241
    utility of, 43–44
gold drugs, 217–228
    anti-HIV, 227
    anti-malaria, 226

antitumor properties of, 223–226
asthma treatment, 227
aurocyanide, 221–222
in chrysotherapy, 218–219
classes of, 218
colloidal metallic gold, 222–223
crystal structures of, 218–219
history of, 217–218
mechanisms of actions, 221
nanoparticles, 226
pharmacokinetics of, 219–220
gold electroplating, 231–271
applications of, 259–271
decorative uses, 260–264
engineering uses, 264–271
hard gold, 268–271
soft gold, 264–268
classification of, 245–246
composite electrodeposition, 257–258
cyanide-plating baths, 247–253
deposition kinetics, 241–245
cyanide electrolyte, 241–243
sulfite electrolyte, 243–244
sulfite-thiosulfate electrolyte, 244–245
thiosulfate electrolyte, 244–245
electrodeposition fundamentals, 234–238
electroforming, 258–259
evolution of, 233*f*
history of, 232–234
noncyanide baths, 253–254
overview of, 231–232
processes, 261–263*t*
pulse electrodeposition, 254–257
thermodynamics, 238–241
gold films, 319–326
burnish golds, 319, 320–321
dissolution of, 32–33
early gold precursors, 319–320
history of, 321–326
liquid bright golds, 319
liquid golds, 321
gold futures, 10
gold leaf, 357–363
in ancient China, 317–318
gilding techniques, 361–362
gold beating, 362–363
history of, 358–361
overview, 357–358
gold mercaptide systems, 327–337
formulations, 335–337
overview, 327
patents, 327
physical properties, 328–331
solubility of, 329–331
structural studies, 331–332
synthetic routes, 328
thermal decomposition studies, 332–334
gold nanorods, 371, 372*f*
gold nanospheres, 370–371
gold peroxide, 54–55
gold potassium cyanide (GPC), 201
gold powder, scanning electron micrograph of, 320*f*
gold resinate, 283–284

gold sponges, 26, 374
gold standard, 305–306
gold(0) complexes, 33–35
gold(I) alkynyl complexes, 76–79
gold(I) chalcogenido complexes, 71–76
gold(-I) complexes, 33
gold(I) complexes, 35–38
antitumor properties of, 223–226
with dicyanides, 79
luminescent, 70–80
gold complexes with dicyanides, 79
gold(I) alkynyl complexes, 76–79
gold(I) chalcogenido and thiolato complexes, 71–76
gold(I) phosphine complexes, 70–71
trinuclear complexes with exobidentate N^C/N^N ligands, 79–80
thermodynamics of electroplating, 238–241
gold(I) phosphine complexes, 70–71
gold(I) thiolate complexes, 71–76
gold(I)-N-(2-mercaptopropionyl)glycine, structure of, 330*f*
gold(II) complexes, 39–40
gold(III) complexes, 40–42
antitumor properties of, 223–226
luminescent, 81–82
thermodynamics of electroplating, 238–241
gold(V) complexes, 42, 43*f*
gold-alloy thermocouples, 400
gold-antimony phase diagram, 169*f*
gold-coated glass, 401
gold-copper brazing alloy, 185–186
gold-germanium phase diagram, 169*f*
gold-germanium soldering alloys, 171–172
gold-indium soldering alloys, 172
gold-lead-tin ternary system, liquidus projection of, 165*f*
gold-nickel brazing alloy, 186–188
gold-on-carbon catalysts, 99
gold-palladium alloys, 305, 401, 402
gold-platinum alloys, 305, 403
gold-silicon phase diagram, 170*f*
gold-silicon soldering alloys, 171–172
gold-silver mercaptides, 337–338
GOLDTECH® BIO2000, 302, 311
gold-tin phase diagram, 168*f*
gold-tin soldering alloys, 168–171
GPG bonding wire, 290*t*
GPG-2 bonding wire, 290*t*
grain refinement, 124–127. *See also* metallurgy of gold
grain growth, 126–127
recrystallization, 126–127
during solidification, 125–126
grain size control, 128
Gravant, Francois, 323
ground coats, 318

H

H[AuCl$_4$].n(H$_2$O), 31
H$_2$S sensor, 115
H9 cells, 227
hard gold, 268–271
hardness, Vickers, 17

hardening, 128–144
   disorder-order transformation, 131–133
   dispersion, 144
   precipitation, 133–144
   solid solution, 129–131
   work, 128
heap leaching, 3
heat conduction, 18
hedging, 1–2, 4–5
Hellot, Jean, 324
Herschel, John, 402
heterogenous catalysis, 95–101
   complete oxidation, 95–107
   reactions of environmental importance, 97–98
   selective hydrogenation, 100–101
   selective oxidation, 98–99
   vinyl chloride, 98–99
   water-gas shift, 99–100
hexanuclear gold(I) complex, structure of, 72*f*
high bond reliability wire, 291–292
high noble dental alloy, 306
high strength bonding wire, 289
high-firing gold formulations, 344–345
HIV infection, 227
hollow nanoparticles, 373–374
hollow-ware, 200
homogenization, 133
Höroldt, Johannes Gregorius, 323
hot melt inks, 349
hydrochloric acid, 31
hydrodechlorination catalysts, 94
hydrogen peroxide, 113
   reduction, 58
   sensors for, 115
hydrogenation, selective, 100–101, 113
hydrous (β) gold oxide, 53, 55
hydroxymethylfurfural (HMF), 108
hypervalent carbon, gold stabilization of, 36*f*
hypervalent nitrogen, gold stabilization of, 36*f*

I

imagen®, 314
imines, enantioselective hydrogenation of, 103
immersion plating, 235
implants, 404–405
impregnation method, 92
incinerator gases, 104
incipient hydrous oxide/adatom mediator (IHOAM), 56–57
incipient wetness (IW), 92
incisors, 296
India, gold jewelry market in, 8
inflation and gold, 9
ink-jet printing, 350
inlays, 299
internally reinforced gold, 312
International Council on Mining and Metals (ICMM), 3
International Cyanide Management Code, 3
investment casting, 124. *See also* lost wax casting
investment in gold, 2, 9–10
   futures, 10
   over-the-counter markets, 10

paper gold, 10
   as protection against inflation, 9
   as safe have in economic/political crisis, 9–10
ion exchange, 92
ionization potentials, 15–16
isocyanates, 91

J

Japan, dental alloy market in, 308
jewelry, 2, 7–8
jewelry manufacturing, 191–211
   blanking, 199
   cable-making technology, 206
   chain-making, 199–200
   coining, 199
   cold forging, 199
   cold forming, 198–199
   computer modeling in, 210–211
   computer-aided design in, 205–206
   drivers for change in, 193
   early history, 192
   electroforming, 200–201
   engineering technologies in, 193
   finishing, 203–204
   hollow-ware, 200
   lasers in, 207–209
   low wax casting, 194–198
   polishing, 203–204
   powder metallurgy in, 209–210
   rings, 204–205
   robotics in, 208
   soldering, 201–203
   stamping, 198–199
   texturing, 203–204
Johnson Matthey Noble Metals, 400

K

karat, 7, 124
karat gold brazing alloys, 179–184
karat gold solders, 175–178
   22-karat, 175–177
   diffusion soldering of, 177–178
knitting of wires, 206, 207*f*
Koch, Robert, 217
Kramers-Kronig relation, 19
Kühn, Heinrich Gottlob, 319, 325, 326–327
Kunckel, Johann, 322–323, 352

L

lactose, 111
laser forming, 208
laser welding, 203
lasers, 207–209
Latin America, gold production in, 4
lattice constant, 18*t*
LC bonding wire, 290*t*
Lewis acid, 33
Lewis base, 33
ligand-to-metal-metal charge transfer (LMMCT), 73–76
liquid bright golds, 319, 321, 325–326

liquid golds, 326–327
    development of, 324–325
    vs. frit-type gold paste, 283–284
liquid-phase homogenous catalysis, 101–103
liquid-phase reductive deposition, 93
Lister, Martin, 323
London Fix, 11
lost wax casting, 194–198
    artificial intelligence in, 198
    casting machines, 195–196
    computer modeling in, 198
    gemstones in, 198
    literature, 194–195
    porosity, 195
    steps in, 194
    technology trends, 196–198
low-alloyed gold, 140–143
lower jaw, 296
lubrication, 403–404
luminescent gold complexes, 69–82
    future prospects, 81
    overview of, 69
luminescent gold(I) complexes, 70–80
    gold complexes with dicyanides, 79
    gold(I) alkynyl complexes, 76–79
    gold(I) chalcogenido and thiolato complexes,
        71–76
    gold(I) phosphine complexes, 70–71
    trinuclear complexes with exobidentate N^C/N^N
        ligands, 79–80
luminescent gold(III) complexes, 81–82

M

machine banding, 348
Macquer, Pierre-Joseph, 324
macrothrowing power, 238
magnetic circular dichroism (MCD), 70
magnetic resonance imaging (MRI), 385
malaria, 226
malleability of gold, 17
maltose, 111
mandible, 296
Matthey, George, 326
maxillary, 296
$(Me_3P)_2Au^{III}I_3$, structure of, 42f
medicine, gold in. See gold drugs
melting, 292–293
melting point of gold, 17
6-mercaptopurine (6-MP), 223
mercury oxidation, 106–107
mercury vapor, detection of, 401–402
mesoporous gold sponges, 374
metal injection molding (MIM), 209–210
metal joining, gold in, 161–188
    advantages of, 161–162
    gold-based brazing alloys, 178–188
    gold-bearing soldering alloys, 166–178
    solid-state joining, 162–163
metallic glazes, 376
metallurgy of gold, 123–155
    bulk metallic glass, 152–153
    color variation, 144–150
    dispersion hardening, 144

fluidity, 150–152
grain refinement, 124–127
metal-matrix composites, 154–155
overview, 123–124
precipitation hardening, 133–144
shape memory alloys, 154
solid solution hardening, 129–133
strengthening mechanisms, 127–144
wetting, 150–152
metal-matrix composites, 154–155
metal-organic gold paste, 283–284
metastable metal surface (MMS), 51–52
methanol oxidation, 61–62
methyl glycolate, 110–111
methyl oxirane, 111–113
methylene oxazoline, 102
methylfuroate, 108
microalloyed gold, 124, 140–143
microalloying, 131
microwave-resistant gold, 344
middle ear implant, 404, 405f
mines, 2–4
minor metals, 345–348
mixed metal catalysts, 94
MMS electrochemistry, 52–53
monolayer (α) gold oxide, 54–55
monolayer protected clusters (MPCs), 73
multicomponent alloys, 143–144
Myochrisine®, 218
Myocrisin®, 218

N

N^C bridging ligands, 79–80
N^N bridging ligands, 79–80
naked gold sols, 98–99
NanoGold Oil™, 403–404
nanoparticles, 369–388
    carbon monoxide oxidation, 59–61
    color of, 23–26
    in decorative gold materials, 340–341
    in dentistry, 314–315
    electrocatalysis of, 57–62
        carbon monoxide oxidation, 59–61
        formic acid oxidation, 61–62
        methanol oxidation, 61–62
        oxygen reduction, 58–59
    formic acid oxidation, 61–62
    gold drugs, 226
    gold nanorods, 371, 372f
    gold nanospheres, 370–371
    industrial applications, 376–379
        catalysis, 379
        high-density data storage, 378–379
        metallic glazes, 376
        optoelectronic devices/switching, 379
        pigments/colorants, 376
        sinter inks, 376
        spectrally selective coatings, 376–378
    in medical diagnostics, 384–386
        magnetic resonance imaging, 384
        optical coherence tomography, 385–386
        photoacoustic microscopy, 385
        two-photon luminescence, 384

mesoporous gold sponges, 374
methanol oxidation, 61–62
nanorods, 371, 372*f*, 378–379, 386–387
nanoshells, 373–374
oxygen reduction, 58–59
semishells, 373–374
sensors/analytical applications, 380–383
colorimetric, 380
electrochemical, 382
refractometric, 380–382
surface-enhanced Raman spectroscopy, 382–383
therapeutic possibilities, 226, 386–388
thin films, 374–376
nanoporous gold, 374
nanorods, 371, 372*f*, 378–379, 386–387
nanoshells, 373–374, 386–387
nanospheres, 370–371
Nanostellar Inc., 105
nanotechnology, gold in, 369–370
Nd:YAG laser, 207
needles for acupuncture, 405–406
neutral electrolytes, 250–251
N-heterocyclic carbenes (NHC), 225–226
Ni-Hg gold layers, 268–271
Nippon Shokubai, 110
Niroyal™ stent, 404
Nitinol wire, 298
noble dental alloy, 306
noncyanide baths, 253–254
nylon precursors, 110–111

O

oil gilding, 361–362
old scrap, 6
oleylamine nanoparticles, 342*f*
onlay, 176
optical coherence tomography (OCT), 385–386
optical properties, 18–26
color of bulk gold, 18–21
color of nanoparticles, 23–26
gold black, 26
gold sponges, 26
thin gold films, 21–23
optical reflector coating, 401
optoelectronic devices, 379
organic light emitting diodes (OLEDs), 81
overaging effect, 132
overcoats, 318
over-the-counter (OTC)markets, 10
oxidants, 31–32
oxidation
complete oxidation, 95–107
selective, 98–99
oxidation states, 238–241
oxygen reduction reaction (ORR), 58–59

P

pad printing, 350
paillons, 202
palate, 296
palladium, 345

paper gold, 10
paper products, 2
Parent, Melchior-Francois, 324
particle induced X-ray emission (PIXE), 346
Pd/Al$_2$O$_3$ catalysts, 107
petroleum refining, 113
pharmacokinetics, 219–220
phosphine complexes, 70–71
photoacoustic microscopy, 385
photographic applications of gold, 402
physical properties, 17–18
physical vapor deposition (OVD), 93
pigments, 376
plasma frequency, 20
Plasmodium falciparum, 226
platform chemicals, 107–108
plating baths. *See* cyanide-plating baths
platinum, 345
point of zero charge (PZC), 92
polarization of light wave, 21
polishing, 203–204
pollution control, 104–107
air cleaning, 104
autocatalysts, 104–106
ethene oxidation, 107
mercury oxidation, 106–107
volatile organic compounds, 107
water pollution control, 107
polymethylemethacralate (PMMA), 267
porcelain teeth, 297–298
porcelain-fused-to-metal (PFM) restoration, 299, 307*t*
porosity, 195
posttreatment, 94–95
powder metallurgy, 209–210
precious metal clays, 210
precipitation hardening, 133–144
Au-Cu-Ag-based alloys, 136–139
Au-Ni-based alloys, 139–140
Au-Pt-based alloys, 139–140
low-alloyed gold, 140–143
microalloyed gold, 140–143
multicomponent alloys, 143–144
overview of, 128
predominantly base dental alloy, 306
preferential oxidation (PROX) systems, 91, 100
premonolayer oxidation, 54
press and sinter technique, 204
price, 11–12
process scrap, 6
producer hedging, 4–5
ProMetal RCT™, 314
ProMetal®, 314
propene oxidation, 111–113
propene oxide, 111–113
pseudoelasticity, 154
Pt cathode electrocatalysis, 58
pulse electrodeposition, 254–257
purple gold, 148–149
Purple of Cassius, 350–357
color generation of colloidal gold, 353–354
in decorative gold materials, 318
gold nanoparticles, 340
history of, 351–353
overview, 350–351

as pigment/colorant, 376
preparation of, 354–357
4-pyridinethiol, 32, 33*f*

R

Raney gold, 374
rapid manufacturing (RM), 205
rapid prototyping (RP), 205
reactions of environmental importance, 97–98
reflectivity curves, 148*f*
refractometric sensors, 380–382
resinate, 319
reversible hydrogen electrode (RHS), 52
rheumatoid arthritis, treatment with gold compounds, 217
rhodium, 346
rings, 204–205
robotics, 209
rupture discs, 402
Rutherford back-scattering spectrometry (RBS), 346

S

Sanochrysin®, 218
scanning electron microscopy (SEM), 80, 346–347
screen printing, 348–350
decals, 349–350
direct printing onto glass, 349
second ionization potential, 15–16
secondary amines, 91
Seebeck effect, 400
Seebeck, Thomas, 400
selective hydrogenation, 100–101, 113
selective oxidation, 98–99
self-burnish, 320
self-burnish gold, 343–344
semimatte gold finish, 320
semishells, 373–374
sensors, 114–115, 380–383
colorimetric, 380
electrochemical, 382
refractometric, 380–382
surface-enhanced Raman spectroscopy, 382–383
shape memory alloys (SMA), 154
shrinkage porosity, 195
silicon, 150–152
silver alloys, 345
silver(I) mercaptide compounds, 337–338
single restorations, 302
sinter inks, 376
sintering, 312
sodium aurothiopropanol sulfonate, 218
sodium aurothiosulfate, 218
soft gold, 264–268
soldering alloys, gold-bearing, 166–178
gold-germanium, 171–172
gold-indium, 172
gold-silicon, 171–172
gold-tin, 168–171
karat gold solders, 172–178
melting points, 166
soldering,
definition of, 161
gold in, 163–178

gold-bearing soldering alloys, 166–178
indium-bearing solders, 165
in jewelry manufacturing, 201–203
thickness of gold surface, 163
tin-containing solders, 164–165
underlayer, 162–163
Solganol®, 218
solid freeform fabrication (SFF), 205
solid solution hardening, 129–133
conventional alloying, 129–131
definition of, 128
disorder-order transformation hardening, 131–133
microalloying, 131
solidification, 125–126
solid-state joining, 162–163
solution annealing, 133
South Africa, gold production in, 3–4
spark plug electrodes, 401
specific heat, 18*t*
spectral encoding, 378
spectrally selective coatings, 376–378
spinnerets, 403
stamping, 198–199
static casting, 195–196, 197*t*
static vacuum-assist machine, 195
steel wheel applicator system, 348
stents, 404
Stern-Volmer equation, 70
Stöltzel, Samuel, 352
strengthening mechanisms, 128–129
succinic acid, 107
sugars
gold catalysts, 98–99
selective oxidation of, 111
sulfite electrolyte, 243–244
sulfite-thiosulfate electrolyte, 244–245, 267–268
sulforesinate, 319, 326
sulfur balsam, 326–327
sulfur-free gold compounds, 338–340
superconducting wires, 400
supercritical CO_2 antisolvent technique, 93
superelasticity, 154
super-Nernstian E/pH shift, 55
super-paramagnetic iron oxide nanoparticles (SPIO), 385
supersaturated solid solution, 133
supply, 1–2
aboveground stocks, 1, 5–7
from new mine production, 1, 2–4
producer hedging, 4–5
supported gold, 91–92
surface electrocatalysis, 56–57
surface electrochemistry, 52–56, 286
EMS electrochemistry, 52–53
FT-AC voltammetry studies, 55–56
hydrous (β) gold oxide, 55
MMS electrochemistry, 53–54
monolayer gold oxide, 54–55
overview of, 51–52
premonolayer oxidation, 54
surface metal atoms, 51–52
surface plasmon polariton (SPP), 21–23
surface plasmon resonance spectroscopy (SPRS), 21
surface-enhanced Raman spectroscopy (SERS), 382–383

surpluses, 2
swarna bhasma (gold ash), 217

T

tarnish films, 300–301
Tauredon®, 218
teeth, 296
telescopic crown, 312*f*
tetranuclear gold(I) alkynyl complexes, 77*f*
tetranuclear gold(I) dithioacetate, 74
textile dye, 376
texturing, 203–204
thermal conductivity, 18*t*
thermal fuses, 402
thermal print heads, 286
thermal properties, 303–304
thermal protection layers, 403
thermocouples, 400
thick film conductors, 283*t*
thick film pastes, 279–286
    applications, 284–286
    electrode, 279–280
    features, 280–282, 283*t*
    future trends, 284–286
    glass frit, 282–284
    gold resinate, 283–284
    sintering behavior, 281*f*
    vs. thin film pastes, 282
    types of, 282–284
thin films, 21–23
    as lubricants, 404
    nanoparticles, 374–376
    vs. thick film, 282
thiolate complexes, 71–76
third ionization potential, 15–16
3D printing, 312–314
three-way catalysts (TWC), 104–105
time-dependent density functional theory (TDDFT), 81
transient liquid-phase (TLP) joining, 177–178
transition temperature, 303
transverse electric mode, 21
transverse magnetic mode, 21
trichloroethene, 107
tungsten carbine wheel applicator, 348
tungsten inert gas (TIG) welding, 203
Turkey, gold jewelry market in, 8
turnover frequencies (TOFs), 90, 102
turnover numbers (TONs), 90, 102
two-photon luminescence (TPL), 384

U

underpotential oxidation, 54
United Chemical Laboratories, 104
United States
    dental alloy market in, 305–308
    gold jewelry market in, 8
unsupported gold, 91
upper jaw, 296
U.S. dollar, and gold price, 11–12
utility variance, 182

V

Vickers hardness, 17
Vincennes porcelain factory, 323–324
vinyl acetate monomer (VAM), 91, 108–110
vinyl chloride, 99, 108–110
Vitallium®, 299
volatile organic compounds (VOCs), 97, 104, 107, 327
vulcanite, 299

W

water gilding, 361–362
water-gas shift (WGS), 99–100, 113–114
weaving of wires, 206
wedge bonds, 287–289
welding
    definition of, 161
    explosive, 162
    laser, 203
    tungsten inert gas, 203
wetting, 150–152
white gold, 7
white gold alloys, 145–147
winding, 293
wire bonding, 287–293
    ball bonds, 287–289
    fine pitch bonding type, 291
    high bond reliability type, 291–292
    high strength type, 289
    production process, 292–293
        annealing, 293
        casting, 292–293
        drawing, 293
        final inspection, 293
        melting, 292–293
        winding, 293
    types of, 290*t*
    wedge bonds, 287–289
wires
    knitting of, 206, 207*f*
    weaving of, 206
Worcester Royal Porcelain Company, 324
work function of gold, 18
work hardening, 128
World Gold Council, 3

X

x-ray diffraction (XRD), 80, 346–347

Y

Y type bonding wire, 290*t*
Yanacocha mine, Peru, 4
Young's modulus, 18*t*

Z

zinc, 150–152, 180
Zsigmondy, Richard, 353